IET TRANSPORTATION SERIES 36

Vehicular Ad Hoc Networks and Emerging Technologies for Road Vehicle Automation

Other related titles:

Volume 1	**Clean Mobility and Intelligent Transport Systems** M. Fiorini and J-C. Lin (Editors)
Volume 2	**Energy Systems for Electric and Hybrid Vehicles** K.T. Chau (Editor)
Volume 5	**Sliding Mode Control of Vehicle Dynamics** A. Ferrara (Editor)
Volume 6	**Low Carbon Mobility for Future Cities: Principles and Applications** H. Dia (Editor)
Volume 7	**Evaluation of Intelligent Road Transportation Systems: Methods and Results** M. Lu (Editor)
Volume 8	**Road Pricing: Technologies, economics and acceptability** J. Walker (Editor)
Volume 9	**Autonomous Decentralized Systems and their Applications in Transport and Infrastructure** K. Mori (Editor)
Volume 11	**Navigation and Control of Autonomous Marine Vehicles** S. Sharma and B. Subudhi (Editors)
Volume 12	**EMC and Functional Safety of Automotive Electronics** K. Borgeest
Volume 16	**ICT for Electric Vehicle Integration with the Smart Grid** N. Kishor and J. Fraile-Ardanuy (Editors)
Volume 17	**Smart Sensing for Traffic Monitoring** Nobuyuki Ozaki (Editor)
Volume 18	**Collection and Delivery of Traffic and Travel Information** P. Burton and A. Stevens (Editors)
Volume 20	**Shared Mobility and Automated Vehicles: Responding to socio-technical changes and pandemics** Ata Khan and Susan Shaheen
Volume 23	**Behavioural Modelling and Simulation of Bicycle Traffic** L. Huang
Volume 24	**Driving Simulators for the Evaluation of Human-Machine Interfaces in Assisted and Automated Vehicles** T. Ito and T. Hirose (Editors)
Volume 25	**Cooperative Intelligent Transport Systems: Towards high-level automated driving** M. Lu (Editor)
Volume 26	**Traffic Information and Control** Ruimin Li and Zhengbing He (Editors)
Volume 30	**ICT Solutions and Digitalisation in Ports and Shipping** M. Fiorini and N. Gupta
Volume 32	**Cable Based and Wireless Charging Systems for Electric Vehicles: Technology and control, management and grid integration** R. Singh, S. Padmanaban, S.Dwivedi, M. Molinas and F. Blaabjerg (Editors)
Volume 38	**The Electric Car** M.H. Westbrook
Volume 45	**Propulsion Systems for Hybrid Vehicles** J. Miller
Volume 79	**Vehicle-to-Grid: Linking Electric Vehicles to the Smart Grid** J. Lu and J. Hossain (Editors)

Vehicular Ad Hoc Networks and Emerging Technologies for Road Vehicle Automation

Amit Kumar Tyagi and Shaveta Malik

The Institution of Engineering and Technology

Published by The Institution of Engineering and Technology, London, United Kingdom

The Institution of Engineering and Technology is registered as a Charity in England & Wales (no. 211014) and Scotland (no. SC038698).

First published 2022

The Institution of Engineering and Technology
Futures Place
Kings Way, Stevenage
Hertfordshire, SG1 2UA, United Kingdom

www.theiet.org

British Library Cataloguing in Publication Data
A catalogue record for this product is available from the British Library

ISBN 978-1-83953-428-7 (hardback)
ISBN 978-1-83953-429-4 (PDF)

Typeset in India by MPS Ltd
Printed in the UK by CPI Group (UK) Ltd, Croydon

Contents

About the authors

Amit Kumar Tyagi is an assistant professor (senior grade) and a senior researcher at Vellore Institute of Technology (VIT), Chennai, Tamilnadu, India. He received his PhD in 2018 from Pondicherry Central University, India. He joined the Lord Krishna College of Engineering, Ghaziabad (LKCE) for the periods of 2009–2010, and 2012–2013. He was an assistant professor and head-research, Lingaya's Vidyapeeth (formerly known as Lingaya's University), Faridabad, Haryana, India in 2018–2019. His current research focuses on machine learning with big data, blockchain technology, data science, cyber physical systems, smart and secure computing and privacy. He has contributed to several projects such as "AARIN" and "P3- Block" to address some of the open issues related to the privacy breaches in vehicular applications (such as parking) and medical cyber physical systems (MCPS). Also, he has published more than 15 patents (nationally and internationally) in the area of deep learning, internet of things, cyber physical systems, and computer vision. Recently, he has been awarded the b paper award for a paper titled "A Novel Feature Extractor Based on the Modified Approach of Histogram of oriented Gradient," ICCSA 2020, Italy (Europe). He has edited more than 10 books for IET, Elsevier, Springer, CRC Press, etc. Also, he has authored book on Internet of Things, intelligent transportation systems, and Vehicular Ad Hoc Networks. He is a regular member of the ACM, IEEE, MIRLabs, Ramanujan Mathematical Society, Cryptology Research Society, and Universal Scientific Education and Research Network, CSI, and ISTE.

Shaveta Malik is an associate professor in the Computer Engineering Department at Terna Engineering College, University of Mumbai, India, where she has more than 12 years of work experience. She received her PhD from Lingaya's University, Faridabad, India. Her research area focuses on image processing, machine learning, deep learning, and artificial intelligence. She has published a number of patents related to machine learning. She received a best paper award for "A Novel Feature Extractor Based on the Modified Approach of Histogram of oriented Gradient" at ICCSA-2020. She has written and edited several books.

Preface

Today Internet of Things (IoTs) is the emerging technology for the world (including next generation society) and science believes that IoTs will change the complete Internet and its structure. To know more about IoTs, we need to know its role, applications, scope, and combination of various other computing environment/ technologies used in IoT for communication as well as Network security or to complete everyday task. Today Government of India promotes Digital India project, under which government wants to promote projects like smart cities, smart grid, smart agriculture, and smart transportation and so on. Such examples belong to Internet of Everything (IoE), i.e., to variants of Internet of Things. IoT and IoE can provide efficient services through its implementation in many sectors and can make people life better and easier. Society 5.0 and Industry 4.0 can be connected together and can be connect together through IoT for betterment of Human being.

Today's IoE are being used in many sectors/areas and make a smart, efficient environment through using many IoT devices together. Building a smart era is quite challenging and have many issues like security, privacy, trust, etc. If we succeed with this, we faced several issues like battery of devices, privacy, security, scalability, compactibility, etc. The coming smart era (the next decade) talked about the role of IoT in driverless cars, smart lock, and many more applications.

This book includes basic information (as much as little information) about IoTs like software, hardware, middle ware, tools used, etc., in detail. This book provides a platform to researchers, readers, scientists, etc., to write their research work and idea towards digital India and IoT technology development. This book also gives exposure to researchers for exploring how IoT is not complete without Artificial Intelligence (AI) and Blockchain Technology. Few Chapters are including in this book with respect to these topics. This book will basically focus on recent advancements on Internet of Things and it's implementations for society.

—Amit Kumar Tyagi
—Shaveta Malik

Acknowledgement

First of all, we would to extend our gratitude to our family members, friends, and supervisors, who stood with us as an advisor in completing this book. Also, we would like to thank our almighty "God" who makes us to write this book. We also thank IET Publishers (who has provided their continuous support during this COVID 19 Pandemic) and our colleagues with whom we have worked together inside the college/university and others outside of the college/ university who have provided their continuous support towards completing this book on Vehicular Ad Hoc Network (VANET)/Intelligent Transportation Systems (ITS).

Also, we would like to thank our Respected Madam, Prof. G Aghila, Prof. Siva Sathya, our Respected Sir Prof. N Sreenath and Prof. Aswani Kumar Cherukuri, for giving their valuable inputs and helping us in completing this book.

—Dr Amit Kumar Tyagi
—Dr Shaveta Malik

Chapter 1

Vehicular Ad Hoc Networks: past, present and future

Abstract

Vehicular Ad Hoc Networks (VANETs) involve the integration of transportation systems with Internet systems, which constitute the major motive for the increase in passenger safety. The Internet of Things (IoTs) contains a mobile ad hoc network component which in turn includes a part called the Vehicle Ad hoc Network. The Internet of Energy (IoE) is a new realm formed by electric automobiles connected to ad hoc vehicle networks. Because various transport systems are being built and numerous applications designed to manage these networks, there are also increasing attacks in this area. As the energy Internet is connected to ad hoc vehicle networks via electric vehicles, there may be a question of the survival of the ad hoc vehicle networks if safety considerations are not relevant. In this survey, several kinds of car network attacks are covered by existing security solutions to address the attacks intelligent transportation system (ITS), with information and communication technologies and wireless embedded sensor devices in today's vehicles, is a realistic and required component of smart cities. The smart transportation system is intended to improve road safety and traffic efficiency and to provide information services. Notifying drivers of dangerous road conditions and giving them knowledge from past traffic will undoubtedly increase driver security and decrease traffic congestion at the right time. Technically, the smart transport system depends on known ad hoc vehicle networks, self-organized wireless networks. Mobile cars in the VANET can do the job of stationary sensors in infrastructure networks. In real-time, they can gather, identify, and communicate traffic information, driving conditions, and any road hazards. VANETs have become a highly active area of research in recent years. VANETs are very attractive to academics and industry because of their unique properties including extremely dynamic topology and predictable movement.

Key Words: Vehicular Ad Hoc Networks (VANET); Intelligent transportation system (ITS); Internet of Things (IoT); Information and communication technologies (ICT); Internet of Energy (IoE); Mobile ad hoc network (MANET).

1.1 Introduction

During the past decade, mobile communication technologies have revolutionized the automobile sector by making it possible to communicate across various devices at any time and from any location. This simple connection allows for the transmission of critical information across flying devices [1]. For the industry, the seamless flow of information in real-time has emerged as a new paradigm. As a result, advancements in information technology and communications have made the concept of communication across mobile devices a simple reality. Among these advancements is the concept of the Vehicle Ad Hoc Network (VANET), which offers up new possibilities for the deployment of safety-related apps. A VANET is an ad hoc network that connects various moving vehicles and other connected devices with relevant information through wireless channels. VANETs are becoming more popular. At the same time, a modest network of cars and other devices that serve as network nodes is being constructed. Any information about a node is sent on to all other nodes in the network [2]. Similarly, after the transmission of their data set, all nodes receive data from other nodes that have been sent to them. After all of the data has been gathered, the nodes synthesis the important information and transmit it to the rest of the network.

When the link between devices becomes strong enough, nodes may be joined and disconnected from the network, indicating that it is an open network. Vehicles outfitted with onboard sensors, such as those being introduced into the market today, make it simple to integrate and fuse the vehicle while also making use of VANET technology, which is becoming more common. MANET is made up of nodes that communicate with one another without the need for a centralized network with networking capabilities. VANET, on the other hand, was a more complex and responsible MANET class or variant to operate in comparison [3]. VANET needs a different set of routing protocols than MANET to provide unrestricted access to nodes. A vehicle communication network links vehicles and allows information to be sent and received to enhance traffic efficiency, monitor road conditions, reduce accidents, detect emergencies, and boost overall network efficiency. With the assistance of several hops, VANET can transmit information to far-flung devices. The VANET in Figure 1.1 may be described by the parameters listed below. A VANET is a subtype of MANET that is used for communication between automobiles and nearby permanent roadside infrastructure. VANET communications include a variety of types, including vehicle-to-vehicle and vehicle-to-infrastructure [4].

When used in conjunction with multihop communication, V2V may be used to provide emergency warnings or to reach coverage nodes. To keep up with the rapid growth in-car communication, several studies have been conducted that have covered all aspects of the technology. These studies have included channel design, appropriate Media Access Control (MAC) scalability, security policies, reliability and latency improvement, Vehicular Ad Hoc Network-Long-Term Evolution (VANET-LTE) integration, as well as routing protocols that ensure good performance and can be adjusted to changes in network topology. Route construction and maintenance protocols in ad hoc networks are very essential since they build and

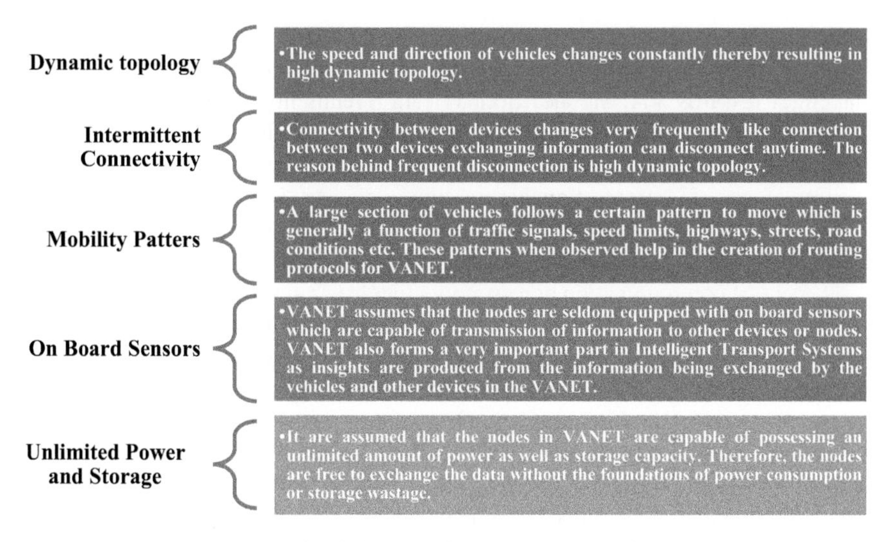

Figure 1.1 The following factors can describe VANET

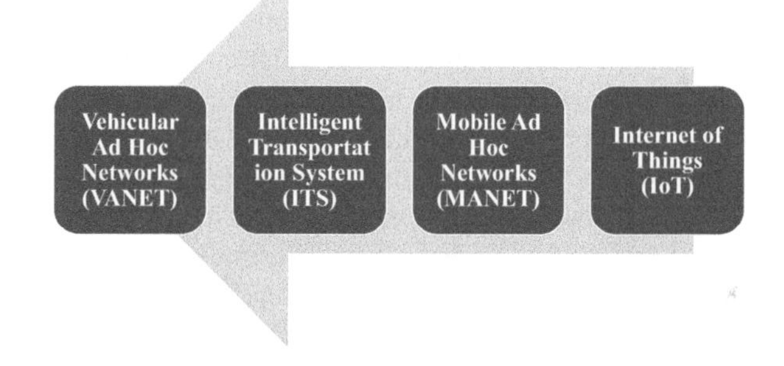

Figure 1.2 Relationship of IoT and VANET: a hierarchical understanding [6]

maintain routes for multi-hop communication while also expanding the service area of the network. A variety of circumstances, including node mobility, interference, and bandwidth constraints, are addressed by the VANET routing protocols, which are intended to cope with the main features and limitations of vehicle grids such as these. In Figure 1.2, the computer displays the connection between the Internet of Things (IoT) and the VANET [5].

The rest of the chapter is arranged according to: Section 1.2 describes VANET history. Section 1.3 describes the VANET literature survey. Section 1.4 explains the Standardization routing protocols of VANET and describes different types of routing protocols. Section 1.5 explains ad hoc network types, and characteristics for VANET, and Section 1.6 explains Frameworks/models of VANET. Section 1.7 describes VANET architectures: previous decade and in current era and Section 1.8 describes VANETs security services, security threats, and attacks in VANET.

Section 1.9 describes VANETs privacy services and Section 1.10 describes Research issues over privacy and security attack, whereas Section 1.11 is Open challenges towards VANET and Section 1.12 Trends in near future for VANET applications and, at last, Section 1.13 describes the conclusion over VANET and summary of the chapter.

1.2 History

VANET is a generic term that refers to the possibility of a vehicle node being linked to another communication node that may be observed via wireless communication in the radio spectrum. In addition, since the vehicles are moving objects, the network architecture may change at any moment, even though, in this particular situation, the location of communication nodes can be anticipated because each vehicle should follow a predetermined route across the city (i.e., roads). We were acquainted with the concept of using vehicle radio communications to improve safety long before digital radio communications were widely available [7]. The patent for radio alerting systems for use in automobiles was filed in 1922 and granted in 1925, and it was based on the idea of peer-to-peer radio communication between identical equipment placed in two separate automobiles that were introduced in the 1920s. Because of its simplicity, the suggested system anticipated certain vehicle safety requirements, which in recent years has increased the demand for communication between automobiles (autonomous vehicles).

This technique was not yet recognized as wireless networking, but it was created to meet one of the most comparable needs to the one we are now discussing, which led to the development and strengthening of VANET technology, which we are currently discussing [8]. However, outside of the military, nothing occurred for more than half a century, until the 1980s and 1990s, when most automobiles were already supplied with a "radio set" as a standard feature. Local radio stations, of course, offered radio broadcasts at the time, informing drivers of severe weather and important events, but only via the speaker system of their vehicles. There was no networking of any sort, just a broadcast, and this was the case. When it comes to private functionality, radio data system (RDS) specifies the number of options, including how to bundle it in unused, internal, or other unspecified groups [9]. Each traffic incident is represented as a binary TMC message, which is sent across the network.

Each message includes an event identification, a location code, an estimate of the duration of the incident, the severity of the issue, and other pertinent information about the incident. Some expressions refer to particular circumstances, such as an accident, while others refer to occurrences that cause significant delays, such as construction projects. Sources of traffic information include, in general, police officers, traffic control centers, traffic cameras, and speed sensors, vehicle floating information, winter driving records, and roadwork data, among others [10]. However, as shown in Figure 1.3, these communications remained unidirectional rather than being distributed across networks.

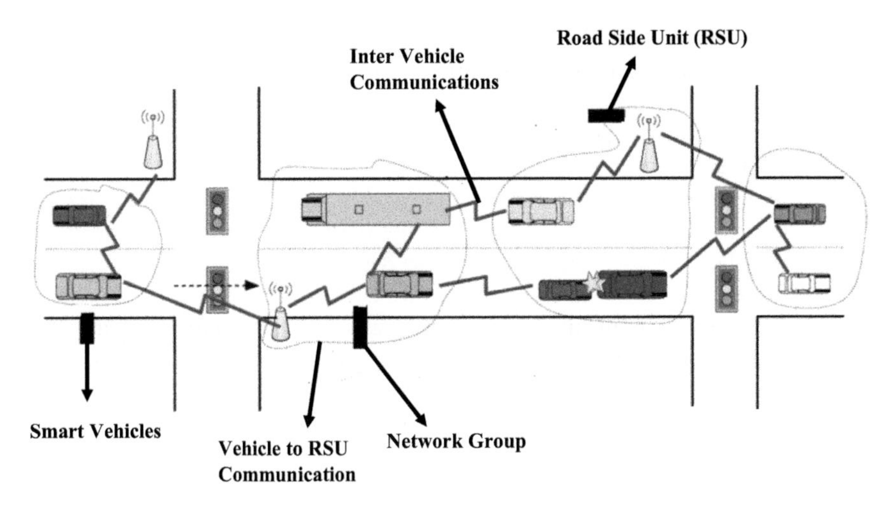

Figure 1.3 VANET basic structure [11]

These were the world's first two-way communications, although in a primitive version, in which the infrastructure analyzed radio-frequency identification (RFID) tags in their most basic form as they passed under the beacons and the tag responded by identifying the tags. More sophisticated systems, such as those manufactured by Philips, may also store and transmit data input and output [12]. The dedicated short-range communications (DSRC) in vehicle system was capable of performing a variety of transactions, and it was anticipated to develop into a natural delivery mechanism for IVHS in the 1990s, given the limitations of the telephone and Internet infrastructure at the time. However, since the vehicles were not linked, there was no automobile network, much alone an ad hoc network, and, as a result, the Downlink beacons, which were expensive and fixed in place, were not used. The connection between the technology and the infrastructure was one of master and slave. Furthermore, the bandwidth and the range were insufficient, and the three main manufacturers of DS RCS standards were at the time attempting to lock the technology inside their protocols, which was unsuccessful (at least two of the three, with the third fighting for non-proprietary protocols).

However, the business case had a basic flaw that needed to be addressed. Although there is no economic rationale, road tolls may be able to pay the necessary infrastructure costs. As a result of one of the countless legal battles that ensued and the eventual compromise and defective standard, several competitors and one of the most prominent experts from that period came together to discuss how to avoid a similar catastrophe in the future world of intelligent transportation system (ITS) communication in the future. Their concept was revolutionary at the time, but they were unaware that similar design problems faced Internet developers at the time, and that their solution was similar to what would become the norm for the internet, namely, separating the applications from the media and introducing online communication and network administration [13].

Because, although the ITS community has been developing these future technologies for more than a decade and a half and has yet to see any major corporate deployments, mobile communications have advanced significantly in recent years. It was about this time when analog cellular phones were phased out and cellular phones were transitioned to the Global System for Mobile Communications, and primarily to the Universal Mobile Telecommunication Systems (UMTS). In the last several years, a gap has developed between the marketing divisions of car manufacturers and their research and development departments. Smartphones are now available to nearly everyone. At least as many cell phones are now accessible in every vehicle as there are people on the road today. Another feature of Bluetooth is that it can link many devices, thus reducing issues with synchronization and calling, to name a few of them [14]. For their owners, they have evolved into sources of amusement and communication and International Organization for Standardization (ISO) standards are shown in Figure 1.4.

Because hand-holding of mobile phones is prohibited by law, almost every vehicle is equipped with Bluetooth or wired capabilities to allow for hands-free use of mobile phones. Several vehicle owners have now requested that car manufacturers utilize this link to offer infotainment for their smartphones. Marketing departments have put pressure on car designers to incorporate these traits into their creations. We are just a few steps away from having this technology available,

Figure 1.4 ISO standards [15]

Table 1.1 Category VANET top-level research [16]

Classes		
Topic examples	**Description**	**Title**
Deployment and field-testing framework	Testbeds	Tool
Connectivity analysis, modelling, management	Mobility issues	MOB
Proposal of a new routing protocol	Routing protocol	ROUT
Location tracking, location estimate correction	Complementary services	SERV
Data collection and message dissemination method	Data management	DATA
Safety, efficiency, entertainment, and environment	Application layer protocols	APP
Protocol design, testing, and verification analysis	Performance comparison analysis	PERF

whether it is linked to the car owner's telephone or integrated inside the vehicle via Bluetooth for ITS services. In Europe, the eCall law – a "silo" system for connecting automobiles to emergency services after an accident – legally requires the installation of a Universal Subscriber Identity Module (USIM) in every light vehicle, which will serve as the vehicle's basic infrastructure and connect the vehicle to emergency services [17]. The VANET top-level research category is shown in Table 1.1.

1.3 Literature survey

VANETs are created in the automobile industry by using the ideas of MANET, which are defined as the spontaneous creation of wireless networks of mobile devices. Initially, VANETs were referred to and introduced in 2001 as ad hoc mobile communication and networking applications in car-to-car applications, in which networking can be formed and information can be shared between cars. Today, VANETs are used in a variety of applications, including transportation. Road safety, navigation, and other route services are delivered via the use of vehicle communications systems, which coexist with the cars themselves in the VANET.

VANET is a critical component of the design of ITS. VANETs are sometimes referred to as intelligent transport networks in certain circles. Some think they have grown into a larger automotive internet that may ultimately evolve into an Internet of autonomous vehicles. While VANETs was first considered to be a straightforward one-to-one application of MANET ideas in the early 2000s, they have now evolved into a research topic in their own right. Although VANET primarily stood for the most often used word, inter-vehicular communication (IVC), its emphasis is on the spontaneous networking component, rather than the usage of infrastructures such as highway side units and cellular networks, as was the case in 2015 [18].

Sherali Zeadally *et al.* (2021): Recently has opened up significant opportunities for intelligent transportation security, convenience, and efficiency solutions for communications, ITSs, and computer systems, among other things. Artificial intelligence (AI) is presently being utilized in a variety of application fields because of the significant potential it has to enhance traditional data-driven methods, among other things. These data are utilized for a variety of purposes, including routing, driver awareness, and anticipating mobility to avoid potentially hazardous situations, thus enhancing the comfort, safety, and quality of the passenger experience overall [19].

Abdelkader *et al.* [2021]: Machine learning is being intensively investigated in 6G automobile networks, which enables the provision of vehicle application services and is a hot topic in the most recent literature study. With the use of vehicle management network strengthening and enhancement learning algorithms, this article offers an in-depth description of an ongoing research effort that is aimed at addressing the challenges associated with automobile communications. Vehicle networks have risen to prominence as a research topic because of their unique characteristics and uses, which include standardization, effective traffic control, road safety, and entertainment systems. In such networks, network entities must decide on how to maximize network performance in the face of uncertainty. Reinforcement learning may be used to effectively address decision-making issues to accomplish this objective. Broad wireless networks, on the other hand, have a vast and complex state, as well as actionable regions. As a result, RL may be unable to identify the optimal approach in a reasonable amount of time [20].

Marjan Kuchaki Rafsanjani *et al.* (2021): VANETs are becoming more popular. These networks are intended to enhance safety, improve urban and road traffic management, and offer passenger services, among other things. Because of challenges such as dependability and privacy, network communications must be safe and secret to be effective. As a result, we need a safe topology to retain trust while yet allowing for cryptography. To be successful, the authentication method must be capable of accurately detecting rogue nodes as well as decreasing latency and costs. One of the primary goals of the proposed technology is to establish secure and robust clusters that will contribute to the overall stability of the network. The result is the integration of trust between cars and trust between vehicles and roadside units and the choice of leaders of clusters based on the expected level of trust for each vehicle. Cluster chiefs, working in conjunction with verifiers, are in charge of keeping track of each vehicle. The cluster heads, on the other hand, ensure that messages are sent most efficiently and securely. Digitally signed messages are encrypted by a trusted authority using the Public–Private key that is provided and then decoded, resulting in each communication having a certificate from the trusted authority [21].

Jyoti Grover *et al.* (2021): VANETs based on software-defined networking (SDN) are being developed, and this is one of the main 5G technological enablers. VANETs, which link vehicles to road units, are capable of providing a broad variety of services. To enhance traffic efficiency while also improving passenger comfort, safety, and entertainment offerings, ITS has adopted this innovative technology. When it comes to managing big and dynamic networks with a stable and integrated policy, traditional VANETs are insufficient. The Open

Network Foundation (ONF) supports the adoption of SDN by promoting, among other things, open standards and logical, centralized network management. When new services and capabilities are offered, VANET may become a more flexible and programmable network. Transmission of data packets over data plane transmission devices is managed by a central control unit, which monitors the operation of the whole network and the transmission of the data packets. SDN enhances VANET's efficiency while also giving safety advantages. On the other side, the integration of new technology and architectural components into the network typically leads to new security problems. This article gives an extensive study of the design and execution features of VANETs, SDNs, and VANETs based on SDN. When utilized in combination with the conventional VANET, the influence of SDN on VANET security is explored. This essay aims to analyze the approaches for providing VANET security solutions using SDN in-depth. [22].

Benedito *et al.* (2021): Vehicle traffic is rising every day in big cities. Despite advances in telecommunications, the amount of connectivity between these vehicles must continue to increase to make mobility smarter. Several experiments were carried out to build a network between cars and the VANET computer system. Air lights can play a vital function in communicating information to cars in VANET environments which can assist in minimize congestion. Intelligent traffic lights have been proposed to change urban junctions and to decrease the usage of idle roadways. Several areas of research were previously studied in the fields of intelligent traffic management, including network latency, pedestrian safety, and driver assistance. Due to the broad variety of problems and solutions, research trends and gaps in intelligent traffic cannot be detected without a complete systematic assessment of the literature. The evaluation revealed research gaps and emerging trends in the field of intelligent traffic management. This document includes references for the finding of mapped articles to make future research more trustworthy. Furthermore, we identify research concerns and recommendations based on the number of publications read and evaluated. An extensive assessment of the research issues is conducted based on the following main themes: safety, the physical safety of persons and processes, and safety of processes [23].

Wu *et al.* (2021): In the context of sustainable economic, environmental, and economic issues, the energy Internet (EI) will enable a future energy system to provide for more flexibility in the physical fields and digitizing and social involvement in data-driven cyberspace. This article gives a comprehensive review of a multi-faceted approach to understanding how EI controls energy, data and information flows to fully grasp how EI handles these flows. The most essential features of the building are separated into three categories: energy-oriented networks, communication networks, and services. These studies also offer a complete overview of the current EI's architecture, technology, standards, services, and platforms [24].

Bhanu *et al.* (2021): Researchers in both the academic and industry sectors have been very interested in VANETs. By using networking technologies and by increasing smart numerical vehicles unexpectedly, traditional VANETs confront a wide range of methodological difficulties in the field of jobs and management. These conflicts are produced by less flexibility, inadequate intelligence, scalability, and particularly faulty connections. ITS must adopt enhanced road transport VANETs to tackle these problems and offer passengers a journey that is enjoyable, unpleasant, and stress-free. The number of cars with wireless sensors is currently growing. Due to these techniques, a wide array of powerful and promising safety technology, well-ordered public road management, driver comfort, and pleasure during trips are all enhanced. It consists of high mobility nodes that are changing constantly in dynamic areas; the dissemination of data from such nodes also creates new differences in design and implementation for researchers as well as for consumers. This results in the efficient clustering of VANET-based data distribution systems and the associated implementing issues. The use of VANET–SDN can enhance the efficiency of the network. One issue in dispute with related vehicles is that information is secure and safe throughout the configuration and that a large number of attacks may be initiated in conjunction with the configuration. On the other hand, confidence and reputation are hard to create with conventional cryptographic approaches. The history and vision of VANETs, location technology, the SDN–VANET approach, and trust management systems will be covered here for a short while, among other things [25].

Jorge *et al.* (2021): The number of mobile apps and services has risen considerably in recent years, making the maintenance of existing network infrastructures problematic. With the introduction of multiple management solutions to manage this expansion along the network chain, it is becoming increasingly difficult to coordinate diverse sections of the whole infrastructure. From its conception, the Zero-touch Network and Service Management (ZSM) concept has been created to automatically manage and configure network resources while still delivering the user experience needed. Mechanical learning (ML) is a crucial technology that many ZSM frameworks employ to offer the network management system with smart decision-making, which is becoming increasingly popular. This paper gives a detailed summary of current ZSM performance improvement strategies in ML. The major processes of standardization and related international initiatives and research efforts in this context are discussed extensively. This dissection shows the upward evolution of the ZSM paradigm. Several standards bodies, for example, have developed reference designs that will serve as the basis for future automated network management and resource orchestration services. Along with these advancements, other approaches to machine learning are employed in several ways to build on future ZSM developments, including multi-lease management, traffic monitoring, and architectural coordination. Several obstacles are emphasized to establish a stable ZMS ecosystem such as the complexity, scalability, and safety of machine learning processes, and future research directions [26].

Ferreto *et al.* (2021): VANET research dates back to the early twenty-first century. The possibility of a wireless network connection between cars has pushed the development of new protocols, devices, and a range of application scenarios. Several simulators were created to address the inherent restrictions of judging these scientific achievements using a real test board at the time. The growth of self-driving vehicles and the advent of new technologies have brought new research problems for virtual independent networks to light (e.g., 5G and edge computing).

As a result, the VANET simulators must be analyzed to determine if they are still able to evaluate these new circumstances. This article presents an updated evaluation of VANET simulators, showing their existing status and their ability to assess novel situations for VANET research [27].

Ali *et al.* (2021): In years, vehicle transmission has become increasingly popular and is currently regarded as a cornerstone of the ITS to manage road traffic effectively and minimize traffic accidents. To allow safety applications, standard families ETSI ITS-G5 and IEEE 1609 require that every car periodically communicates awareness throughout the region. With increasing vehicle density, a wide range of messages can be replaced under dynamic settings in short periods, resulting in rapid congestion of the radio channel and a worsening of security essential services. The European Telecommunication Standards Institute (ETSI) has created the decentralized congestion control (DCC), which has different performances based on the algorithm employed to decrease channel congestion caused by transmission properties (i.e., message rate, transmission power, and data rate). In this article, a study of the DCC standardization attempts and an analysis of current congestion reduction methods and algorithms are shown [28].

Kaul *et al.* (2021): A large number of resources are available to vehicles at the nodes of VANETs, allowing for the transmission of a reliable and trustworthy intelligent transmission system in a timely and secure manner. VANET may make use of these vehicle resources to improve the driving experience, reduce traffic congestion and accidents, and promote the use of contamination-free ITSs in a variety of other new applications, according to the authors. To support and promote these activities, the concept of a vehicle cloud has received a great deal of attention. In addition to allowing new kinds of collaboration with vehicles and producing a broad variety of applications and services, the deployment of VANET's cloud resulted in the extension of the VANET's and ITS' perimeter. The scope of this research included a comprehensive examination of all conceivable features of the VANET cloud. Initial research has focused on topics such as architecture and the communication types that are supported in the VANET cloud environment. Now that the necessity for the VANETs cloud has been explained, it is time to summarize the reasons behind this. A comprehensive comparison between the VANET cloud and its parent network, i.e., VANETs, may be used to demonstrate the similarities and differences between the two networks. We will be concentrating our efforts in this post on the search for different apps that are supported by the VANET cloud. It has been determined that there are many problems associated with the security, implementation, and usage of the VANETs cloud. Finally, this paper discussed the use of VANET clouds in a variety of future technologies, including the IoT, software-defined networks, and cognitive radio [29].

Shawal *et al.* (2021): A VANET is a group of mobile or stationary wireless networks. VANETs play an essential role in providing safety and comfort for drivers in the car. You are responsible for delivering intelligent traffic management and real-time event information. VANETs have taken care of applications, such as safe driving, smart sailing, emergency response, and entertainment for vehicles. However, the increasing number of connections presents a range of new security and security problems, which may be fatal for both the host and the linked features. Violating national privacy as well as identifying automobiles is a major impediment to the usage of interactions with compelled vehicles. The phrase “location privacy” means both the vehicle (driver) and the privacy of the location. When a vehicle transmits a message, none but authorized bodies should be aware of the real identity or position of the vehicle. During processing, location secrecy is a key design element of VANET operations, given the necessity for validation of all communications provided by the vehicle. The peculiarity of this essay is that it addresses operational and security problems related to the privacy of an individual located in VANETs. This chapter also includes a detailed analysis of the many attacks, identity theft, and manipulation, and other advanced ways of privacy on VANET systems. By using the material in this article, researchers may learn a great deal about privacy issues and various security concerns faced by VANETs and critically assess the present VANET privacy protection technology [30].

Behzadan *et al.* (2021): The intelligent transport system (ITS) aims at integrating sensing, control, analysis, and communication technologies into transportation infrastructure, mobility, comfort, safety, and efficiency. Carmakers continue to build smarter automobiles while road and infrastructure upgrades have changed the traveling experience. The utilization of new technologies, as well as research and development in ITS, are achieving greater efficiency and dependability. Although each year better cars are created with more attention on passenger and pedestrian safety, new technology and an expanded ITS link characterize evil actors. New issues about privacy are generated by the data collected by passengers and their patterns of travel in clever cities with integrated public transport systems. In this article, we offer a thorough classification of information technology systems security and privacy problems [31].

Shirin *et al.* (2021): IoT has made it possible to link things and share information in recent years, thanks to the proliferation of IoTs across a variety of sectors. With the help of this function, the author may satisfy the needs of various industries. Smart transport makes use of the Internet of Vehicles (IoV) as a vehicle communication solution to improve efficiency. Traffic control applications and services are improved to improve road safety. The author identified and proposed a taxonomy for IoT services, applications, and infrastructures. To ensure high-quality apps and services, we have prioritized them based on driving needs as well as non-functional criteria. This article summarizes the present status of IoT in architectures, services, and application development. It may serve as a starting point for developing solutions to traffic congestion issues in urban areas. Applied to the creation and administration of intelligent cities, this research is beneficial. These precautions need the employment of IoV features such as real-time operation, application accident prevention, and complex user data management [32].

1.4 Standardization routing protocols of VANET

VANET’s route protocols are divided into five categories: a routing protocol based on topology, a routing protocol based on a position, a cluster-based protocol for

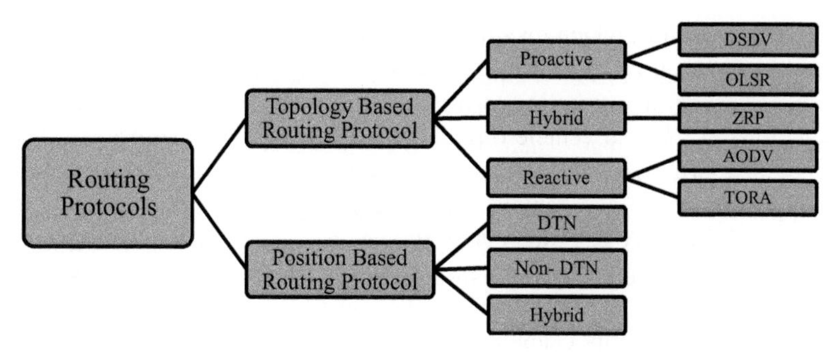

Figure 1.5 Types of routing protocols [34]

rutting, a geo-positioning protocol, and the radio transmission routing system. These protocols are distinguished by the most suited area/application [33]. Figure 1.5 describes the types of routing protocols.

1.4.1 Topology-based routing protocols

These routing techniques employ information about the network connection to deliver packets. They are also proactive and reactive. A pre-active routing protocol means that independent of communication requests, routing information such as the following forwarding hop is maintained in the background. The advantage of the proactive routing protocol is that it provides minimum latency in real-time applications since the target path is retained in the background. A table is built and held in an ode. Each entry in the database is therefore the next-hop node to a particular destination. It also allows the preservation of idle data channels that limit the bandwidth available. It only maintains the present routes and lowers the network pressure. Routing involves the route search phase during which query packets are flooded in the searching network, and when the route is identified, this phase is finished. All kinds of reactive routing protocols are AODV, DSDV, OLSR, TORA, etc. [35].

1.4.2 Position-based routing protocols

Position-based routing includes the routing algorithm. They share the role of choosing the next transmission hopes with information on the geographical location. The packet is transmitted without map information to the nearest one-hop neighbor. Position-based routing is a benefit since it is not required to build and maintain the global route from the source node to a target node. The position-based routing is separated into two different kinds [36].

1.4.2.1 Position based on Greedy V2V protocols

The message moves in greedy strategy to the distant neighbor and an intermediate node to the next goal. The Greedy method requires that the intermediate node be located, located, and destined. This is why these protocols are sometimes called min delay protocols to send data packets to the destination as fast as feasible. GPCR, CAR, and DIR are several kinds of greedy location-based V2V protocols [37].

1.4.2.2 Greedy perimeter coordinator routing

Greedy perimeter coordinator routing (GPCR) is based on the assumption that the urban road is a natural planner scheme. For its functionality, GPCR does not require an external static street map. There are two components of the GPCR: A restricted Greedy transmission technique, a routing algorithm repair strategy. A GPCR uses a greedy transmission strategy based on its location and routes to crossroads. GPCR does not use an external static road map, therefore nodes cannot be identified at the intersection. GPCR uses a heuristic approach to locate and mark nodes at crossings as coordinators [38].

(i) Neighbor table approach

The beacon nodes routinely transmit information about their location and the last location of each neighbor by listening to the beacon messages of their position and the position of the neighbor as data on their position.

(ii) Correlation coefficient

In that instance, the node utilizes its location data and its immediate neighbors position information to identify the coefficient. This strategy is preferable to the approach in the next table. The technique can avoid dependency on the external road map by utilizing this strategy.

(iii) Connectivity aware routing protocols (CAR)

CAR techniques identify routes toward a destination and have a specific purpose in the cache of successful routes between diverse source and destination pairs. The position of the destination route is also forecasted if the location changes. Hello, beacons with vector data are routinely issued by CAR protocol nodes. Piggybacked beacons can also be stored on bandwidth and congestion waste packets. When the distance between the nodes in the following table exceeds the threshold value, the entries expire.

(iv) Protocol on routing based on the diagonal crossroads

The DIR protocol establishes numerous diagonal connections between the vehicle's source and destination. The DIR protocol is developed into a geographical protocol in which the source car geographically distributes data packets to the first diagonal crossroad, second diagonal crossroad, and so on. Auto-adjustment involves dynamically choosing a sub-route with a shorter delay in data packet between two close diagonal crossings for the transfer of data packets. Self-adjustment denotes DIR. To decrease the data packet, the route delay is automatically picked with the lowest subpath delay. The DIR protocol may automatically adjust the routing path to preserve the lower packet latency.

(v) Protocols on tolerance delays

Finding a message node in an urban environment where the vehicle is thick is not a difficulty, but few automobiles work in rural circumstances or cities at night, and it is tough to build end-to-end highways. In such conditions, certain concerns must subsequently be taken into account in sparse networks. Delay protocols MOVE, VADD, and SADV are tolerant in various types.

(vi) Assisted vehicle for data delivery (VADD)
VADD utilizes a forward-looking approach to enable car packets to be delivered over sparse networks when the node enters the broadcast range and therefore allows a sparse network packet to be carried through a relay. VADD requires each car's position and an external static road map. E Packet is composed of three modes: traversing, right, and destination. Each mode depends on the node position of the packet. The crossing is utilized when the packet reaches an intersection where routing choices can be used to pass the packet to a vehicle with one of the several connected routes. The current node is mode Right on a route where the packet may only go in two ways into the present node or the opposite end.

(vii) Static node-assisted adaptive routing protocol (SADV)
In sparse communications networks, SADV focuses on reducing latency. S ADV also adjusts dynamically to different traffic volumes by letting every node measure the time required for the transmission of a message. S ADV implies that every car has access and knows where it is via GPS to an external static road plan. Three major components are available: SNAR, Multipath Data Diffusion, and Link Delay Update. S ADV consists of three components. Two modes are used to operate SADV: "Road Mode" and "Crossroad Mode." The SNAR uses the optimal pathways given by the graph in the route map. By monitoring the delivery time between static nodes, LDU maintains the delay matrix dynamically. MPDD enables routing multipath.

1.4.2.3 Cluster-based routing

It is recommended that cluster-based routing be carried out in clusters. A node group identifies as a part of the cluster and the cluster is allocated an ode in front of the cluster. For large networks, there can be great scalability, but network latency and overhead are achieved when highly mobile VANET clusters are established. A cluster-based architecture of the virtual network must be built with clustering nodes for scalability [39].

1.4.2.4 Broadcast routing

In the field of VANET, broadcasting is common in terms of sharing, traffic, weather and emergency, vehicle road conditions, advertising, and advertising. The various routing protocols for broadcasting include BROADCOMM, UMB, VTRADE, and DV-CAST [40].

1.4.2.5 Geo cast routing

Mainly Geo cast routing is a multi-cast routing. The goal is to deliver the packet from the originating node to all other nodes in a certain geographical position (Zone of Relevance ZOR). In geo-casting vehicles beyond ZOR, unnecessary fast reactions are not advised. It generally establishes a forwarding area for packet flooding to reduce the overhead message and network congestion that flood packets cause everywhere. Unicast routing may be used to move the packet to the target location. One risk of Geo cast is the splitting of networks and unwanted neighbors that might prevent the transmission of correct messages. Geo-casts include IVG, DG-CASTOR, and DRG routing techniques [41].

1.5 Types of ad hoc network and characteristics of VANET

Currently, wireless communication technologies for phones and the Internet are common in the industrialized world. Many people use these technologies daily to talk, to social networking, to read news and e-mail, to surf the world, to play and love things. Including smartphones, tablets, computers, all wireless cards allowing the user to communicate with others [42]. The normal situation in the preceding usage scenarios is a client who utilizes a communication gadget to access a service provider's network. Therefore, the network operator needs to build some type of infrastructure to deliver wireless coverage to network users. Although this strategy is popular and incredibly effective, the availability of communication infrastructure cannot blindly depend on communication communications. For example, natural calamities or military campaigns could have destroyed it. To coordinate the effort, rescue teams and the military still need to contact. In other circumstances, direct connections without crossing other elements are the ideal approach to provide a critical delay service. For example, consider a vehicle that breaks down and transmits a warning message to the cars that approach. To avoid a collision, the message should be reached as soon as possible by neighboring cars to ensure that direct contact with vehicles is better than the intermediary part of the infrastructure (which adds more delay). In other cases, the usage of money-saving communication over the network infrastructure is merely more beneficial. Thus, if a voice call can be established between the closest terminals without the operator network, the user will not bear the charge [43].

1.5.1 Types of ad hoc network

These specialties differ in node mobility, power processing, criticalities of energy efficiency, etc. Although they all uphold the same core principles for wireless multi-hop communications, they have enough specifics to ban effective one-size-fit solutions [44]. Figure 1.6 shows with their applications the ad hoc network types.

1.5.1.1 Mobile ad hoc network

This is the most common ad hoc type of network. A Mobile ad hoc network (MANET) is made up of an arbitrary collection of wireless models. It is meant to be a spontaneous, unplanned network. Node mobility and wireless signal fluctuations change the topology of the network over time. Battery nodes are regularly operated; energy efficiency is consequently one of the most significant design problems. Computing and storage resources may be restricted, although not always. Because a MANET provides an excessively broad environment, network behavior might be of little benefit. Other ad hoc networks, sometimes in military, rescue, or similar networks, are generally employed in MANET.

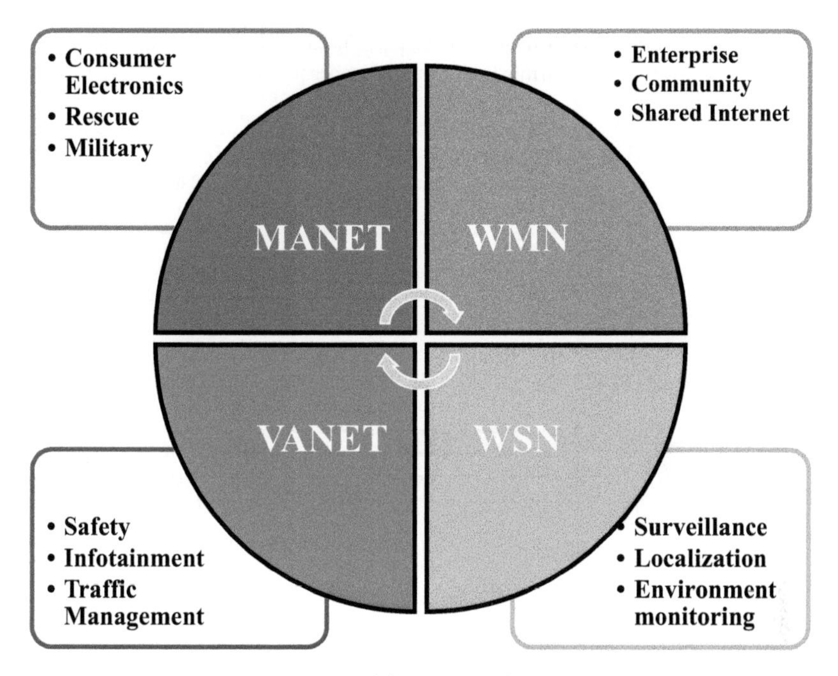

Figure 1.6 Ad hoc network types [45]

1.5.1.2 Wireless mesh network

This is a unique situation in the ad hoc network where nodes are static base stations that interact with multi-hop routes. They offer a backbone for the communication of mobile consumer devices between mesh nodes. Note that between mesh nodes an ad hoc network is created. Although static, the architecture of the network is dynamic, because wireless signals are fundamental in making the network independent in a different environment. Mesh routers are frequently equipment devoted to ongoing electricity. Energy, computing, and memory resources are typically not an issue. In contrast to generic MANETs, a wireless mesh network (WMN) can be produced or not. A WSN usually consists of simple wireless sensor-based devices. Each sensor has very limited energy, memory, and computational capabilities. Efficiency is consequently of critical significance in this sort of network. A WSN can therefore be composed of hundreds or even thousands of nodes. The target is a sink device that processes observed data in many applications [46].

1.5.1.3 VANET

It is a particular example of a MANET consisting of automobiles. This knowledge has a major influence on the assumptions that a VANET may make. Vehicles cannot roam free because they must meet existing road and road topologies and traffic conditions. They can achieve huge speeds and produce extremely unstable network architecture sometimes. Moreover, a VANET might be made up of multiple nodes. Cars have no energy limitations and can be fitted with important

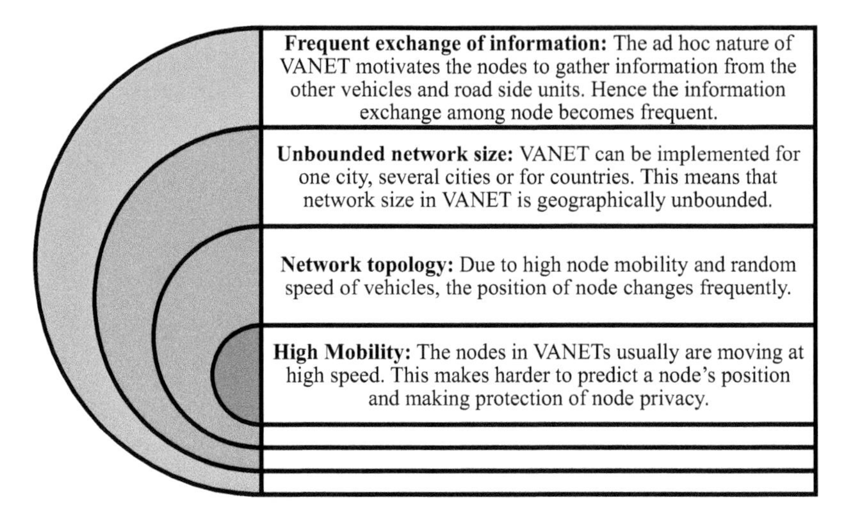

Figure 1.7 VANET characteristics

communication and processing capacity. However, many technological barriers must be solved before the service is supplied [47].

1.5.2 VANET characteristics

VANET is a MANET application but has its features which may be summarized as seen in Figure 1.7 [48].

1.6 Frameworks/models of VANET

Since VANETs have a potentially huge scale, it needs extensive development to introduce new technologies into VANETs and the experimental implementation is highly expensive [49]. In general, before the market introduction, two crucial and necessary procedures are taken:

- simulation analysis and assessment;
- field-test analysis and verification.

1.6.1 VANET models

VANETs are a large and complex global model of a system consisting of four sub-models for each aspect: a vehicle and driver model, a traffic flow model, and a communication model, as shown in Figure 1.8 [50].

1.6.2 Simulation methods

Simulation is undoubtedly a necessary step forward before new technologies are used in VANETs. A VANET simulation includes two separate components: a traffic simulator and a network simulator described in Figure 1.9 [51].

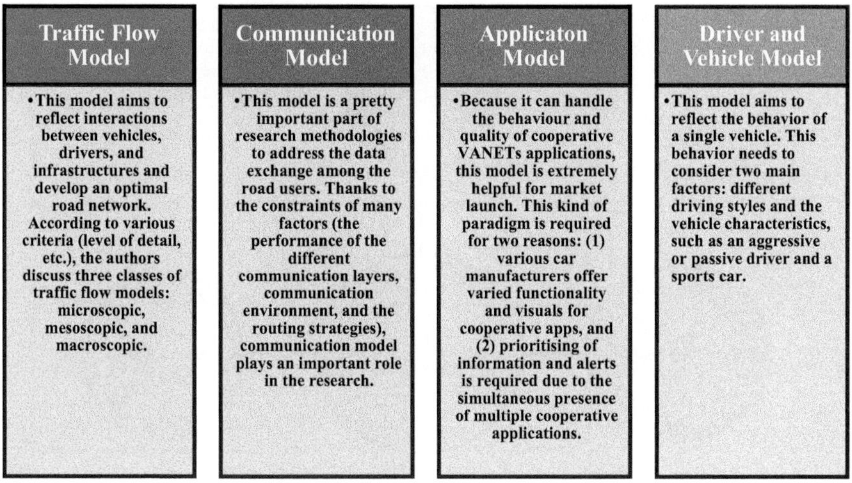

Traffic Flow Model

- This model aims to reflect interactions between vehicles, drivers, and infrastructures and develop an optimal road network. According to various criteria (level of detail, etc.), the authors discuss three classes of traffic flow models: microscopic, mesoscopic, and macroscopic.

Communication Model

- This model is a pretty important part of research methodologies to address the data exchange among the road users. Thanks to the constraints of many factors (the performance of the different communication layers, communication environment, and the routing strategies), communication model plays an important role in the research.

Applicaton Model

- Because it can handle the behaviour and quality of cooperative VANETs applications, this model is extremely helpful for market launch. This kind of paradigm is required for two reasons: (1) various car manufacturers offer varied functionality and visuals for cooperative apps, and (2) prioritising of information and alerts is required due to the simultaneous presence of multiple cooperative applications.

Driver and Vehicle Model

- This model aims to reflect the behavior of a single vehicle. This behavior needs to consider two main factors: different driving styles and the vehicle characteristics, such as an aggressive or passive driver and a sports car.

Figure 1.8 Different VANET models

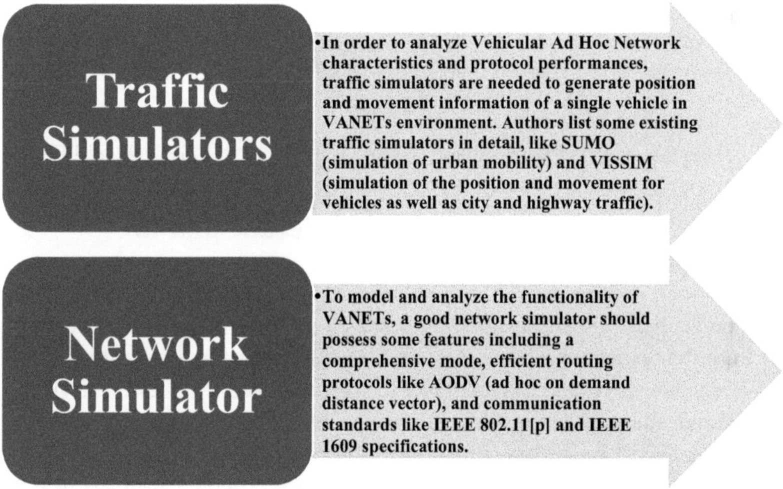

Traffic Simulators

- In order to analyze Vehicular Ad Hoc Network characteristics and protocol performances, traffic simulators are needed to generate position and movement information of a single vehicle in VANETs environment. Authors list some existing traffic simulators in detail, like SUMO (simulation of urban mobility) and VISSIM (simulation of the position and movement for vehicles as well as city and highway traffic).

Network Simulator

- To model and analyze the functionality of VANETs, a good network simulator should possess some features including a comprehensive mode, efficient routing protocols like AODV (ad hoc on demand distance vector), and communication standards like IEEE 802.11[p] and IEEE 1609 specifications.

Figure 1.9 Description of types of simulation methods

1.6.3 Field operational testing

Although the method of simulation contributes greatly to VANET research, it does not reflect cars' real environment. Researchers have been lured to the field of operational testing (OTT) to tackle these challenges, which attempt to study and evaluate these applications in a wider range of realistic environments.

These testing can move the VANET system closer to the market and generate economic value. On the contrary, FOT data can make network models more realistic and performance-enhancing. The characteristics of FOT are described in Figure 1.10 [52].

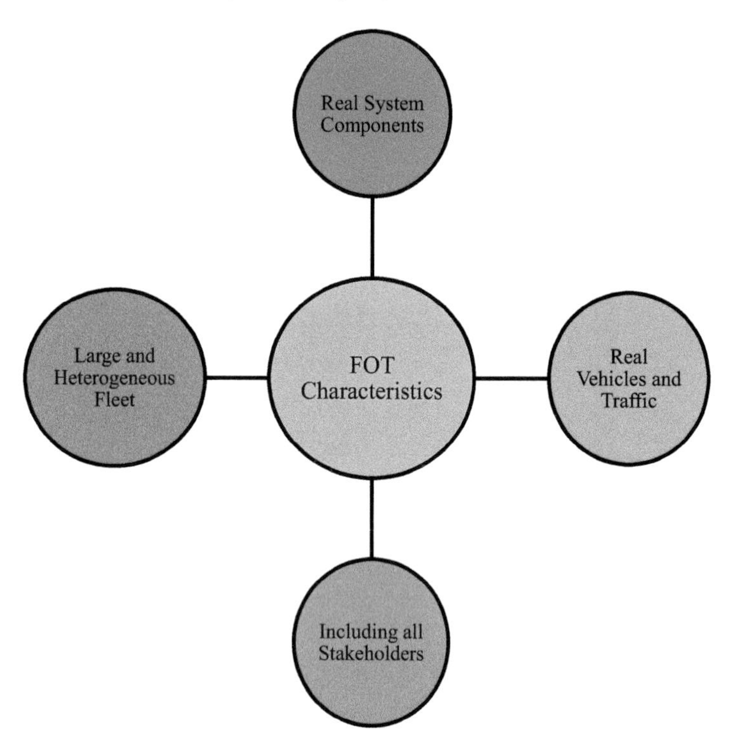

Figure 1.10 Field operational testing (FOT) characteristics

1.7 VANET architectures: previous decade and in current era

Drawing on the list of the most significant research subjects in the VANET (extracts and blends from the "request for papers," a series of eight research topics that span a wide variety of challenges related to automotive communications) (top-level categories). On the other hand, the APP class stores all documents relating to user applications (e.g., intersection collision avoidance, road congestion notification, and multimedia streaming). Therefore, if the new MAC or routing protocol was proposed, the paper was named MAC-PHY or ROUT instead of PERF, and its performance evaluation [53] was followed. TOOL was documented that detailed a new tool, platform, structure, or architecture. All experimental research, usually dealing with deployment and field tests, has also been included for simplicity. The ROUT class is a new protocol routing paper, whereas the MOB class focuses on mobility issues such as mobility modeling and clustering. Over the last decade, academia and business have sought to ensure the efficiency and success of wireless communication within vehicles. The applicability of network coding techniques in VANETs might be a significant factor for this trend. Second, we have the SERV class, composed of papers on supplementary services, whether it is the Quality of Service (QoS), security techniques, or location services. This section covers the network system architecture of ad hoc vehicles [54].

1.7.1 Principal components

VANET is possible for organizations that may be split into three domains: a mobile domain, an infrastructure domain, and a larger domain by IEEE 1471-2000 and the architectural standard ISO/IEC 42010. The vehicle area contains all vehicle kinds, including cars and buses. The Mobile Domain encompasses all sorts of mobile devices, including personal browsers and smartphones. In the infrastructure sector, there are two domains: the roadside infrastructure domain and the central infrastructure domain. The field of road infrastructure includes roadside organizations such as traffic signals. Infrastructure management centers like TMCs and vehicle management centers are the main area of infrastructure. The architecture of VANET nevertheless changes from region to region. There are three sections in the system design: in-car, ad hoc, and infrastructure. In general, these connections are wired and sometimes wireless. However, the ad hoc realm comprises automobiles with On Board Units (OBUs) and Roadside Units (RSUs). An OBU may be seen as a mobile ad hoc node and RSU is also a static node [55].

1.7.2 Architecture of a communication

VANETs may categorize communication types into four kinds. In-vehicle communication, increasingly required and crucial for research in VANETs, refers to the domain of the vehicle [56]. Figure 1.11 explain the many forms of VANET architectures of communication

- The vehicle communication system can assess vehicle performance, in particular driver weariness and somnolence vital to drivers and public safety.

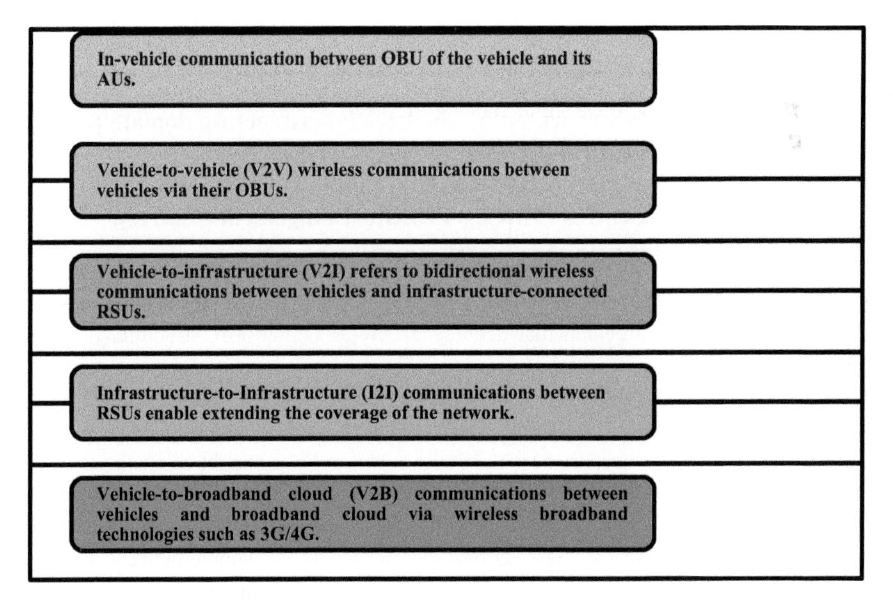

Figure 1.11 Different types of VANET communication architecture

- Vehicle communication (V2Vs) can provide drivers with a data exchange platform to convey information and warning messages to aid drivers.
- Another intriguing area of research in VANETs is the relationship between vehicle-to-road infrastructure (V2I). V2I connectivity enables drivers to update traffic/weather in real-time and offers monitoring and environmental sensing.
- Vehicle-to-vehicle broadband communication (on cloud) means vehicles can connect themselves using 3G/4G wireless broadband technology. Since there may be more information, data monitoring and entertainment on the broadband cloud, this type of connection is beneficial for active driver support and vehicle tracking.

Note that vehicle-to-building (V2B) charging is a fruit, like vehicle-to-grid (V2G), of large battery capacities developed for the Electrical Vehicles (EVs).

1.7.3 Layered architecture for VANETs

The open systems interconnection (OSI) model, which puts identical communication into one of seven logical layers, is well known to most readers. VANETs may usually be designed from area to region and hence they differ in their protocols and interfaces. IEEE 1609 offers an intermediary level of the standard family for flexible support of VANET security applications, whereas other protocols do not allow security applications at the intermediate level of the protocol stack. In an example, three extremely dependable protocols do not offer security applications with services for network layer services and transport layer services: TCP, IPv6, and UDP [57].

Mobile domain: Includes car domains and mobile gadgets. The first contains all vehicle kinds (e.g., cars, trains, buses). This includes all kinds of mobile devices (e.g., smartphones, laptops, smart watches). Figure 1.12 illustrates VANET kinds of mobile domains.

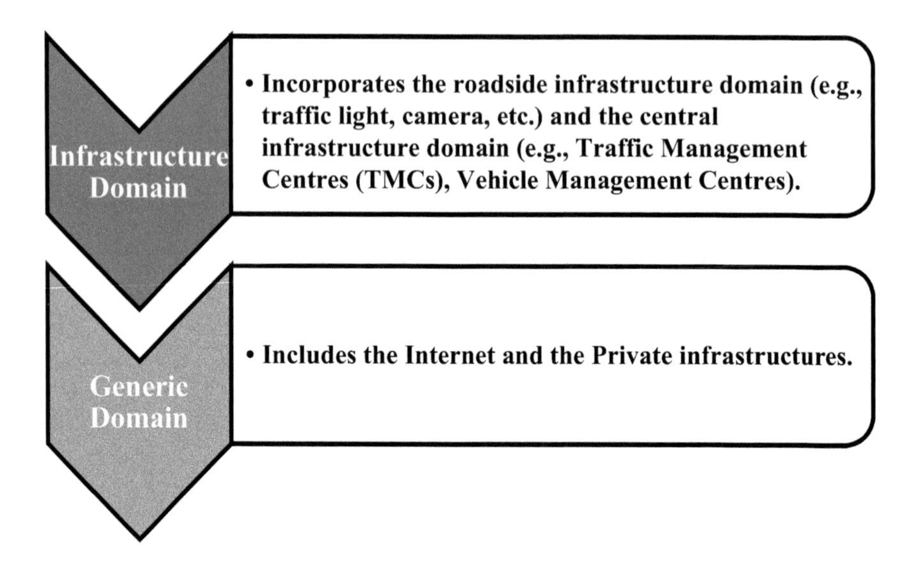

Figure 1.12 Types of mobile domain in VANET

The European architectural standard for VANETs is slightly different. The CAR-2-X communication system is based on the CAR-2-CAR Communication Consortium. The architecture of the C2C Communication System comprises the following areas:

- *The domain of the in-vehicle* consists of a single AU and an Onboard Unit (OBU). An AU is a specialized device that can be a car element or a separate mobile device such as a smartphone, laptop, etc. It runs one or more programs that benefit from OBU communication. The AUs and OBUs are permanently linked by cable or WiFi.
- *The ad hoc domain* consists of OBUs and stationary RSUs deployed on a certain roadside location. With short-range wireless communication devices, OBUs may communicate with each other directly or multi-hop to allow for ad hoc communication between vehicles. An RSU is a stationary device accessible through the Internet or the infrastructure network.
- *Infrastructure domain* access includes HSs and RSUs. If either RSU or HS does not offer Internet connectivity, OBUs can run mobile radio networks such as HSDPA, WiMax, and 4G, for example.

1.8 VANETs security services

MANETs has presented a new safety concern, which the researcher considered to be a major difficulty addressing security issues, including fewer key issues, mobility, poor wireless connectivity, and driving problems [58]. The driver also has to inform the traffic conditions in a defined time range accurately. Because of its unique characteristics, the VANET is particularly vulnerable to attacks. For VANET security it is important to state the criteria that the system must comply with the appropriate network operation. Threats or assaults in VANETs may not be able to meet these conditions. The main safety needs are grouped into five main areas: availability, secrecy, authentication, data integrity, and non-repudiation [59]. These security services are addressed as seen in Figure 1.13.

- *Availability*: Due to its direct connection with all security applications, availability is the most crucial element of security services. Availability is primarily responsible for functional management and its safety must guarantee that network and other applications stay functional in the event of malicious or defective situations.
- *Confidentiality*: Secret information based on certificates and shared public keys guarantees that the specified recipient has access to data while external nodes are prevented from accessing the data until the designated user has received confidential data.
- *Authentication:* In VANET, authentication plays a key role. It stops VANET from fighting suspicious network entities. Related mode information such as user identification and sender address are necessary.
- *Non-repudiation*: it guarantees that the recipient of the message does not refuse to transmit and receive the dispute sender.

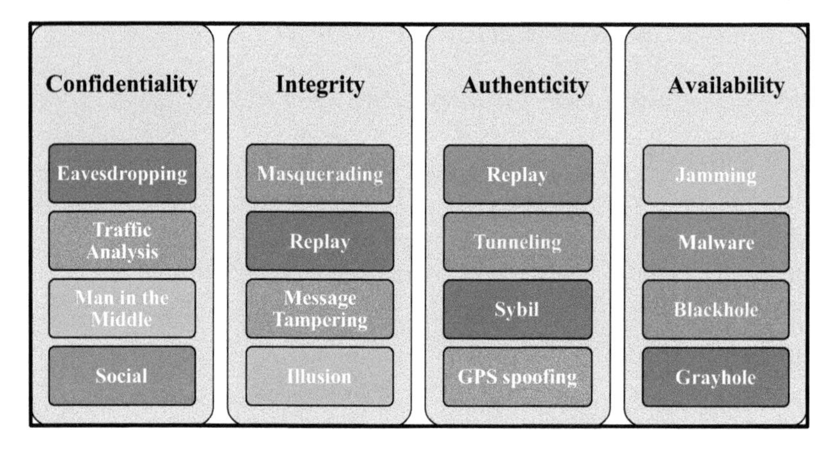

Figure 1.13 Security services [60]

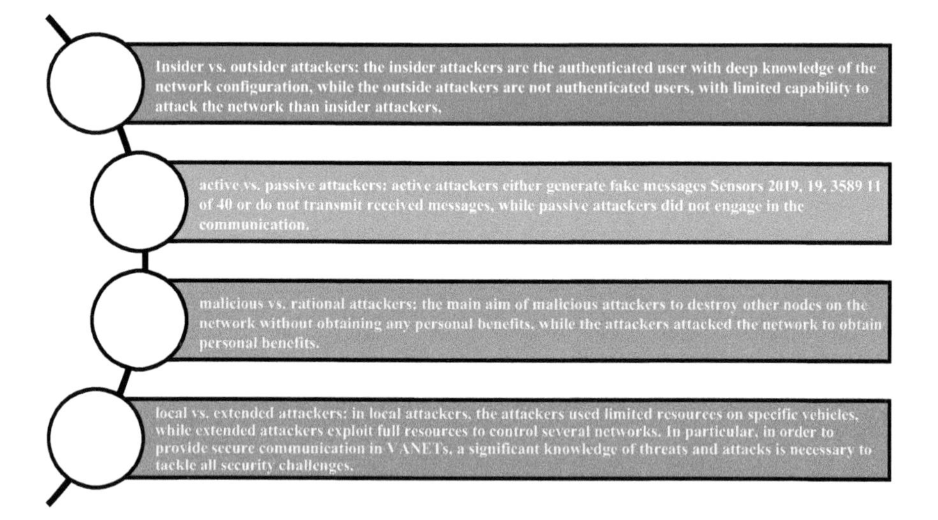

Figure 1.14 Different types of threats in VANET

Security threats and attacks in VANETs

Due to their fundamental characteristics, such as their rapid mobility and frequent disconnection VANETs are more vulnerable to assault. Due to the high mobility of vehicles, the network topology of VANETs is rapidly developing [61].

The vulnerability of VANETs increases as a result of the attacker's ability to insert false information into the message being broadcast. VANETs are exposed to a wide range of potentially hazardous assaults as a result of these difficulties. In VANETs, the attacker's goal is to cause widespread disruption throughout the network to further his interests. Figure 1.14 depicts the classification of attackers in VANETs, which are divided into four groups [62].

1.9 VANETs privacy services

The desire for privacy is addressed in a variety of ways in different nations. For the sake of crime prevention, several nations require drivers to be identified by their license plates. Others, on the other hand, may put a required privacy policy on the system in their nations. Furthermore, the necessity for privacy is one of the most important factors in gaining public approval for VANET implementation [63]. Network communication should remain anonymous, meaning that a message should not expose any information about the sender. In addition, the communication delivered should be safeguarded if an unauthorized observer comes into contact with it. Furthermore, the behaviors of a sender should be unlikable to the person who is receiving the message. A greater level of anonymity is envisaged in certain plans, in which the identities of cars transmitting notifications are concealed from the public, as well as from the government. However, complete anonymity may allow for the possibility of misbehavior since attackers would be able to act intentionally without fear of being discovered [64].

Figure 1.15 illustrates several privacy considerations that should be taken into account while developing a privacy-preserving architecture for VANET.

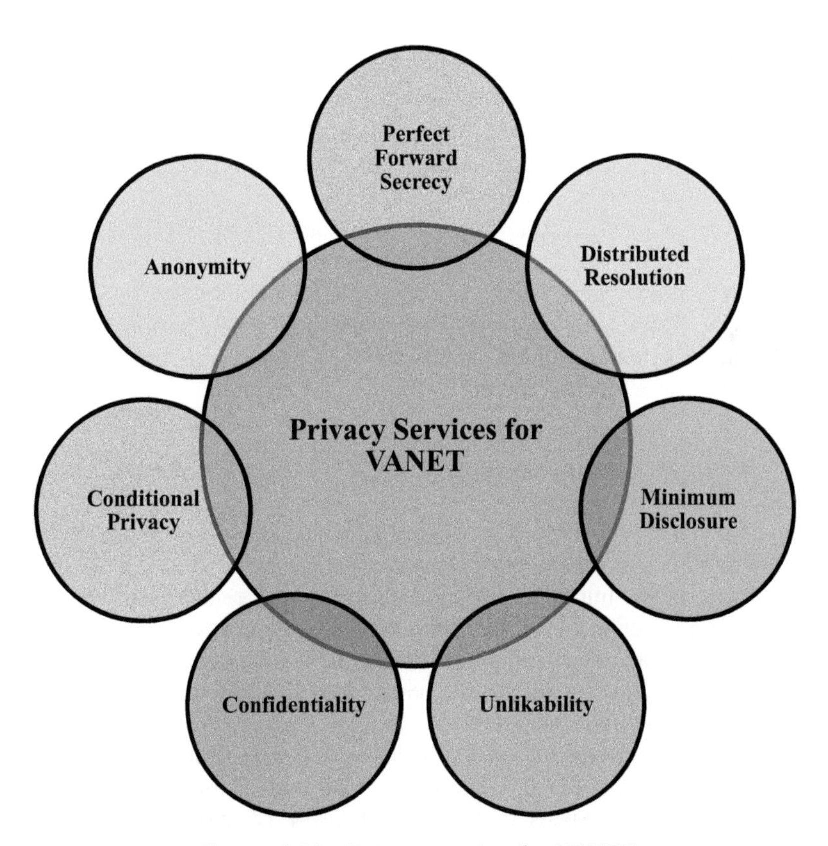

Figure 1.15 Privacy service for VANET

Anonymity: The sender of communication should be indistinguishable or anonymous when a collection of senders is involved. VANET must give anonymity to senders/drivers to protect the privacy of those who transmit messages. As a result, it should not be able to link the content of a communication to the identity of the person who sent the message. However, this creates a contradiction between the need for responsibility and the need for anonymity. Conditional anonymity is therefore required to meet both the security and privacy criteria.

- *Unlikability*: An adversary is unable to determine if the Items of Interest (IOI) (messages, behaviors, and/or subjects) utilized in vehicular networks are linked or unrelated in a sufficient amount of detail. It is important to note that the unlikability of the sender to a particular message might be referred to as anonymity, because the sender's anonymity may be compromised as a result of this.
- *Message confidentiality*: is ensured by this security service, which prohibits the exposure of message contents to unapproved parties to protect the privacy of the user.
- *Minimum information disclosure*: During a communication, a user should only give the bare minimum of information. The information about the user that is exposed during a transaction should be kept to a bare minimum, in other words, no more information than is necessary for the task. The information gathered should be tailored to the unique requirements of the individual involved.
- *Conditional privacy*: There is no question that a driver benefits from the traffic-related communications that are automatically transmitted by other nearby cars. These communications, on the other hand, contain private information about the sender, such as the identification of the car (plate license number), its location, and its direction. Individuals are apprehensive about disclosing this sensitive information to unauthorized third parties. As a result, a safe method should prohibit an unauthorized person from learning the combination of true identity and other sensitive information.

1.10 Research issues over privacy and security attack

Today, newer and more powerful onboard apps may store a wide variety of personal information and trajectory information that can be used to disclose people's behaviors, habits and traces, and whereabouts [65]. Before deploying the VANET communication architecture, the aforementioned risks must be addressed. Otherwise, the confidence, reliability, and individual acceptance of the system would probably be low because of the potential of attackers altering the signals or tracking automobile movement.

Table 1.2 provides an overview of assaults and compromised services. During the past several years, various approaches have been introduced to tackle the security and privacy problems created by the Internet. Most focus more on two key characteristics of VANETs: communication and network architecture.

Table 1.2. Summary of attacks and compromised services [66]

Compromised services	Attacks	Category
Availability	Timing attack	Content and user privacy
Authentication	Snooping attack	
Authentication Non-repudiation Integrity	Masquerade	
Authentication Availability	Sybil	
Integrity Authentication	Bogus information	Content protection
Integrity Authentication	Replay	
Authentication Confidentiality Integrity Nonrepudiation	Man-in-the-middle	Infrastructure protection
Confidentiality	Wormhole	
Availability	Black hole and gray hole	
Availability Authentication	DOS	

1.11 Open challenges towards VANET

VANETs are one of the foundational technologies for a transportation system that is efficient, safe, instructive, and entertaining, among other things. Nowadays, individuals spend a substantial portion of their time in automobiles or other forms of transportation [67].

Higher levels of safety can be achieved in this setting by dependably sharing crucial events. Better efficiency is accomplished by reducing traffic congestion and pollution while also increasing the predictability of journey time. VANETs make use of two sorts of communications: beacon messages and safety messages [68]. In contrast, safety messages aid automobiles on the road through the transmission of emergency information to avert accidents and to protect persons from life-threatening conditions. VANETs employ RSUs as gateways that function as an on-the-go virtualization layer that allows users to access cloud services. Further, then the present tendency, future VANETs, and their applications will include new developing technologies that will add new functionality to their existing features. Figure 1.16 depicts some of the most significant problems that future VANETs will face.

1.12 Trends in near future for VANET's applications

As an access point or router, the RSU may be used to store and provide data when needed. It can also be used to store and deliver data when needed [70]. Vehicles are

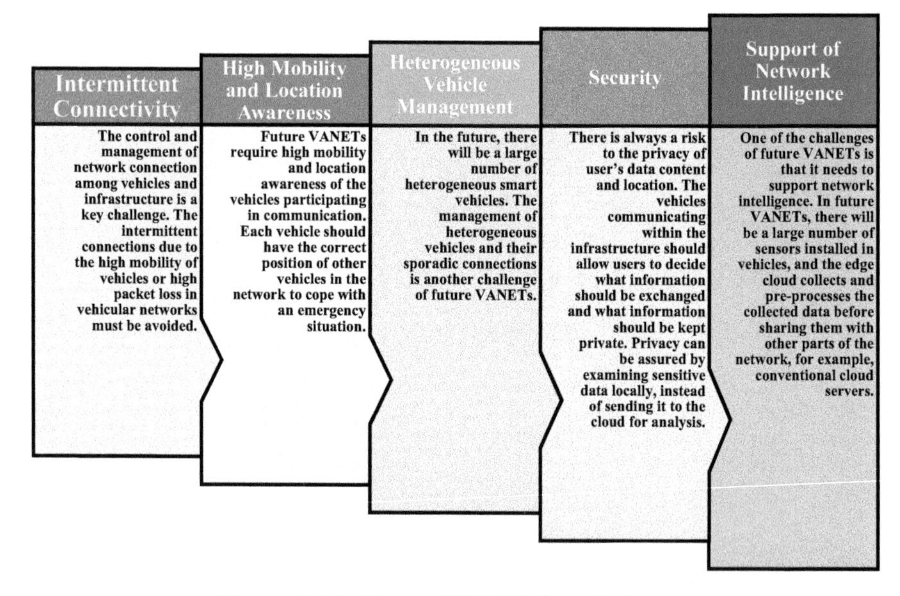

Figure 1.16 Future VANETs key challenges

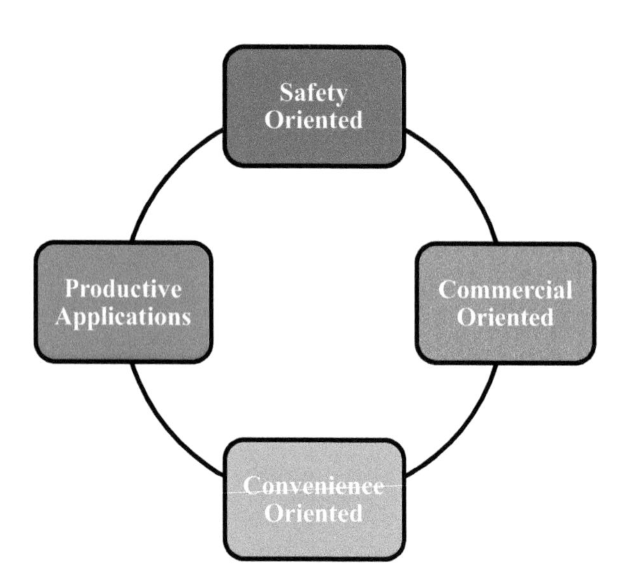

Figure 1.17 Applications of VANET

responsible for uploading and downloading all data from RSUs. According to Figure 1.17, Four types of applications may be divided into vehicle-to-vehicle traffic, vehicle-to-home, and routing-based applications. Applications are divided into four categories. The applications of VANETs are classified according to whether they are V2I or V2V communications [69].

1.12.1 Safety applications

As seen in Figure 1.18, safety applications include the monitoring of the surrounding road, oncoming cars, the road surface, road bends, and other factors.

1.12.2 Commercial applications

Commercial apps offer entertainment and services to the driver, such as Web access, music, and video streaming, and other features and functions. Commercial applications are classified as shown in Figure 1.19.

1.12.3 Convenience applications

The convenience application largely works with traffic management to improve traffic efficiency by increasing driver comfort. As illustrated in Figure 1.20, the comfort applications may be categorized.

1.12.4 Productive applications

It is deliberately called productive since this application is extra to the apps listed before. Productive applications as seen in Figure 1.21.

1.12.5 Future work

Since mobile devices are recognized and used by us in our everyday lives, the future of VANETs is surely safe. It formed part of the government's efforts. The cameras, also known as Concept II, are manufactured by Tele-Traffic UK and are

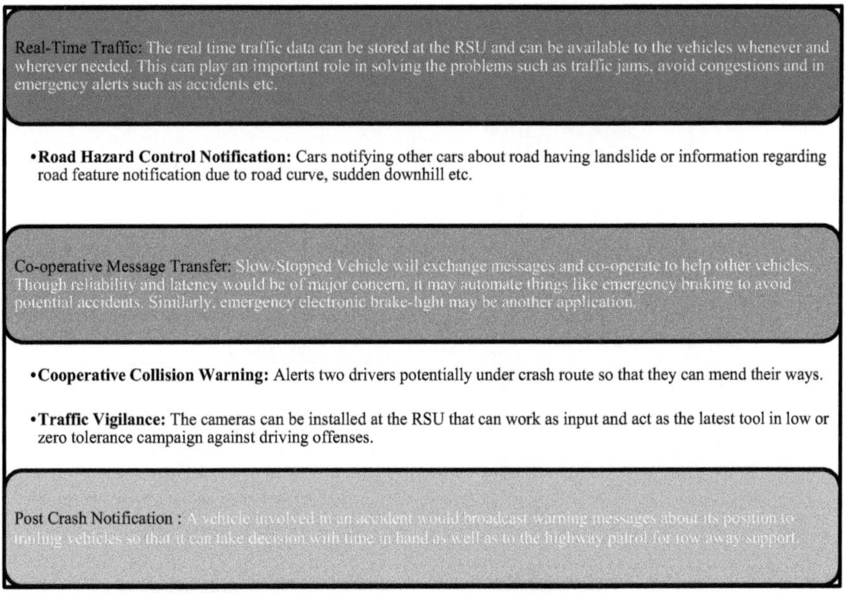

Real-Time Traffic: The real time traffic data can be stored at the RSU and can be available to the vehicles whenever and wherever needed. This can play an important role in solving the problems such as traffic jams, avoid congestions and in emergency alerts such as accidents etc.

- **Road Hazard Control Notification:** Cars notifying other cars about road having landslide or information regarding road feature notification due to road curve, sudden downhill etc.

Co-operative Message Transfer: Slow/Stopped Vehicle will exchange messages and co-operate to help other vehicles. Though reliability and latency would be of major concern, it may automate things like emergency braking to avoid potential accidents. Similarly, emergency electronic brake-light may be another application.

- **Cooperative Collision Warning:** Alerts two drivers potentially under crash route so that they can mend their ways.
- **Traffic Vigilance:** The cameras can be installed at the RSU that can work as input and act as the latest tool in low or zero tolerance campaign against driving offenses.

Post Crash Notification : A vehicle involved in an accident would broadcast warning messages about its position to trailing vehicles so that it can take decision with time in hand as well as to the highway patrol for tow away support.

Figure 1.18 Road safety applications

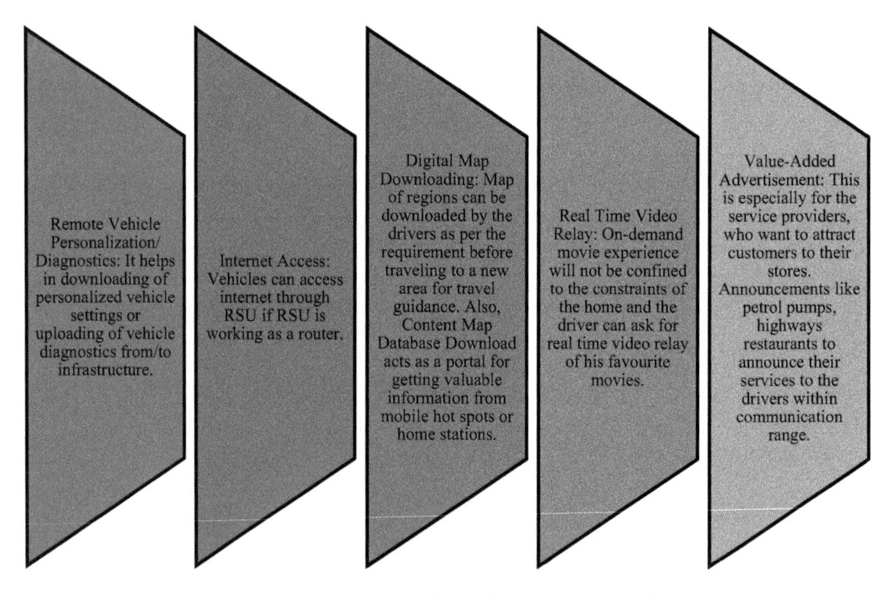

Figure 1.19 Commercial applications classifications

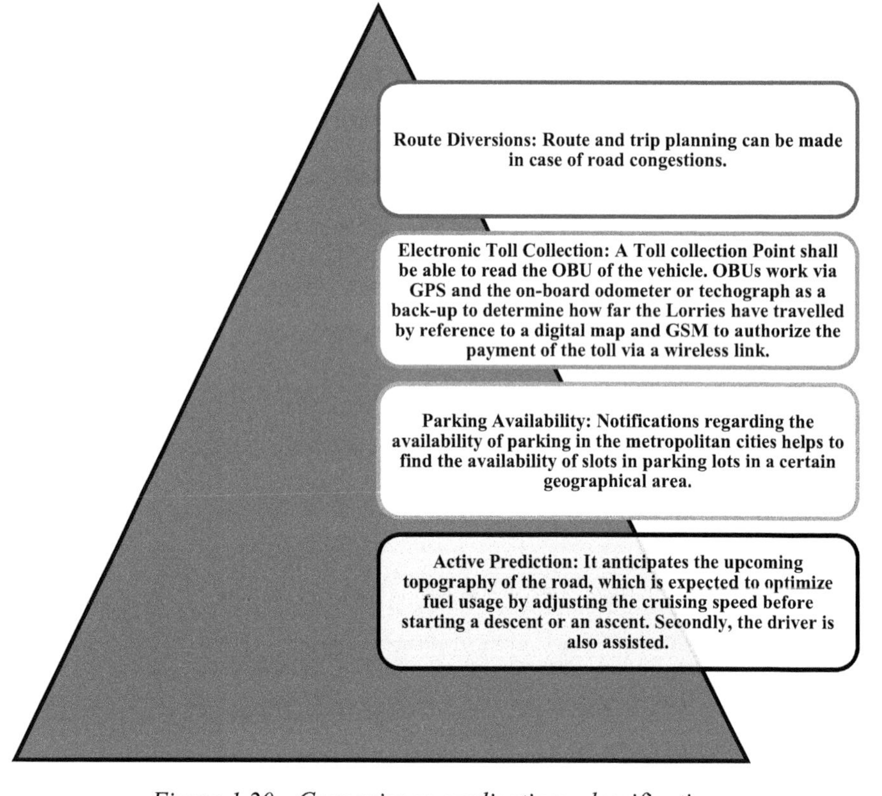

Figure 1.20 Convenience applications classification

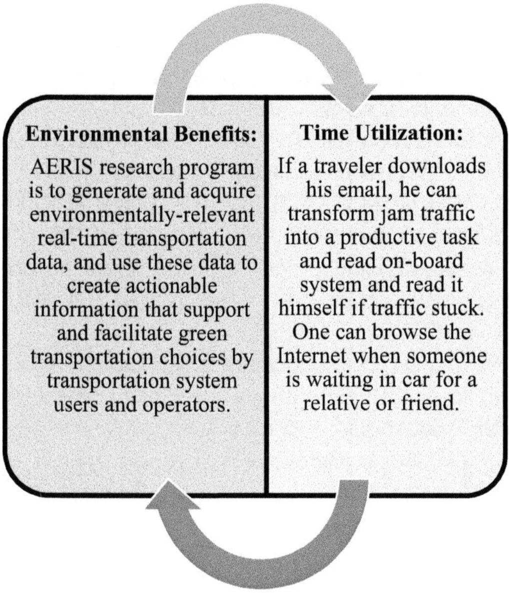

Environmental Benefits:
AERIS research program is to generate and acquire environmentally-relevant real-time transportation data, and use these data to create actionable information that support and facilitate green transportation choices by transportation system users and operators.

Time Utilization:
If a traveler downloads his email, he can transform jam traffic into a productive task and read on-board system and read it himself if traffic stuck. One can browse the Internet when someone is waiting in car for a relative or friend.

Figure 1.21 Productive applications classifications

currently the last weapons in the UK Dorset Police's zero-tolerance campaign for traffic violations. Similarly, various attempts to employ VANETs in terms of traffic safety and efficiency are being taken in other countries. There are still many more problems that will have a big influence on the future of VANETs.

1.13 Conclusion

The automation provided by VANET, as well as the introduction of V2V and V2I connections, will allow for the establishment of an efficient and safe future transportation system. After that, it will show a state-of-the-art VANET-driven development of traffic management and monitoring applications, platooning, junctions, VRUs, and other related features and functions. The present situation has prompted the consideration of several potential VANET-based applications in the future. These applications address critical aspects of mobility, such as safety, efficiency, long-term sustainability, and user comfort and convenience. Several applications are based on other technical advancements, such as the IoT, smart cities, and smart grids. This chapter provides an overview and a prediction of the existing transport infrastructure for future VANET applications, as well as a discussion of the future transport infrastructure. Future VANET apps may be able to provide suggestions for a completely automated and cooperative transportation system if they are designed properly. Ad hoc vehicles have network architecture, standards, and protocols, which are covered in this chapter, as well as the characteristics mentioned in Section 1.5.

References

[1] Lee, M., and T. Atkison. "Vanet applications: past, present, and future." *Vehicular Communications* 28 (2021): 100310.

[2] Günay, F.B., Öztürk, E., Çavdar, T., Hanay, Y. and Khan, A.R. "Vehicular ad hoc network (VANET) localization techniques: a survey." *Archives of Computational Methods in Engineering* 28.1–33 (2020). 10.1007/s11831-020-09487-1.

[3] Soleymani, S.A., Goudarzi, S., Anisi, H., Zareei, M., Abdullah, H. and Kama, Ni. "A security and privacy scheme based on node and message authentication and trust in fog-enabled VANET." *Vehicular Communications* 29 (2021): 100335.

[4] Javadpour, A., Rezaei, S., Sangaiah, A.K., Adam, S., and Shadi, M K. "Enhancement in quality of routing service using metaheuristic PSO algorithm in VANET networks." *Soft Computing* (2021): 1–12.

[5] Bangui, H., M. Ge, and B. Buhnova. "A hybrid machine learning model for intrusion detection in VANET." *Computing* (2021): 1–29.

[6] Verma, A., Saha, R., Kumar, G., and Kim, T.-h. "The security perspectives of vehicular networks: a taxonomical analysis of attacks and solutions." *Applied Sciences* 11.10 (2021): 4682.

[7] RadhaKrishna Karne, T.K. "Review on VANET architecture and applications." *Turkish Journal of Computer and Mathematics Education (TURCOMAT)* 12.4 (2021): 1745–1749.

[8] Chen, J., and Z. Wang. "Coordination game theory-based adaptive topology control for hybrid VLC/RF VANET." *IEEE Transactions on Communications* 69.8 (2021): 5312–5324.

[9] Ren, M., Zhang, J., Khoukhi, L., Houda, L., and Véronique, V. "A review of clustering algorithms in VANETs." *Annals of Telecommunications* 76 (2021): 581–603.

[10] Aboud, A., H. Touati, and B. Hnich. "Handover optimization for VANET in 5G networks." *2021 IEEE 18th Annual Consumer Communications & Networking Conference (CCNC)*. IEEE, 2021.

[11] Yeun, C.Y., M. Al-Qutayri, and F. Al-Hawi. "Efficient security implementation for emerging VANETs." *UbiCC J* 4.4 (2009).

[12] Pal, R., R.K. Then, and F. Siddiqui. "Optimal cognitive channel selection using deep learning in CR-VANET." *Proceedings of Integrated Intelligence Enable Networks and Computing*. Springer, Singapore, 2021. pp. 803–810.

[13] Min, M., Wang, W., Xiao, L., Xiao, Y. and Han, Z. "Reinforcement learning-based sensitive semantic location privacy protection for VANETs." *China Communications* 18.6 (2021): 244–260.

[14] D. Silva, E., Tomás, J.M.H., de Macedo, and A.L.D. Costa. "Design of a context-aware routing system model with mobility prediction for NDN-based VANET." *DCE21 13 Symposium on Electrical and Computer Engineering: Book of Abstracts*; 2021.

[15] Patel, N.J., and Rutvij H.J. "Trust-based approaches for secure routing in VANET: a survey." *Procedia Computer Science* 45 (2015): 592–601.

[16] Cavalcanti, E.R., Rodrigues, A., Spohn, M., Cézar, R., and Fabiano, A. "VANETs' research over the past decade: overview, credibility, and trends." ACM SIGCOMM Computer Communication Review 48.2 (2018): 31–39.

[17] Sajini, S., Mary, E.A., and Janet, J. "Improving The Security of Data Communication in Vanets Using ASCII-ECC Algorithm." (2021). 10.21203/rs.3.rs-597331/v1.

[18] Mao, M., Yi, P., Hu, T., Zhang, Z., Lu, X., Lei, J. "Hierarchical hybrid trust management scheme in SDN-enabled VANETs." *Mobile Information Systems* 2021 (2021).

[19] Mchergui, A., T. Moulahi, and S. Zeadally. "Survey on artificial intelligence (AI) techniques for vehicular ad-hoc networks (VANETs)." *Vehicular Communications* (2021): 100403.

[20] Mekrache, A., Bradai, A., Moulay, E., and Dawaliby, S. "Deep reinforcement learning techniques for vehicular networks: recent advances and future trends towards 6G." *Vehicular Communications* (2021): 100398.

[21] Mirsadeghi, F., M.K. Rafsanjani, and B.B. Gupta. "A trust infrastructure-based authentication method for clustered vehicular ad hoc networks." Peer-to-Peer Networking and Applications 14.4 (2021): 2537–2553.

[22] Sultana, R., J. Grover, and M. Tripathi. "Security of Sdn-based vehicular ad hoc networks: state-of-the-art and challenges." *Vehicular Communications* 27 (2021): 100284.

[23] Cunha, B., Brito, C., Araújo, G., Gomes, R., Soares, A., Francisco, and Silva, F. A. "Smart traffic control in vehicle ad-hoc networks: a systematic literature review." *International Journal of Wireless Information Networks* (2021): 1–23.

[24] Wu, Y., Wu, Yanpeng, Guerrero, Josep & Vasquez, Juan C. "A comprehensive overview of a framework for developing sustainable energy internet: from things-based energy network to a services-based management system." *Renewable and Sustainable Energy Reviews* 150 (2021): 111409.

[25] Bhanu, C. "Challenges, benefits and issues: future emerging VANETs and cloud approaches." *Cloud and IoT-Based Vehicular Ad Hoc Networks* (2021): 233–267.

[26] Gallego-Madrid, J., Sanchez-Iborra, R., Ruiz, P.M., Skarmeta, A.F. "Machine learning-based zero-touch network and service management: a survey." *Digital Communications and Networks* (2021).

[27] Weber, J.S., M. Neves, and T. Ferreto. "VANET simulators: an updated review." Journal of the Brazilian Computer Society *27.1* (2021): 1–31.

[28] Balador, A., Cinque, E. and Pratesi, M. "Survey on decentralized congestion control methods for vehicular communication." *Vehicular Communications* 33 (2021): 100394.

[29] Sharma, S., and A. Kaul. "VANETs cloud: architecture, applications, challenges, and issues." *Archives of Computational Methods in Engineering* 28 (2021): 2081–2102.

[30] Khan, S., Sharma, I., Aslam, M., Khan, M.Z., and Khan, S. "Security challenges of location privacy in VANETs and state-of-the-art solutions: a survey." Future Internet 13.4 (2021): 96.

[31] Hahn, D.A., A. Munir, and V. Behzadan. "Security and privacy issues in intelligent transportation systems: classification and challenges." *IEEE Intelligent Transportation Systems Magazine* 13.1 (2021): 181–196.

[32] Abbasi, S., Rahmani, A.M., Balador, A., and Sahafi, A. "Internet of vehicles: architecture, services, and applications." *International Journal of Communication Systems* 34.10 (2021): e4793.

[33] Hamdi, M.M., Al-Dosary, O.A.R., Alrawi, O.A.S., Mustafa, A.S., Abood, M.S., and Noori, M.S. "An overview of challenges for data dissemination and routing protocols in VANETs." *2021 3rd International Congress on Human-Computer Interaction, Optimization and Robotic Applications (HORA)*. IEEE, 2021.

[34] Gillani, M., A. Ullah, and H.A. Niaz. "Trust management schemes for secure routing in VANETs—a survey." *2018 12th International Conference on Mathematics, Actuarial Science, Computer Science and Statistics (MACS)*. IEEE, 2018.

[35] Aljabry, I.A., and G.A. Al-Suhail. "A simulation of AODV and GPSR routing protocols in VANET based on multimetrices." *Communications* 7 (2021): 9.

[36] Shah, P., and T. Kasbe. "A review on specification evaluation of broadcasting routing protocols in VANET." *Computer Science Review* 41 (2021): 100418.

[37] Mahdi, H.F., M.S. Abood, and M.M. Hamdi. "Performance evaluation for vehicular ad-hoc networks-based routing protocols." *Bulletin of Electrical Engineering and Informatics* 10.2 (2021): 1080–1091.

[38] Smiri, S., Abdelali, B., Adil, B.A., Azeddine, Z., and Rachid, B.A. "Implementation and QoS evaluation of geographical location-based routing protocols in vehicular ad-hoc networks." *WITS 2020*. Springer, Singapore, 2022. pp. 515–526.

[39] Yelure, B., and S. Sonavane. "Performance of routing protocols using mobility models in VANET." *Next Generation Information Processing System*. Springer, Singapore, 2021. pp. 272–280.

[40] Shrivastava, P.K., and L.K. Vishwamitra. "Comparative analysis of proactive and reactive routing protocols in VANET environment." *Measurement: Sensors* (2021): 100051.

[41] Belamri, F., S. Boulfekhar, and D. Aissani. "A survey on QoS routing protocols in vehicular ad hoc network (VANET)". *Telecommunication Systems* (2021): 1–37.

[42] Hamdi, M. Maad, Y.A. Yussen, and A.S. Mustafa. "Integrity and authentications for service security in vehicular ad hoc networks (VANETs): a review." *2021 3rd International Congress on Human-Computer Interaction, Optimization and Robotic Applications (HORA)*. IEEE, 2021.

[43] Al-Absi, M.A., Al-Absi, A.A., Sain, M., and Lee, H. "Moving ad hoc networks—a comparative study." *Sustainability* 13.11 (2021): 6187.

[44] Ros, F. Scalable Data Delivery in Mobile Multi-hop Ad-hoc Networks – Application to Gateway Discovery Protocols, 2011. AAMAS'06 May 8–12

2006, Hakodate, Hokkaido, Japan Available at: https://ri.cmu.edu/pub_files/pub4/yu_bin_2006_1/yu_bin_2006_1.pdf.

[45] Deshpande, A. "Review of effective trust management systems in VANET environments." *International Journal of Grid and Distributed Computing* 14.1 (2021): 1771–1780.

[46] Akter, S., Rahman, S., Bhuiyan, Z., and Mansoor, N. "Towards secure communication in CR-VANETs through a trust-based routing protocol." IEEE INFOCOM 2021 – IEEE Conference on Computer Communications Workshops (INFOCOM WKSHPS). IEEE, 2021.

[47] Feng, X., Shi, Q., Xie, Q., and Liu, L. "An efficient privacy-preserving authentication model based on blockchain for VANETs." *Journal of Systems Architecture* 117 (2021): 102158.

[48] Kaur, R., Ramachandran, R., Doss, R., and Pan, L. "The importance of selecting clustering parameters in VANETs: a survey." *Computer Science Review* 40 (2021): 100392.

[49] Kamoi, R.N., Alves, P. Jr, Lourenco and Verri. "Platoon grouping network offloading mechanism for VANETs." *IEEE Access* 9 (2021): 53936–53951.

[50] Vanitha, N., and G. Padmavathi. "Support of mobility models for the decentralized multi-layer UAV networks assisting VANET architecture (DMUAV)." *International Journal of Engineering Research and Technology* 13.12 (2020): 5219–5226.

[51] Sharma, S., Ajay, K., Suhaib, A., Surbhi, S. "A detailed tutorial survey on VANETs: emerging architectures, applications, security issues, and solutions." *International Journal of Communication Systems* (2021): e4905.

[52] Fourati, L.C., S. Ayed, and M.A.B. Rejeb. "ICN clustering-based approach for VANETs". *Annals of Telecommunications* (2021): 1–13.

[53] Moni, S.S., and D. Manivannan. "A scalable and distributed architecture for secure and privacy-preserving authentication and message dissemination in VANETs." *Internet of Things* 13 (March 2021): 100350.

[54] Suresh Kumar, K., Radha Mani, A.S., Sundaresan, S. and Ananth Kumar, T. "Modeling of VANET for future generation transportation system through edge/fog/cloud computing powered by 6G." *Cloud and IoT-Based Vehicular Ad Hoc Networks* (2021): 105–124.

[55] Zhang, S.Y., Lagutkina, M., Akpinar, K.O., and Akpinar, M. "Improving performance and data transmission security in VANETs." *Computer Communications* 180 (2021): 126–133.

[56] Tyagi, A.K. and Sreenath, N. "Providing safe, secure and trusted communication among vehicular ad-hoc networks' users: a vision paper." *International Journal of Information Technology and Electrical Engineering*, 5.1 (2016): 35–44.

[57] Tyagi, A.K., Rekha, G., and Sreenath, N. (eds). "Opportunities and challenges for blockchain technology in autonomous vehicles." IGI Global, 2021. http://doi:10.4018/978-1-7998-3295-9.

[58] Shivanand, B., Shrikant, S.T., Geetha, D.D., and Sunilkumar, S. "A survey on security and safety in vehicular ad hoc networks (VANETs) cloud."

Cognitive Informatics and Soft Computing. Springer, Singapore, 2021. pp. 309–319.

[59] Tyagi, A.K. and Aswathy, S.U. "Autonomous Intelligent Vehicles (AIV): research statements, open issues, challenges and road for future." *International Journal of Intelligent Networks*, 2 (2021): 83–102. https://doi.org/10.1016/j.ijin.2021.0

[60] Hou, L., Yao, N., Lu, Z., Zhan, F., and Liu, Z. "Tracking based mix-zone location privacy evaluation in VANET." *IEEE Transactions on Vehicular Technology* (2021).

[61] Alharthi, A., Q. Ni, and R. Jiang. "A privacy-preservation framework based on biometrics blockchain (BBC) to prevent attacks in VANET." *IEEE Access* 9 (2021): 10957–10969.

[62] Jiang, X., Yu, F.R., Song, T. and Leung, V.C.M. "Resource allocation of video streaming over vehicular networks: a survey, some research issues, and challenges." *IEEE Transactions on Intelligent Transportation Systems* (2021).

[63] Nandy, T., Mohd Yamani, I.I., Rafidah, Md N., *et al.* "A secure, privacy-preserving, and lightweight authentication scheme for VANETs." *IEEE Sensors Journal* 21.18 (2021): 20998–21011.

[64] Hemalatha, R. "A survey: security challenges of vanet and their current solution." *Turkish Journal of Computer and Mathematics Education (TURCOMAT)* 12.2 (2021): 1239–1244.

[65] Nair, M.M. and Tyagi, A.K. "Privacy: history, statistics, policy, laws, preservation and threat analysis." *Journal of Information Assurance & Security* 16.1 (2021): 24–34.

[66] Potluri, S., Mohammad, G.B., Mohanty, S.N., Vaishnavi, M. and Sahaja, K. "Secure intelligent framework for VANET: cloud-based transportation model." *Cloud Security: Techniques and Applications* (2021): 145.

[67] Kanellopoulos, D., and F. Cuomo. "Recent developments on mobile ad-hoc networks and vehicular ad-hoc networks." *Electronics* 10 (2021): 364.

[68] Saivichit, S.T.C. "Using a Distributed Roadside Unit for the Data Dissemination Protocol in VANET With the Named Data Architecture." IEEE Access. 6, 32612–32623 (2018).

[69] Mihret, E. Tilahun, and K.A. Yitayih. "Operation of VANET communications: the convergence of UAV system with LTE/4G and WAVE technologies". *International Journal of Smart Vehicles and Smart Transportation (IJSVST)* 4.1 (2021): 29–51.

[70] Sathyapriya, A., Sathiya, K., Sneha, T.M., Rohit Raja, D. and Manikandan, T. "Automatic speed control system in vehicles using VANET". *Advances in Smart System Technologies*. Springer, Singapore, 2021. pp. 719–726.

Chapter 2

MAC layer, Hybrid MAC protocols in Vehicular Ad Hoc Networks (VANETs)

Abstract

The rapid advancements of network technology over the past decade have accelerated the development of state-of-the-art software, hardware and transmission technologies. This has led to the promotion of several diverse networks in multiple technological domains that differ based upon their application requirements. A particular network type that has attained the spotlight recently is the Vehicular Ad Hoc Networks or VANETS. The promising potential of VANETS in improving human transportation, by providing digital vehicle infrastructures and customized passenger comfort, has been responsible for the development of vehicular networks as a domain of extensive research and standardization. Routing, broadcasting, Quality of Service (QoS), and security are the areas specified in focus for quite a number of VANET research work. This chapter discusses the role of Medium Access Protocols in VANETS. MAC protocols clearly explain the method by which the nodes share the underlying channel. As there is no VANET standard, earlier used one was IEEE 802.11a and 802.11b for the purpose of MAC layer access technologies. Here in this discussion, we have taken up designing an effective MAC so that timely safety messages can be sent reliably, since it is a serious matter of human lives involvement in VANET. Two major goals of VANETs are to improve road safety and to increase transportation efficiency. An unfailing and effective MAC is the need to achieve the two important goals. Delivery success rate, delay, throughput, bandwidth utilization, fairness, and overhead of the transmitted packets are to be balanced critically by the MAC protocols. A synopsis of development of the development of MAC protocols and its standards have been elucidated in this chapter.

Key Words: Global position system; Mobile ad hoc networks; Vehicle-to road infrastructure; Medium access control

2.1 Introduction

High mobility, high storage capability, good battery power backup, high processing power, and inbuilt global position system (GPS) are the valuable properties of Vehicular Ad Hoc Networks (VANETs). They are one type of mobile ad hoc networks (MANETs). They are an integral part of intelligent transport system (ITS). They are used for efficient services including road side safety, reduce/avoid traffic density. On board unit (OBU) on each vehicle has communication devices. They also have GPS receiver. Out of the three types of communication in VANETs, which are vehicle to vehicle, vehicle to road side unit (RSU) and RSU to RSU, RSUs are allocated at the roadside, for supporting vehicular communications. Dedicated short-range communication (DSRC) in VANETs (75 MHz spectrum) are in the range of 5.850–5.925 GHz. These are allocated by United State Federal Communication Commission. Channel 178 is used as the control channel for safety and control message communication along with other six channels are also used for communication of service messages. The in-vehicle domain refers to in-vehicle communication, which is becoming increasingly necessary and crucial in VANETs research. A vehicle's performance, particularly driver weariness and drowsiness, can be detected by an in-vehicle communication system, which is crucial for driver and public safety. Vehicle-to-vehicle (V2V) communication can be used to provide a data exchange platform for drivers to share information and warning messages, hence increasing driver assistance.

Another useful study subject in VANETs is vehicle-to-road infrastructure (V2I) communication. V2I connection allows drivers to get real-time traffic and weather reports, as well as environmental sensing and monitoring. Vehicle-to-broadband cloud (V2B) communication refers to the use of wireless broadband methods such as 3G/4G to connect between automobiles. This type of connectivity will be important for active driver assistance and vehicle tracking because the broadband cloud may include more traffic information and monitoring data as well as infotainment. Details of MANETs differ from VANETs though they both support a big range of applications, from one hop information dissemination of say, cooperative awareness messages (CAMs) to multi-hop dissemination of messages over vast distances. VANETs move in an organized manner. The roadside equipment interactions will be accurate. Vehicles are mostly restricted in their movement range, as they are constrained to follow a paved highway. Some typical applications of Vehicular Ad Hoc Networks include:

- *Electronic brake lights* – they enable a driver on road to react quickly to braking vehicles before him, though they are not visible due to intermittent vehicles.
- *Platooning* – driver can follow a leading vehicle enabled by wireless information on acceleration and braking. This forms electronically coupled "road trains."
- *Traffic information systems* – drivers get obstacle reports updated on live basis, which use VANET communication. This is enabled through vehicle's satellite navigation system.
- *Road transportation emergency services* – to reduce time lapse and severe delays in emergency scenario, VANET communications/networks manage the road safety warning and status information dissemination. They speed up emergency rescue operations and save lives.

- *On-the-road services* – transportation highway in future is expected and viewed to be information-driven and wirelessly-enabled. Advertisement of services to the driver regarding shops, gas stations, etc. available on the roadway and sale notifications are also enabled through VANETs, making its potential useful for commercial application also.

How are the vehicular nodes made to communicate without associating with the access point or base station coordinator? Without joining the network, they communicate, playing an important role, irrespective of access technology is enabled by the MAC layer. Multiple stations effectively share regulation of access to the common wireless medium. No central system is required to manage the channel access or assignment of resource, giving decentralized operations made possible by MAC in a vehicular environment. There are many instances where specific frequency bands are allocated for vehicular movements, which are further categorized into channels specifically for vehicle communication. MAC protocols for VANETs, as per physical division of spectrum resources, are classified as single channel MAC protocols and multi-channel MAC protocols.

- Single channel MAC protocols: They use tools like carrier sensing, time division multiple access (TDMA), code division multiple access (CDMA), and space division multiple access (SDMA) for allocation of resources for multiple nodes. There is a possibility of packet collisions here, to a certain level, which is reduced by using multi-channel MAC protocols.
- Multi-channel MAC protocols: They use multiple transceivers or channel switching mechanisms and support packet transmission on multiple channels and promote effective use of channel resources. The possibility of packet collisions in single channel is reduced by this protocol, thereby improving the efficiency of network throughput.

When a vehicle is in mobility, the dynamic topology change, short connection life, severe broadcast environment, high node density and different types of traffic quantity and quality, pose a heavy demand on the resources of medium access control (MAC) layer for vehicle networks. The design and functions of the MAC (viz channel access rules, schemes of prioritization, types of frames and formats) as specified by the IEEE 802.11p and ETSI ITS-G5 standards, and the multi-channel operation will be described in this chapter. Other areas of discussion will be main evaluation tools (field-tests, simulations, analytics) which will scrutinize the MAC performance.

2.2 Description of background and related work

One of the most well-known specializations of MANETs is VANET (Vehicular Ad Hoc Networks) [1]. Vehicles in VANETs communicate with one other or with the infrastructure (vehicle-to-vehicle, V2V) (vehicle-to-infrastructure, V2I). The term “vehicle-to-everything” (V2X) has recently gained popularity. Vehicular networking has piqued the interest of both academics and industry since its debut.

Several international industrial and governmental consortia, such as the European Commission, have been formed. There were issues in routing protocol for data routing among vehicles, called V2V (vehicle to vehicle communication) or V2I (vehicle to road side infrastructure) [2]. Wireless communication has been a vital technology in VANETs to enable the development of a wide range of services and applications [3]. The present routing methods in MANETs are really not available for most application scenarios in VANETs due to the features of VANETs, such as high dynamic topology and intermittent connectivity. As a result, researchers make every effort to improve existing algorithms as well as create new ones in order to assure communication dependability. Routing approaches can be classified into three categories based on the number of senders and recipients associated: geocast/ broadcast, multicast, and unicast [1, 4]. Three routing protocols were compared for different parameters and were simulated on Network Simulator-2(ns-2). Previous Research works have tested out the following methodologies: Ad hoc on-demand distance vector routing (AODV), destination sequenced distance vector (DSDV), and dynamic source routing (DSR). The power supply restraints or time sync problems should not affect an effective VANET MAC protocol [5].

The protocol should cope with topology changes there by improving vehicular technology efficiency. There are difficulties in hilly environments, with many curves and turns. The long lasting Geocast is supporting elimination of transmission problems of the 802.11p. Also, as ZigBee nodes reduce the transmission period within vehicles, the routing of packets has seen a big improvement in this model [6]. The design will include the HP ProLiant DL580 Gen8 Server, the core for VANET components. The channel access is managed by deploying RSUs or cluster heads (CHs) by centralized MAC protocols. Both RSUs and CHs are effective in access-scheduling function [7]. They can manage channel resources on both single channel and multiple channels. Assignment of timeslots, codes or frequency bands for all the coverage vehicles. Overhead of neighbor discovery is limited that too in a disseminated manner. The ratio of CCHI to SCHI in alternating access channel modes is dynamically adjusted, thereby improving the performance. Conventionally, the goal of safety applications has been to prevent accidents, and this has been the driving force behind the development of automotive ad hoc networks. Crash avoidance applications, for example, have a high demand for V2V or V2I communication [8]. Vehicles equipped with various sensors continuously gather traffic data and monitor the environment, and cooperative vehicular safety programs can alter real-time traffic information and send/receive warning signals via V2I or V2V communication to improve road safety and prevent accidents [9].

SCHS are reserved by the variable service reservation intervals. This makes the load balancing between the control channel (CCH) and SCH possible. Adaptive intervals are achieved, number of rounds are optimized thereby maximizing the throughput of network though multiround elimination on CCH. Real-time adjustment of CCHI and SCHI, based on network and traffic flow conditions has become important, as per multi-channel MAC protocols with adaptive variable intervals for VANETs. Dual transceivers deployment has made it possible to sense safety messages on CCH and service messages on SCH, also receiving safety-related messages. The channel access supervision is efficiently managed by cluster-based MAC protocols due to spectrum of new solutions possible [10]. Vehicles in one area can be classified into a cluster and a CH is identified to work as Cluster Manager to manage channel access and timeslot allocation. The CHs minimize the communication link interruption, vehicles relative motion and communication links remain uninterrupted for longer periods [11]. Adaptive learning tools incorporated in the fuzzy logic inference system (FIS) are used to organize vehicles into more stable and non-overlapping clusters [12]. This improves the life time of CH and its members. Timeslots and channels have to be mapped to vehicle identification ID, different vehicles [13] are assigned disjoint channels and timeslots, thereby delivering transmissions without contention.

2.3 Layers used for VANETs

Vehicle mobility, short connection lifetimes, severe propagation environments, high node density, and heterogeneous traffic nature and quality needs all require careful consideration when designing the MAC layer for automotive networks. Most readers are familiar with the open systems interconnection (OSI) concept, which divides communication functions into seven logical levels. In this architecture, the session and presentation layers are eliminated, and each layer can be further divided into sublayers [14]. The physical and electrical features of the network are defined by the PHY layer. It is in charge of the circuitry that modulates and demodulates RF signals. Sending and receiving RF frames is handled by the MAC layer. Further, figure 2.1 shows the *Layer architectures of VANETs*.

The attributes of dedicated short-range communication (DRSC) physical layer are as follows: 802.11p OFDM, range of 5.89 GHz Bank (5.89–5.91), 10 MHz wide channel in the WAVE. Basic data rate is approx. 3 Mbps. Default data rate of 6 Mbps. The DSRC physical layer attributes from transmission power control to using multiple (or single) antennas and from channel estimation to channel selection, contribute to the

<table>
<tr><td rowspan="5">Vehicular Network</td><td>Application Type</td><td>• Safety application
• Intelligent transport application
• Comfort application</td></tr>
<tr><td>Quality of Service</td><td>• Non-real-time
• Soft-real-time
• Hard-real-time</td></tr>
<tr><td>Scope</td><td>• Wide area
• Local</td></tr>
<tr><td>Network Type</td><td>• Ad hoc
• Infrastructure based</td></tr>
<tr><td>Communication Type</td><td>• V2I
• V2V</td></tr>
</table>

Figure 2.1. Layer architectures of VANETS

limitations imposed on the scalability of the network. Multi-path delay spread and mobility in the multi-path environment, makes the communication quite difficult. Frequency selective fading caused by the delay spread and time selective fading caused by the mobility are the challenges. When getting into the cause of delay spread, non-line of sight (NLOS) conditions and highways having large Doppler spread appear as the major ones. The following are physical layer challenges in VANETs, apart from channel estimation error and lack of time-interleaving.

- Single and dual radio: Within the default scheme, addition of another radio to improve performance of safety communication does not work, though there are some other advantages to support usage of dual radio.
- Propagation model: Different models taking into consideration various topographies of landscapes, under the VANETs falling under the propagation model is better. For example, free space model is used for highways. However, in cities, since there will be many buildings and hoardings, the space model will result in shadowing and multi-path fading. In rural areas, trees, mountains will pose different problem of reflections. So, the propagation model takes into considerations the various landscapes.

Note that selection of best channel and data rate through game theoretic approach has been arrived at through a simulation study for the physical layer's channel/selection channel. Advanced channel estimation techniques are applied to get precise channel state information (CSI). Diversity techniques are used to bring down fading and interference to a minimum level fair channel access selection and scalability of VANETs depend on the MAC protocol to a critical level. As an effective and trustworthy MAC protocol is the dire necessity, researchers focus on effectivity using the multi-channels and performance modeling of the EDCA mechanism, so that performance of the MAC layer can be analyzed through an investigative model. At present, the IEEE 802.11p MAC along with the 1609 Multi-Channel Operations is now used as the MAC standard for vehicular communications [15] under the WAVE technology control overhead surges due to recurrent and dynamic topology change in the attributes of the VANET stochastic modeling studies are aiding to design an effective protocol, for better performance of the MAC layer. The performance modeling and investigation demands fulfillment of conditions such as saturated/unsaturated, access categories count, back off counter freezing, collision-both internal and external, nodes movement and computational complications. Transmission and collision probabilities can be calculated through Markov chains method. At present, most of the modeling are being done for the IEEE 802.11p enhanced distributed channel access (EDCA) mechanism. As a perfect model will give clarity to MAC performance, continued studies are required. Dynamic adjustment of the contention window (CW): support of the packets received correctly should adjust the CW size and detect the traffic of the network through the received packets sequence. Simulation has proved that many algorithms help in dynamic adjustment of CW with better reception rates. Congestion in CSMA/CA can be avoided through this strategy of CW adjustment. This strategy is used as default to avoid congestion.

- Congestion control: DCC adjusts the physical layer parameters (called as cross layer approach). This happens through using MAC layer as channel sensing mechanism. Algorithms for MANETs are several, but as they are not well matched for VANETs (due to dynamic network), the congestion control solutions provide a high throughput with two types of CC algorithms – proactive and reactive. Another one has now emerged as Hybrid Algorithm. The basis of all these algorithms is that the drivers of vehicles should work with the intention of controlling the congestion of vehicles in the area. Transmission power adjustments and message rate adjustment are the two important factors on which the algorithms focus.
- Routing: This is important for any network efficiency. The ideal routing protocol enables movement of data through series of hops, selecting the option of better route, shorter and less crowded. Data transmission [16] should consider status of the network in VANET, as the nodes connect and disconnect with the network regularly. The environment is temporary and keeps changing, a dedicated path may not be available. Network performance may be impacted if regular routing algorithms are used. The capability of the protocol should be such that it does localized operations, whenever routing decisions are based on locally available information. So, node availability is not required to know the entire network topology. As a result, control overhead is reduced. The difficulties in routing layer are Route data management.

2.4 Protocols used for VANETs

Routing protocols have established themselves as the industry standards for data transmission in networks. In a network, efficient routing protocols make dynamic routing selections (refer Figure 2.2).

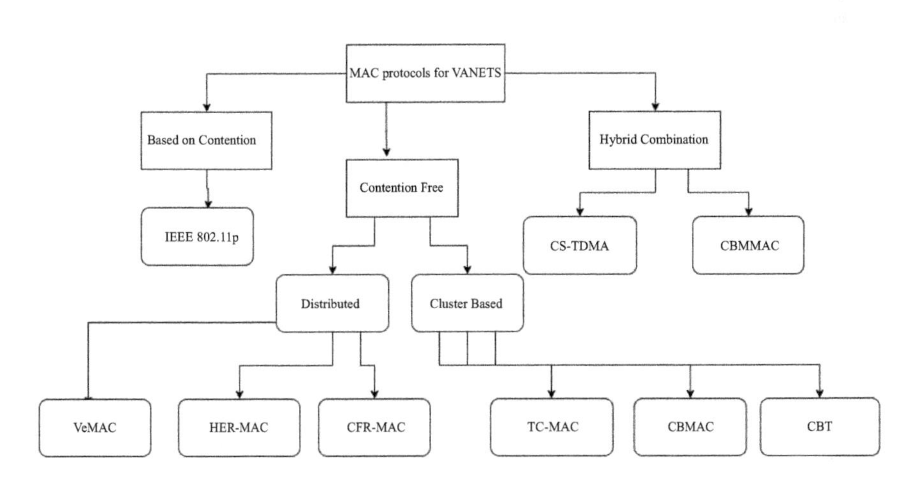

Figure 2.2. MAC protocols in VANETs

Topology-based routing protocol: Proactive and reactive topology-based routing protocols are two types of topologies-based routing protocols. When compared to position-based routing protocols, topology-based routing systems have restricted performance [17]. During the routing decision process, most topology-based routing algorithms require extra node containing topology information.

Proactive routing protocols: Tables describing the topology are maintained by proactive routing protocols. The tables in these protocols are updated on a regular basis, and data is sent from one node to another. Because of their nature, proactive routing protocols are often known as table-driven protocols. In proactive protocols, there are two forms of updating available: periodic update and triggered update. Broadcasting update tables wastes network power and bandwidth. When nodes are added to networks, the table size is increased in proactive protocols, which increases the demand.

Destination sequence vector router routing: DSDV provides loop-free routes, employs a single source to destination approach, and employs the distance vector shortest path method. There are two types of packets sent by the protocol: incremental and full dump. Full dump packets provide routing information, while incremental packets contain updates. Full dump packets consume more bandwidth, while incremental packets are sent more often, increasing network overhead. Due to bandwidth use and update methods, the DSDV protocol is not appropriate for big networks.

OLSR stands for optimized link state routing protocols, which is a point to point and proactive routing protocol based on the classic link-state algorithm. For route establishment or route maintenance, it employs a technology known as multipoint relaying, which optimizes the message and flooding process. Multipoint relay (MPR) is a method that reduces the number of active relays required to cover the neighbors [18]. The protocol was created to ensure the accuracy and stability of data routing in a network.

Ad hoc On-Demand Distance Vector Routing (AODV): Based on demand and requirements, DSDV and DSR algorithms are used in AODV protocols. The protocols work with routing tables to start the discovery process. The packet broadcast through the source in the discovery technique is a route request (RREQ) packet, and the neighbor nodes forward the packet to their neighbors until an active route is located and the maximum number of hops is reached. Before transmitting the packet to their neighbors [19], the packets are unaware of the active route for the requested destination. Many studies have concluded that AODV performance and efficiency are best measured by three metrics: packet delivery ratio, routing overhead, and path optimality [20]. Many MAC protocols are suggested for VANETs, as they decide as to how all the vehicles get access to the channel and such decisions are critical in achieving the target of effective and congestion free channel access. This is important for betterment of safety/non-safety connected services for the transportation system.

The issues faced in this regard are many, such as continuously changing environment for network, nodes movement on the higher side, coordination in multi-channel environment, issue of hidden & exposed nodes and nearby channel disturbance. Figure 2.3 showcases the WAVE protocol stack.

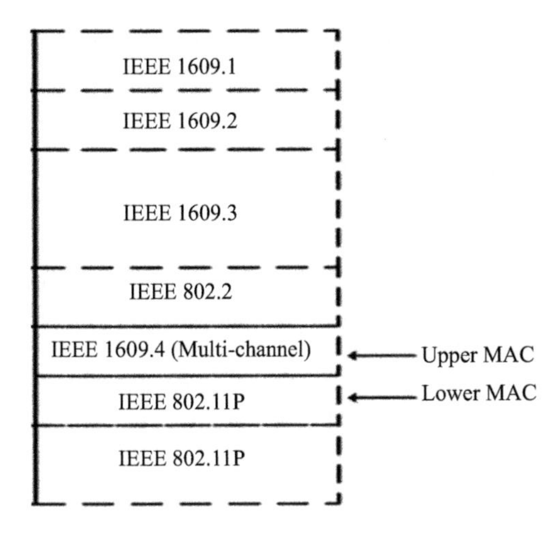

Figure 2.3. WAVE protocol stack

The division and sharing of MAC protocols that are based on channel partitioning can be accomplished with high performance with uniform load distribution. There are different methodologies through which channel partitioning can be achieved. These methods include namely, code-division multiple access, frequency-division multiple access, and time-division multiple access. Multiple node enablement in the same frequency division is characteristic of time division multiple access. The transmission is assigned to one of the time slots (generated from different time frames). The timing of the slots is adapted to the requirements of the node assigned to it. The communication from the nodes occurs consecutively and every node adheres to its slot. The most significant advantages attained through the usage of these aforementioned protocols is maintaining maximal fairness and reduced interference. However, there is some significant inefficiency associated with the protocol when handling low loads.

Vehicular self-organized MAC (VeSOMAC) was a wireless networking protocol for inter-vehicular transmissions. The design of deterministic and short delay bounds for inter-vehicular information transmission has been enabled through a time-division multiple access (TDMA) self-configuration. The movement and location of the vehicles often hold an integral role in determining the time division multiple access slot time slot to ensure minimal delay. The allocation of TDMA slots with respect to its physical location is made to be in the same sequential order.

The introduction of a multi-channel MAC protocol for vehicular networks called VeMAC is built upon a TDMA backbone scheme structure helped combat the interferences that may arise from multiple vehicle network overlap or through the changes caused from vehicle mobility. GPS synchronization is carried out in all vehicles. All vehicles are equipped with two different transceivers with one assigned to the control channel and the other one assigned to the service channel. The VeMAC comprises of multiple service channels for example, WAVE and one control channel.

The messages that are transmitted from the vehicle through the control channel can be categorized into control messages and high priority messages.

2.5 Standardization of layers and protocols

The process of standardization and the establishment of uniform standards, globally for all products in a particular domain is essential to ensure that quality and safety is ensured. The use of standards also helps eliminate incompatible heterogeneity and helps different systems to function in harmony. There are several wireless standards that enable vehicular networks to operate at different levels (V2V, V2I, and infrastructure to infrastructure communication). The establishment of network communication standards for VANETs is motivated by the need to improve transportation efficiency and road safety. The available standards range from basic wireless technology to state of the art 4G technology [21]. The predominant wireless access methodologies used in VANETS originate from cellular access namely WAVE and 2G/2.5G/3G/4G. The dominance and governance power held by countries play a major role in determining the standardization of protocol stacks for VANETS. In terms of the automotive industry power, United States, Japan, and Europe are the forerunners in propagating VANET standardizations. IEEE-1609 WAVE wireless access are employed in the United States for ensuring compatibility with even single radio-based vehicles. The protocol stack is highly secure and lightweight in nature. Most technical activities related to VANETS is conducted and made popular by the technical subcommittees of the IEEE Communication Society. The concentrations of the aforementioned society accommodate future telematics applications, ITS-based services, V2V and V2I communications, standards, intersection management technologies communications enabled vehicle/road safety, and real-time traffic monitoring.

Several initiatives have been taken by the US government and its Department of Transportation to test the beneficial attributes of WAVE standard technology. The road safety improvements and threat reduction was also well documented in the government's initiative project conducted in Ann Arbor, Michigan in the year 2012. The final report published by The US National Highway Traffic Safety Administration (NHTSA) stated that vehicle to vehicle networks have been explored extensively and are suitable for mass deployment across the United States. The only factor holding back the technology was the requirement to equip a significant portion of the vehicle fleet before deployment. VANETs work based upon intercommunication and information exchange and hence, there is a need for a minimum threshold of vehicles to be a part of the network. United States Law makers and transportation regulators were also convinced in 2014 that the adoption of V2X (vehicle to something) networks will increase user comfort, security and safety [22]. While most parts of the United States and the world have implemented VANETs for mainstream traffic control and communication, rapid measures are being undertaken to increase the employment of VANETs for the domain of personal and public transportation.

2.6 Possibilities towards development of new protocols

Based upon empirical understandings and comparison of existing implementations, we can understand that the employment of an efficient MAC protocol can help guarantee fairness. It is however essential to determine whether the protocol will be able to provide increased Quality of Service (QoS) to the user. VANETs assign the highest importance to the transmission of safety messages. Next-generation VANETs, on the other hand, will have unique needs for self-driving cars with high flexibility, reduced latency, real-time applications, and connectivity that ordinary cloud computing may not be able to meet. As a result, combining fog computing with traditional cloud computing for VANETs is suggested as a viable solution to a number of existing and future VANET difficulties. Fog computing can also benefit from the addition of a software-defined network (SDN), which gives network flexibility, easy implementation, and global information. Multi-channel protocols help transmit information packets through the employment of multiple channels. The channels are completely utilized when multi-channel protocols are used for the transmissions. They also ensure high efficiency by making the interval mode dynamic and adaptive to the network's requirements. Furthermore, because centralized MAC protocols may acclimatize to high dynamic networks and permit direct multi-hop relaying transmissions by using RSUs or CHs, they are used under the assumption of an unstable link and high vehicle mobility. The ideal interval ratio, clustering, multi-channel coordination, and QoS compliance are some of the open difficulties in future development. To boost the overall throughput, innovative methods and infrastructure should be considered in MAC protocol design, such as the employment of numerous transceivers and adoption of the multiple-input multiple-output (MIMO) methodology.

2.7 Challenges in VANETS

The unique characteristics of VANETs necessitate alternative communication paradigms, security and privacy techniques, and wireless communication technologies when compared to MANETs. Network connections, for instance, may not be steady for an extended period of time. Researchers have looked into making better use of existing infrastructure, such as roadside units and cellular networks, in order to increase communication performance. Although some specific VANET issues have been overcome, many major research challenges remain unsolved. As a result, in order to address these issues, academics will need to dig deeper. The primary challenges will be summarized in the following discussion. From a more theoretical standpoint, nothing is understood about the fundamental limitations and prospects of VANETs communication. We feel that preventing accidents and reducing resource consumption are both significant theoretical research concerns. The original IEEE 802.11 standard fails to meet the criteria for reliable network connectivity, and the IEEE 802.11p protocol's present MAC parameters are inefficiently setup for a large number of cars. As a result, more research on standards is required. Although many effective routing protocols and algorithms have been presented by researchers, such as Cognitive MAC for VANET (CMV) and Greedy Traffic-Aware Routing (GyTAR), the critical challenge is to design good routing protocols for VANETs communication with high vehicle mobility and high dynamic topology

The most essential challenge in VANETs communication is the administration and regulation of network connections among vehicles and between vehicles and network infrastructures. The primary problem in vehicular communication design is to achieve adequate delay performance under the restrictions of high dynamic topology, channel, and vehicular speeds. Designing cross-layers among original layers is one option for supporting real-time and multimedia applications. Cross-layer protocols, which operate at many layers, are commonly used to establish priority across distinct flows and applications. After studying the performance measures, we need to discuss the importance of cross-layer architecture in VANETs communication in a collaborative environment. The authors regard VANETs to be a form of cloud known as mobile cloud computing (MCC) and provide a broadband cloud in vehicular communication. As a result, in the context of vehicular clouds, collaboration between vehicular clouds and Internet clouds is possible. Mobility is required to provide real-time and multimedia applications. The mobility of vehicular networks causes the topology to change quickly. Furthermore, vehicle mobility patterns on the same road will change/ show strong connections. The writers of highlight the notion that mobility is important in modeling and creation of vehicular protocols. Many examples can be found in reference solutions with major disadvantages and the "key pair/certificate/signature" is still used in the mainstream solution.

2.8 Conclusion

Over the last several years, improvements in wireless communication technologies and technical innovations in the auto-mobile sector have sparked a surge in research interest towards the topic of VANETs). V2V and V2I communications are enabled by wireless access technologies such as IEEE 802.11p in a vehicular network. Through the development of intelligent transportation systems (ITSs), this advancement in wireless communication is predicted to boost road safety and traffic efficiency in the near future. As a result, governments, the automobile industry, and academia are working together on numerous active research initiatives to develop VANET standards. The architecture, standards, and protocols of vehicle ad hoc networks were defined initially in this study, followed by the features of protocols. MAC protocols were primarily discussed across this chapter. The existing technology, standards, and layers involved in the functioning of MAC protocols has been discussed extensively. VANET has become an intriguing sector of mobile wireless communication due to the typical collection of VANET application areas, such as car collision warning and traffic information distribution. This chapter depicts the various operational and performance aspects of each protocol for various applications, using various MAC layer technologies. The absence of compatibility between protocols, which puts their heterogeneity to the test in various ways, causes the homogeneity paradigm to break down. Robustness, distance coverage, data transfer rates, and energy efficiency are all characteristics that contribute to protocol diversity, allowing it to suit a wide range of applications.

References

[1] Paul B., Ibrahim Md., Bikas Md., and Naser A. "Vanet routing protocols: pros and cons." *arXiv preprint arXiv*:1204. 1201 (2012).

[2] Rani P., Sharma N., and Singh P. K. "Performance comparison of VANET routing protocols." In *2011 7th International Conference on Wireless Communications, Networking and Mobile Computing* (pp. 1–4). IEEE, 2011.

[3] Park C.K., Ryu M.W., and Cho K.H., "Survey of MAC protocols for vehicular ad hoc networks." *Smart Computing Review*, 2012;2(4):286–295.

[4] Amit Kumar Tyagi, S U Aswathy, Autonomous Intelligent Vehicles (AIV): Research statements, open issues, challenges and road for future, International Journal of Intelligent Networks, Volume 2, 2021, Pages 83–102, ISSN 2666-6030. https://doi.org/10.1016/j.ijin.2021.07.002.

[5] Ma Y., Yang L., Fan P., Fang S., and Hu Y., "An improved coordinated multichannel MAC scheme by efficient use of idle service channels for VANETs." In *Proceedings of the 2018 IEEE 87th Vehicular Technology Conference* (*VTC Spring*), Porto, Portugal, 3–6 June 2018; pp. 1–5.

[6] Mao Y., Yan F., and Shen L., "Multi-round elimination contention-based multi-channel MAC scheme for vehicular ad hoc networks." *IET Communications* 2017;11:421–427.

[7] Nguyen V., Pham C., Oo T.Z., Tran N.H., Huh E.N., and Hong C.S., "MAC protocols with dynamic interval schemes for VANETs." *Vehicles Communications* 2019;15:40–62.

[8] Song C., Tan G., and Yu C., "An efficient and QoS supported multichannel MAC protocol for vehicular ad hoc networks." *Sensors* 2017;17:2293.

[9] Hafeez K.A., Zhao L., Mark J.W., Shen X., and Niu Z., "Distributed multichannel and mobility-aware cluster-based MAC protocol for vehicular ad hoc networks." *IEEE Transactions on Vehicular Technology*. 2013;62:3886–3902.

[10] Shahin N. and Kim Y.T., "An enhanced TDMA cluster-based MAC (ETCM) for multichannel vehicular networks." In *Proceedings of the 2016 International Conference on Selected Topics in Mobile & Wireless Networking (MoWNeT)*, Cairo, Egypt, 11–13 April 2016; pp. 1–8.

[11] Batth R.S., Gupta M., Mann K.S., Verma S., and Malhotra A., "Comparative study of TDMA-based MAC protocols in VANET: a mirror review." In *International Conference on Innovative Computing and Communications* (pp. 107–123). Springer, Singapore, 2020.

[12] Wu J., Zhang L., and Liu Y., "On the design and implementation of a real-time testbed for distributed TDMA-based MAC protocols in VANETs." *IEEE Access* 2021;9:122092–122106.

[13] Zain I.F.M., Awang A., and Laouiti A., "Hybrid MAC protocols in VANET: a survey." 2017 In *Vehicular Ad-Hoc Networks for Smart Cities* (pp. 3–14). Springer, Singapore, 2017.
[14] Ma M., Liu K., Luo X., Zhang T., and Liu F., "Review of MAC protocols for vehicular ad hoc networks." *Sensors* 2020;20(23):6709.
[15] Nguyen V., Khanh T.T., Oo T.Z., Tran N.H., Huh E.N., and Hong C.S. "A cooperative and reliable RSU-assisted IEEE 802.11 p-based multi-channel MAC protocol for VANETs. *IEEE Access* 2019;7:107576–107590.
[16] Akhter A.F.M., Ahmed M., Shah A.F.M., Anwar A., and Zengin A., "A secured privacy-preserving multi-level blockchain framework for cluster based VANET." *Sustainability* 2021;13(1):400.
[17] Lyu F., Zhu H., Zhou H., *et al.* "MoMAC: mobility-aware and collision-avoidance MAC for safety applications in VANETs." *IEEE Transactions on Vehicular Technology* 2018;67(11):10590–10602.
[18] Singh G.D., Tomar R., Sastry H.G., and Prateek M., "A review on VANET routing protocols and wireless standards." In *Smart Computing and Informatics* (pp. 329–340). Springer, Singapore, 2018.
[19] Ibrahim B.F., Toycan M., and Mawlood H.A., "A comprehensive survey on VANET broadcast protocols." In *2020 International Conference on Computation, Automation and Knowledge Management (ICCAKM)* (pp. 298–302). IEEE, 2020.
[20] Nguyen V., Pham C., Oo T.Z., Tran N.H., Huh E.N., and Hong C.S., "MAC protocols with dynamic interval schemes for VANETs." *Vehicular Communications* 2019;15:40–62.
[21] Hussain R., Lee J., and Zeadally S., "Trust in VANET: a survey of current solutions and future research opportunities." *IEEE Transactions on Intelligent Transportation Systems* 2020;22(5):2553–2571.
[22] Buinevich M., Izrailov K., Stolyarova E., and Vladyko A., "Combine method of forecasting VANET cybersecurity for application of high priority way." In *2018 20th International Conference on Advanced Communication Technology (ICACT)* (pp. 266–271). IEEE, 2018.

Chapter 3

Opportunistic routing protocols for Vehicular Ad Hoc Network (VANETs) and intelligent transportation system (ITS)

Abstract

The innovative technology Vehicle Ad Hoc Network (VANET) has received a great amount of interest in recent years. Due to rapid changes in topology and frequent disconnections, the efficient protocol for data routing between automobiles known as vehicle to vehicle (V2V) or vehicle to infrastructure (V2I) is hard to design for roadside infrastructure and automotive communication. The current VANET routing methods cannot deal with all traffic situations. This emphasized the development of an excellent routing mechanism. It is therefore important to determine the advantages and disadvantages of routing protocols that may be utilized to design or construct new routing protocols. In this article, the pros and disadvantages of VANET communication routing methods are addressed. An ad hoc car network that can set up all network devices to act as a host and a network router. Only nodes help each other to deliver information. Ad hoc vehicle networks were mostly temporary and less accessible infrastructure networks. The channel location and network connection instability diminish mobility and resource boundaries in the performance unit. Improve the relevant layering methods in other performance units used to stack communication through information exchange at certain other protocol levels. Everything is good ad hoc protocol routing for AODV. While the VANET is helping to provide efficient logistics and transport solutions, several difficulties must be addressed if an appropriate solution is to be found.

Key Words: Vehicular Ad Hoc Network (VANET); intelligent transportation system (ITS); Internet of Things (IoT); carrier sense multiple access (CSMA); Quality of Service (QoS)

3.1 Introduction

Vehicle using connected things/automotive Internet application is the ad hoc Vehicular Ad Hoc Network (VANET) that delivers driving safety and comfort (using (Internet of Things) IoT devices). Vehicles in a VANET serve various functions. They provide VANET statistics on local traffic as a source of data. They also serve as intermediary nodes to transfer packets to other vehicles. It might also be a database that receives the information [1]. Further infrastructure such as route devices improves VANET resilience through the relay of packets. However, owing to economic, technological, regulatory, and other constraints, the quantity and positioning of roadside devices are limited. A high number of information relays may therefore be carried out across the VANET, particularly when the distance between the source and destination is vast. The huge number of relays is connected to the danger of effective communication, as cars are very mobile and automotive connectivity may be stopped unexpectedly. Current research tends to jeopardize connection bandwidth and network capacity. However, limited network resources in VANETs play an important role in the network service experience [2]. To ensure that information transmission is reliable at VANET, an information transmission system has to be designed that decreases information distribution through frequent connection failures and enhances road safety and durability. Although tiny connections permit fewer interruptions, they create a means of increasing the transmission time and bandwidth efficiency through several short connections. There is a balance between robustness and speed in the VANET packet routing [3].

The term connection is a possible trade balance approach. A route may be constructed with long links that offer stability throughout the transmission of the packet knowing the duration of the connection. Furthermore, since a link lifetime may be expected, a substitute route can be set up to assist the flow of data. The movement of cars is confined to fixed roads and speeds in a VANET. These limitations might provide important information to forecast the life lifetime of the relationship. This research offers an effective predictive strategy for data transmission to balance the reliability and efficiency of data transmission in VANETs [4]. Our system determines the length of the connection between two motorcars fitted to the transmission kit based on information about the speed and position of the vehicle in the regular contact package. Our method uses a statistical tool called a connecting tool that detects the effects of data transfer in relative speed and distance vehicles on efficiency. The utility value is used to optimize the path of the selection of the transmission node. VANET is a MANET variant that provides a wireless network for cars and cars. It is an autonomous wireless network in which VANET nodes are utilized for information exchange and sharing as servers or clients. The VANET network architecture may be broken down into cellular/WLAN, ad hoc, and hybrid categories. Applications from VANET include automobile collision alerts, driver assistance, driver support, cooperative driving, cooperative cruise monitoring, information streaming, Internet access, vehicle location, parking, and autonomous cars [5].

3.1.1 No sheltered boundaries

VANET has no structure, no restrictions, no topology established, and no network nodes. If necessary, it can connect more than one vehicle/node to any node/vehicle. Every emergency can happen on a road like an ambulance that has to go to a hospital and cannot get to the hospital more quickly. In this case, it joins other road

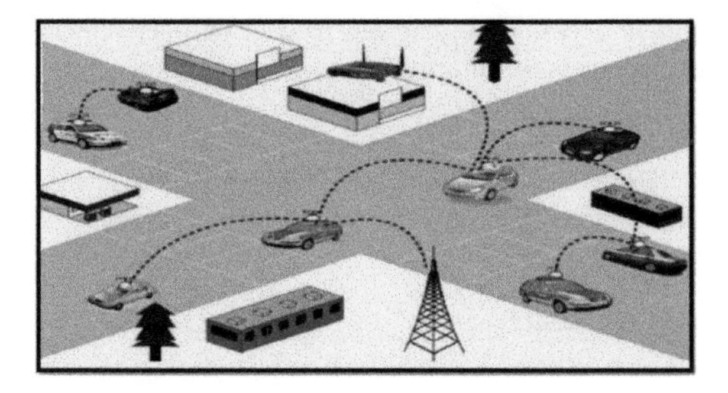

Figure 3.1 VANET design [6]

cars and attempts to locate a path with less traffic. This is a VANET benefit, but it is frequently a drawback because there is no zone or boundary where it works. An intruder can stop message processing and intervene because any node/vehicle that makes VANET untransparent for assaults can participate and since it has no defined topology, unpredictability, and dynamism that makes the messages defenseless and Figure 3.1 shows VANET design.

3.1.2 Compromised node

The nature of VANET is dynamic and allows any node/vehicle to connect or exit the network as needed. To share information or ask inquiries, you will connect to the network and instantly disengage the network when the necessity is done. If you want information or inquiries to trade. The changing nature of the nodes achieves this dynamic feature of VANET. The VANET connection is relatively short-lived, therefore it is not a lengthy network. At the same time, the network is positive and negative since the damaged node/car in the network is very hard to trace, making the network reputable [7]. The harmful nodes can infiltrate the network, interrupt the network, engage actively in the network, or, depending on what kind of assault VANET can helplessly, and Figure 3.2 shows the needed flow chart.

3.1.3 No fundamental management

A normal network always includes an important node or network management facility that covers information exchange, network failures, and node features that complicate network hacking. Sometimes this core node is also called a basic node. But VANET does not have this fundamental structure. There is no center node to supervise the VANET network process as the design does not include a certain network architecture alone, it also makes it easy to access the network and intercept the transaction for the destroying node [8].

3.1.4 Scalability problem

The nature of the exchange and the type of nodes can be readily foreseen in a normal network as they communicate via media that makes their activities in the

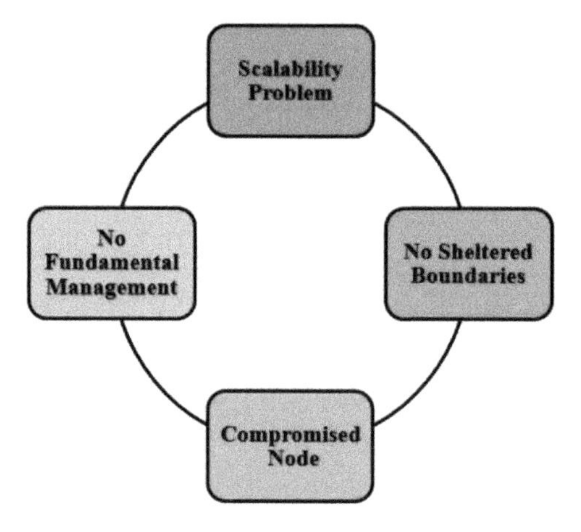

Figure 3.2 Needed flow chart

network predictable and scalable. VANET is unpleasant. The VANET node is movable and dynamic, which allows connecting and communicating with any network node that creates network reliability problems. In other words, any damaging node can connect and intercept communications through the network since it is dependent on blind confidence [9].

The rest of this chapter is organized as follows. In Section 3.2, research works related to opportunistic routing protocols for VANETs. Section 3.3 describes routing protocol, types, and challenges of VANET and in Section 3.4, opportunistic routing schemes in VANET have been described. Section 3.5 describes the network layer used in VANET and Section 3.6 describes urban VANET routing protocol, whereas Section 3.7 describes the novel routing framework for VANET. Section 3.8 describes VANET receiver-based forwarding system, and Section 3.9 describes opportunities toward efficient routing protocols and standardization in VANET. Finally, Section 3.10 concludes this chapter and identifies some potential future works.

3.2 Literature survey

There are a few techniques for data transfer depending on VANET receivers. To facilitate video streaming over the VANET environment, a reactive receiver solution known as VIRTUS was created. VIRTUS enables the transmission of large data speeds between cars without the requisite roadside infrastructure. Nodes are conveyed here based on their current position and future estimates. The VIRTUS transmission area is directed towards the goal. The receiving nodes calculate waiting time depending on geographical progress and the dependability of the connections. VIRTUS delivers adequate video streaming performance and also minimizes car network transfer volumes. RPBL is another lightless routing method based on the recipient in which the nodules select their waiting time on the path towards their destination [10].

Azimi Kashani *et al.* (2020): Ad hoc vehicle networks create a wide range of technologies including vehicle safety, traffic control, and intelligent transport systems. Given the rising mobility of automobiles and their inhomogeneous distribution, an efficient routing system seems vital. Ascertain sections of a route are crowded and others are not packaged, the routing protocol should be in a position to make dynamic judgments. On the other hand, the VANET network environment is vulnerable when data is transmitted. Broadcast routing could increase its efficiency, similar to opportunistic routing, then other protocols. In this article, a novel logic of opportunity routing ,protocol describes the decision-making process of packet transmission through a fuzzy logic system with three packet progress input parameters, local density, and several packets sent. Re-diffusion procedures use the value of these parameters as inputs to tackle a multi-casting problem for the fuzzy logic system with the congested and scattered locations in mind [11].

Hong Huang *et al.* (2020): The growth of mobile and IoT devices and the improvements made in wireless communication have highlighted the necessity for heterogeneous hybrid networks to have new communication paradigms. Researchers have indicated an opportunity to harness the opportunities offered by these various networks. Although there are many recommendations for various opportunistic routing systems, very few have examined futility to evaluate the network's wireless connection status to design reliable and faster paths to destinations. The AGO-FQ, the new asymmetric link for Fuzzy logic Q-learning, allows long-term transmission connections that assign forwarding candidates to a specific site. The proposed routing procedure employs useless logic to decide if a wireless connection is useful, by recording several network parameters, bandwidth availability, link quality, node power transfer, and remote progress. The proposed routing process uses a Q-learning algorithm to select the best candidate based on his smooth logical evaluation [12].

Juan Huan *et al.* (2020): Opportunistic routing in ad hoc car networks can increase reliability and performance efficiently. Opportunistic routing, however, also poses security problems. Malicious nodes, for example, may simply be combined into applicant node groupings which can impede network speed. The paper offers a node-based notion of trust to solve the problem of malicious nodes in the opportunistic transmission of candidates. The suggested confidentiality model uses customization and filtering techniques to rule out harmful suggestions in the computation via the dynamic weight calculation process, which integrates a direct trust with indirect trust [13].

Carlo Augusto *et al.* (2020): V2X technology is built on heterogeneous and secure applications based on V2V-typical broadcast messages and V2I/I2V-typical unicast messages. We have devised an Opportunistic Routing Algorithm (RSSI) for mobile ad hoc network infrastructure to support unicast communications from vehicle infrastructure such as rescue operations and to fulfill the VANET routing problem (IRONMAN). IRONMAN uses opportunistic routing decisions based on measured RSSI by roadside units instead of traditional GPS-based solutions (RSUs). With a genuine testbed, the author shows that IRONMAN offers standard ad hoc network routing options based on Linux, such as BATMAN and HWMP, which deliver almost optimal functionality without adding any routing costs [14].

Alshehri *et al.* (2020): Recently, the rapid development in wireless network communication applications has lowered network capacity with an increasing number of nodes. Existing routing solutions such as extensive delays, network infrastructure needs, limited traffic patterns, high technological complexity, and efficient bandwidth use cannot meet this capability constraint. The scalability of wireless network capacity is therefore an important topic. This chapter describes a new, scalable opportunistic routing method for a large multi-hop wireless network. Different network metrics analyze the performance of the routing protocols for accurate analysis. In conjunction with the advances in wireless communications capabilities of these devices, new paradigms for these heterogeneous hybrid networks were required for mobile and IoT systems to thrive. Researchers have indicated an opportunity to harness the opportunities offered by these various networks. Although there are many recommendations for various opportunistic routing systems, very few have examined futility to evaluate the network's wireless connection status to design reliable and faster paths to destinations. The proposed routing procedure employs useless logic to decide if a wireless connection is useful, by recording several network parameters, bandwidth availability, link quality, node power transfer, and remote progress. The proposed routing process uses a Q-learning algorithm to select the best candidate based on his smooth logical evaluation [15].

Pang *et al.* (2021): In contrast to regular ad hoc mobile networks (MANET) and ad hoc vehicular networks, the UAV system has two characteristics (VANET). First, the architecture of the UAV network changes more often and more rapidly. Second, the connections between UAVs vary considerably in time and space. Existing hop-by-hop routing techniques cannot adapt to the highly dynamic network environment and do not take full advantage of the transmitting possibilities for all different UAV network connections. In comparison with the Hop-by-Hop routing protocols, GPHLOR may take maximum advantage of the transmission chances for each connection regardless of hop distance and connecting quality, utilizing the concept of despair. Based on information on the geographical position of each node, GPHLOR chooses relays but does not care about topology so that they may be adapted to common topologic changes. GPHLOR also enables every node to compute its shipping priority according to its location information disseminated. It does not rely on frequent network data monitoring and delivery, thereby lowering overhead procedures. The simulation data showed that GPHLOR offers improved packet delivery rates, end-to-end latency, throughput, and overhead compared to conventional UAV routing protocols [16].

Parameswari *et al.* (2021): VDTNs are an innovative DTN solution for vehicle communications amid demanding delays and unpredictable connectivity situations. This method uses the transport bundles in the warehouse to reach their destinations by driving cars with short-distance wireless devices in asynchronous form, hop by hop. The VDTN architecture relies on out-of-band signals and employs an IP for VDTN with separate control and data plans. This article provides an Opportunist OPRNET routing protocol to work on global location data routing decisions and to fuse a hybrid approach between multiple copying and single copying methods. It also improves vehicle communication performance by minimizing energy use.

The findings of our proposal for simulation demonstrate that OPRNET is practicable and can be regarded as a highly important vehicle communication technology, even if it incorporates appropriate interference and QoS support technologies. OPRNET seeks to improve network capacity such as storage, latency, and energy use while expanding distribution opportunities and reducing overhead and delays [17].

Sharvari *et al.* (2021): Emergency communication is essential to enable rescue and search efforts during disasters. The UAV networks can be used, under such circumstances, to supplement wrecked cellular networks in larger areas. But routing for these UAV networks is not a problem because of high UAV mobility, inconsistent connectivity between UAVs, changing UAV topology (3D), and constraints on resources. Although several UAV routing strategies are available, no integrated approaches to inter-UAV coverage, collision, and routing have been adopted too yet. Analytical formulas for probability and probability of collision are obtained, and a minimum and maximum distance between UAVs is calculated for empiric computation. The next step is a novel multi-hop opportunistic 3D routing approach (MO3DR) with inter UAV coverage and crashes limits to optimize the expected progress of all data packets [18].

Somayeh Razaghi Kariznoi *et al.* (2021): multi-change wireless sensor networks are widely used by enhancing the Internet of things applications as a promising technology. In the light of power restrictions in WSNs, extending the network life is an issue that needs to be addressed clearly in every design – medium access and routing protocols. Adaptive sleep-wake scheduling based on the usage of nodes and energy level in the proposed medium access control (MAC) protocol enables the sender to prevent repeated data transmission synchronization and competition. In addition, the proposed approach prevents priority channel saturation depending on the number of competing nodes and channels. This balances network load across many channels. Moreover, an opportunistic routing region is limited to improving road conditions that balance the loads between various rivals. The simulation results show that the approach proposed increases performance over the equivalent works (MORR protocol and SLOR protocol) [19].

Qin *et al.* (2021): Design of an optional offload data routing in-car opportunistic networks using sparingly distributed static relay systems. The objective is to enhance the data download ratio while meeting the data delay criteria, and also to minimize the average delivery time. The author demonstrates the efficiency of static relay encounters to enhance data download speed based on the vehicle trajectory. A greedy technique for optimal positioning of the relay is proposed. The author decreases the computational complexity of the data download technique and also the relay placement approach [20].

3.3 Routing protocol, types, challenges in VANET

In a routing protocol, the interaction between two communication components is specified. It comprises setting up a route, conveying information, and acting to sustain or recuperate from a failure of the route. Wireless communication's primary

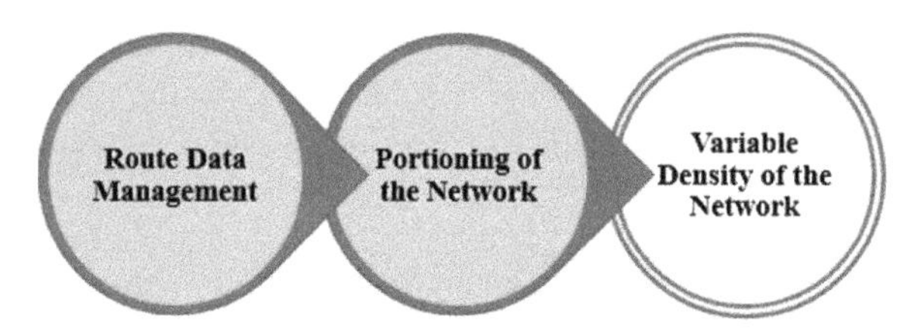

Figure 3.3 Routing layer challenges

objective is to minimize the time of communication while using the lowest network resources. Figure 3.3 shows routing layer challenges [21].

MDVNET routing protocols have been examined for several uses in-vehicle networks. Routing systems are divided into four types, i.e., unicast, multicast, broadcast, and MDVNET-based routing protocols. Routing plays a very significant part in each network's overall operation. The best routing protocol transmits the data along the shortest, least congested path to several hops [22].

Before transmitting the data, the network state in VANETs should be considered as nodes regularly join and depart the network. The topology is uncertain and routes are not occasionally available. The usual routing methods are not suited for communication with the car; the network performance is thus constrained. When routing choice is based entirely on nearby information, the Protocol should be able to conduct localized activities [22]. It removes the need for the node to know the entire network topology and therefore reduces overhead control. Figure 3.4 shows types of routing protocols in VANET and Figure 3.5 shows the ideal routing protocol for VANETs.

3.3.1 Unicast protocol for routing

It refers to a transmission independently from one communication device to another. The MDVNET unicasting major objective is to transfer packages from a single-source vehicle via wireless single-hop/multi-shop communications through a shop-by-hop strategy or a "store-and-forward" technique to another single destination vehicle. The major distinction is whether intermediate vehicles are redirecting or transporting incoming packets to the appropriate routing algorithm. Unicast routing algorithms were investigated in the current investigation in two ways: greedy and opportunistic [23]. The source vehicle sends packets to its outermost (the nearest intermediate) neighbors and transfers packets to its outermost neighbor (the second intermediate vehicle hop) with greedy protocols for unicast routing until the target vehicle obtains these packets. Figure 3.6 shows source to destination data packets.

This indicates that future decisions will be based on the geographical information in greedy Unicast routing algorithms for automobiles. In basic communication settings, the greedy unicast routing protocols have proved useful, for example in using "Hop-by-Hop" techniques for automobiles on straight highways. But greedy storage and forward procedure are best suited for city zones with a

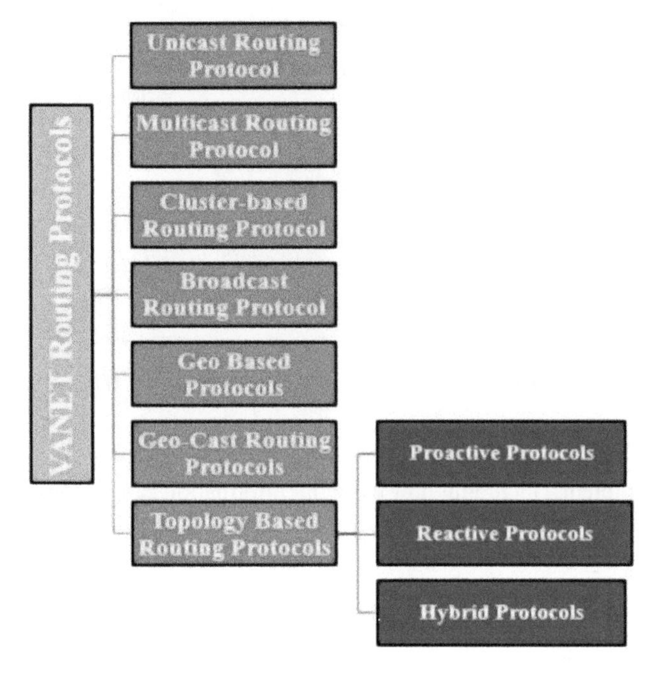

Figure 3.4 Types of routing protocol in VANET

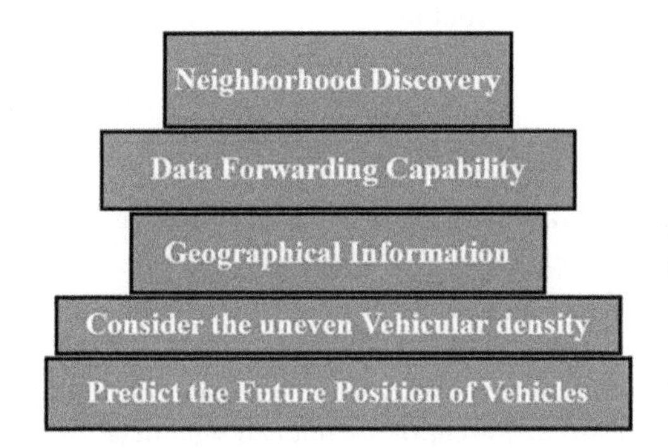

Figure 3.5 Ideal routing protocol for VANETs

plenty of crossings so that applications without safety that enable delay tolerance may be implemented. Opportunistic unicast routing algorithms allow wireless networks with frequent connections like vehicle networks to be operated. Originally, MANETs with homogeneous mobile node settings allowed more opportunistic unicast routing methods. Relays are chosen for special utility features, such as cars with different speeds, for the creation of appropriate opportunistic unicast routing protocols for networks with diverse setups. In the selection of intermediate vehicles, an opportunistic street routing strategy was devised, which

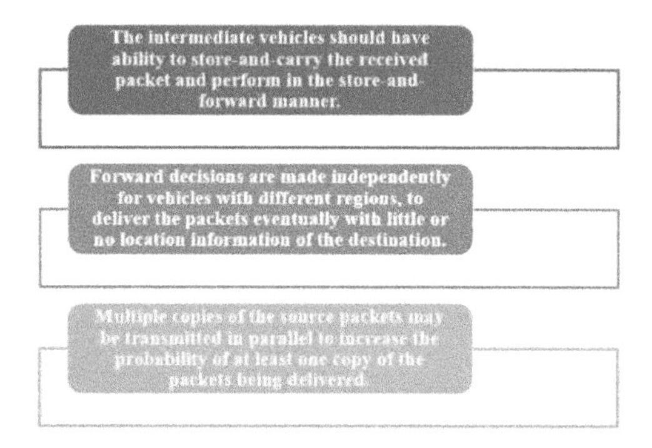

Figure 3.6 Source to destination data packets

took into consideration sign fading and mobility. Unicast routing systems have significant problems, including how essential information can be conveyed rapidly, how communication delays and packet losses may be avoided, and how routing conflicts can be resolved in various unicast routing applications [24].

3.3.2 Multicast routing protocol

There are two types of multi-group communications: single-group and multi-group. Multicast routing is useful for inter-vehicle communication in situations like roadblocks, traffic jams, calamities, and geographical advertising. Geofact is one of the most often used multicast routing techniques in MANETs, where geographical locations define the group of destinations. The Intervehicle Geocast protocol alerts all automobiles on a roadway of potential hazards based on their position, speed, and direction. The Geocast approach may be improved by incorporating more data, such as mobility and trajectory connections. The major goal is to overcome network connectivity, mobility unpredictability, and package duplication [24].

3.3.3 Broadcast routing protocol

It utilizes a communication method to transmit a message to all receivers at once. This important data is sent to other cars to encourage safe and cooperative driving applications and to distribute the aforementioned vital data for unicasting or multicasting. There is no denying that broadcasting is feasible. However, while designing a broadcast protocol for vehicle networks, two key issues must be addressed: broadcast storm and disconnected network difficulties. A broadcast storm occurs when many vehicles submit duplicate broadcast packets at the same time. During a broadcast storm, close automobiles may experience channel congestion and packet collision. Vehicle networks are actively at work to overcome this. With these methods, vehicles rebroadcast packets received with a probability of one or less. A car's chance of retransmission depends on its distance from the sender in these three techniques, thus only GPS or signal strength packets from cars in the single-hop area are required. In addition to distance, data on local topology and space distribution are used to pick retransmission vehicles [25].

3.3.4 Cluster-based routing protocol

This is a routing system where automobiles are organized into several sets (clusters) by specific criteria that pick a vehicle from each set to become a cluster head (CH). Note that three routing protocol types of unicast, multichannel, and broadcast may be used to develop all routing techniques for vehicle networks. The protocol based on the clusters, in particular, is a blend of unicast, multicast, and broadcast rather than a unique routing mechanism. For example, a communications mode can be a CM unicasts package for their CHs, a CH transfers packages to all its members or multichannel road infrastructure or other CHs, and a one-channel package for the infrastructure for one CH or multi-channel. Research has shown that the clustering process can increase the scalability and reliability of the routing protocol by grouping automobiles based on relative speeds and geographical distribution data [26].

3.3.5 Geo-based protocols

A source interacts with the destination in these protocols through geographic location and network address. Calculate the large balancing routing protocol (LBRP) and set the route according to the location of the node. The routing tables are therefore not necessary. The procedure has three components: lighting, placement, and transmission. The problem is that the Global Positioning System requires assistance. Furthermore, when the car enters the area like a tunnel, the satellite signals become faint [27]. However, it gives optimal performance for the road conditions. Its characteristics also include high mobility efficiency. Figure 3.7 shows a comparative study of different protocols.

3.3.6 Geo-cast routing protocols

The geo-cast protocol has two primary areas (e.g., zone of relevance (ZOR) and zone of future (ZOF)). ZOR is the area given to the regional nodes. The main objective of this protocol is to facilitate communication between the ZOR vehicle. In addition, if the source truck decides to talk with the vehicle, but not ZOR, the truck will be ZOF, and every truck entering ZOF will be responsible for transferring the information to another ZOR. The link might often be broken down owing to the frequent zone shifts that are unfavorable [29].

3.3.7 Topology-based routing protocols

The above category has two protocol kinds, proactive and reactive [30]. In addition, ad hoc routing protocol classification is given in Figure 3.8.

3.4 Opportunistic routing schemes in VANET

An appropriate updating technique for transmission of beacon signals into an opportunistic routing system for certain features of VANETs, including position, speed, and direction [31]. Two rules underpin this arrangement:

(i) estimate the lifespan of the link between automobile beacon messages after the predicted time expires;

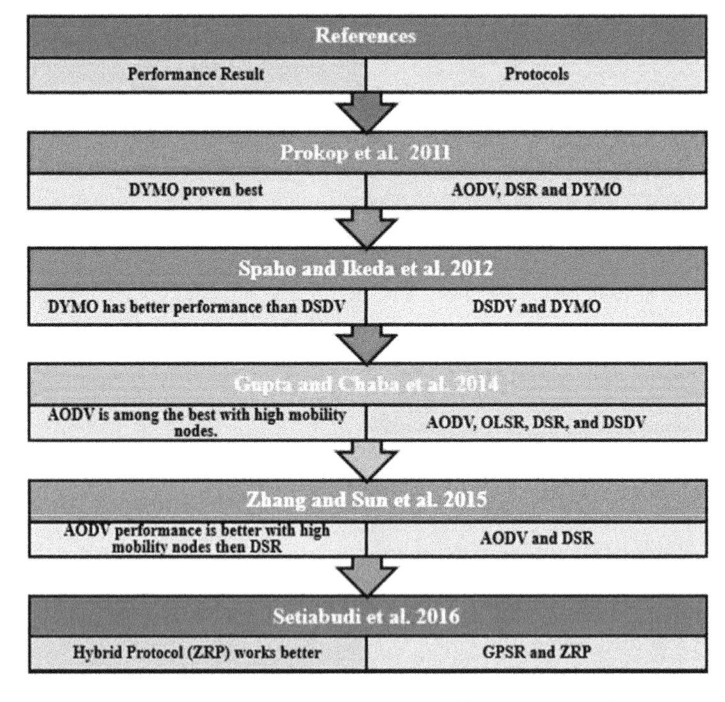

Figure 3.7 Comparative study of different protocols [28]

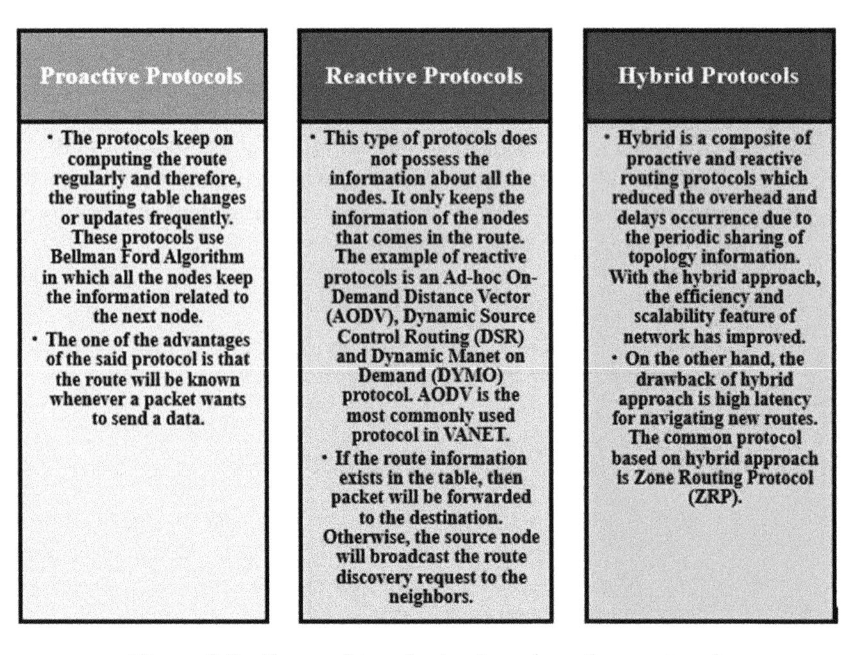

Figure 3.8 Types of topologies-based routing protocols

(ii) a beacon message should be sent if the transmission received of the data packet is altered sequentially to preserve topological correctness.

The dependable and effective opportunistic routing (PRO) of VANETs is based on estimations of probability. To evaluate the candidate's performance, this approach anticipates changes in the signal-to-interference-plus-noise (SINR) ratio and the packet queue (PQL) of the objective vehicle nodes.

3.4.1 Routing systems based on position

Two location-based solutions to the minimum local problem of greedy algorithms [32]. Created two protocols for CSVNs:

(i) GF fully spread with the relay system
(ii) GF with virtual roadside units with local minimal requirements during the early design process.

In PGRP, the weight of the vehicle is based on features such as the direction and angle of the node. Hello is used by PGRP to forecast the position of each vehicle node. After a short time, PGRP transmits packets by vehicle location. PGRP has two major problems:

(i) Estimate the vehicle node's future position using GPS and Hello packets.
(ii) Support the dynamic speed rate length and the Hello packet reaction.

RRP employs the road side unit (RSU) and picks blocks to send to the target car via data packets. The shortest path between the routes is picked from the source node to the destination node. Whenever a flag node is received, the data packet uses its digital map to select the block nearest to the target node and transfers the packet to a vehicle node on the block. Pass CAR is a passive routing clustering solution that originally uses the clustering technique. For each node in the routing process, a priority is then calculated. The fundamental concept of the Pass CAR was to select the appropriate vehicle nodes for the CH or the gateways to forward the packets in the road discovery phase. Pass CAR evaluates the appropriateness of nodes in an election plan based on factors such as reliability, stability, and durability. A geographical link guaranteed by quality in urban VANETs. Each road segment is allocated weight to identify the ideal route depending on the delay and connection in each road sector. The traffic segment can be picked one by one for the ideal routing way to prevent local maximum and data congestion based on weight data. The data packet will subsequently be sent via the designated segment of the route with a dependable greedy technique. A logical fusion system routing protocol that could help to evaluate conflicting parameters. It comprises different aspects, such as position, direction, quality, and achievable performance of the vehicle to determine the perfect relay node [33].

3.4.2 Hybrid opportunistic routing methods based on positions

GeOpps uses pathways of the navigation system for predicted nodes that get closer to the final destination node. It finds the shortest path from the source node to the destination node and the end-to-end latency from the source to the target node. Routing continues until the target node is sent to the data packet. The fundamental disadvantage of GeOpps is that this protocol requires VANET navigational information. LSGO uses the combination of geographical location and connection status information to select the data transmitter. This protocol is intended to ensure the delivery of packets over a highly dynamic network. It also attempts to improve the reliability of the transfer of data. This method also minimizes the number of transmissions and delays in transmission. The key components of the protocol are the evaluation of the connection quality, the selecting process for a group of application nodes, and the selection of a data communication node prioritization technique. In this strategy, the transmitter uses concurrent packet receipts supported by the wireless media's broadcasting feature and carries out opportunistic transmission through a selection of neighboring nodes receiving the packet. Opportunistic geographic and route-based VANET routing. ORRIS takes characteristics such as location topography, motion vectors and road flows into consideration. ORRIS analyses the history of traffic flow densities in the opposite direction to estimate traffic flux density. Decisions on data transfer are based on the geographical topography and the flow of dispersed cars. Some strategies assessed opportunities or positions to choose and prioritized applicant relay nodes. In this article, we provide an OPBR routing system that combines the opportunity and position-based routing for the design and elimination of expired connections for the applicant relay node selection procedure. On the other hand, the system presented includes different acceptable and adequate characteristics to select a relay node, a priority algorithm, and remove the expired connecting mechanism [34].

3.5 Network layer used in VANET

After VSN has been introduced and deployed, adverts are then periodically relayed by the VSN nodes to their neighbors, recognizing their identity and number of Hops from the sink. All nodes are positioned in the initial place so that they are away from the sink node. The corresponding sink node is added to the routing tables and the nodes that receive the publicity. Moreover, their hop numbers are recorded as a hop from the sink. When a node distributes an ad, it reveals its number of hops from the sink rather than from endlessness. In many cases, the routing table containing the sink address and addresses from which they are acquired is updated. The shopkeeper is only kept at her routing table when she receives an advertisement of a sink from several neighbors.

3.5.1 Vehicular networks applications

Communication networks for moving vehicles are often designed to share information and to enable a wide variety of cooperative applications [35,36].

Safety applications: The transmission of safety information can offer security services, substantially decrease traffic accidents and effectively protect the lives, health, and property of commuters. Once safety data from other automobiles are collected, the drivers take preventive actions to enhance driving safety or are informed in advance of projected dangerous situations to avert transportation accidents. One type of vehicle safety information comprises information on traveling countries such as current position, speed, and direction in real-time. This type of information is essential not only for drivers or automated driving systems to pass and change routes and to prevent collisions but also for platoon-based independent vehicles to keep the string stability of the pads and to ensure that driving is essential.

Non-safety applications: To increase passenger comfort by exchanging information between mobile cars, valuable services such as traffic management and help for entertainment may be given. Similar to certain security systems, most traffic management software is intended to eliminate traffic bottlenecks to improve traffic flows and save passenger time. Another common example of traffic management systems is electronic toll collection. The collection of electronic tools can help to get information about the identification of the car and to eliminate the delay in collecting the tolls so that passengers can save time.

3.5.2 Vehicular networks characteristics

Wireless communication is carried out largely in-vehicle networks utilizing the MANET principles, i.e., wireless communication spontaneously for data exchange. In addition to similar traits, such as self-organization and management, short to medium-range, omnidirectional transmission, and low bandwidth, the moveable nodes of a vehicle network have unique properties [37].

Detrimental features: These features provide barriers or problems to vehicle network connectivity, including high mobility, severe retardation limitations, and sophisticated communications settings.

Fast mobility: Vehicle networks offer even higher mobility than mobile ad hoc network (MANETs) due to their high movement speed. High mobility typically results in wireless communication lines that are often disconnected and lower the efficient communication time between vehicles. In addition, it will also dynamically modify the network structure and pose additional challenges to the information transmission between cars.

Strict time constraints: Information transmission is needed for some car network applications, such as safety applications and some entertainment apps, to prevent traffic accidents, to protect the lives of travelers, and to maintain the quality of information services.

Complicated settings of communication: In three communications contexts, vehicle networks are commonly employed. The first is one-dimensional communication scenarios such as the road traffic scenario. Whilst the automobiles are continuously traveling quicker inroads than in other locations, this environment is reasonably straightforward due to the apparent direction and typically fixed speed.

Profitable features: These qualities are useful for wireless communications in-car networks including low energy and track predictions.

Low-energy constraints: While automobiles always have sufficient electricity, vehicle networks, like conventional mobile communication networks, have no power constraints. Each vehicle may also offer significant sensing and computing capabilities, including data storage and processing, as it can supply energy while continuing to communicate with other automobiles.

Driving route forecast: cars on networks are confined to road use under normal conditions, so that the driving route may be forecast for themselves or other vehicles if the information on the roadmap and speed of vehicles is available.

3.5.3 Multi-layer VANET challenges

The physical (PHY) and MAC layer problems discussed and detailed the PHY chain estimation and carrier sense multiple access (CSMA) behavior in high node density and possible congestion control methods along with multi-canal operations. The performance utilizing single and multiple radios and some simulation results were presented. The study also underlines the development of the DSRC from the standpoint of PHY and MAC layer. Multiple access self-organized time division (STDMA) solves the scalability challenge of VANET and has demonstrated promising results in terms of scalability and reliability compared to CSMA [38].

3.5.4 Multi-layer challenges

3.5.4.1 Layer of physical

The physical layer in VANETs is a well-investigated field. From power control through multiple (or single) antennas and from estimate to channel selection. Many factors of the physical layer contribute to the limiting of the network's scalability. The multi-path environment makes communication exceedingly difficult because of the multi-path latency and mobility. The delay spread results in a selective decline in frequency and selective decline in mobility. Non-line of sight (NLOS) circumstances leads to a significant delay spread because of dispersal and the Doppler dispersion of roads [39]. This leads to two distinct challenges: the miscalculation of the channel and the absence of time. In VANETs, the physical layer difficulties include as shown in Figure 3.9.

3.5.4.2. Medium access control (MAC) layer

In fair channel access, the MAC protocol plays a key role. The scalability of VANETs is also strongly dependent on the MAC protocol, thus it is very vital to create an efficient and reliable MAC protocol. The MAC IEEE 802.11p is now utilized with 1609. The multi-channel operations are used as the MAC standards for WAVE vehicle communication [40]. Most current research focuses on the effective use of the EDCA Mechanism's multichannel and performance modeling to build an analysis model to investigate the behavior of the MAC layer. In the MAC protocol for VANETs, the following should be borne in mind as shown in Figure 3.10.

3.6 Urban VANET (UVANET) routing protocol

UVANET has numerous routing mechanisms to effectively transport data to the target. Implementation of UVANET routing protocols based on topology is

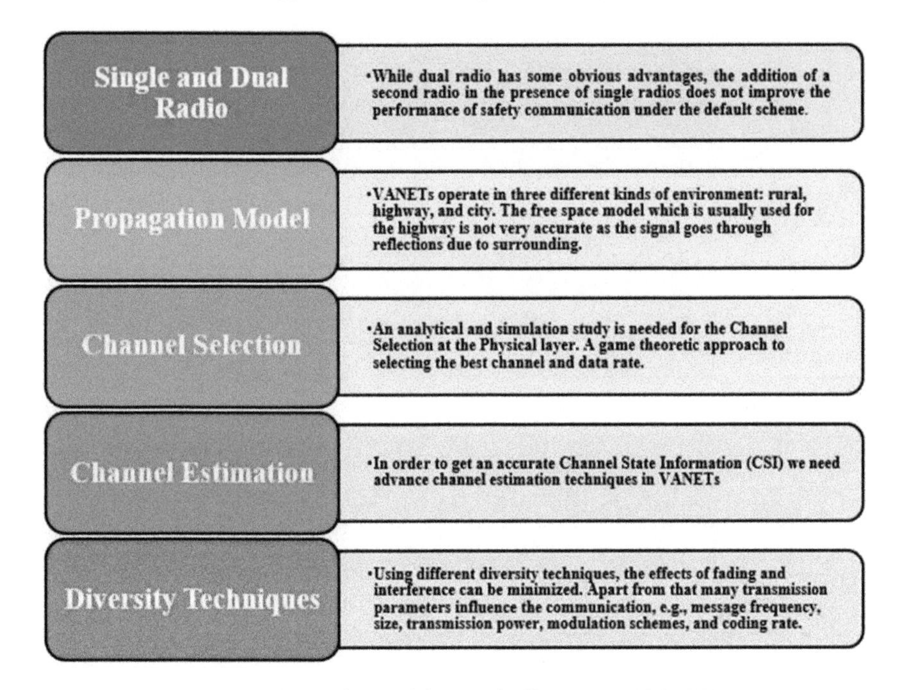

Figure 3.9 Physical layer challenges in VANETs

problematic due to frequent changes in topology. UVANET uses position-based routing protocols that merely provide data on the location of the destination and the surrounding position if road repair is not necessary. The GSR protocol is meant to effectively convey data via the fastest route to the destination. The greedy perimeter stateless routing (GPSR) protocol provides data to cars nearest to the target as soon as possible. The perimeter mode is used to recover from the optimal local problem. The GPCR protocol sends the data via the shortest path using the Dijkstra algorithm. The coordinator node is used to decide the next path for data transfer. The Aware Road and Traffic (A-STAR) anchor-based protocol transmits data over major routes [41]. The main streets are crowded cars. Greedy Traffic-Aware Routing (GyTAR) transmits smart data to the forward traffic destination. The cell data packet (CDP) is used for the transmission of density information between cars from one junction to another. When a vehicle reaches a junction with the data, weight is calculated at each nearby crossroad. Then the data is moved to the highest weight. The IBR protocol transmits data to a route that is least late in the transmission of packets. The diagonal cross-based routing (DIR) protocol gives data via diagonal crossings to the destination. The data is transmitted at the diagonal intersection of that path, which takes a minimum of time to transfer a packet. With cars outside of the communications range, the P-GEDIR protocol provides data to the target. The IGRP sends data to the Internet gateway across the junctions. The route is determined by delay, QoS, and bandwidth. UVANET enhanced

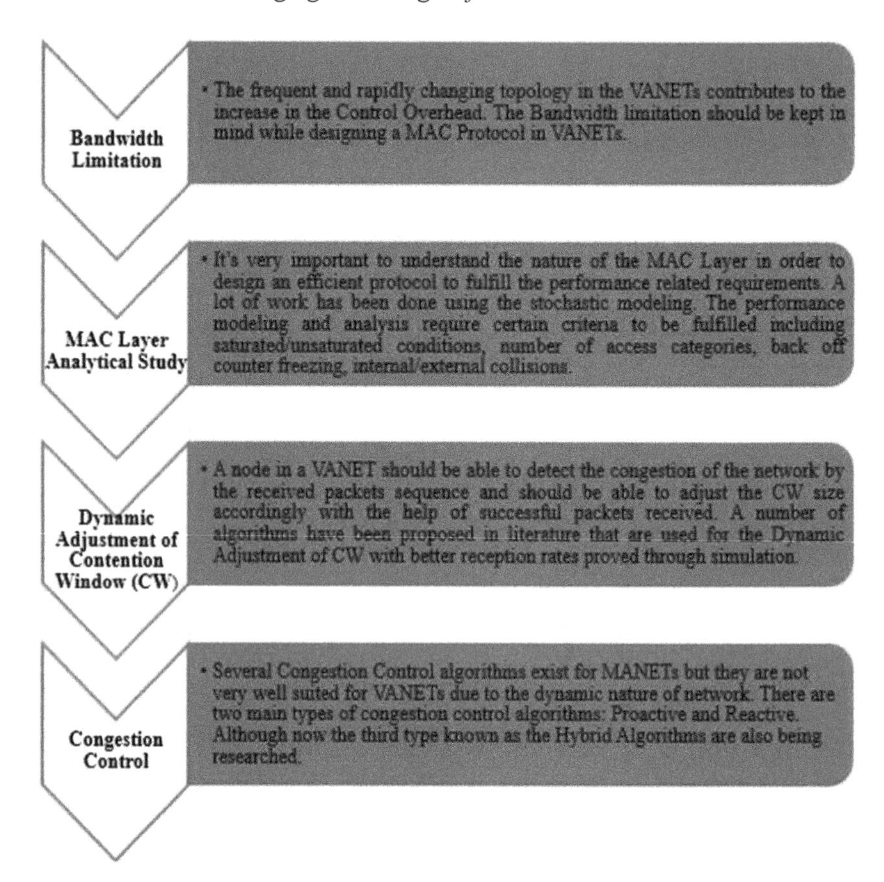

Figure 3.10 MAC protocol for VANETs

state routing technology for effective data transmission to the destination. A greedy backbone routing strategy to transfer data with the fewest intersection nodes to the destination. The static intersection nodes are set to transmit the data on the following route at the intersection. UVANET hybrid routing system that reduces overhead routing and improves scalability. It uses reactive routing and location to successfully transmit data to the target. A reliable method of routing automobiles by vector speed grouping. This reduces the breakdown of the link between cars. These are numerous UVANET protocols used to route the data to the target in a minimum time framc. UVANET is provided with a fly-overs path-based data transmission routing system. The main focus of this study will be communication between ground vehicles, communication between ground vehicles, and flyover communication. This protocol transfers the data in a very short time to the destination. A routing system based on route selection, where protective nodes control communication at the crossings, is described [42].

3.7 Novel routing framework for VANET

VANET provides information on the state of the channel for a range of circumstances, which may be very valuable for making routing decisions. Each of these approaches is for a specific set of conditions and is chosen based on the specified situation. Some quality channel information, such as signal fading or signal ratio, may be beneficial for selecting routes and some of these data are provided below. Path selection can be simplified and improved by employing the MAC layer that uses locally accessible information. This requires the creation of a VANET link-layer routing protocol. The performance of this approach is mostly controlled by the decision threshold value. However, a number cannot be chosen that will achieve great performance in all circumstances. In addition to the node density, the ideal value also impacts geographic dispersion patterns and wireless channel quality. In automobile networking, broadcasting strategies suited for vehicle networking should make changes in these parameters adaptable. In actuality, wireless signals in the system unexpectedly interact with one other, such that messages that seem not deterministic are received [43].

Wireless signals can interfere and cause messages to be lost at one or both nodes when two nodes simultaneously broadcast messages. Although only one node is broadcast, it fades, the phenomenon of interfering with various sections of the same signal on several channels impairs communication. If the reliability of communication is low, multi-hop wireless protocols must be able to work successfully even if the dependability of communication is inadequate. The threshold is used to determine which routes are excellent and which ones are bad based on channel circumstances. Routes that fall below a certain threshold are disallowed. Function thresholds also included channel-level characteristics such as the signal-to-noise ratio (SNR), the packet retry rate, and the channel availability percent for channel quality assessment [44]. It is proposed that a path calculation massaging framework be used that includes four types of paths calculation massaging, namely, the paths request, response, error path calculation massaging, and path answer recognition as shown in Figure 3.11.

3.7.1 Framework for the implementation of ITS in public transportation

A theoretical framework is provided to develop expectations of the process of integration, implementation, and adaptation for the new integrated system of public transport based on technology adoption in the Free State. The ITS application to improve safety and mobility while at the meme time reducing harmful environmental impacts using traffic signals and controls seeks to facilitate a multi-modal surface transport with a synchronized transportation system of all types, configurations, and likely driver reactions to improve safety and environmental performance at the same time. However, the arguments have demonstrated that the complete effect of ITS components can only be achieved through the integration of ITS components. To achieve this, ITS uses a variety of communication and technological features, including infrastructure, Wi-Fi (Wide Area Network), Bluetooth, GIS, data acquisition and communications systems, camera systems and artificial view systems, moving and classificatory detection systems, vehicle, and digital mapping systems and others. Providing eco-friendly mapping systems [45].

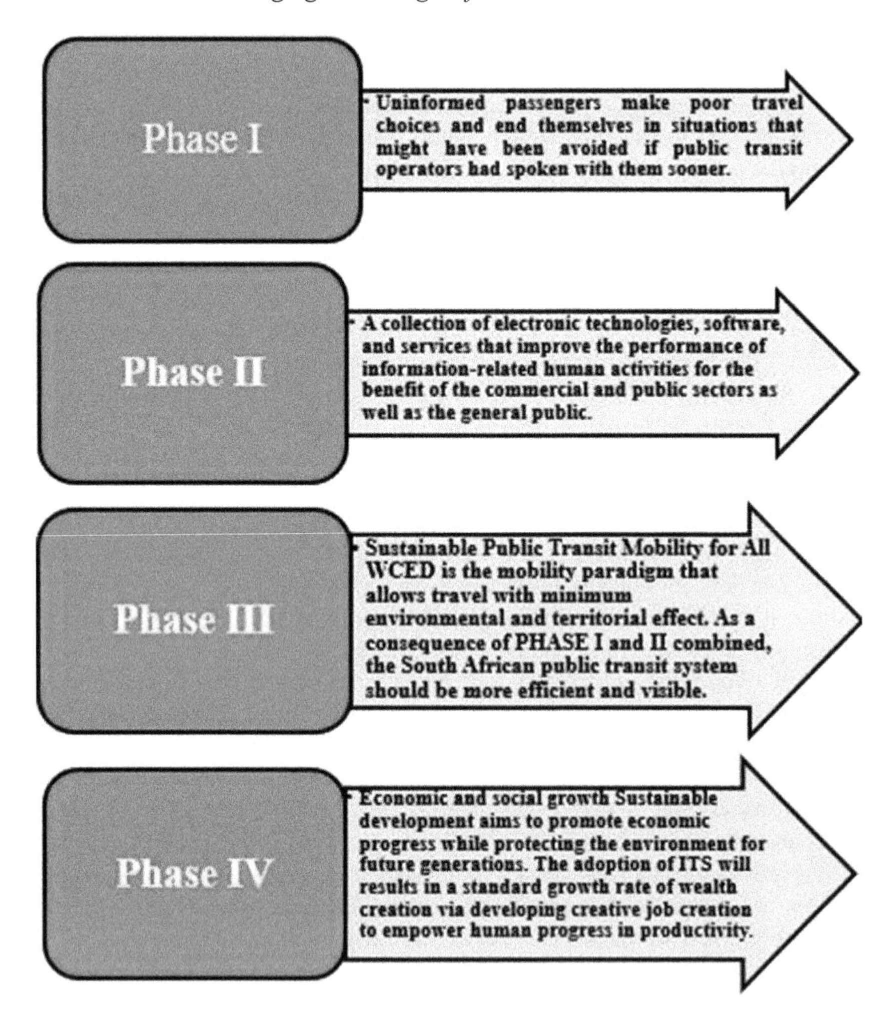

Figure 3.11 ITS implementation framework in 4 phases

3.8 VANET receiver-based forwarding system

There are several recipient-based VANET data transfer mechanisms. VIRTUS has been developed as a reactive receiver-based solution to allow video streaming across the VANET environment. VIRTUS allows high-speed connections without road infrastructure between cars. According to your current position and future locations, the receiving nodes communicate. The shuttle area VIRTUS goes to the destination. The reception nodes calculate the time to wait according to geographical development and stability. In automobile network, video streaming VIRTUS gives acceptable performance and reduces the number of transfers [46].

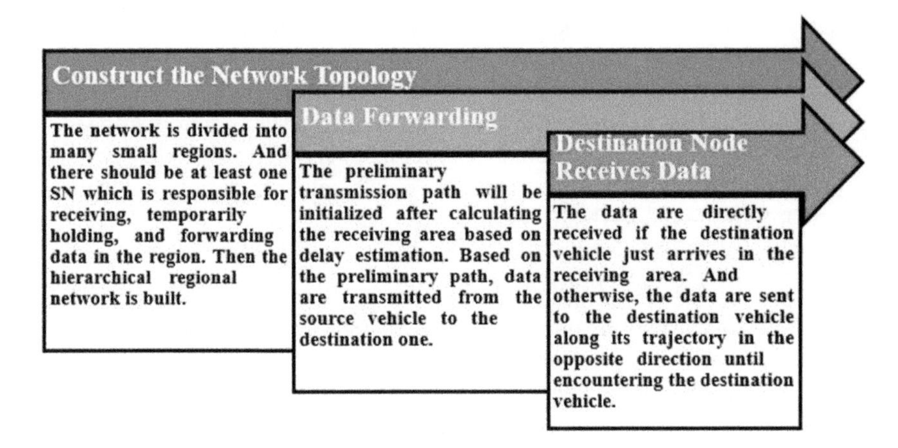

Figure 3.12 THAF—trajectory-based hierarchical adaptive forwarding steps

3.8.1 Efficient forwarding strategy based on prediction

Like prior research, our technology is used to forecast the connection life for position and speed data of surrounding cars. The problem is that with increasing network density the number of hops for packet transmission from source to destination increases. To tackle this problem, the efficiency of packet transmission is reduced [47].

The vehicle selects a neighbor with low relative speed and long distances from the sender as the next hop transfer node, taking into account the forecasted link life and connection utility. This data transmission strategy is designed to achieve stable connections and lower latency from end to end. In addition, we explore a variant of the connection tool that is based on neighboring two-hop data to boost the dependability of the communication. This ensures that a trustworthy connection to continue packet forwarding is found in the next-hop node specified by the sender. Generally speaking, because of the relative speed higher, the link between vehicles in the opposite direction is shorter than between vehicles in the same direction [48]. Cars should not be used on the reverse lines as the next forwarders without knowing where the target node is. The following three stages include the protocol for trajectory-based hierarchical adaptive transmission as shown in Figure 3.12.

3.9 Opportunities toward efficient routing protocols and standardization in VANET

The VANET technology is fascinating and an ITS solution is feasible. Many researchers have tried to create VANET routing protocols reliable, scalable and efficient and improve their QoS. Communicating between vehicle knots that allow drivers to take proper decision-making requires a great deal of reliability; therefore, developing a routing protocol to ensure certain QoS is one of the main challenges for car networks, given that VANET has special features such as reduced mobility, high knot speed, and high-dynamic topology. QoS is considered one of VANET's hardest tasks. The change in network topology precludes assured QoS settings and incorrect routing status information. To accomplish QoS, researchers evaluated several levels of the VANET protocol stack, with a major effort to create a strong vehicle network routing protocol and increase QoS inside the VANET [49].

3.9.1 QoS in VANET networks

QoS refers to its capacity to provide assured service levels for applications at a network level. QoS is evaluated based on supported VANET network applications. In VANET applications, certain restrictions, such as road safety, comfort, and traffic monitoring can be discovered. The major constraint for road safety and traffic surveillance is the real-time accuracy of information. Comfort applications nevertheless demand constant communication. QoS for VANETs co-operates with and coordinates various network components such as the QoS protocol, resource reserve system, and MAC layer [50].

3.9.2 QoS routing in VANET

QoS refers to its ability to deliver guaranteed service levels for network-level applications. QoS is assessed based on VANET network applications supported. Certain constraints can be identified in VANET applications, such as road safety, comfort, and traffic monitoring. The main restriction on road safety and traffic monitoring is the accuracy of the information in real-time. Nevertheless, comfort applications require regular contact [51].

3.9.3 QoS routing parameters for the improvement of protocols

The key QoS parameters are utilized to optimize and improve routing system performance [52]. Used to determine the optimal next node or path between the source and destination node, the QoS routing parameters:

Delay: Latency is the key factor in protocols for routing. The time of access, transmission time, propagation time, and processing time are all concerned. Access delay known as queuing delay is the average time a node tries to transmit a packet. The time to send an internet packet was called a transmission delay.

Distance: The geographical distance from the next node to the present node is shown. Distance is the most common fundamental parameter.

Link reliability: Communication lines in high-dynamic networks such as VANET are particularly prone to disconnect. Therefore, considerable attention must be paid to the routing dependability of these networks. The dependability of links refers to the probability that during a particular period a direct link between two nodes is still accessible.

Hop count: The number of hops or hop counts reflects the number of connections between resources on the route between sources and destination nodes.

Security: QoS-protected routing algorithms are a key component of QoS services and secure wireless networks. Security attacks create significant difficulties with safe routing, as routing control messages may be reduced by attacking the routing process and by QoS for the entire network.

Neighbor nodes: There are several one-hop nodes in a node. These one-hop nodes are known as neighboring nodes. By giving Hello message to you, you update your information on the current position, time, speed, and direction.

Mobility: The high mobility of moving nodes such as VANETs typically results in changes in network architecture and fragmentation. For these reasons, it is difficult to route packets over the network. By considering mobility, routing performance may be considerably improved.

3.9.4 Metrics of QoS evaluation

In addition to the core measures for the evaluation of routing protocols [53] like end-to-end delay (E2ED), packet delivery ratio (PDR), packet loss (PL), network load (NL), normalized routing load (NRL), normalized routing load (NRL), normalized overhead load (NOL), average routing replay ratio (ARRR), link failure (LF), etc. as shown in Figure 3.13.

E2ED: This refers to the time it takes for the packet to send a source to the destination. The delay with network congestion rises and has a significant impact on QoS.

PDR: The ratio is between the number of data packets delivered successfully to the destination and the number of packets communicated through the source.

PL: The failure of one or more packets being transmitted to the destination is caused by data transmission faults or network congestion. Packet loss is calculated as a percentage of the packets lost relative to the packets transmitted.

Bandwidth: Bandwidth measures the rate at which packets may be sent through a network or network path.

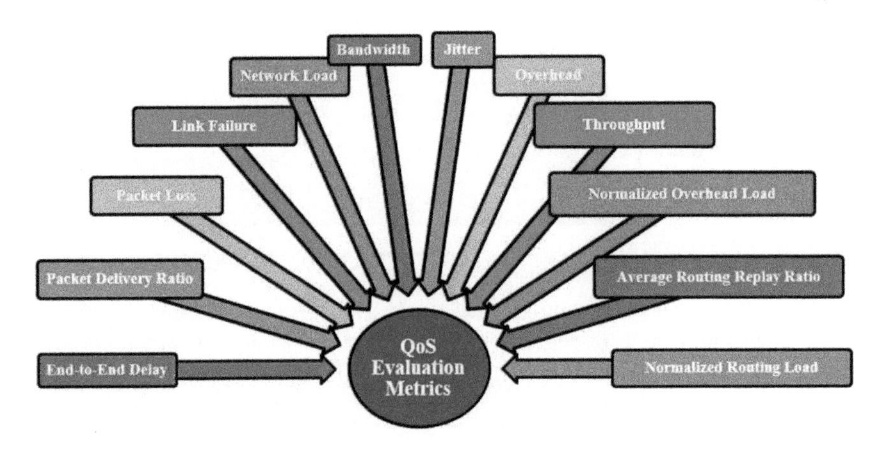

Figure 3.13 QoS evaluation metrics

Throughput: This refers to the average number of bits transmitted on a communication link by time slot. Node mobility hops number, and transmission range affects data.

Jitter: The gap between the longest time from end to end to the least time from end to end (the variance of the delay). This is due to the difference in successive delays in packet queuing.

Overhead: Routing and data packets are usually required to share the same network capacity, and network routing packets are therefore considered overhead.

NL: The ratio of automobiles receiving a duplicate message and the general acceptance of a packet is indicated.

NRL: It shows the proportion of packets that the data packet delivers to the destination and each hop is measured individually.

NOL: This is the ratio of the total number of routing packets to the number of data packets delivered correctly. The increased bandwidth required for routing packets is displayed by NOL.

ARRR: It refers to the average rate of route response of all network nodes when the route requests generated by the all-source node on all route requests are intended.

LF: During the routing procedure, the average number of connection failures is reported. This graphic shows how the routing protocol works to prevent connection failures.

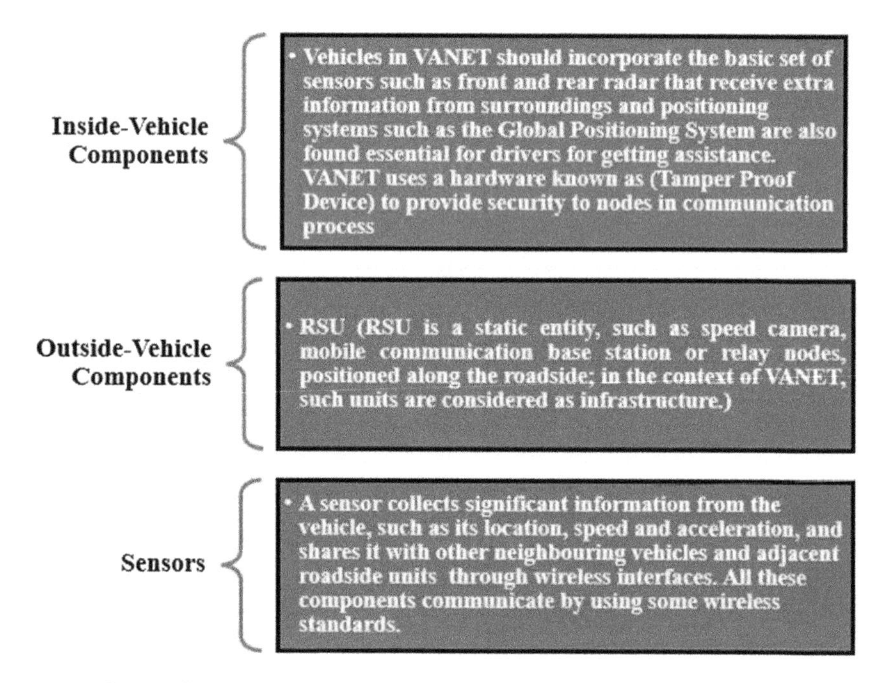

Figure 3.14 VANET system, main communicational components

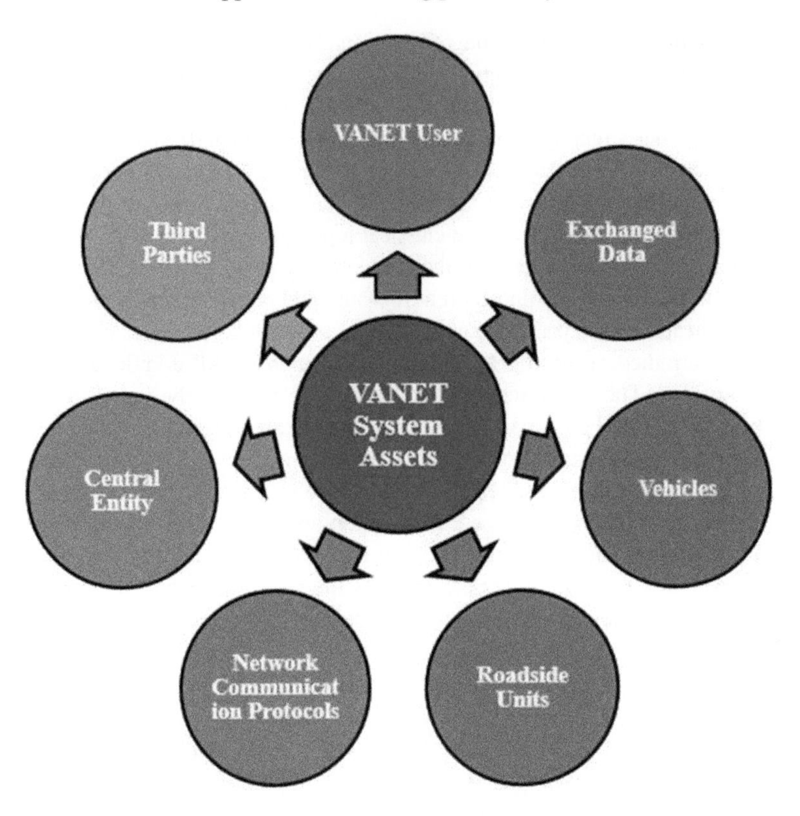

Figure 3.15 VANET system assets

3.9.5 Standardization in VANET

In the VANET system, the main communicational components are [54]:

1. Inside-vehicle components
2. Outside-vehicle components
3. Sensors

And described in Figure 3.14.

VANET system assets

An asset is a crucial network component. The failure of this asset causes harm to the overall network, which has a major impact on its users [55–58]. Thus, the process of identification and security of assets and the access control of network users is a critical stage in VANET security, since the VANET system assets are shown in Figure 3.15.

3.10 Conclusion

The VANET network is dynamic, so it is relatively hard to disturb the connection, i.e., the path between the source node and target node. Connecting firms are not reliable, thus the road needs to be reliable. In the automobile world, ITS handles challenges of traffic and road safety. VANET is one of the competent ITS systems for transport, road safety, and logistics. It may also be used in logistics and transport to cope with flow issues relating to material mobility information. VANET needs specific attention for road and safety optimization, with rising vehicle numbers, speed, and high bandwidth requirements for new applications. The literature demonstrates that the VANET routing protocols emphasize classical topological routing. The SINR and the transmission zone are the criteria for the contention qualifying depending on the recipient. Only those with SINR nodes on a threshold and an angle of transmission under the angle indicated in the package can compete for transmission rights if their waiting period is determined according to geographical progress. All additional nodes that refuse the packet. The suggested approach gives one route to the target, minimizing network congestion when no transmission timeouts are required. The method also picks a forwarding node nearest to the destination, leading to fewer expectations and better routing efficiency. However, in the case of transmission timeouts, the possibilities of several routes are enhanced. In addition, excessive transmission might substantially increase the latency from one end to the other. Therefore, the current and planned simulations for future work will assess the performance of the proposed system in various scenarios. The protocol design and limitation issues were examined in a multi-faceted way. Protocols presently in use and new protocols were also constructed and comparisons were produced that give prospects for the research in the field of VANET. VANET will allow VANET architecture via the Internet of Vehicles (IoV) and Vehicle Clouds in its future technologies (VC). In addition, vehicle communications are being shifted towards cellular communication standards, which will give more opportunities for study in the future, with the development of LTE-D2D technologies.

References

[1] Bozorgzadeh E, Barati H, and Barati A. "3DEOR: an opportunity routing protocol using evidence theory appropriate for 3D urban environments in VANETs." *IET Communications* 14(22) (2021): 4022–4028.

[2] Bangotra DK, Singh Y, Selwal A, Kumar N, and Singh PK. "A trust-based secure intelligent opportunistic routing protocol for wireless sensor networks." *Wireless Personal Communications* (2021): 1–22.

[3] Wu W, Wang X, Hawbani A, Liu P, Zhao L. and Al-Dubai AY. "FLORA: fuzzy based load-balanced opportunistic routing for asynchronous duty-cycled WSNs." *IEEE Transactions on Mobile Computing* (2021). DOI: 10.1109/TMC.2021.3074739.

[4] Das D, Bapat J, and Das D. "Connectivity and collision constrained opportunistic routing for emergency communication using UAV." *arXiv preprint arXiv:2103*.16117 (2021).

[5] Derakhshanfard N, and Soltani R. "Opportunistic routing in wireless networks using a bitmap-based weighted tree." *Computer Networks* 188 (2021): 107892.

[6] Rahnamaei YS, Rahnamaei B, Barekatain, and Raahemifar K. "TIHOO: an enhanced hybrid routing protocol in vehicular ad-hoc networks." *EURASIP Journal on Wireless Communications and Networking* 2019(1) (2019): 1–19.

[7] Li N, Yuan X, Martinez-Ortega J-F, and Diaz VH. "The network-based candidate forwarding set optimization for opportunistic routing." *IEEE Sensors Journal* 21(20) (2021): 23626–23644.

[8] Vadicherla P, and Vadlakonda D. "Study on energy-efficient routing protocols scheme in heterogeneous wireless sensor networks (network and mobility)." *Materials Today: Proceedings* 47(15) (2021): 4955–4958.

[9] Thyagarajan J, and Kulanthaivelu S. "A joint hybrid corona based opportunistic routing design with quasi mobile sink for IoT based wireless sensor network." *Journal of Ambient Intelligence and Humanized Computing* 12 (2021): 991–1009.

[10] Wang L, Chen Z, and Wu J. "An opportunistic routing for data forwarding based on vehicle mobility association in vehicular ad hoc networks." *Information* 8(4) (2017): 140.

[11] Azimi Kashani A, Ghanbari M, and Rahmani AM. "Improving the performance of opportunistic routing protocol using fuzzy logic for vehicular adhoc networks in highways." *Journal of AI and Data Minining* 8(2) (2020): 213–226.

[12] Alshehri A, Abdel-Hameed AB, and Huang H. "FQ-AGO: fuzzy logic Q-learning based asymmetric link aware and geographic opportunistic routing scheme for MANETs." *Electronics* 9(4) (2020): 576.

[13] Su B, Du C, and Huan J. "Trusted opportunistic routing based on node trust model." *IEEE Access* 8 (2020): 163077–163090.

[14] Grazia CA, Klapez M, and Casoni M. "IRONMAN: Infrastructured RSSI-based opportunistic routing in mobile adhoc networks." *2020 16th International Conference on Wireless and Mobile Computing, Networking and Communications (WiMob) (50308)*. IEEE, 2020.

[15] Alshehri AM. *Opportunistic Routing Schemes for Large-scale and Heterogeneous Multi-hop Wireless Networks Using Directed Energy Links and Fuzzy Logic Q-Learning*. Diss. New Mexico State University, 2020.

[16] Pang X, Liu M, Li Z, Gao B, and Guo X. "Geographic position based hopeless opportunistic routing for UAV networks." *Ad Hoc Networks* 120 (2021): 102560.

[17] Parameswari S. "Opportunistic routing protocol for resource optimization in vehicular delay-tolerant networks (VDTN)." *Turkish Journal of Computer and Mathematics Education (TURCOMAT)* 12(11) (2021): 3665–3671.

[18] Sharvari NP, Das D, Bapat J, and Das D. "Connectivity and collision constrained opportunistic routing for emergency communication using UAV." *CoRR* (2021). https://arxiv.org/pdf/2103.16117.pdf.

[19] Abedi O, and Kariznoi SR. "Load-balanced and energy-aware opportunistic routing with adaptive duty cycling for multi-channel WSNs." *The Journal of Supercomputing* 77(2) (2021): 1038–1058.

[20] Qin X, Guanglun H, Baoxian Z, and Cheng Li. "Sparse relays assisted opportunistic routing for data offloading in vehicular networks." *ICC 2021-IEEE International Conference on Communications*. IEEE, 2021, 1–6.

[21] Huang Z, Xu Y, and Pan J. "TSOR: Thompson sampling-based opportunistic routing." *IEEE Transactions on Wireless Communications* 20(11) (2021): 7272–7285.

[22] Wang Z, Chen Y, and Li C. "Opportunistic routing in mobile networks."*Opportunistic Networks: Fundamentals, Applications and Emerging Trends* (2021): 243.

[23] Chen Y, Li C, and Wang Z. "Opportunistic routing in mobile networks." In *Opportunistic Networks*. London: CRC Press; 2017. pp. 243–298.

[24] Vesco A, Scopigno R, Casetti C and Chiasserini C-F. "Investigating the effectiveness of decentralized congestion control in vehicular networks." (2021): 1314–1319. 10.1109/GLOCOMW.2013.6825176.

[25] Li N, Yan j, Zhang Z, Martínez-Ortega J-F and Yuan X. "Geographical and topology control-based opportunistic routing for ad hoc networks." *IEEE Sensors Journal* 21.6 (2021): 8691–8704.

[26] Wang J, Mei A, Tang X, and Shi B. "Social-based link reliability prediction model for CR-VANETs." In Liu Z, Wu F, and Das SK (eds), *Wireless Algorithms, Systems, and Applications. WASA* 2021. Lecture Notes in Computer Science, 12937 (2021). Springer, Cham. https://doi.org/10.1007/978-3-030-85928-2_30.

[27] Shugran MAA. "Applicability of overlay non-delay tolerant position-based protocols in highways and urban environments for vanet." *International Journal of Wireless & Mobile Networks (IJWMN)* 13(2) (2021).

[28] Raj RV, and Balasubramanian K. "Trust aware similarity-based source routing to ensure effective communication using a game-theoretic approach in VANETs." *Journal of Ambient Intelligence and Humanized Computing* 12.6 (2021): 6781–6791.

[29] Hamdi MM, Yussen YA, and Mustafa AS. "Integrity and authentications for service security in vehicular ad hoc networks (VANETs): a review." *2021 3rd International Congress on Human-Computer Interaction, Optimization and Robotic Applications (HORA)*. IEEE, 2021.

[30] Abbas A, Krichen M, Alroobaea R, Malebary S, Tariq U, and Piran MJ, "An opportunistic data dissemination for autonomous vehicles communication." *Soft Computing* 25 (2021): 11899–11912.

[31] Omar N, *et al.* "Route beaconing (RouteBea) process in GreedLea routing protocol for Internet of Vehicle (IoV) network environment." *Journal of Physics: Conference Series*. 1878; Second International Conference on Emerging Electrical Energy, Electronics and Computing Technologies 2020, 28–29. Available at: https://iopscience.iop.org/article/10.1088/1742-6596/1878/1/012057 October 2020, Melaka, Malaysia.

[32] Guillen-Perez A, Montoya A-M, Sanchez-Aarnoutse J-C, and Cano M-DA "A comparative performance evaluation of routing protocols for flying ad-hoc networks in real conditions." *Applied Sciences* 11(10) (2021): 4363.

[33] Alzamzami O, and Mahgoub I. "Geographic routing enhancement for urban vanets using link dynamic behavior: a cross-layer approach." *Vehicular Communications* 31 (2021): 100354.

[34] Hamdi MM, Al-Dosary OAR, Alrawi OAS, Mustafa AS, Abood MS and Noori MS, "An overview of challenges for data dissemination and routing protocols in VANETs." *2021 3rd International Congress on Human–Computer Interaction, Optimization and Robotic Applications (HORA)*. IEEE, 2021, 1–6.

[35] Tyagi AK and Sreenath N. "Vehicular ad hoc networks: new challenges in carpooling and parking services." in *Proceeding of International Conference on Computational Intelligence and Communication (CIC)*, Volume 14. International Journal of Computer Science and Information Security (IJCSIS), Pondicherry, India, pp. 13–24.

[36] Ram A. "An efficient routing protocol for vehicular ad-hoc networks." *International Journal of Sensors Wireless Communications and Control* 11(4) (2021): 472–481.

[37] Sachidhanandam P, and Balasubramanie P. "Elevated ensemble dynamic energy-aware routing optimization based energy management and network life-time improvement in WSN." *Wireless Personal Communications* (2021): 1–13.

[38] Patil AP, and Hurali LCM. "Analysis of routing protocols for software-defined vehicular ad hoc networks." *International Journal of Networking and Virtual Organisations* 24(2) (2021): 161–181.

[39] Rahmani AM, Tehrani ZH, and Souri A. "Evaluation of energy consumption in routing protocols for opportunistic networks." *Telecommunication Systems* (2021): 1–33.

[40] Mukherjee A, Dey N, Mondal A, Debashis De, and Crespo RG. "iSocialDrone: QoS aware MQTT middleware for social internet of drone things in a 6G-SDN slice." *Soft Computing* (2021): 1–17.

[41] Kabbaj S, Rahman AU, Malik AW, Baba AI, Sri Devi R. "Time-bound single-path opportunistic forwarding in disconnected industrial environments." *Vehicular Communications* 27 (2021): 100302.

[42] Vanitha N. "Binary spray and wait for routing protocol with controlled replication for DTN based multi-layer UAV ad-hoc network assisting VANET." *Turkish Journal of Computer and Mathematics Education (TURCOMAT)* 12(10) (2021): 276–2782.

[43] Hawbani A, Wang X, Al-Dubai A. *et al.* "A novel heuristic data routing for urban vehicular ad hoc networks." *IEEE Internet of Things Journal* 8(11) (2021): 8976–8989.

[44] Khezri E, and Zeinali E. "A review on highway routing protocols in vehicular ad hoc networks." *SN Computer Science* 2(2) (2021): 1–22.

[45] Chen W, Su J, Cui C, and Chen B. "Topology control routing strategy based on message forwarding in apron opportunistic networks." *Peer-to-Peer Networking and Applications* 14 (2021): 3605–3618.

[46] Debnath A, Basumatary H, Dhar M, Debbarma MK, and Bhattacharyya BK. "Fuzzy logic-based VANET routing method to increase the QoS by considering the dynamic nature of vehicles." *Computing* 103 (2021): 1391–1415.

[47] Khamer L, Labraoui N, Gueroui AM, and Abba Ari AA "Enhancing video dissemination over urban VANETs using line of sight and QoE awareness mechanisms." *Annals of Telecommunications* 76 (2021): 759–775.
[48] Tyagi AK and Sreenath N. "Providing trust enabled services in vehicular cloud computing," 2016 *International Conference on Research Advances in Integrated Navigation Systems (RAINS)* (2016): 1–7.
[49] Mahdi HF, Abood MS, and Hamdi MM. "Performance evaluation for vehicular ad-hoc networks based routing protocols." *Bulletin of Electrical Engineering and Informatics* 10(2) (2021): 1080–1091.
[50] Shiddharthy R. "A Selective reliable communication to reduce broadcasting for cluster-based VANET." *Turkish Journal of Computer and Mathematics Education (TURCOMAT)* 12(3) (2021): 4450–4457.
[51] Belamri F, Boulfekhar S, and Aissani D. "A survey on QoS routing protocols in vehicular ad hoc network (VANET)." *Telecommunication Systems* (2021): 1–37.
[52] Costa LPP, Marcondes CAC, and Senger H. "Non-cooperative vehicular density prediction in VANETs." *International Conference on Computational Science and Its Applications*. Lecture Notes in Computer Science, vol. 12952 Cham: Springer; 2021.
[53] Srivastava A, Prakash A, and Tripathi R. "Improved store-carry-forward scheme for information dissemination in unfavorable vehicular distribution." *Advances in VLSI, Communication, and Signal Processing*. Singapore: Springer; 2021. pp. 657–666.
[54] Zhou Chunyue, Hui Tian, Yaocong Dong, Baitong Zhong, "An energy-saving routing algorithm for opportunity networks based on asynchronous sleeping mode." *Computers & Electrical Engineering* 92 (2021): 107088.
[55] Talukdar MI, and Hossen MS. "Reactive and proactive routing strategies in mobile ad hoc network." *Opportunistic Networks*. London: CRC Press; 2021. pp. 37–54.
[56] Lu Y, Wang X, Yi B, and Huang M. "The reliable routing for software-defined vehicular networks beyond 5G." *Peer-to-Peer Networking and Applications* (2021): 1–15.
[57] Khalid K, Woungang I, Sanjay K, Jagdeep Singh D. "Reinforcement learning-based fuzzy geocast routing protocol for opportunistic networks." *Internet of Things* 14 (2021): 100384.
[58] Ghori MR, Sadiq AS, and Ghani A. "VANET routing protocols: review, implementation, and analysis." *Journal of Physics: Conference Series*. 1049 (2018). IOP Publishing.

Chapter 4

Forwarding and remote networking in Vehicular Ad Hoc Networks

Abstract

Short-distance communication technology is used by Vehicular Ad Hoc Networks (VANETs) to connect vehicles and infrastructure. As a result of this, new apps can now be made available. Attackers can, however, take advantage of the dynamic network design to compromise the security and privacy of cars and message forwarding. Based on trust evaluation, this study proposes a socially aware security message forwarding mechanism. As a means of protecting vehicle privacy, this pseudonym evaluates the exchange conditions to ensure that the attacker cannot track the vehicle at all times while it's on the road. The road side units deliver the message to numerous relay cars and an application provider via the trust gradient to several different receivers. In light of the data, the suggested method appears to have the potential to improve message forwarding success rates, reduce message leakage risks, and better safeguard the privacy of users on VANETs. As wireless communication and the automotive sector continue to advance, interest in VANETs has risen dramatically in recent years. This advancement in wireless communication is expected to have a positive influence on vehicle safety and efficiency when it is incorporated in intelligent transportation systems (ITS). VANETs standards are being developed through collaboration between many organizations, including the federal government, the automotive industry, and academia. VANETs uses include accident warnings and sharing of traffic information. VANETs, as a result, have turned into an interesting area of study for many persons interested in mobile wireless communication. You will learn about VANETs' current state-of-the-art and the problems and opportunities that lie ahead. You will also learn about recommendations for moving forward towards the long-awaited ITS.

Keywords: NDN (named data networking); Dedicated short-range communications (DSRC); Vehicle to infrastructure (V2I); Intelligent transport system (ITS); Vehicle to vehicle (V2V); Vehicular Ad Hoc Networks (VANETs)

4.1 Introduction

Vehicle Ad Hoc Networks (VANETs) have gotten a lot of interest lately due to recent developments in in-vehicle connectivity and dedicated short-range communications (DSRC) technology. In cars equipped with on-board units, there are two ways to communicate: vehicle to vehicle (V2V) and vehicle to infrastructure (V2I). VANETs are the acronym for Vehicle Ad Hoc Networks (VANETs) (V2I). Intelligent transportation systems (ITSs) contain services such as emergency vehicle warnings, curve speed warnings [1], pedestrian crossing warnings, and more (ITS). This technology is critical to ITS because it allows drivers to have a more enjoyable driving experience while also increasing traffic safety. Nevertheless, because VANETs use open wireless media, a large number of self-driving cars make up the vehicle sub-networks, putting them at risk. Many assaults, such as the tracking of the attacker, the theft or modification of messages, the fraudulent delivery of messages, and so on, put the privacy of vehicles in danger [2]. As a result, the VANET's security is compromised. VANET adoption is influenced by communication security as well. It is possible to cause a serious car accident by sending the wrong message on the road. The traffic control center made a mistake due to incorrect traffic flow messages, and attackers may pretend to be ambulances and approach traffic signals for help [3]. Message and vehicle privacy information are accessible to channel eavesdroppers and relay automobiles. VANET's efficiency is influenced by the routing protocol used. Traditional VANET communication security employs authentication mechanisms for symmetric and asymmetric key management.

However, because the cars must be verified by Trusted Authority (TA), this method cannot be used for large-scale vehicle communication under strict symmetric key management systems. It is for this reason that asymmetrical key management authentication systems that use individual encryption-decryption keys are becoming increasingly popular in terms of authentication methods, PKI and ID both have their advantages and disadvantages [4]. Authentication frameworks based on ID signature and ID cryptography are presented as a remedy to the high communication and processing costs associated with PKI authentication systems. Digital signature technologies simplify the certificate administration procedure as well. Traditional VANET key approaches like those listed above, however, have numerous drawbacks. To prevent key reliability from being compromised by an attacker or being lost during transmission, the vehicle's subnets must be changed often. Once the attacker has hacked or interfered with the signals, the target vehicle will be unable to decode them even with the legitimate private key. To overcome these security issues, it is challenging to rely just on message encryption, signing, and authentication Asymmetric key management is also less flexible than the secret-sharing approach [5]. As a result, a message encrypted with the public key can only be decoded by the vehicle specified as the destination, and no other vehicles are involved. The encryption-carry-decryption cycle must be repeated if the message is sent to a large number of vehicles, reducing transmission efficiency and reducing message usage. The use of socially aware security message routing methods can increase VANET safety. If used, vehicles can communicate using a pseudonym instead of their real identity, and the pseudonym is updated each time the two pseudonyms are exchanged. To fragment the resulting message, the source vehicles use top-secret sharing methods like secret key sharing. A trust evaluation on encounter vehicles should be performed to pick selected fragment message forwarding vehicles to ensure that as many fragment messages as possible reach the application provider [6]. Figure 4.1 shows the chapter's major contributions.

While maintaining the confidentiality of communication, message forwarding may be made more adaptable and effective. Computer-intensive duties can be sent to mobile devices such as smartphones and smart watches instead of the vehicle's onboard units (OBUs). Utilizing computation, storage, and power capabilities of serving edge nodes from mobile devices in VANETs minimizes latency, conserves energy, and safeguards user privacy. For the most part, mobile device-to-OBU communication has to be protected from spoofing attacks, especially rogue edge assaults, due to the high mobility of OBUs and the large-scale topology of the VANET network structure on edge computing networks [7]. The rogue edge attackers can exploit the Universal Software Radio Peripherals to intercept radio signals from adjacent radio sources, such as cars. As a result, they can send out spoofing signals to generate erroneous reports of VANET intrusions. Cryptography, trust, certificates, and physical authentication are the current basis for VANET authentication. By utilizing PHY features like received signal strength indicators, packet receipt signals, and channel responses, a simple security protocol can be employed to verify wireless communications. The mobile device's authentication accuracy is improved thanks in part to ambient WiFi signals. Mobile devices and OBUs in a vehicle can detect ambient radio

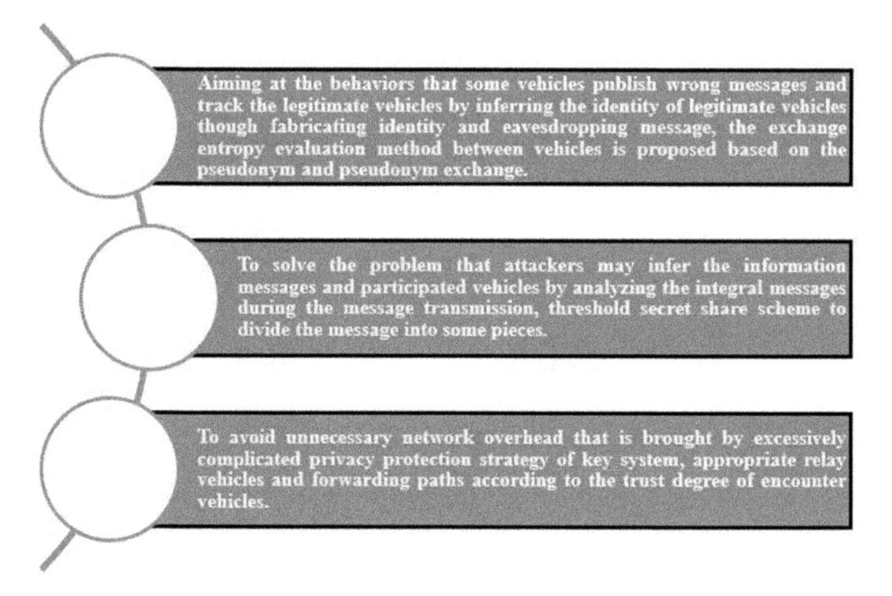

Figure 4.1 Main contributions of this chapter

signals and extract PHY features like packet arrival intervals and RSSIs from those signals thanks to PHY authentication. VANET mobile devices generate various numbers of priority packets based on the type of application being used [8].

4.1.1 Authentication scheme in VANET

Since of this, authentication is critical in VANETs because it helps to confirm that the data is coming from a trustworthy source. VANETs can authenticate their users using symmetrical or asymmetrical keys, depending on the implementation. A public-private cryptographic key pair, implemented using public key infrastructure, protects data being transported across a network. Authentication is achieved using PKI by signing the communication with the sender's private key and confirming it with the receiver's public key [9]. To authenticate and secure vehicle-to-vehicle communication, many VANET techniques use PKI. Establishes trust in V2V communication through member identification with a non-interactive ID mechanism. A blind signature-based approach is used to keep automobiles and RSUs anonymous when communicating. Securing communication between cars with a symmetric key could lessen the asymmetric key's large processing burden. Technique for preserving conditional privacy that is quite effective. In ECPP, a short-lived anonymous key is used to communicate between the RSUs and onboard devices. OBU will authenticate the RSU first before requesting with its true ID and pseudo-ID if it needs a short-term public key certificate from the RSU. Temporary public key certificates are issued by the RSU. With this strategy, each OBU will have fewer anonymous keys to store. Numerous ways in the authentication process make use of symmetric keys. Secure, loss-tolerant stream-based authentication. Figure 4.2 depicts how VANET Authentication Schemes are organized [10].

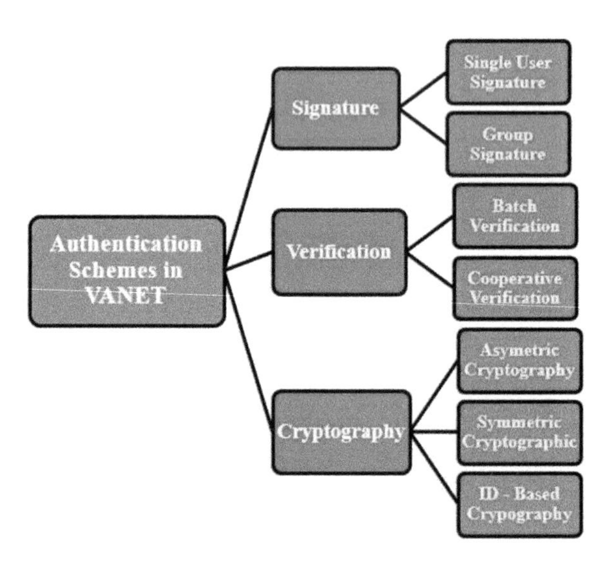

Figure 4.2 Classification of VANET authentication schemes [11]

4.1.2 Rogue nodes detection

Hostile nodes masquerading as trustworthy nodes are known as rogue nodes. The malicious node could be a router, a fog node, or an Internet of Things (IoT) device. There are several ways to identify "rogue nodes," which are unreliable [12]. One method is to use a threshold anonymous announcement (TAA). Before receiving information obtained through the TAA system, a car verifies that it was provided by unique vehicles. To avoid concerns with double-messaging, the TAA checks for many signatures from different vehicles for each message. There must be enough messages received beyond a defined threshold before a message is accepted. TAA's base is one-time anonymous authentication, while the backbone is direct anonymous attestation. Safety in VANETs is maintained through some rules and policies/guideline. For detecting fake messages, the REST-Net monitoring and plausibility tests can be utilized. In REST-Net, rules are used to specify inappropriate behavior and notify users of it. REST-Net monitors the vehicle's behavior both before and after the message is delivered to ensure its legitimacy. If the vehicle's behavior does not match the behavior stated in the notification, the message will be deemed incorrect. An in-vehicle controller's controller area network (CAN) is injected with an undesired packet using deep neural network topology. Training and detection are the two phases of the process. This will be done by analyzing the CAN packet for statistical characteristics that can be used to classify packets as normal or attack [13]. This is referred to as training. The presence or absence of CAN features determines whether a packet is considered normal. In the event of a packet loss, the node is responsible for the loss or duplication. Each solution relies on the verifier vehicle, which is the one with the lowest amount of distrust in the entire system. The verifier maintains track of the number of packets received and sent back and forth by the vehicles. Using the verifier, you may determine whether or not a vehicle sent the same packet more than once, or whether or not an invalid packet was passed. When a car initially joins the network, it is given a mistrust value. If that figure exceeds a predetermined threshold, the vehicle is deemed rogue and is yanked from service. The mistrust value rises if the verifier observes something out of the ordinary. Vehicle communication type and following behavior are taken into account while using this detection method to identify rogue vehicles.

4.1.3 Security in fog computing

It is important to understand the security and privacy issues associated with fog computing before employing one. There are several issues specific to fog computing, some of which have been passed down from cloud computing. As a security measure, you must address issues like rogue node identification and authentication to maintain your network free of malicious nodes. The study's goal is to build trust among fog nodes. Fog computing has piqued the interest of researchers since it offers more advantages than cloud computing presently does. Fog computing is utilized in an IoT setting, whereas fog computing is used in a VANET context [14]. Fog nodes and users at the network's edge employ blockchain technology and a secret sharing mechanism to authenticate each other. To verify the authenticity of a fog node, the fog node must maintain a blockchain (fog nodes would be able to establish mutual authentication with each other). The fog node verifies the end user's identity using a secret sharing approach. To authenticate users and fog nodes without using the cloud, a cloud broker and blockchain information would be needed. Brokers provide fog nodes with a verification parameter, which they utilize to verify the user's identity. When the broker receives the public key, he verifies the fog node's certification before giving it out to the user. Users must first sign up for an account in the cloud before they can use it for ultimate authentication. New user credentials will be provided for the fog-level authentication procedure by the cloud. A foggy node's publicly assigned key is compared to the broker's validation public key by an end-user to verify its authenticity. The fog node would authenticate the user by decrypting and confirming the credentials with polynomial interpolation using the fog public key. Fog computing-based cyber-attack prevention for Internet of Things devices [15].

4.1.4 Network model

VANETs make it possible for automobiles to communicate with one another through the usage of messages. To convey information to other surrounding vehicles, such as speed, position, acceleration, and brake condition, automobiles are equipped with a variety of sensors [16]. These sensors include GPS, Radar, and OBU. V2V and V2I communication are used in both multihop and broadcasting tactics to convey messages. To maintain the present state and behavior of the network, each vehicle modifies its speed, acceleration, and other characteristics in response to information received from its neighbors. Honest nodes provide authentic messages to their neighbors whereas rogue nodes inject bogus data before broadcasting it to the network. A dynamic fog layer identifies rogue nodes by comparing and analyzing the vehicle speed in the beacon messages, which detects nodes broadcasting false signals. A dynamic fog layer is created by guard nodes using OBUs from all automobiles because individual OBUs are resource-constrained. As opposed to traditional computer systems that are limited in processing power, fog computing provides a large amount of processing capacity to the guard node, making it faster and more scalable at any vehicle density. It is not necessary to have a large number of guard nodes to spot a rogue node now. Figure 4.3 explains how to pick a guard node and how it works [17].

4.1.5 Traffic flow model

To simulate traffic flow on highways and in cities, Greenshield uses its traffic flow model concept. According to real-world examples, the Greenshield model is an effective model to use when trying to predict traffic flow patterns. The density falls in direct proportion to vehicle speed due to their inverse relationship. The rogue nodes' network nodes create a false traffic collision situation while the network is under attack by sending low-speed beacon signals to any vehicle in the vicinity [19]. False signals emanate from the red area, which is home to many malicious nodes. Vehicles begin to slow down as if an accident is imminent after receiving

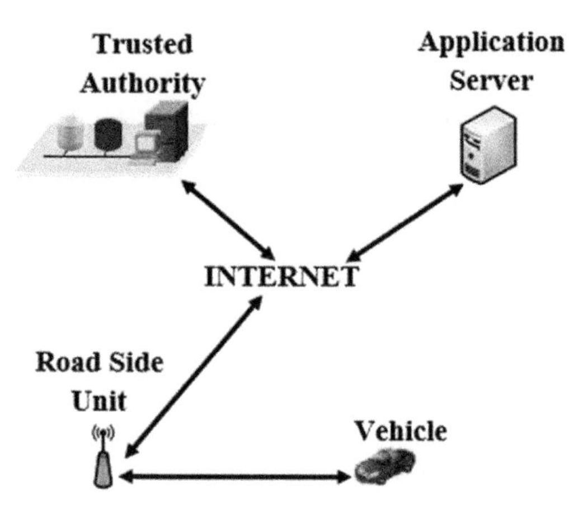

Figure 4.3 VANET network model [18]

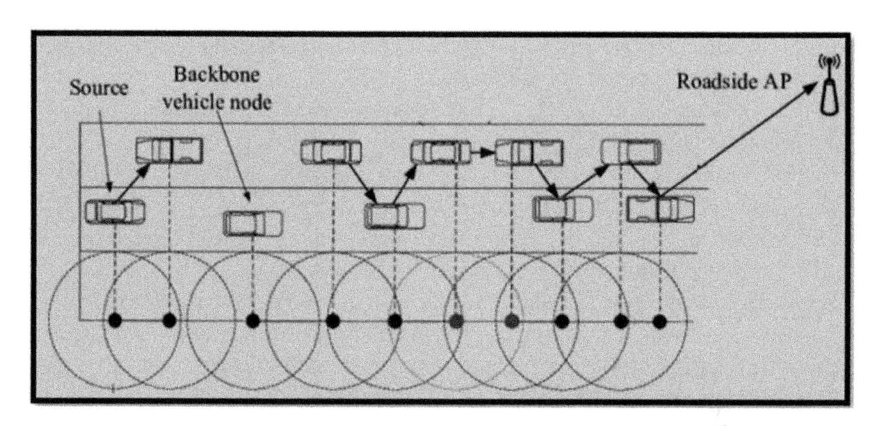

Figure 4.4 VANET traffic model [20]

beacon messages from the rogue nodes. Due to the rogue nodes' communication range, vehicles have not started braking yet (see Figure 4.4).

Afterward, this chapter is structured as follows: Section 4.2 introduces the related/background work. Section 4.3 discusses VANET rogue edge detection using learning-based techniques. VANET's transport layer is covered in detail in Section 4.4. VANET forwarding techniques based on Geo Networking are described in Section 4.5. Sections 4.6 and 4.7 include performance evaluations of vehicular named data network (VNDN) forwarding strategies and VANET wireless access standards. Section 4.8 discusses networking standards in VANET issues and future prospects. A worm propagation model for the web Vehicular Ad Hoc Network (WVANET) is described in Section 4.9, and the chapter's conclusion is discussed in Section 4.10.

4.2 Related/background work

With regards to VANET security, no stone goes unturned. It is critical to identify rogue nodes that send out fake beacon messages to keep the network safe. VANETs use broadcasting or multihop procedures to send messages to other vehicles on the network. Users can save their lives by using information gleaned from other automobiles. Nodes in dynamic vehicular networks that broadcast misleading information about traffic congestion or accident reports to nearby autos could potentially cause chaos by creating traffic gridlock [21].

Tanmay Kasbe *et al.* (2021): Automobile non-deterministic mobility and high energy and velocity make routing a difficult problem to solve. They are self-contained systems connected to the Internet as a part of the infrastructure. These problems make it difficult to build VANET routing protocols that are both efficient and scalable in the real world. We wanted to provide an in-depth look at broadcasting and routing, security, QoS, and entertainment while exploring the wireless protocols utilized in the VANET. To evaluate the VANET's accuracy and performance, routing protocols have been deployed. This study compares suggested swarm intelligence routing protocols with several other possibilities using well-known ad hoc routing protocol baselines [22].

B. Ramakrishnan *et al.* (2021): Vehicle-to-vehicle (V2V) communication via a remote system is a typical VANET function. By remotely giving explicit services to its nearby vehicles, it encourages basic improvements in health regardless of how it is communicated. According to the findings of this study, CEOs in VANETs have confidence in one another. The framework ensures vehicle confidence and reliability, and as a result, system overhead is decreased. You can use sidekicks to pick dependable vehicles and a decent route utilizing a flower pollination method that opposes your own (e.g., integrating the principles of connection quality and neighbors trust estimation). Recurrence, order, and setting-based aspects of message transmission are used to strengthen the steering process based on trust [23].

Basavaraj Mathapati *et al.* (2021): Since VANET relies on a computer, control, and communication technology, its position-based routing is particularly sensitive. Using an IoT-enabled VANET application raises questions about security, privacy, and integrity, among other things. Security solutions have evolved over the previous two decades, with cryptography-based and trust-based approaches dominating. Modern intelligent trust-based procedures outperform earlier cryptography-based methods due to their increasing computational complexity. Using trust-based techniques, it is possible to deal with VANET's well-documented security issues. Position-based routing that takes vehicle behavior into account is what we arc after for VANET communications. Attack trust evaluation and artificial intelligence are used in RPBVANET's resilient position-based VANET protocol to keep hostile vehicles out. To begin, ARPVP's periodic self-trust evaluation algorithm finds suspect nodes by looking at several trust factors. This is done throughout the project's initial development phase. The second phase of ARPVP's position-based route generating algorithm was created using an AI technique called ant colony optimization [24].

Bouziane Brik *et al.* (2021): Many academics have recently been inspired by the Internet of Cars, which has recently been an active study issue since it provides efficient and reliable wireless connections between automobiles and everything, to design new communication paradigms that fulfill new vehicle needs in smart cities. 5G and DSRC, such as DSRC, are the most promising mobile communication technologies for vehicles. This research provides a dependable data forwarding method for the Internet of Cars dubbed DFS-IoV to support quick data transfer, smooth and integrated communication between vehicles. Urban regions with a wide range of network characteristics will be used to test the suggested approach [25].

T.N. Shankar *et al.* (2021): The virtual autonomous network (VAN) is an extension paradigm for connecting moving vehicles to wireless transmission devices within a specific geographical area without the use of permanent infrastructure. Several of the model's most important vehicles are hidden in a dimly lit area of the radio spectrum, making it difficult to locate them. This study proposes a fuzzy-based multi-hop broadcast protocol for highly mobile VANETs as an alternative to standard message distribution methods. The transmission range is reduced as the number of intermediate forwarding points is reduced. To prioritize the delivery of emergency messages, the proposed protocol uses numerous nearby nodes that match three criteria: the distance between cars, the number of nodes, and signal strength [26].

Liang *et al.* (2021): A growing number of people are considering VANETs as a way to communicate beyond regional boundaries. The communication range of these systems can be expanded by connecting VANETs and satellite networks. Using VANETs and satellite networks together, this research proposes ways to increase the transmission performance of the networks. The control of VANET transmission from end-to-end using satellite networks multi-hop VANETs bandwidth-delay analysis in heterogeneous networks and a transmission control model for cross-regional contexts long delays and high bit errors rates in a heterogeneous network can be dealt with using congestion avoidance and slow start transmission methods to create a congestion window in the model. Congestion avoidance increases data transmission capacity while reducing packet losses. VANETs and satellite networks have been used as testbeds, and the findings demonstrate that the system can deliver good transmission performance. Network technologies such as VANETs and satellites. This study's findings suggest that VANET data can be transmitted across different networks without a problem [27].

Rituparna Chaki *et al.* (2021): Consequently, technological improvements have benefitted healthcare systems in both urban and rural areas. However, because of accessibility issues, this method frequently fails to connect India's rural residents with modern healthcare. In a growing country like India, where people are spread out across vast distances, universal health care is impractical. To address this issue, a new approach to network communication is required. This proposal's primary goal is to find a new way to communicate health-related information to the general public. As long as we do not require a nationwide communication network, healthcare services like these will remain more affordable [28].

N. Vanitha *et al.* (2021): During natural disaster recovery, the unmanned aerial vehicle network (UAVN) is a wireless communication network that transmits images or video taken by the UAV back to the aircraft on the ground. The UAV network has a wide range of applications, but the most common one is searching and rescue. After a disaster, this study focuses on an ad hoc UAV network built on the delay tolerant network (DTN). Notoriously dynamic UAV networks suffer from packet forwarding issues due to unbalanced connections. To provide a binary spray and to wait for a routing system with controlled replication, DTN-based decentralized multilayer UAV networks are deployed. When working in a multi-layer UAV ad hoc network, next layer UAVs can supply a well-organized store. End-to-end delays will be reduced, and the packet delivery ratios will be improved while packet loss ratios will be reduced, on all DTN UAVs and the ground vehicle. This will reduce packet loss ratios. To avoid message duplication and ensure single-copy data forwarding, the proposed BSnW has a high packet delivery ratio [29].

Zhou Shuang *et al.* (2021): VANET performance can be improved with an effective routing protocol. However, due to difficult road conditions, intermittent connectivity between vehicles, and frequent network structure changes, building an efficient data packet transmission route in an urban context remains a big challenge. There are numerous intersections to choose from depending on the chance of connecting to different sections of the road as a starting point, and the author describes how to determine whether a two-way lane will have a connected road segment when there are traffic lights in their V2V communication routing system for urban VANETs. To discover the shortest route between two points, the author develops an optimization model based on the likelihood that two road segments will be connected. Following the optimum road, a path has been established, this study examines how many other neighboring vehicles are within the communication range of the packet-transporting vehicle. It then accounts for their combined positions, speeds, and directions [30].

N.M. Kumar *et al.* (2021): VANET technology communication networks use ad hoc networks, which are flexible communication hubs. Vehicle-specific system settings discourage the use of the most recent telematics technology. Using dynamic well-being software that does not sufficiently handle security issues exposes the general public to dangers. A single, clear vision guides the security program's projections. Several specialized urban co-ops are taking a look at using VANET. The raging information storm hurts the presentation's overall message. AODV, DSR, DYMO, TORA, and TIHOO are among the protocols being used to address and analyze the problem. With an upgraded fluffy and cuckoo method, an improved VANET security protocol can handle the location of the steadiest path between a source and destination hub. Information about specific courses is kept anonymous using the AESP-fuzzy VANET framework [31].

G. Padmavathi *et al.* (2021): Using unmanned aerial vehicles (UAVs) is more important in today's diseased world. aiding with an unmanned aerial vehicle Using U2U communication, the VANET is designed for disaster relief and search-and-rescue missions. You do not have to be there when a drone (also called an unmanned aerial system or UAS) takes off and flies. Nearly every business makes use of unmanned aerial vehicles. Ad hoc networking between UAVs makes it possible to create flying ad hoc networks, which alleviates some of the issues associated with a UAV network that is completely reliant on infrastructure. Ad hoc networks for UAVs are examined and discussed in this study to determine which designs are most appropriate. As well as using mobility measurements such as network diameter and average clustering coefficient with selected mobility models, these designs' network performance is also assessed by using the OLSR routing protocol [32].

Mistareehi *et al.* (2021): VANETs are expected to have a substantial impact on future ITS design and implementation. While a car is traveling, onboard equipment collects data that can help drivers avoid traffic congestion and provide useful information. However, OBU is not available to all car owners just yet because it is not compatible with all vehicle makes and models. We developed a hardware implementation for rural areas to make use of OBU. To be effective, OBU installations don't have to be time-consuming or expensive. According to our findings from testing, the model we've provided is capable of quickly sending and receiving signals (such as safety messages) to nearby automobiles as well as the access point and the final destination [33].

SuYu *et al.* (2021): This study proposes a new approach to enhancing the vehicular network's data transmission performance and security by utilizing machine learning techniques for data processing. 5G cellular networks and other data transfer mechanisms are combined in the new technology. Researchers used 5G equipment to collect data across Berlin for three months. Large volumes of data were analyzed and classified using position-based routing protocols and SVM algorithms. These techniques were utilized in real-time to identify non-line-of-sight conditions so that data may be transferred safely without loss or degradation. The study's findings show that the proposed technique works well with massive amounts of data and can be utilized to improve the efficiency of urban VANET networks and increase data transmission security [34].

I. Jasmine *et al.* (2021): Because of VANET's novel topology, which alters the mobility of nodes, data transmission in the vehicle environment will be more efficient. Density, mobility, and distribution are all key factors to consider when it comes to emergency message broadcasting. Problems occur when an emergency message must be broadcast over the airwaves in real-time. Send and receive messages and collect data from network neighbors are made possible by VANET, which makes use of beacons to do so. Messages cannot be delivered if there is a broadcast storm caused by many cars sending messages at the same time. To maintain message dependability while also enhancing transmission efficiency, we use the adaptive scheduled partitioning and broadcasting approach described in this chapter. Black widow optimization is used to calculate the size of each division schedule based on the transmission density of the network. The use of adaptive scheduled partitioning and broadcasting reduces message redundancy and delay by forwarding new messages and selecting the appropriate partition. The performance research makes use of currently available methodologies for determining efficiency, redundancy, collisions, and latency [35].

Ilyong Chung *et al.* (2021): Even in the age of the Internet of Things, VANET research is a hot topic. VANETs have been the subject of numerous investigations utilizing a variety of methods and methodologies. Using a wireless heterogeneous vehicular communication system is difficult because of the network's sophisticated and dynamic connection architecture. Rapid data exchange among cars in a particular area depends on VANET's spontaneous V2V message transmission capabilities. Long-distance vehicle V2V message transmission within the range of various RSUs has not been completely explored due to short interaction intervals and significant vehicle mobility in connection. V2V and V2R (vehicle-to-RSU) communication is considerably impeded by environmental limits such as physical barriers and signal interference, resulting in significantly reduced wireless connection in actual VANET deployments. Communication between V2R devices. Using unmanned aerial vehicles (UAVs) as support facilities improves the VANET's transmission quality and availability. An authorized UAV group association is used in the study's basic design. It is possible to distribute V2V messages far and wide since every RSU along the path has a V2V connection [36].

4.3 Learning-based rogue edge detection in VANETs

Nodes must be able to be correctly identified and differentiated from one another to function properly in any network while yet preserving privacy. In other words, a certificate authority (CA) validates the validity of every node in the network. By confirming the validity of each vehicle with a certificate authority, it is assumed that each vehicle has acquired a valid certificate and public/private key pair (Pseudonyms – PNs). Using the keys, routine communications may be encrypted, and the required public keys can be used by others to authenticate and decode them if they have the necessary permissions. Every automobile is believed to have a sufficient number of key pairs to allow it to live for a long length of time, and these key pairs are changed regularly to ensure that the vehicle remains anonymous [37].

Figure 4.5 shows symbols and notation summary and the reason for this is that the keys are delivered at a regular interval and not too soon, preventing short-term linkability. Since nodes can change their PIN, the most recent messages sent by that node will remain linked to the network even if their PIN changes. Further, Bob wants to provide Alice with real-time data such as images, videos, and information from virtual reality in addition to verifying whether or not the edge testing is Alice rather than an Eve node operating outside the vehicle and attempting to deceive Bob. The edge node can reduce processing delays while also conserving mobile device computation energy by providing computing and caching resources. Bob, Alice, and Eve's radio environment vary over time due to the constant movement of OBUs. Vehicles carrying mobile devices and edge nodes are assumed to travel at a speed denoted by V(K) 1 during time slot K, without sacrificing generality. Let R (K)(I) denote this. Let (I) represent the RSSI reading for Alice's *i*th signal sent at time slot K. In time slot K, the rogue edge (Eve) is assumed to be traveling in a different vehicle with a speed of V(K) 2. Eve intends to deceive Bob with fictitious

$C_F^{(K)}$
• Spoofing Detection Mode Cost at Time Slot K
$I^{(K)}$
• Importance of Bob's Sensing Data at time slot K
$U^{(K)}$
• Utility of Bob at Time Slot K
$\alpha_{F/M}$
• Weight factors between the false alarm rate and the miss detection rate at time slot K
$p_{F/M}^{(K)}$
• False alarm rate/miss detection rate at time slot K
δ
• Learning Discount factor of Bob
• Learning rate of Bob
• Spoofong rate at time slot K
$V_{1/2}^{(K)}$
• Speed of the target/attack OBU at time slot K
$F^{(K)}$
• Ambient Signal feature of the Edge Node under test at time slot K
$R^{(K)}(I)$
• RSSI of the Ith packet sent by the edge node under test at time slot K
$\hat{R}(I)$
• RSSI of the Ith packet sent by the edge node under test at time slot K

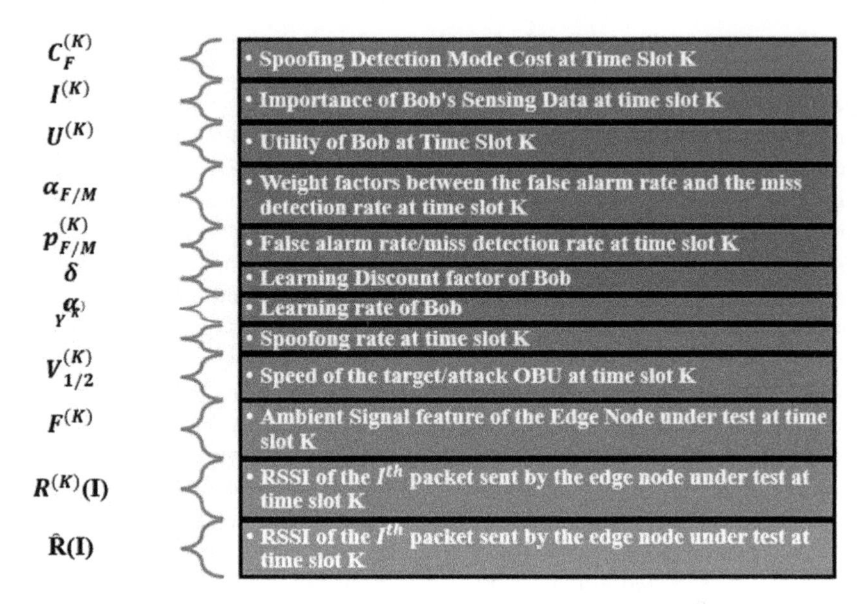

Figure 4.5 Symbols and notation summary [38]

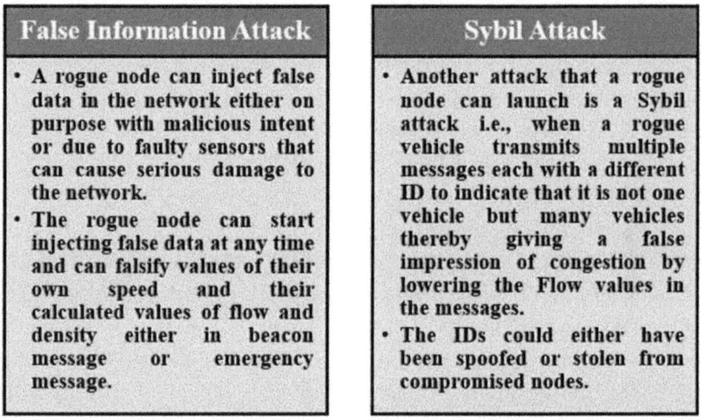

False Information Attack
• A rogue node can inject false data in the network either on purpose with malicious intent or due to faulty sensors that can cause serious damage to the network.
• The rogue node can start injecting false data at any time and can falsify values of their own speed and their calculated values of flow and density either in beacon message or emergency message.
Sybil Attack
• Another attack that a rogue node can launch is a Sybil attack i.e., when a rogue vehicle transmits multiple messages each with a different ID to indicate that it is not one vehicle but many vehicles thereby giving a false impression of congestion by lowering the Flow values in the messages.
• The IDs could either have been spoofed or stolen from compromised nodes.

Figure 4.6 Different types of attacks

edge computation results or illegal access advantages by sending a spoofing message with Alice's MAC address. Eve must provide Bob with the PHY-layer feature of the ambient radio signals in the spoofing signals in this PHY-layer authentication framework [39].

Attack model

There are different types of attacks that can take place in VANETs as shown in Figure 4.6.

4.4 Transport layer in VANTEs

Open systems interconnection (OSI) is a communication paradigm that divides communication functions into seven logical layers. The session and presentation levels are not included in this design, but each layer can be further split into sublayers within the framework of this architecture. Every region has its own set of protocols and interfaces because VANET design changes from one location to the next. There are specific protocols and standards for automobile use in the DSRC system. Protocols that are still under development will be used at different levels of the protocol stack, according to this piece of writing. When it comes to non-safety applications, IPv6, TCP, and UDP are the most dependable protocols. IPv6 is a network protocol that also offers transport-layer services. The job of the transport layer is to ensure that the full message is delivered securely and intact from one procedure to the next. The transport layer is in charge of ensuring the integrity and sequencing of message transmission [40]. Segments are produced at the transport layer by grouping multiple bytes and using the transmission control protocol (TCP).

VANET needs to send a lot of information. The employment of various TCP versions in a dense network architecture helps to keep traffic moving. Data is sent from the application layer of the source host to the transport layer protocol of the destination host via the user datagram protocol (UDP) and TCP, among others. The TCP is used for communication that relies on connections, but the UDP is utilized when the connections are bad [41]. TCP transmits communications control information to aid in the management of traffic flow, error control, and congestion. From beginning to end, it offers a dependable, congested-controlled Internet connection. SCTP, or stream control transmission protocol, is a novel transport-layer protocol that is both dependable and message-oriented. TCP comes in several flavors, including SACK, Westwood, and Westwood, to name a few. All of these things are targeted toward making mattresses perform better. The Internet Engineering Task Force advocated using the SACK option in Reno TCP version 2, which is now known as SACK TCP. SACK TCP is a security-conscious tweak. To provide a transport layer with interchangeable protocols and the ability to dynamically and intelligently select network interfaces and protocol settings, extremely sophisticated machinery is needed. Figure 4.7 depicts a few of the several pieces of equipment that require maintenance. There are several times when a program's preferences are supposed to take precedence over a system's control features. There are numerous instances where a system policy and an application preference must work together, and a more robust API that provides applications with a wider range of network mechanisms to choose from and configure will necessitate meaningful interaction between the API and the policy manager of the underlying system. The API is the subject of this article. As a result, we seek out related work to gain a deeper understanding of how things operate behind the scenes [42].

4.5 Forwarding in VANETs: Geo Networking

Safety, efficiency in traffic, and informational apps are all built on automotive connectivity. For these apps to work successfully, information must be transmitted across

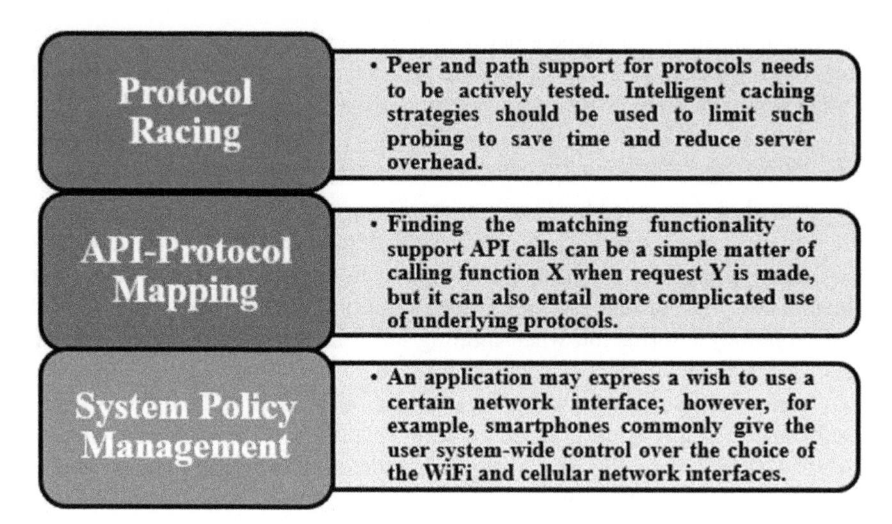

Figure 4.7 Machinery needs to take care of these three parameters

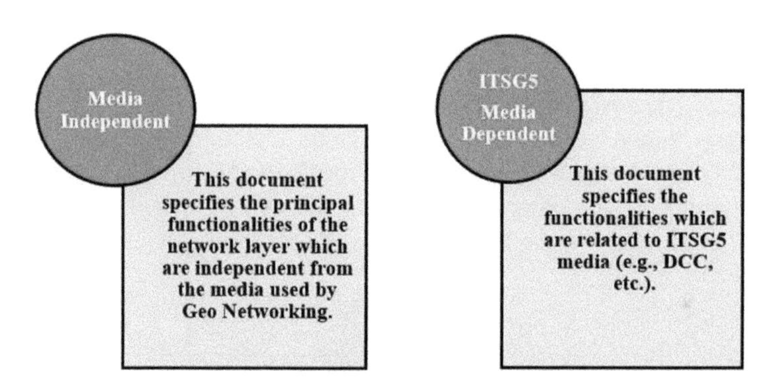

Figure 4.8 Two major documents

specific geographic regions. Researchers and standardized in Europe have been working on Geo Networking for VANET forwarding for the last decade. The several Geo Networking addressing strategies are described. are discussed. In terms of Geo Networking's protocol features, they are quite comprehensive. Last but not least, let's have a look at security, duplicate packet identification, and Geo Networking. Figure 4.8 displays the release of two important documents in quick succession [43].

In most vehicle applications, communication between vehicles within or at certain geographic regions is required. All vehicles entering the traffic bottleneck should be alerted if there is a backup or collision. Vehicles departing the area will be unable to make use of this data any longer. As a result, vehicles and roadside equipment can connect wirelessly and on-demand thanks to this innovation in communications technology. As part of the basic ITS station architecture, geo networking is included since it allows ITS applications for traffic efficiency, entertainment, and safety. Broadcasting safety-related messages to a specific geographic location may necessitate the usage of multi-hop communication or Internet application unicast communication [44].

4.5.1 Routing based on positions

Data is routed or forwarded by a communication network from a source to a destination. In most cases, data packets are forwarded by intermediary nodes on behalf of the sender. The job of a routing protocol is to figure out the best route to your destination for your current situation. Several routing systems take advantage of network topology information when figuring out the best method to connect two nodes on the network. The topology is determined by the connections that allow for direct communication. A VAN's topology, on the other hand, is dynamic. A link's value fluctuates greatly since it exists for such a short period before disintegrating. The spread of link state changes is huge when determining end-to-end pathways using link-state information. Finding the whole path may be impossible if the procedure takes longer than the typical lifetime of communication connections [45]. Instead of computing the complete path all at once, nodes could make forwarding decisions for each communication step (intermediate forwarder).

4.5.2 Addressing

The address of a data packet identifies the recipient. Geo Networking uses modified methodologies to address the specific information dissemination requirements of vehicle safety and traffic efficiency applications. As shown in Figure 4.9, most modern networks employ one of four different addressing methods.

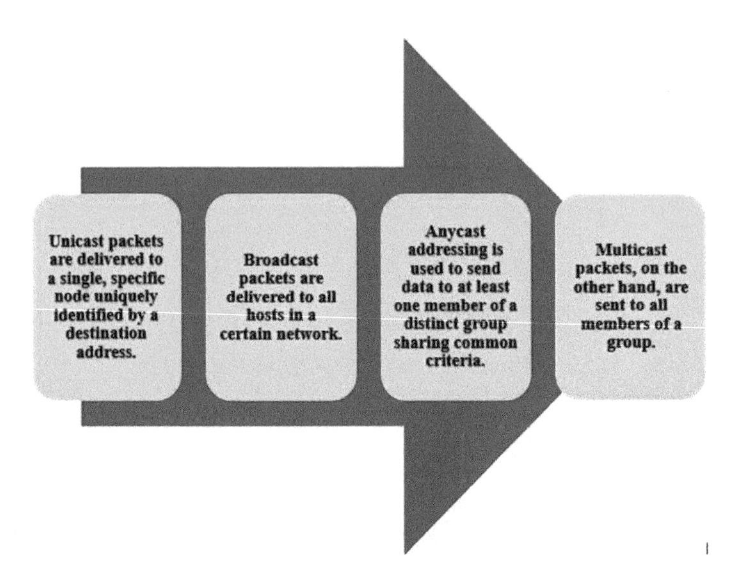

Figure 4.9 Four different addressing schemes

Geo Unicast

Geo Unicast focuses on a single node. This is comparable to IP-based networks' typical unicast. But in addition to the node identity, the packet is routed based on the node's location.

Geo Broadcast/Geo Anycast

Anycast addresses at least one node, whereas Broadcast addresses all nodes within a specific target zone, respectively. Higher-level requests specify and supply the required destination region to lower-level requests. Origin: The packet's origin may or may not be in the location where it will be received. Geo Broadcast/Geo Anycast is identical to traditional multicast/anycast operations if all nodes in the target area are considered to be members of a multicast or anycast group [46].

4.6 VNDNs forwarding strategies

Named data networks (NDNs) convert the Internet from a host-to-host link to content delivery. NDN's name-based search makes it possible to locate and retrieve any piece of data without the need for an IP address. Also, the Interest/data packet paradigm is used for the sharing of NDN architecture in this system [47]. However, the request packet is equally as critical to NDN's design because it represents the data. Figure 4.10 depicts VANET networking based on NDN.

4.6.1 VANET benefits of NDNs forwarding

By addressing the routing inefficiencies of present IP-based networks and ensuring effective data distribution, NDN's forwarding and routing functions will achieve their primary goal. Figure 4.11 illustrates and discusses the enormous advantages of stateful and adaptive forwarding for vehicular communication [49].

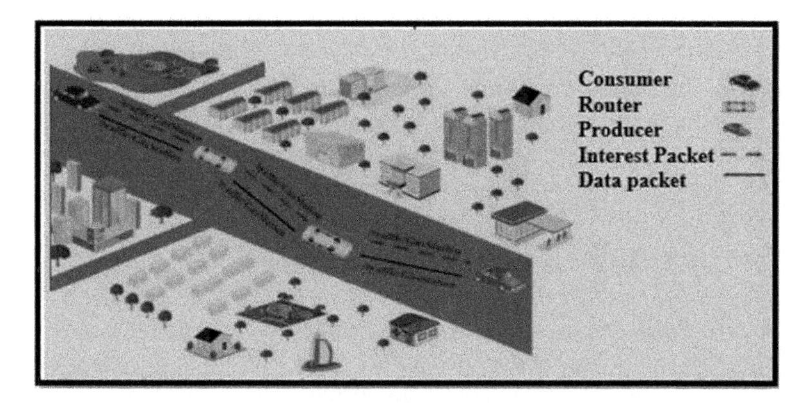

Figure 4.10 NDN-based VANET networking [48]

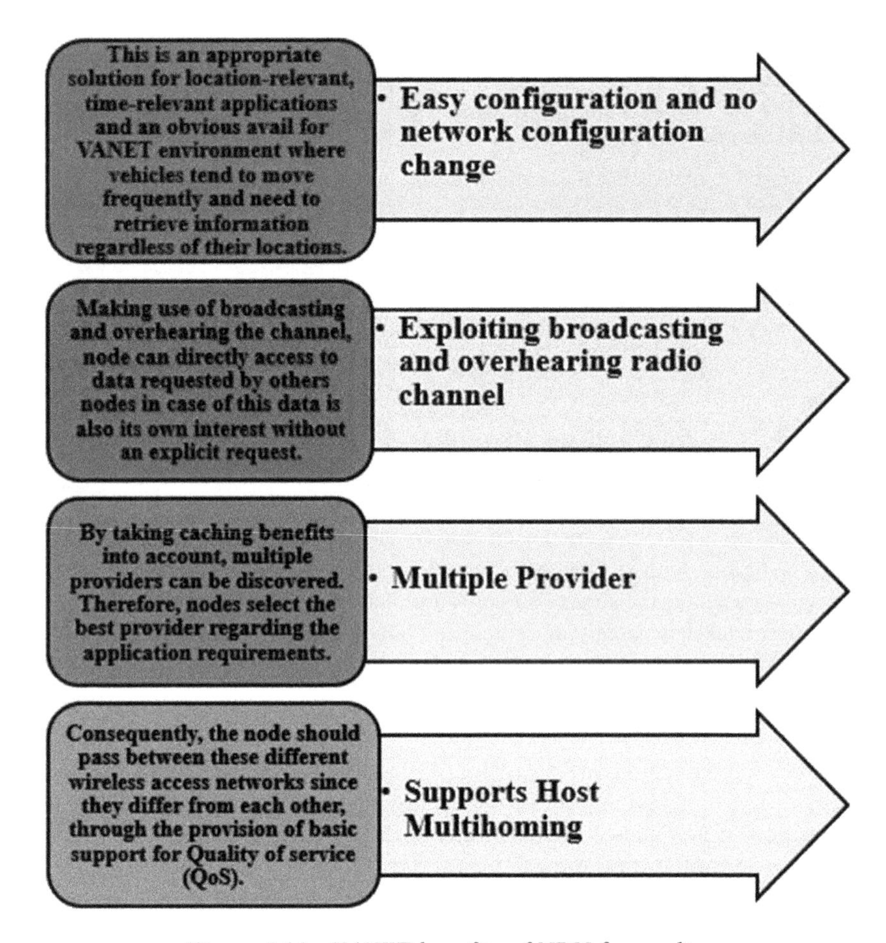

Figure 4.11 VANET benefits of NDN forwarding

4.6.2 VANET encounter NDNs forwarding challenges

Content may be distributed more quickly and efficiently in a vehicle setting with NDN. When it comes to vehicle data dissemination, NDN concepts face several serious roadblocks, such as network partitions and the broadcast storm problem. Figure 4.12 illustrates these issues succinctly [50].

4.7 VANET wireless accessibility standards

The usage of standards makes it easier to develop new items and to compare them with one other. For vehicles to communicate from V2V, V2I, or infrastructure to infrastructure, many wireless protocols are available that provide radio access for vehicles. It is the goal of these communication standards to increase traffic safety while also increasing efficiency and driver and passenger comfort. To the best of

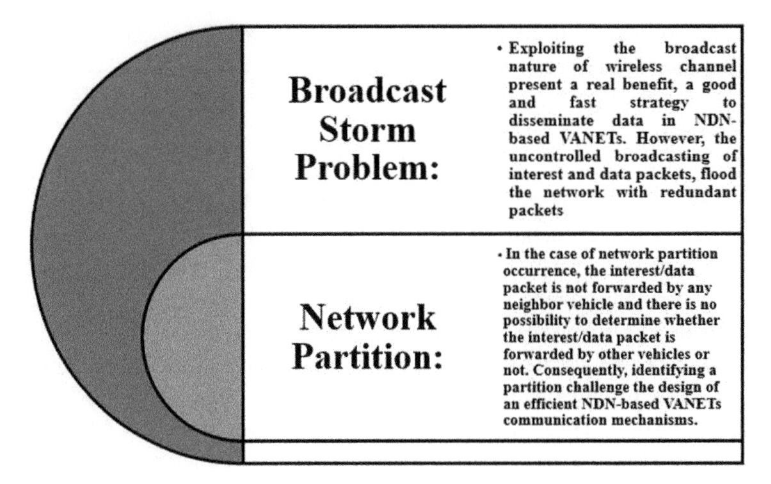

Figure 4.12 VANET face NDN forwarding challenges

our knowledge, no previous studies have collected standards like 4G cellular access in one study. DSRC and WAVE are examples of industry-standard protocols [51].

4.7.1 Vehicle-based cellular access (2G, 2.5G, 3G, and 4G)

Cellular network access relies on repurposing the restricted radio frequency spectrum that has been granted to various services. Data transfer rates are available on GSM-based cellular networks, which date back to the early 1990s. After Finland became the first country to use the second-generation (or 2G) cellular technology in 1991, other countries followed suit. Both TDMA and FDMA mechanisms are utilized by GSM to transport data, with TDMA accounting for the vast majority of all GSM connections [52]. When it comes to GSM, you will see both TDMA and FDMA in action. The European Telecommunications Standards Institute (ETSI) created GPRS, or the 2.5G standard, in 2000 to aid in the efficient transfer of data at high bandwidth and maximum data rate. However, the GPRS transmission rate was inadequate for delivering multimedia data. As a result, mobile telecommunications networks of the third generation (3G, also known as UMTS) were created in 2008. Since the development of HSDPA in 2005, 3G has been able to send data at a higher rate. To achieve high-speed continuous data transfer rates, GSM offers a standard for cellular networks that do not include these two standards. 3G and 4G, the most recent cellular communication standards, are currently under development. To improve mobile data transmission speeds while maintaining high data transfer rates, 4G is an important goal to pursue. 4G networks all across the world have been using the long-term evolution (LTE) standard since 2010. With 4G, data transfer will be faster, users will have greater mobility, and switching between networks will be seamless. To put it another way: at this time, 4G appears to fall short on all counts. The primary reason for using a cellular network is to take advantage of pre-existing infrastructure. Base station participation in the mobile environment has the drawback of greatly delaying data transmissions. DSRC has solved this issue by eliminating the requirement for a communication base station [53].

4.7.2 DSRC

DSRC intends to improve communication network efficiency by promoting low overhead operations. Many different types of data can be sent over the airwaves by a car's wireless communications system. A couple of instances are notifications concerning hazardous road conditions or accidents. High data transfer speeds can be achieved quickly while employing DSRC [54]. The DSRC radiofrequency spectrum has been licensed, however, it is unrestricted at this time. Because the FCC does not charge for its spectrum, using it is cost-free. However, it requires a license to limit the amount of spectrum that may be used while also helping to alleviate congestion. Thus, all radio stations must follow the FCC's laws and regulations when it comes to the use of particular frequencies or channels. One channel is devoted to the control signal, with the remaining six channels going to the service signals. Messages of the utmost importance and management data are sent over the control channel. The control channel is being monitored while service channels are being switched. In the literature, you will find DSRC standards utilized by various countries, including the USA, Europe, and Japan. Data transmission speeds, coverage, and modulation vary among these standards depending on the type of communication and frequency bands used. However, in Japan, OBU uses half-duplex transmission while RSU uses full-duplex, as is the case in the US and Europe [55].

4.7.3 Wireless access in vehicular environment (WAVE)

Low-latency in terms of overall network setup costs, WLAN is less expensive, but it also necessitates the purchase of additional hardware such as wireless adapters and routers. The IEEE 802.11 wireless networking standard enables Wi-Fi connectivity. The Media Access Control Layer of 802.11 was enhanced by IEEE Standard 1609.4-2006 to allow WAVE. Broadband wireless access solutions from diverse vendors can work together thanks to the IEEE 802.16e standard. It is possible to use WAVE on both a desktop and a mobile device. OBU or RSU can function as both a service provider and consumer, depending on the platform. WAVE discusses RSU-based apps that give multiplexed access to the OBU [56].

4.8 Issues and future perspective related to networking standards in VANET

When compared to other kinds of mobile networks, vehicular networks have unique traits and behavior. Compared to MANETs and WANETs, vehicular networks have significantly greater processing, transmission power, and predictability. The findings of this study will be useful to academics working on standard routing protocols for VANETs, which may be used in both urban and rural settings [57]. Vehicle networks encounter a few obstacles that can be used for future research in this sector, as detailed below:

Data storage and administration make large-scale projects feasible: With vehicle ad hoc networks, the scale of the network is not constrained like it is with mobile and wireless ad hoc networks, which have a finite number of cars they may include. Data generated by automobile networks, particularly those with a large number of vehicles, is vast. Sending and storing these files are both necessitated by using the VANETs, which are huge and dynamic while simultaneously creating massive volumes of data, which have never been more difficult to collect and administer.

Network topology, partitioned network, and connectivity: Interconnectivity, network structure, and a partitioned network all have a role. The term "vehicular networks" refers to ad hoc networks that don't include any moving vehicles. These networks are unique since they are continually shifting places. Because nodes join and leave the network regularly, and the underlying networks are repeatedly partitioned, the network's topology changes frequently. As a result, the network is constantly evolving. Although this dynamic nature of traffic can create large gaps between vehicles in a densely populated region, it can also create networks with many isolated nodes.

A high degree of mobility, along with a wide range of network densities: The vehicle network is characterized by vulnerability and extreme configurations in a dynamic environment. A roadway or a city may be an appropriate place to use it. In this situation, the VANET is more dispersed and less interconnected. There are hence measures necessary for the detection of disconnections. Designing protocols for medium access control can help prevent data corruption and transmission errors. For high data transfer, vehicles in both scenarios must adapt their behavior to changing network densities.

Security and privacy: Other types of classic wireless ad hoc networks, such as mesh networks, have similar security and privacy concerns to VANET. Because of frequent changes in network architecture, the large network size, and the high mobility patterns as well as a wide range of software applications and services running on VANET systems, VANET security presents a unique and fundamental problem in this context. Authentication, nonrepudiation, and personal privacy are all challenges that need to be addressed in the car industry. Despite much research, valet network security remains a concern. All of these diverse vehicle networks necessitate new, highly secure communication methods.

Standardization of protocols: VANET connection may now be used by many different types of vehicles because of the standardization of protocols. Since the network runs smoothly when all nodes adhere to the same communication protocol, it is critical. Industry, government, and academics can only accomplish standardization and standard procedures by working together on the same topic at the same time.

Disruptive tolerant communications: When sparse networks are used, reliability and delivery times suffer, as does latency. Some systems use the Carry-and-Forward strategy, which increases the overall delivery time of information, to enable fast delivery and dependability of the information they deliver. New data communication mechanisms customized to various vehicle ad hoc networks should be investigated to minimize or eliminate these problems.

Need for a benchmark for testing: To properly test the car ad hoc network, a benchmark is needed. There are therefore sections of a constantly changing network. An inadequate or overburdened network may need a protocol transition from time to time. It is critical to follow globally accepted standard protocols. Furthermore, to the best of our knowledge, no existing scientific literature benchmarks have been used to compare the proposed procedures' performance in this domain.

Localization systems and geographical addressing: To make use of localization systems and geographic addressing, many applications require a geographic address that can carry out data transmissions. The future position of a vehicle can be predicted based on the vehicle's mobility pattern and driving style. However, keeping track of and maintaining the geographic addresses that cars and apps utilize presents a significant challenge.

4.9 Worm propagation model for Web Vehicular Ad Hoc Network (WVANET)

In recent years, scientists have become increasingly interested in how worms propagate between nodes and even beyond networks. Finding the network's weak points is important to infect the nodes. Attackers can exploit the network's weaknesses by employing several target-finding strategies. Figure 4.13 depicts the three stages of worm propagation in which a node may be at any time [58].

4.9.1 Worm propagation model for WVANET worm propagation

Because every node has the same probability, the web car ad hoc network is homogeneous. A homogeneous network can be illustrated by a fully linked graph. There is an interconnected topology because every node in the WVANET is

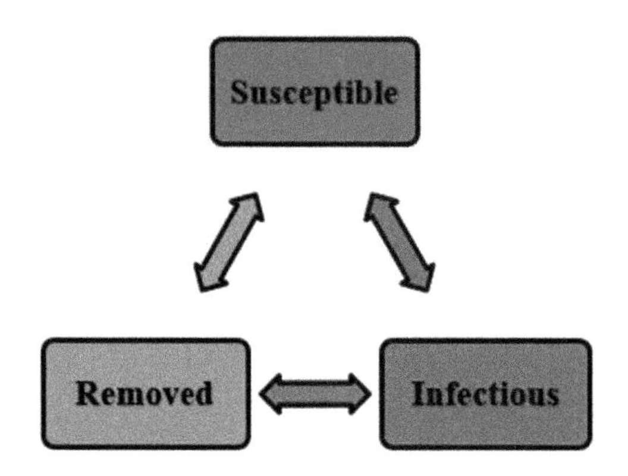

Figure 4.13 Three states during worm propagation

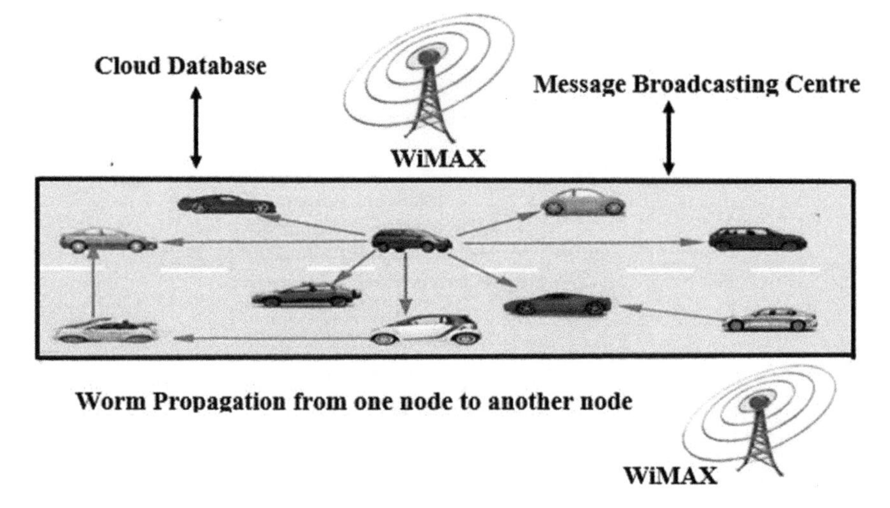

Figure 4.14 Worm propagation architecture of WVANET [59]

connected to every other node. Each infected weak host has a 50% chance of infecting another weak host on the network if there are several sick ones. WVANET nodes should be immune to worm infection as a result. It is possible to interact among the WVANET nodes because each one is connected to the others, as shown in Figure 4.14.

4.9.2 Behavior of worm propagation model for WVANET worm

When a node is infected, the worm examines the IP addresses of other nodes that have communicated with it. As soon as it discovers an IP address that is not secured, the worm will move on to the next node. For starters, the worm scans for recently connected machines to see whether it can infect those as well. To begin infecting another machine, it must first be located. The infection spreads like wildfire over the network until all vulnerable nodes are infected. Remember that the worm only scans for IP addresses that have recently communicated with the infected host.

Worm detection model for WVANET

When an infected node rejoins an ad hoc web network in a car, it must be rendered immune to attack until it is cleaned up. Worms must be detected and eliminated on each infected node regularly to maintain their safety. All other nodes must be safeguarded by eliminating any inaccessible nodes.

Worm detection model for WVANET worm detection scanning process

To keep the remaining nodes safe, a worm-infected node must be repaired or removed from the network entirely. Thanks to established standards, worm-vulnerable nodes may be identified and remedied. Figure 4.15 depicts the many types of regulations that have been framed.

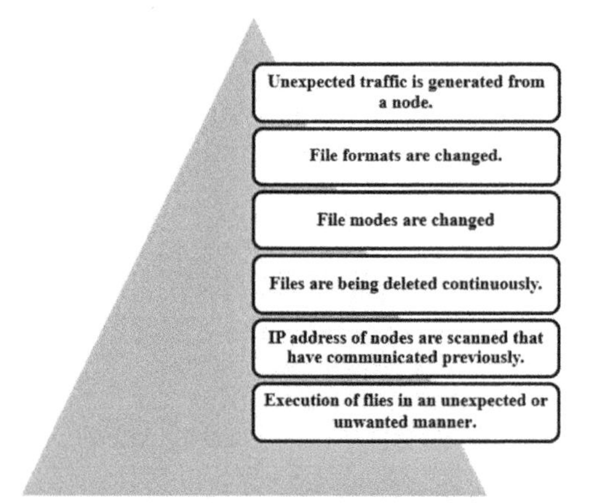

Figure 4.15 Various rules for WVANET worm detection scanning process

The node will be temporarily disabled if any of the rules are breached. To top it all off, the node is constantly monitored for possible worm attacks and remedied accordingly. It is possible for nodes that have been automatically forbidden from interacting with the particulars node to reconnect. To keep the network safe, no data can be transferred from the temporarily blocked nodes to the other nodes. A node's worm status is determined after it has been shut down. Worms are eliminated and the temporary lock is reset to keep the network architecture's communication flowing. During this time, all nodes are being checked for worm assaults and are being monitored and examined often [60].

4.10 Conclusion

As a consequence of significant investments from a wide range of sectors, such as government agencies and the automobile and navigation safety industries, VANET is no longer a far-off dream. Opportunities, regions of application, and prospects for VANETs are rapidly developing, encompassing a wide range of service types with a wide range of demands and goals. Many new and distinct open research topics, such as developing wireless networks, dependable message distribution, and event detection, make VANET research increasingly intriguing. Several crucial issues are now being explored and debated in vehicular communication. In the end, these prospective technologies would be integrated into the various worldwide vehicular communication programmers and standards currently being developed. A more efficient and secure transportation system can be developed with VANETs, which are enabled by automated cars and V2V and V2I communications. There is a BSA for driving assistance applications that will be released by the EU soon. VANETs are discussed as a possible answer to future transportation concerns in this chapter. Next, advanced state-of-the-art technology using VANET-based traffic management and control apps is presented for platooning, junctions, and VRUs. VANETs-based applications are being evaluated in light of the current circumstances. Many variables are taken into consideration when developing these transportation-related apps, such as safety, efficiency, sustainability, and comfort. There are several new technologies on the horizon, including smart cities, smart grids, and the Internet of Things (IoT). Because it gives both an overview of the current transportation system and insight into possible future VANET-based applications, this chapter accomplishes a double goal.

References

[1] Sultan, S, Javaid, Q, Malik, AJ, Al-Turjman, F and Attique, M. "Collaborative-trust approach toward malicious node detection in vehicular ad hoc networks." *Environment, Development, and Sustainability* (2021): 1–19.

[2] Tyagi, AK and Sreenath, N. "Preserving location privacy in location based services against Sybil attacks." *International Journal of Security and Its Applications*. 9(12) (2015): 189–210.

[3] Amari, H, Louati, W, Khoukhi, L and Belguith, LH. "Securing software-defined vehicular network architecture against DDoS attack." *2021 IEEE 46th Conference on Local Computer Networks (LCN)*. IEEE, 2021, 653–656.

[4] Tyagi, AK, Aswathy, SU, Aghila, G, and Sreenath, N. "AARIN: Affordable, Accurate, Reliable and INnovative Mechanism to Protect a Medical Cyber-Physical System using Blockchain Technology." *IJIN* 2 (2021): 175–183.

[5] Singh, P, Raw RS, and Khan SA. "Link risk degree aided routing protocol based on weight gradient for health monitoring applications in vehicular ad-hoc networks." *Journal of Ambient Intelligence and Humanized Computing* (2021): 1–23.

[6] Mekki, T, Jabri, I, Rachedi, A, and Chaari, L. "Software-defined networking in vehicular networks: A survey." *Transactions on Emerging Telecommunications Technologies* (2021): e4265.

[7] Machado, C, and Westphall CM. "Blockchain incentivized data forwarding in MANETs: strategies and challenges." *Ad Hoc Networks* 110 (2021): 102321.

[8] Mao, M, Yi, P, Hu, T, Zhang, Z, Lu, X, and Lei, J. "Hierarchical hybrid trust management scheme in SDN-enabled VANETs." *Mobile Information Systems* 2021 (2021): 1–16.

[9] Prakash, J., Sengottaiyan N, and Anbukaruppusamy TKP. "Dynamic routing for the various topologies in a VANET." *Annals of the Romanian Society for Cell Biology* (2021): 20486–20499.

[10] Dibaei, M, Zheng, X, Xia, Y, *et al.* "Investigating the prospect of leveraging blockchain and machine learning to secure vehicular networks: a survey." *IEEE Transactions on Intelligent Transportation Systems* 23(2) (2022): 683–700.

[11] Manvi, SS, and Tangade S. "A survey on authentication schemes in VANETs for secured communication." *Vehicular Communications* 9 (2017): 19–30.

[12] Greco, C, Pace, P, Basagni, S, and Fortino, G. "Jamming detection at the edge of drone networks using multi-layer perceptrons and decision trees." *Applied Soft Computing* 111 (2021): 107806.

[13] Hasan, MM, Rahman, MA, Sedigh, A, and Khasanah, AU, *et al.* "Search and rescue operation in flooded areas: a survey on emerging sensor networking-enabled IoT-oriented technologies and applications." *Cognitive Systems Research* 67 (2021): 104–123.

[14] Abbasi, F, Zarei M, and Rahmani AM. "FWDP: a fuzzy logic-based vehicle weighting model for data prioritization in vehicular ad hoc networks." *Vehicular Communications* (2021): 100413.

[15] Hamdi, MM, Yussen YA, and Mustafa AS. "Integrity and Authentications for service security in vehicular ad hoc networks (VANETs): a review." *2021 3rd International Congress on Human-Computer Interaction, Optimization and Robotic Applications (HORA)*. IEEE, 2021.

[16] Channakeshava, RN, and Sundaram M. "A study on energy-efficient communication in VANETs using cellular IoT." *Intelligence Enabled Research*. Springer, Singapore, 2021. 75–85.

[17] Devi, YS, and Roopa M. "Performance analysis of routing protocols IN vehicular ad-hoc networks." *Materials Today: Proceedings* (2021).

[18] He, D, Zeadally, S, Xu, B, and Huang, X. "An efficient identity-based conditional privacy-preserving authentication scheme for vehicular ad hoc networks." *IEEE Transactions on Information Forensics and Security* 10(12) (2015): 2681–2691.

[19] Chen, C, Huan, L., Xiang L., Jianlong Z., Hong W., and Hao W. "A geographic routing protocol based on trunk line in VANETs." *Digital Communications and Networks* (2021).

20] Liu, Y, Niu, J, Qu, G, Cai, Q and Ma, J. "Message delivery delay analysis in VANETs with a bidirectional traffic model." *2011 7th International Wireless Communications and Mobile Computing Conference*. IEEE, 2011, 1754–1759.

[21] Dutta, N, Deva Sarma, HK, Jadeja, R, Delvadia, K and Ghinea, G. "Integrating Content Communication into Real-Life Applications." *Information-Centric Networks (ICN)*. Springer, Cham, 2021. pp. 169–194.

[22] Shah, P, and Kasbe T. "A review on specification evaluation of broadcasting routing protocols in VANET." *Computer Science Review* 41 (2021): 100418.

[23] Selvi, M, and Ramakrishnan B. "Secured message broadcasting in VANET using Oppositional Flower Pollination Algorithm (OFPA)." *2021 Third International Conference on Intelligent Communication Technologies and Virtual Mobile Networks (ICICV)*. IEEE, 2021.

[24] Maranur, J, and Mathapati B. "ARPVP: attack resilient position-based VANET protocol using ant colony optimization." (2021).

[25] Azzaoui, N, Korichi A, and Brik B. "A reliable data forwarding technique for 5G internet of vehicles networks." *Innovative and Intelligent Technology-Based Services for Smart Environments–Smart Sensing and Artificial Intelligence*. CRC Press, 2021. pp. 142–148.

[26] Basha, SK, and Shankar TN. "Fuzzy based multi-hop broadcasting in high-mobility VANETs." *IJCSNS* 21(3) (2021): 165.

[27] Zong, L, Wang, H, Bai, Y, and Luo, G. "Cross-regional transmission control for satellite network-assisted vehicular ad hoc networks." *IEEE Transactions on Intelligent Transportation Systems* (2021), doi: 10.1109/TITS.2021.3106018.

[28] DasGupta, S, Choudhury S, and Chaki R. "VADiRSYRem: VANET-based diagnosis and response system for remote locality." *SN Computer Science* 2 (1) (2021): 1–13.

[29] Vanitha, N. "Binary spray and wait for routing protocol with controlled replication for DTN based multi-layer UAV Ad-hoc network assisting VANET." *Turkish Journal of Computer and Mathematics Education (TURCOMAT)* 12(10) (2021): 276–2782.

[30] Zhou, S, Demin, L, Qinghua, T, Yue, F, Chang, G and Xuemin, C. "Multiple intersection selection routing protocols based on road section connectivity probability for urban VANETs." *Computer Communications* 177 (2021): 255–264.

[31] Ganesh, A., Ayyasamy S, and Kumar NM. "Performance and analysis of advanced and enhanced security protocol for vehicular ad hoc networks (VANETs)." *Wireless Personal Communications* (2021): 1–21.

[32] Vanitha, N, and Padmavathi G. "Support of mobility models for the decentralized multi-layer UAV networks assisting VANET architecture (DMUAV)."

[33] Mistareehi, H. "Message dissemination scheme for rural areas using VANET (hardware implementation)." *2021 Twelfth International Conference on Ubiquitous and Future Networks (ICUFN)*. IEEE, 2021, 120–125, doi: 10.1109/ICUFN49451.2021.9528820.

[34] Zhang, SY, Lagutkina, M, Akpinar, KO, and Akpinar, M. "Improving performance and data transmission security in VANETs." *Computer Communications* 180 (2021): 126–133.

[35] Devi, MR, and Jeya IJS. "Black Widow Optimization Algorithm and Similarity Index-Based Adaptive Scheduled Partitioning Technique for Reliable Emergency Message Broadcasting in VANET." (2021).

[36] Tan, H, and Chung I. "RSU-aided remote V2V message dissemination employing secure group association for UAV-assisted VANETs." *Electronics* 10(5) (2021): 548.

[37] Lopes, R, Luís M, and Sargento S. "Real-time video frame differentiation in multihomed VANETs." *Wireless Networks* 27(4) (2021): 2559–2575.

[38] Xiao, L, Zhuang, W, Zhou, S, and Chen, C. "Learning-based rogue edge detection in VANETs with ambient radio signals." In *Learning-based VANET Communication and Security Techniques. Wireless Networks*. Springer, Cham. 2019. https://doi.org/10.1007/978-3-030-01731-6_2.

[39] Sindhwani, M, Singh, R, Sachdeva, A, and Singh, C. "Improvisation of optimization technique and AODV routing protocol in VANET." *Materials Today: Proceedings* 49 (2021): 3457–3461.

[40] Tyagi, AK and Aswathy, SU. "Autonomous Intelligent Vehicles (AIV): Research statements, open issues, challenges and road for future." *International Journal of Intelligent Networks*, 2 (2021): 83–102.

[41] Shobana, G, and Arockia XAR. "Detection mechanism on vehicular adhoc networks (VANETs): a comprehensive survey." *International Journal of Computer Science & Network Security* 21(6) (2021): 294–303.

[42] Chiluveru, R, Gupta N, and Teles AS. "Distribution of safety messages using mobility-aware multi-hop clustering in vehicular ad hoc network." *Future Internet* 13.7 (2021): 169.

[43] Kaja, H. Survivable and Reliable Design of Cellular and Vehicular Networks for Safety Applications. Diss. University of Missouri-Kansas City, 2021.

[44] Bendigeri, KY, Jayashree DM, and Kumbalavati SB. "Wireless sensor networks and its application for agriculture." *Intelligent Data Communication Technologies and Internet of Things*. Springer, Singapore, 2021. pp. 673–687.

[45] Ravi, B, and Thangaraj J. "Stochastic traffic flow modeling for multi-hop cooperative data dissemination in VANETs." *Physical Communication* 46 (2021): 101290.

[46] Saravanan, M. "Improved authentication in vanets using a connected dominating set-based privacy preservation protocol." *The Journal of Supercomputing* (2021): 1–22.

[47] Peixoto, MLM, Maia, AHO, Mota, E, *et al.* "A traffic data clustering framework based on fog computing for VANETs." *Vehicular Communications* 31 (2021): 100370.

[48] Ahed, K, Benamar, M, Ait, Ayoub, AL, and Rajae, O. "Forwarding strategies in vehicular named data networks: a survey." *Journal of King Saud University-Computer and Information Sciences* (2020). 10.1016/j.jksuci.2020.06.014.

[49] Raja, M. "PRAVN: a perspective on road safety adopted routing protocol for hybrid VANET-WSN communication using balanced clustering and optimal neighborhood selection." *Soft Computing* 25(5) (2021): 4053–4072.

[50] Badreddine, C, Abderrahim, HB, and Mohammed, E. "Blackhole attack detection in a vehicular ad-hoc network using statistical process control." *International Journal on Communications Antenna and Propagation (IRECAP)* 7 (2017): 208.

[51] Wang, X, Weng Y, and Gao H. "A low-latency and energy-efficient multimetric routing protocol based on network connectivity in VANET communication." *IEEE Transactions on Green Communications and Networking* (2021).

[52] Nayak, BP, Hota, L, Kumar, A, Turuk, AK, and J. Chong, PH. "Autonomous vehicles: resource allocation, security and data privacy." *IEEE Transactions on Green Communications and Networking* 6(1) (2022): 117–131.

[53] Balamurugan, R, and Hariharan MM. "VANET based accident alerting system." *2021 5th International Conference on Trends in Electronics and Informatics (ICOEI)*. IEEE, 2021.

[54] Zhang, X, Wang, W, Mu, L, Huang, C, Fu, H, and Xu, C. "Efficicnt privacy-preserving anonymous authentication protocol for vehicular ad-hoc networks." *Wireless Personal Communications* 120 (2021): 3171–3187.

[55] Zhao, L, Bi, Z, Lin, M, Hawbani, A, Shi, J, and Guan, Y. "An intelligent fuzzy-based routing scheme for software-defined vehicular networks." *Computer Networks* 187 (2021): 107837.

[56] Alaya, B, Khan, R, Moulahi, T, and Khediri, SE. "Study on QoS management for video streaming in vehicular ad hoc network (VANET)." *Wireless Personal Communications* (2021): 1–33.

[57] Kaci, A, and Rachedi A. "Named data networking architecture for the internet of vehicles in the era of 5G." *Annals of Telecommunications* (2021): 1–13.

[58] Wang, S, Zhang Q, and Chen G. "V2V-CoVAD: a vehicle-to-vehicle cooperative video alert dissemination mechanism for internet of vehicles in a highway environment." *Vehicular Communications* (2021): 100418.

[59] Milton Joe, M., Ramakrishnan, B., Karthika Bai, S, Shaji, R. S. "Modelling and Detection of Worm Propagation for Web Vehicular Ad Hoc Network (WVANET)." *Wireless Pers Commun* 109 (2019): 223–241. https://doi.org/10.1007/s11277-019-06561-1

[60] Hongzhi, Guo, Xiaoyi, Zhou, Jiajia, Liu, Yanning, Zhang. Vehicular intelligence in 6G: Networking, communications, and computing, Vehicular Communications, 33, 2022, 100399, ISSN 2214-2096, https://doi.org/10.1016/j.vehcom.2021.100399.

Chapter 5

Unmanned Aerial Vehicles (UAV) relay in Vehicular Ad Hoc Network (VANETs) against smart jamming with reinforcement learning/ deep learning

Abstract

The focus of today's study is on Vehicular Ad Hoc Networks (VANET). Existing VANET security technologies are still being assessed to ensure that the network is protected from attacks and breaches. The security and privacy of VANETs are interwoven with the networks' ability to withstand disasters. This chapter takes a close look into wireless networks, with a particular emphasis on creating research approaches for securing VANETs (virtual area networks). This study, which goes into great detail about the dangers of wireless network jamming, has demonstrated a wireless network jamming attack. The types of jammers used in wireless ad hoc networks, as well as the locations of jammers, are the two most important elements of jamming techniques in these networks. VANETs are a topic that has piqued the interest of many people, who are actively researching them. The development of self-driving and semi-autonomous vehicles will be impossible without the use of VANETs, which increase both safety and comfort. The fact is that, whether or not security concerns are prevalent in VANETs, they provide significant challenges. It is the goal of VANET to create a network of autonomous vehicles to increase driver comfort and safety while also increasing efficiency. The transmission of information in smart vehicles is expanding at a rapid pace, and all of these vehicles currently communicate with one another via internet services. The majority of the functionality of these vehicles is dependent on data control and the environment's cooperative awareness messages. This puts VANETs in danger of a wide range of cyberattacks as a result of their vulnerability.

Keywords: Deep belief network (DBN), Cooperative awareness messages (CAMs), intrusion detection system (IDS), Vehicular Ad Hoc Networks (VANETs), denial of service (DoS)

5.1 Introduction

Vehicular Ad Hoc Networks (VANETs) support V2V and V2I communications, enabling to build unmanned driving and supporting the growing usage of onboard units

(OBUs) gadgets while also enhancing transmission safety. There is a vulnerability in the VANET due to OBU mobility and a large-scale dynamic network with fixed roadside units (RSUs). A jammer can prevent OBUs and servicing RSUs from transmitting continuously [1]. An intelligent radio jammer can monitor and analyze the VANET's ongoing communications and policies by using smart radio devices, giving them control over the jamming frequencies and signal strengths. Before starting an attack, they can coerce the VANET to use a specific communication technique. The use of unmanned aerial vehicles (UAVs) to deliver the OBU message improves VANET anti-jamming communication [2]. By their line of sight (LOS) connectivity and shorter deployment times, UAVs have better channel states and a lower eigenvalue for path loss than RSUs at fixed locations on the ground that may be substantially jammed. UAVs have also made it easier to communicate with OBUs and RSUs while using them. To increase the signal-to-interference-plus-noise-ratio (SINR) of the OBU signals and hence to minimize the bit error rate (BER), UAVs broadcast the OBU message again when the serving RSUs are jammed or interfered with the BER.

The application of game theory to wireless network anti-jamming power control has been studied. VANET jamming resistance is still an open subject, according to what we know thus far. UAV relay game with anti-jamming features is designed utilizing an OBU channel model with route loss, log-normal shadowing, and Rayleigh fade to select whether or not to deliver an OBU message to another RSU [3]. The game's Nash equilibrium (NE) is determined using the jamming model and the UAV channel model, revealing how jamming impacts VANET transmission and the UAV-assisted VANET's BER for the OBU message. Using techniques like Q-learning and reinforcement learning, you can figure out the best UAV team strategy after a while of playing. Prior knowledge about jamming or VANET models need not be required when using policy hill climbing (PHC). This relay method, in contrast to Q-learning strategies, creates more uncertainty in the choice, making it more difficult for jammers to figure out how to intercept it. Our recommendation is to deploy PHC-based UAV relays to increase the anti-jamming capabilities of VANET. Since the jammer cannot anticipate the perfect relay policy in this architecture, an attack on the UAV will fail. The use of the hot-booting technique increases the jamming resistance of the VANET and speeds up the learning process. Instead of starting with a random search of the standard PHC and leaving the Q-function with a null matrix, the hot-booting approach prepopulates the Q-function with data from similar cases [4].

Using a hot booting PHC-based relay, we were able to reduce the BER of the OBU message while also increasing the value of the UAV. The following people contributed to the project: We devised a UAV relay game to test the VANET's defenses against high-tech jamming. The game's NEs and the circumstances under which they occur are described. The sophisticated jamming can be defeated without knowing anything about the UAV channel model or the jamming model by using a hot-booting PHC UAV relay. It is predicted that by employing this method, the BER of an OBU message will drop while the message's utility would rise. To make driving safer over the road network, we may get compromised the security and privacy of the VANETs [5]. VANET intrusion detection networks have systems in place to detect security threats and network attacks, both from within and outside the vehicle. An intrusion detection system (IDS) can pick up on any unexpected or suspect network activity. Cyber analytics are used to support IDS in their search for harmful activity and anomalies. Systems based on abuse are only able to detect known intrusions with a high storage need while anomaly-based intrusions arrive at the rate of one every second. Numerous approaches to intelligent intrusion detection exist, including data mining and decision trees, genetic algorithms, and artificial neural networks (ANNs). The combination of deep learning (DL) and reinforcement learning (RL) in deep reinforcement learning (DRL) produces more precise and reliable function approximations in scenarios with high dimensions or infinite states [6]. Many fields take inspiration from DL, which heavily utilizes ANNs.

Each node in the neighboring layers has a weighted relationship to all of the other nodes. Nodes employ weights and a basic function to compute output values based on input from other nodes. ANNs have become popular inductive approaches due to their great degree of flexibility and nonlinearity, which is driven by data. The most advanced kind of RL to increase the learning rate of the DRL algorithms is DeepMind. DeepMind used DL to build DRL, which is a more advanced version of the RL technique [7]. During the real-time learning process, the learning events will be stored and used to train the neural network. By providing real-time feedback, the latter will assist the agent in making better judgments more quickly. To train the neural network, DRL leverages new experiences obtained through real-time interactions with the surrounding environment instead of DL. The Q-values can be stated as a table for a small number of discrete states and actions. Deep neural networks (DNNs) are necessary, for example, to estimate Q-values in continuous state spaces. Using DQL as an example, a DNN translates continuous state space to a collection of action Q-values. Action zones that are too large or continuous should be avoided. When using this approach, you'll have an endless supply of things to do [8]. In this technique, a critic DNN learns the Q-values for state action tuples using the soft actor-critic approach and then uses that knowledge to predict the Boltzmann distribution across the predicted Q-values of available actions. Figure 5.1 shows UAV relay in VANETs against smart jamming with RL.

The rest of the chapter is organized as follows. We review the related work in Section 5.2 and present the DL and its importance over VANET in Section 5.3. We present the types and characteristics of DL models in Section 5.4 and identify smart jamming in Section 5.5. Solving jamming problems with RL/DL is presented in Section 5.6. Challenges toward DL over VANET are described in Section 5.7. Open issues and future trends with RL/DL are described in Section 5.8, while a conclusion is drawn for this work in Section 5.9.

5.2 Related/background work

In the last few decades, automatic cars have become increasingly popular, especially in Europe. The internet connects all of the vehicles, allowing data to be sent and received back and forth. An accident may occur if a hacker tries to steal data

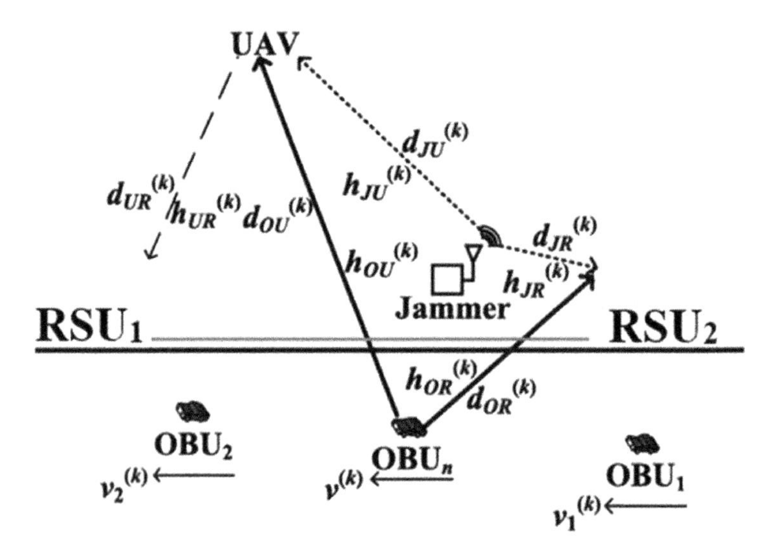

Figure 5.1 UAV relay in VANETs against smart jamming with reinforcement learning [9]

via a car-to-car connection and the transmission is delayed [10]. The following is an example of an algorithm of this type: To classify approaches to VANET intrusion detection and detect different types of attacks, these studies use a variety of machine learning algorithms, including K-nearest neighbor (KNN) and support vector machine (SVM). Data is classified and predicted using SVM and KNN, two machine learning algorithms. This study uncovered denial of service (DoS) and fuzzy assaults as potential threats. The company created a multilayer perceptron neural network as a distributed IDS to detect intruders or attackers on an Internet of Vehicles (IoV) network. Confusion matrices are built by combining forecasts, classification data, and other reports [11]. Mobile ad hoc networks can be compromised by DoS, U2R, R2L, and probe attacks. This research examines several attacks in greater detail. IDSs can identify rogue nodes in automobile ad hoc networks and the incorrect information they send to the network administrator by using an anomaly-based detection technique. RSU was utilized to evaluate the system because its communication range encompassed the whole test zone. RSUs placed within a communication range are utilized by each node to gather meta-information for use in global parameter flow calculations. As a result, no one will know what happened to the car or where it went. Rogue nodes can be located with the use of an intrusion detection system, also referred to as an IDS [12]. DNNs and probabilistic belief networks have been used to build an IDS. Detection and documentation of network-based hazards are the goals of these studies.

Rajesh *et al.* (2021): Anti-jamming methods that depend on frequency hopping are unsuitable for use in mobile, on-demand networks, where OBUs are required to connect. This research makes use of UAVs to boost VANET communications' resistance against sophisticated jamming (UAVs). By keeping an eye on how the OBU and UAV communications are working, these UAVs can adjust their communication methods accordingly. UAV delivers an OBU message to a different RSU when the RSU's radio transmission is considerably obstructed or interfered with (RSU). The anti-jamming UAV relay game simulates the battle between UAVs and jamming devices with intelligence. Before deciding on the jamming intensity, the jammer evaluates the UAV strategy according to the relay game, which decides whether or not the UAV will send an OBU message between RSUs. To minimize OBU message BERs while also optimizing the VANET's advantages, a relay strategy was designed. The scenario is being analyzed with Q-learning [13].

Wang *et al.* (2021): One of the many possible results of a jammer attack on a network of UAVs is the loss of communication. For two reasons: The source UAV cannot simply increase its transmit power due to an energy constraint, and the destination UAV's mobility makes it difficult to design a network architecture that will operate for both devices. This study proposes a distributed RL framework based on energy efficiency for improving communication quality while simultaneously reducing overall network energy consumption in the presence of a greedy jammer for UAV networks under jamming attacks, as demonstrated by experiments. The relay UAVs can autonomously choose their broadcast strength based on past state-related information, rather than relying on the jammer's shifting target or the trajectories of other UAVs, as is the case with earlier methods. It is possible to conceal the precise location and current battery level of an UAV. We offer an anti-jamming relay based on deep RL for UAVs that makes use of portable computer equipment such as the Raspberry Pi. We investigate the NE and performance bonds for the power control game that has been specified [14].

Barbora Buhnova *et al.* (2021): New research issues for the building of smart, sustainable cities have emerged as a result of the rapid development of intelligent transportation systems (ITSs). To improve traffic management and road safety in urban areas, a new transportation technology called VANET allows cars to connect and share data more efficiently. The ITS must find new ways to incorporate new network technologies to provide more dependable and efficient services for passengers and drivers alike. UAVs are an example of how VANET and UAVs are linked. As part of the VANET–UAV partnership, UAVs aid vehicles in expanding network coverage while efficiently avoiding obstacles (e.g., buildings) and providing high data delivery rates (UAVs). Despite their growing popularity, drones and the VANET continue to raise major security issues that need to be addressed immediately. It is now possible to stop cyberattacks on vehicle data using advanced machine learning techniques on VANET and UAVs, such as DL. This chapter discusses recent advances in machine learning, including work on virtualized networks and network IDSs utilizing UAVs. Some of the most significant research challenges mentioned in the literature are also explored, as well as viable ITS security solutions [15].

Ouahouah *et al.* (2021): UAVs have recently piqued the interest of academics and industry leaders alike due to the vast array of exciting new applications that have emerged. Because limits on UAV operations are constantly changing, the majority of UAV applications rely on visual line of sight (VLOS). There is widespread agreement among the commercial UAV sector that commercial UAV operations should be expanded to encompass urban and inhabited areas, rather than merely VLOS only. The usage of BVLOS UAVs is still severely controlled. Problems will certainly arise when employing UAVs to detect and avoid moving and stationary objects in the environment. To identify and avoid a physical collision, an intelligent component must be installed on the UAV or at the multiaccess edge computing (MEC) to receive the data acquired from multiple UAV sensors, process it, and then make the appropriate option. To capture the sensing data, a variety of sensors, such as Lidar, depth cameras, video, and ultrasonic, should be utilized in conjunction. Using probabilistic and DRL, this research provides collision-avoidance algorithms that are both efficient and effective. Depending on the capacity and workload of the UAV, the suggested algorithms can either be executed on the UAV or at the MEC, respectively. It makes no difference whether or not the user has prior knowledge; our algorithms have been devised and developed to be applicable in any situation. The use of UAVs necessitates the testing of solutions in a hostile environment because they fly at random over a limited region with no association. When these strategies were evaluated in both familiar and unfamiliar situations, the results showed that they were effective at avoiding collisions while using less energy [16].

Li *et al.* (2021): UAV networks are especially sensitive because sophisticated jammers can choose their jamming strategy based on the current channel situation. RL algorithms have difficulty converging quickly because of the state space's enormous dimensionality, even though they have the potential to give UAV networks intelligence. This study can reduce the state space that an agent must traverse because of the usage of domain information, and it accelerates the algorithm's convergence as a result. Ultra-high-altitude UAVs can efficiently explore state space at extreme altitudes due to the aircraft's inertial and free-space signal attenuation laws. As a result, a pretraining virtual environment has been developed that contains receiver performance indicators as well as subjective assessments of the activity's worth. The algorithm is more realistic than others in its class because it is based solely on data that can be viewed in real time [17].

Xiao *et al.* (2021): UAVs are particularly sensitive to jamming attacks when utilized for video capture, processing, and transmission. This is due to the changing topology and limited energy of UAVs. UAVs have a new anti-jamming video transmission approach developed by these researchers, which uses RL to select video compression quantization settings and channel coding rates, modulation, and power control tactics to guard against jamming attacks. A video compression and transmission policy for UAVs is chosen based on observed video task priority, UAV controller channel state, and jammer interference power received by the vehicle in this method. When this technology is used, the UAV can produce greater video quality while using less power. As a safe RL-based solution to lessen the danger of video transmission failure, DL is also proposed to speed up the UAV learning process and reduce the likelihood of video transmission failure. The processing complexity of the UAV, as well as its optimal utility in various conditions, may be determined through simulations. By reducing transmission latency and UAV energy consumption, innovative solutions not only improve the video quality but also save money [18].

Jin Shang *et al.* (2021): An architecture for self-organizing UAV networks is being investigated in this study that takes topology construction and power adjustment into consideration simultaneously. With the least amount of information sharing, the backhaul rate can be increased while avoiding negative power competition by each scattered UAV. The core of our solution is DRL. This is our backhaul system: transmission target selection (TS), a power control (PC), and multichannel power allocation (PA). For each of the three submodules, a different set of DRL algorithms and incentives are used to solve the problem. Deep-Q learning is a fantastic choice for TS since it uses fewer relays and can ensure the backhaul rate. A deep deterministic policy gradient is utilized to solve PC and PA to identify the optimal solution to meet traffic demand with necessary fine-grained transmission power. With less malevolent power rivalry, the backhaul rate will go up significantly, which is significant. Simulations show that the proposed framework outperforms DQL and the max–min strategy in terms of overall system effectiveness and performance gains [19].

Sicong Liu *et al.* (2021): UAVs have received a great deal of attention from researchers in recent years, owing to their rapid development in communications, networking, and sensing applications. When UAVs are employed for military surveillance and environmental monitoring, their communication processes are not as secure as they should be because of jamming attacks on the aircraft. A jammer can prevent the controller from getting sensing data by delivering jamming signals during the drones' communication. A jammer can also drain the drone's battery or prevent the drone from following the authorized sensing mission waypoint [20].

Pankaj Verma *et al.* (2021): WPLS, or wireless physical layer security, is the cutting-edge technology in mobile network security today, and it will continue to be so in the future. Several factors contribute to the variety and resilience of the WPLS paradigm, including antenna selection (AS), physical layer authentication (PLA), and relay node selection. Because of heterogeneity, ultra-density, and high mobility, wireless network security is becoming more challenging to maintain. As the complexity of wireless networks continues to grow, machine learning has lately emerged as a promising solution for reducing that complexity. That is why the focus of this study's PLA, AS, and relay node selection is on intelligent WPLS introduction rather than on other aspects of the network. Begin by becoming familiar with WPLS and ML, and then discover how they are related to one another. Improve WPLS by going back and reviewing how to relay node selection, authentication, and the application of machine-learning all operate together in concert. Several significant challenges confronting intelligent WPLS were also addressed, including the Internet of Things (IoT), device-to-device communication, cognitive radio, and UAVs with non-orthogonal multiple access. It also looked at the various applications of the technology in wireless networks. The closing section of the appendix delves into the specific machine learning algorithms used by WPLS researchers. This essay is intended to inform and educate readers on the current status of WPLS and intelligent WPLS technology, which is our primary purpose in writing it [21].

Wooyeol Choi *et al.* (2021): A flying ad hoc network (FANET) is comprised of a group of UAVs that collaborate to perform missions. FANET is a subset of the many different forms of wireless ad hoc networks, and it is an essential one. Today, FANETs are being used to accomplish a wide range of commercial and civilian goals, including traffic congestion management, remote data collection, weather forecasting, and product delivery. FANETs must overcome several significant challenges, such as adaptive routing protocols, flight trajectory selection, energy limits, and self-deployment before they can be fully realized. FANETs will be required to address these concerns. For the past few years, several scholars have been focusing on finding solutions to these issues. Because of the great mobility and unpredictability of the FANET network, changes in its topology are a considerable challenge to implement. To overcome these disadvantages, researchers have implemented RL algorithms in FANETs. Examples of RL applications in FANETs include the routing protocol, trajectory selection, relaying, and charging, all of which were examined in detail and compared and contrasted in this article. In addition, we highlight potential study directions as well as unresolved research challenges [22].

5.3 Deep learning and its importance over VANET

When using deep learning algorithms, there are numerous “layers” of neural network algorithms to consider, and each layer transfers an optimized version of the data from the previous layer to the next layer [23]. When dealing with datasets of a few hundred characteristics or columns, the majority of machine learning approaches work admirably well. Using this strategy becomes time-consuming or even impossible due to the enormous number of features in an unstructured dataset such as one generated from a photograph. DL algorithms learn more and more about a picture as it passes through the many layers of a neural network, which is a process known as reinforcement learning. By merging the low-level qualities of the early layers with the higher-level properties of the later layers, a more holistic image is generated [24]. Some objects, such as a dog or a tree, may be recognized using a middle-layer method, while others, such as a dog or a tree, may be detected using an advanced-layer method.

5.3.1 Why is DL important?

DL is particularly useful when working with unstructured data because it is capable of processing a hugc number of features at the same time. DL techniques, on the other hand, maybe overkill for simpler tasks because they require a great amount of data to be useful [25]. DL models for image recognition can be trained with access to more than 14 million photos, as demonstrated by the ImageNet dataset, as an example. A DL model can get overfitted if the input data is too simple or incomplete. As a result, the model may fail to generalize well to new data. Using alternatives (such as decision trees or linear models) that have larger datasets and more features

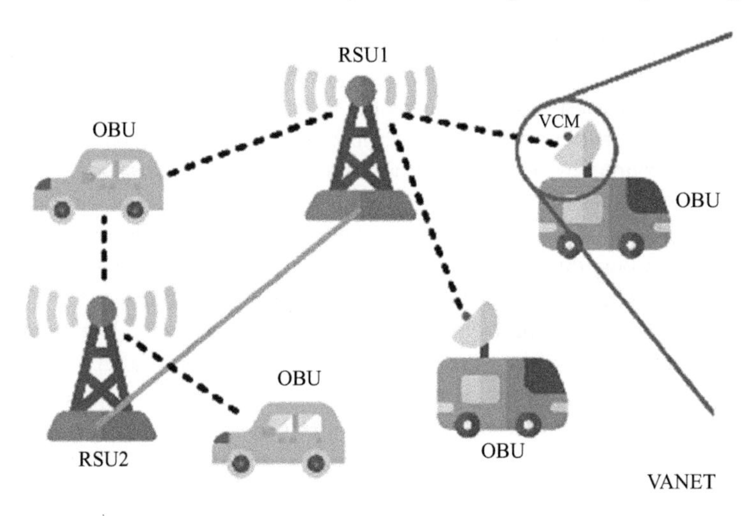

Figure 5.2 DL-based intrusion detection method in VANET [27]

will be necessary if you want to figure out why consumers depart or discover fraudulent transactions. DL can be applied to a variety of tasks, including multiclass classification, even when the dataset is small and well-structured [26]. Figure 5.2 shows a DL-based intrusion detection method in VANET.

VANETs are classified as centreless, self-organized, multihop, and dynamic, to name a few characteristics. VANETs make it possible to communicate from vehicle to vehicle (V2V), vehicle to roadside (V2R), and vehicle to human (V2H) through the use of satellite communications [vehicle-to-thing (V2X)]. When used in conjunction with vehicle-mounted electronics and apps, ITS can minimize traffic congestion and accident risk [28]. Automobiles employ peripheral devices or multihop propagation to communicate with other cars, roadside equipment, or other people when traveling on the road. Increased vehicle mobility improves the adaptability and expansibility of the network, but at the expense of topology maintenance and maintenance. The one-hop transmission range can be exceeded by a wide range of vehicles, resulting in the loss of VANET routing links in the process. Whenever there is a large number of cars on the road, a broadcast storm occurs because the limited spectrum resources are being taxed to their maximum capacity [29]. As the quantity and density of IoT devices have increased, the efficiency of V2X communication has suffered, making it increasingly difficult to communicate between linked devices. Unauthorized users are prevented from accessing a frequency band reserved by standard spectrum allocation methods when the “fixed allocation mode” is enabled.

5.4 Deep learning models – types and characteristics

During their broadcasting of messages to other vehicles, pedestrians, and roadside equipment, automobiles must contend with channel conflict in a specific region. In this case, the problem is divided into two sections, which we will discuss separately [30]. First and foremost, there is a plan in place for moving data from one location to another. Data is broadcast to all cars using the protocol, allowing them to all benefit from the same information once it has been implemented. To distribute information regarding road conditions and accident warnings, the sender must use our G-hop protocol to locate paths that are relatively stable while keeping the number of relays as low as possible to avoid a "broadcast storm," as described above. It cannot be overstated how important this is. Second, due to the enormous number of vehicles (broadcasters) participating in the forwarding strategy described above, there is still the possibility of collisions occurring. Consequently, the issue of a high number of vehicles transmitting at the same time has been handled in this way.

Modern learning techniques, which make use of the vehicle's daytime running lights, can help drivers make better channel selections, lowering the danger of being involved in a collision with another vehicle [31]. It is possible to apply the G-hop and Learning algorithms in real-world systems by utilizing mobile edge computing (MEC) and software-defined radios. The G-hop broadcast protocol employs mobility groups to route cars to the position of the relay, which is where the protocol gets its name. We can avoid channel contention by transmitting messages from multiple autos at the same time using the most appropriate Digital Signature Algorithm (DSA) mechanism, Learning techniques, as described above. G-hop is used by a vehicle to locate the next relay vehicle before following the required instructions to select channel assignments. MEC servers installed at RSUs can collect data, maintain topology, and perform offline learning calculations for required learning algorithms, all of which are beneficial to the organization [32]. As long as a single Multi-access Edge Computing (MEC) server is capable of serving all of the vehicles on a given stretch of road, a large number of MEC servers can be deployed to gather data and update network parameters simultaneously. Figure 5.3 shows the fusion of medical imaging and electronic health records using DL.

5.4.1 G-hop: a group-based multihop broadcast protocol

According to research, cars tend to travel in groups with similar mobility characteristics, such as speed, and messages are more consistently conveyed within the same

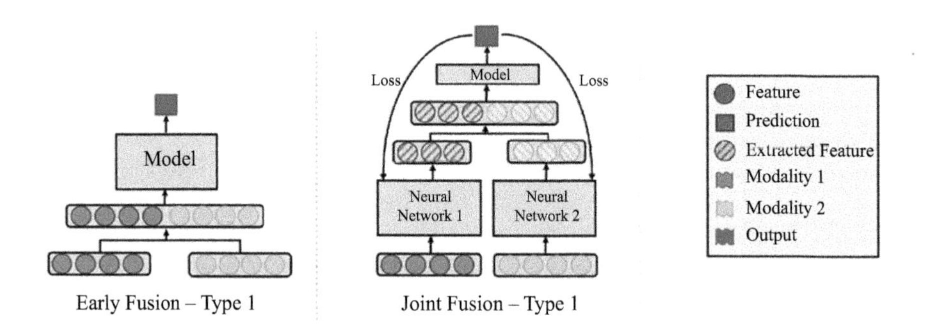

Figure 5.3 Fusion of medical imaging and electronic health records using DL [33]

group when they travel in groups of similar mobility characteristics. The speed and behavior of the vehicle are quite dynamic. The DFS algorithm, which divides the cars into groups based on mobility characteristics, is used to generate a broadcast protocol based on the groups formed by the Depth-First Search (DFS) algorithm [34]. Vehicles will be grouped based on features that they have in common. The initial step is to select a forwarding vehicle within the same group, followed by selecting a forwarding vehicle across groups. The vehicle is given precedence as a result of this. The high speeds cause cars traveling in the opposite direction to have difficulty making connections with one another. As a starting point for our mobility model, let's define the terms listed below as follows. The car itself serves as a node of origin for the data. It is a vehicle or piece of equipment that is used to transport information. A relay node is a vehicle that is used to transmit and receive communications (not the source nor the target node). People are drawn to nodes such as autos, which encourages them to form clusters around them. To facilitate communication between groups, it is necessary to utilize a centralized mechanism [35]. In a network, nearby nodes are vehicles that are near the central node. When a vehicle receives and sends a message, the waiting time is indicated by the letter T [36].

5.5 Identifying smart jamming

It is preferred for long-term evolution (LTE) networks and future releases due to its strong noise and fading resistance. Prior cellular networking technologies lacked the advantages of LTE, which can be attributed in large part to the orthogonal frequency-division multiple access (OFDMA) modulation scheme [33]. Improvements in data speeds and coverage as well as energy efficiency and latency are just a few of the advantages. However, despite the advantages of OFDMA, LTE networks remain vulnerable to attacks from active radio nodes. In most cases, an attacker will first transmit a signal intended to confuse or disrupt the victim before attempting to obtain personal information from them (a jamming node). Layer-1 DoS attacks, for example, are frequently referred to as DoS attacks on communication networks of this nature. It has been a long time since wireless communication networks have not been protected from jamming assaults [37]. As communication technology grows more prevalent in our daily lives, it is projected that the reliability of information flow would improve, enhancing the effectiveness of jamming assaults in the process. There are several different sorts of jamming nodes that can be used to launch these attacks. Identifying and understanding the operation of a jamming node is crucial to lowering the intensity of the interference and its consequences. Figure 5.4 identification of smart jammers.

Identification of jammers and detection of jammers are two distinct but related techniques. To identify jammers, network metrics or signal characteristics are typically used, as will be discussed in further depth in the following section. A variety of statistical approaches may be used by identification algorithms, depending on the sort of assault being detected and investigated [39]. Aside from algorithms for identification and detection, machine learning technologies have a positive impact on them. Deep convolutional neural networks have been the subject

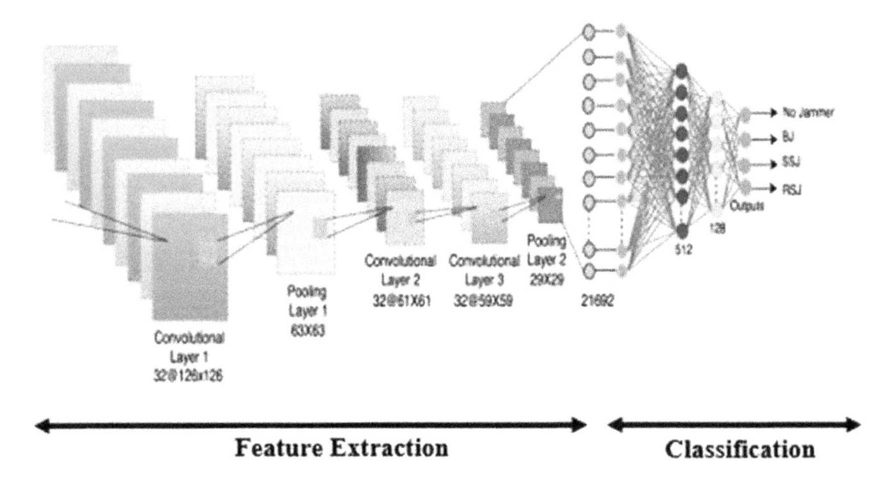

Figure 5.4 Identification of smart jammers [38]

of extensive research because of the demands of real-time communication, which necessitates speedy responses. Only a few of the numerous usual uses include location, modulation classification, channel decoding, and waveform identification [40]. The use of time–frequency transformation (TFT) preprocessing to capture jamming effects on the received signal is a common step in all of these detection and classification techniques. The spectrogram TFT approach is one of the most often used methods in signal analysis, and it is also one of the most complexes. A spectrogram created by applying a short-time Fourier transform (STFT) to the signal represents the signal on the time–frequency plane. In addition to having a predetermined window size, the spectral representation of the signal has a fixed time–frequency resolution, which is reflected in the spectral representation. Due to the expansion of communication networks in recent years, new types of jammers have emerged, some of which are capable of launching assaults over a very narrow frequency band or for extremely short periods [41]. Because jamming nodes are buried in the time–frequency plane, traditional TFT approaches such as STFT are no longer effective at detecting them. Attackers who use smart jammers have already been quantified in terms of their influence, which has been done through numerical and measurement research. Smart jammers can monitor transmissions and determine the most vulnerable sections of transmitted packets, with the possibility to target reference signals, thanks to lower hardware costs and more readily available software. Smart jammers get an advantage over legal nodes by making use of these observations. Many anti-jamming defenses mechanisms have been presented, such as game repeating algorithms and frequency hopping algorithms, to name a few notable examples [42]. If these network strategy algorithms are to be effective in combating an attack, it is critical to be able to identify the type of jammer that is being utilized. Therefore, before deploying jamming prevention/anti-jamming systems, it is necessary to determine who is jamming the signal.

We develop a wavelet transformation-based preprocessing method to improve the identification and classification of jamming devices. It is common to practice employing wavelet transformations in a wide range of signal processing methods, particularly in the audio and video domains. Wavelet transforms, like the Fourier transform, are used to describe signals by describing them with a linear set of basic functions. By employing steady oscillation, the essential functions of the Fourier transform are made sinusoidal rather than time–frequency localized, resulting in the Fourier transform [43]. The usage of wavelets, on the other hand, allows for the detection and identification of jammers because they are time-limited and located on the time–frequency plane. By applying a rectangular window to the signal before performing the Fourier transformation, the STFT may detect temporal variations in the signal. All of the signal's time–frequency resolution will be identical since STFT uses the same rectangular window for all of its locations. Because wavelet transformations and STFTs have different window sizes, it is possible to achieve multiresolution on different signal positions using wavelet transforms. Because intelligent jammers change the characteristics of the signal, a resolution mechanism that can adapt to signal fluctuations is required. It is possible to jam the LTE downlink channel and collect downlink channel observations by continually processing the received baseband signal with a stationary monitoring node (MN). An attack on the base station's (BS) signal is carried out by a mobile jammer while the MN and cell are synchronized. MN is utterly oblivious to the coming jammer attack on his network [44]. In addition to interference with synchronization signals, LTE downstream transmission is vulnerable to two fundamental faults caused by clever jammers: interference with signals required to determine the jammer's location and interference with signals required to synchronize with the jammer.

5.6 Solving jamming problem with RL/DL

With the help of a variety of various ways, it is feasible to recognize jamming attacks and countermeasures in real time. Regardless of how effective the procedures and results of some approaches are, there are flaws in other approaches as well. The potential concerns associated with each of them will be examined in further detail later on. It is hard to determine what type of attack a node has been subjected to while using the JAM mapping protocol [45]. Additionally, the approach does not appear to be capable of appropriately detecting reactive jamming. Because of the message mapping (JAMMED/UNJAMMED), this adds to the network's stress by increasing traffic. According to the Ant system, if jamming is detected in an area before the agents have formed a tabu list, this technique will fail to work as expected and will result in failure. In addition, the use of Ant agents to generate the tabu list increases the amount of time it takes. Aside from that, it utilizes an abnormally large amount of RAM. Network parameters are used to compute the performance of the network to determine whether a certain node or piece of the network is jammed. Besides appearing to be inaccurate when it comes to measuring network parameters, the technologies utilized in jamming detection tend to be prone to flaws [46]. Before implementing this method, it is necessary to duplicate all data across all replicated BSs. Allowing for the possibility that all of the duplicated BS's data has been updated, after some time if a previously operational BS becomes unresponsive before the upgrade, data will be lost.

There are various possible explanations for this. Evasive approaches have a high cost in terms of travel time and reconfiguration of networks. When employing multipath routing, there are several paths to take to go to your destination from your starting point, but none of them are blocked by anything else. This method of channel surfing is inefficient since it demands the synchronization of two nodes, which is not always possible [47]. Because of these variables, overhead must be taken into consideration while transferring or reconfiguring the network. The spatial retreat requires the two nodes to remain in synchronization with one another to prevent jamming of a specific communication channel between them by shifting them to a different, safer place. To ensure consistency, this solution analyzes RSS and location information provided from various nodes as part of the consistency check. They will be unable to communicate with one another if any of these nodes are subjected to jamming attacks. The fuzzy interference technique is used for information warfare jamming detection because it is well-suited for jamming detection in general. Densely distributed networks outperform sparsely distributed networks when using the "2 means clustering of neighborhood nodes" technique, according to the results. As a result, networks with fewer near neighbors should avoid employing it altogether [48]. Depending on the network density, channel hopping can be used sparingly or extensively, depending on the situation. The hopping approach is straightforward to put into action. Because carrier sense time is used as the metric, reactive jammers will not be detected in the network and will not be identified. Figure 5.5 describes the anti-jamming problem solution: using a DRL approach.

5.6.1 Open research challenges

This conclusion is based on the fact that no universal anti-jamming approach has been created to counteract all types of jammers to date, leading us to this point. Even while jamming is less difficult to implement, developing a means for detecting and countering it is more complex to do. Anti-jamming is becoming a

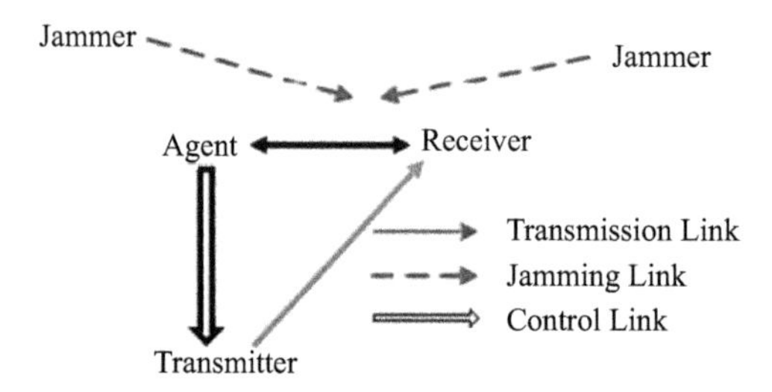

Figure 5.5 Anti-jamming problem solution: using RL approach [49]

more difficult challenge as new wireless network technologies are developed (e.g., vehicle network, WiMax). This section contains information on several significant research topics, including energy-efficient jamming detection, jammer classification-based detection, and anti-jamming in IEEE 802.11n and wireless mobile networks, among others [50].

5.6.1.1 Energy-efficient jamming

When examining simple jamming detection and countermeasure systems, it becomes evident that there is a lack of a well-drafted reactive approach that has been well considered. Incorrect radio connections or interference can both cause packet loss, but an effective detection method should be able to distinguish between the two. Low-power jamming techniques, such as reactive jammers, are widely used nowadays. A low-power jamming detection approach, on the other hand, does not exist.

5.6.1.2 Detection based on jammers

When attempting to identify jammers, there are several different sorts of structured jamming attacks that can be used in a variety of different methods to accomplish this. According to our research, the acts of a jammer can be recognized and classified into several categories. Top-down approaches can be used to derive the features of jammers, as can be seen in the detection methods described above. Decide whether the jammer is basic or complex before proceeding on. The second level categorizes the jammer based on how it operates: proactive, reactive, function-specific, or smart-hybrid, depending on how it operates. The implementation of a top-down plan appears to be less difficult than the implementation of a bottom-up strategy [51].

5.6.2 Anti-jamming in wireless mobile networks

Anti-jamming methods have been developed and tested on static networks for the majority of jamming detection systems. Because mobile network jammers can travel and interfere with algorithms that are used to detect and locate them, combatting jamming has become more difficult to achieve. When it comes to movable nodes, spatial retreats appear to be the only approach being utilized. The effectiveness of wireless mobile networks continues to be a challenge, particularly when overhead costs are taken into consideration [52].

5.7 Challenges toward DL over VANET

A variety of RL and DRL algorithms have been implemented to deal with the dynamic and sophisticated V2X paradigm problems. RL-based V2X research has a bright future, thanks to advancements in the new 5G paradigm in processing, storage, and RL/DRL. In this case, there are several challenges to overcome [53]. When making travel arrangements, keep the following points in mind: The DRL framework requires training and evaluation. As a result, the applicability of the

DRL paradigm in real-world systems has been questioned. A stochastic model is commonly used to create the simulated dataset since it simplifies the real system while omitting nuanced trends. The DRL framework's training and performance evaluation require the development of a more efficient method for creating simulation data. Both the level of mathematical complication and the level of synchronization required are quite difficult. The increasing amount of traffic flow state/control actions that must be taken into account make managing traffic signal timing one of the most difficult challenges for RL to conquer. RSU GPUs would have to do increasingly more intricate computations and operations to assist machine learning algorithms if they are to provide real RL-based congestion management solutions in VANET [54]. Edge computing facilities are required to operate RL-driven V2X applications. Multiagent RL/DRL is utilized for dynamic het nets. A vehicle's network is constructed from nested IoT devices and networks that are continually evolving to meet service requirements and networking conditions that are always changing. The RL/DRL agents for each entity must be compact and versatile to respond fast to changes in the network environment. Because of this, learning spaces for states and actions may be constrained, lowering the effectiveness of the convergent policy. The network environment gets even more complex as a result of the expanded state space created by multiple agent interactions, resulting in longer training times for learning algorithms. VANET-based DL faces numerous problems, as observed in Figure 5.6.

Fairness and optimization are diametrically opposing concepts in the TLC world. Agent fairness policies are critical in the real world. For example, all vehicles passing through a junction would have the same priority under a fair TLC system. In trying to increase traffic while yet being fair, you'll hit a brick wall. Two examples of optimization measures are a reduction in latency and an increase in throughput [56]. Fairness and optimization will remain a challenge in the future, although artificial intelligence technologies like reward functions and other similar tools can help.

5.8 Open issues and future trends with RL/DL

The RL/DRL architecture has an impact on a wide range of vehicle applications [57]. According to the research team, "current RL/DRL-driven vehicular networks demonstrate that they are merely scratching the surface" of what they are capable of, given the limits of current RL/DRL techniques and the constantly changing requirements of vehicle networks themselves. Figure 5.7 depicts the current state of RL and DL as well as possible future directions.

5.8.1 Autonomous and semi-autonomous vehicles

Self-driving cars and semi-autonomous vehicles demand a new level of dependability and reliability. In light of the current state of 5G development, there is little chance that it can satisfy your requirements anytime soon. 6G wireless networks, made feasible by breakthroughs in hardware, software, and new communication

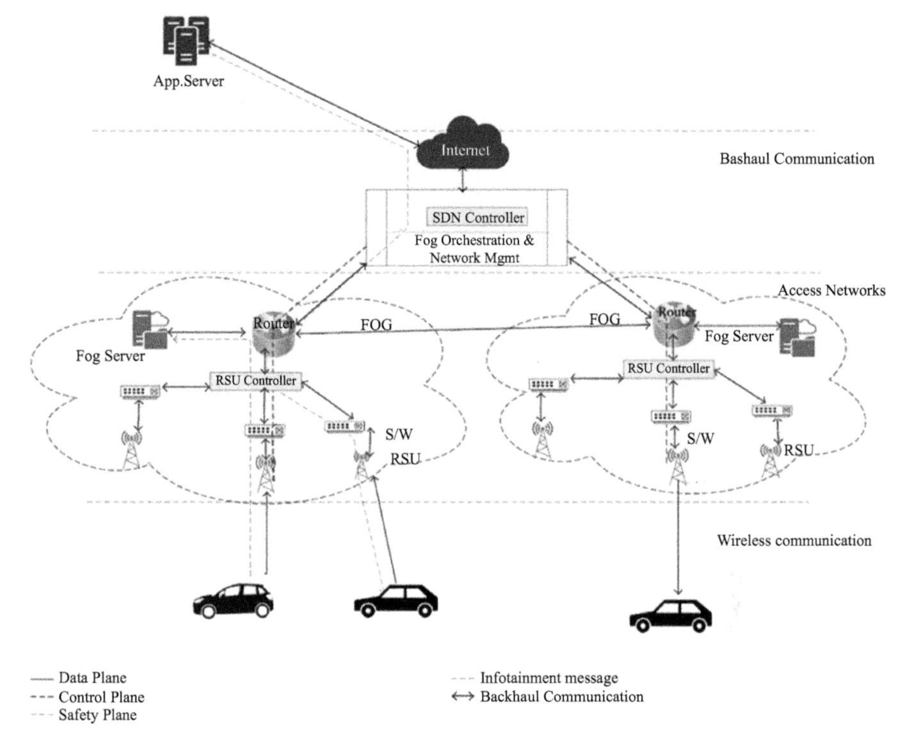

Figure 5.6 Challenges toward DL over VANET [55]

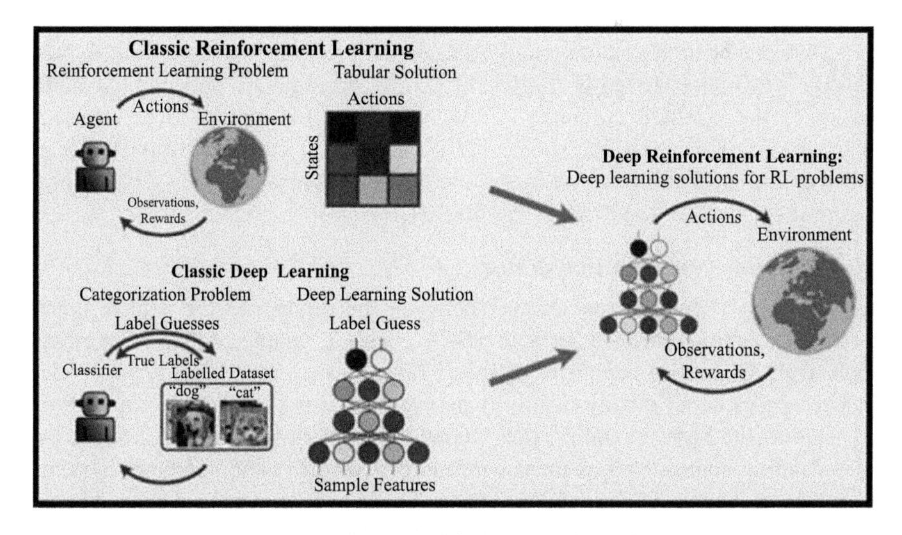

Figure 5.7 Open issues and future trends with RL/DL [58]

possibilities, will be required to get connected vehicles up and running. Numerous research initiatives are devoted to autonomous vehicles, which are continually being developed [59]. Accidents caused by careless driving must be prevented as well as emissions reduced. It has been recently discovered that RL and DRL may be utilized for teaching novice drivers how to perform autonomous driving tasks as lane holding, lane changing, merging into ramps, overtaking, and motion planning using the technology. The DRL systems, DQN and DDAC, keep the vehicle in the lane and maximize the average velocity for discrete activities (DQN) and continuous actions (DDAC).

The Department of Defense Acquisition Council is referred to as DDAC. Because of Q-Learning's help, we learned how to change lanes by just accelerating and decelerating. While working with DQL, researchers honed the most efficient driving technique for dynamic ramp merging. This approach increased the long-term rewards the most [60]. One of Q-multipurpose learning's RL policies uses decision-making that is dependent on the opposing vehicle's willingness to cooperate with the agent for overtaking reasons. Autonomous driving's biggest challenge in the future will be to improve MARL approaches, which have received minimal research funding to now. It has the potential to be extremely advantageous in the context of high-level decision-making and coordination among autonomous vehicles. This algorithm has been employed in artificial vision system investigations in the past. On the other hand, self-driving car communication is separate from the vehicle's motion planning and control systems. Improving telecommunications while also controlling autonomous vehicles is indeed a difficult task. Radio links between self-driving cars can be broken if they are flown incorrectly or separated by a large distance. If not concurrently, RL/DRL should be able to deal with each of these problems [61]. This appears to be a worthwhile topic for future investigation on our part. When autonomous mode fails or when human help is required, semi-autonomous mode (also known as teleoperated driving) is utilized. They can be used as a temporary solution until fully autonomous vehicles are available. With teleoperated driving, vehicles can be remotely operated by radio. This gives the onboard computer of the vehicle direct control over steering, acceleration, and braking operations. Ultra-low latency communication between the driver and the vehicle is essential if the driver is confronted with a potentially dangerous scenario and must react rapidly [62].

5.8.2 Brain–vehicle interfacing

There is no need for physical contact between the driver and the vehicle when operating a brain-controlled vehicle (BCV) with a brain–computer interface. People with disabilities benefit from BCVs because they give them greater independence and a better quality of life by giving them a new method to drive autos [63]. Due to the high reliability, low latency, and massive data rates needed for brain–machine connections, present wireless connectivity, and computer systems are unable to handle BCV. 6G-V2X must be able to recognize human driving behavior if it is to alter it. Previous successful initiatives have demonstrated that BCV is viable. There is no scalable solution for brain–machine interactions that demand great coverage, availability, speed, and low latency for the benefit of users because of the existing BCV implementation's requirement for a wireless connection. The RL and DRL frameworks could be used in future research to better allocate resources to these kinds of applications shortly. But this research did not investigate how AI in BCV processes EEG headset signals obtained via ANN into movement directions (forward, backward, left, or right) (i.e., how they are converted from one to the other). To date, the vast majority of research on BCV has relied primarily on simulation to corroborate its conclusions [64]. As a result, real-world testing is required to assess whether BCV is indeed useful in the real world.

5.8.3 Green vehicular networks

Edge computing and artificial intelligence will be used by 6G automobile network nodes, which will require a substantial amount of energy. Throughput, bandwidth, and the sheer quantity of ubiquitous wireless nodes will make them a significant energy drain on the global grid. The development of energy-saving, emissions-reducing, and pollution-cleaning green vehicle communication systems may place considerable restrictions on present networks [65]. This is a completely new area of study. Faster transmission rates and lower power consumption are the two biggest obstacles to green communications in automobile networks. Better user experience and energy savings at the BS can only be achieved by modernizing the current communication technique. Few pieces have used RL/DRL before, thus there is not much to draw inspiration from in the way of past work. Because of this, more research may be done to determine the role of the green vehicle routing problem (GVRP) in a changing environment, giving the current study more legitimacy. The GVRP of new electric automobiles necessitates the construction of charging facilities [66]. New energy vehicle routing may now be studied in conjunction with charging station locations for the first time, thanks to this discovery.

5.8.4 Integrating unmanned aerial and surface vehicles

With the introduction of 6G, we aspire to create a global, all-encompassing network that is capable of operating in a variety of environments. In this approach, it will be feasible to establish a globally distributed mobile broadband communication infrastructure that is always on and always available. Unlike manned aircraft, UAVs do not have any personnel in the control room [67]. They are becoming increasingly popular as a result of technological advancements and a reduction in their costs. They are being employed by corporations such as Amazon and DHL to deliver packages via the airborne delivery system. There are a plethora of potential uses for UAVs, including providing traffic signals from the air and shooting images from above. An unmanned surface vehicle (USV) is a boat that can operate independently on the water's surface without the assistance of a human operator.

Scientists have recently developed a strong interest in remotely piloted vehicles because of the multiple advantages they provide in various fields, such as environmental monitoring, natural resources discovery and exploration, waterborne transit efficiency and safety, and so on [68]. When it comes to path planning and collision avoidance, for example, the quick mobility of UAVs is advantageous, as researchers observed when they investigated path planning while utilizing RL/DRL technology. When designing a path, approaches such as DRL path planning and USV collision avoidance with DRL are essential. This is especially true for USV guidance and navigation systems. With the emergence of UAVs, having a low-latency and ultra-reliable network will become increasingly more vital, increasing the necessity of traffic control and resource sharing management systems.

5.8.5 Security enhancement with blockchain

The growth of the IoV has been substantially hindered by the necessity for vehicle identity and privacy protection in 6G networks. Particularly, if the authority in charge of providing identifying information has a good reputation, this is true. Blockchain technology may be able to help with this problem. Among the many ways blockchain can be used in the IoT is to enhance network security and enable self-driving communications [69]. Because of its ability to provide security, intelligent architecture, and flexible resource sharing, the AI/blockchain combination is immensely popular in the industrial IoT and IoT arena. Data and multimedia material could be sold near moving cars with DRL and permissioned blockchain connectivity, which changed the game. While the permissioned blockchain is managed by the nodes, the data is stored on-board each vehicle. DRL improved the performance of the blockchain-enabled IoT to boost transaction throughput while protecting the underlying blockchain system's security and ultimate decentralization. Several research projects have focused on the topic of blockchain application security. To enhance IoT security and trust management, several privacy and security concerns must be addressed first. Key sharing and certification storage can be done without the use of standard client–server protocols using a decentralized authentication system built on a consortium blockchain and numerous trusted authorities. Vehicle privacy can be protected via pseudo-random IDs; nevertheless, synchronizing many roadside control units is still a problem [70].

5.9 Conclusion

Autonomous and semi-autonomous vehicle security is now mainly reliant on clever intrusion detection technologies, which are becoming increasingly common. Several sorts of attacks on vehicles, networks, and devices pose a threat to the development and usage of self-driving cars. When an attacker has access to or modifies system data, the likelihood of a catastrophic event increases significantly. This is especially true in the context of the VANET network. Through the use of intrusion detection, accidents can be avoided and the general public kept safe. Because it analyzes and categorizes VANET signals, the IDS is a successful solution for VANET security problems in that it successfully identifies attacks. Previously, artificial intelligence and machine learning were used to detect intrusions on the VANET, but the accuracy of the detection was low. In contrast, the use of a DL algorithm increased the accuracy and efficiency of the system.

DL algorithms have been enhanced as a result of the alterations made to them, and the new algorithms incorporate many additional aspects. Our investigation into the problem of VANET V2X channel contention is presented in this chapter. The problem may be solved in two steps, which we have devised for your convenience. A G-hop protocol based on mobility groups was the first item we proposed to route V2X messages through fewer but more efficient relay nodes, which we called "G-hop 2.0." As soon as we had selected a relay node, we had to deal with the problem of channel contention, which happens when a high number of vehicles are competing for a limited quantity of radio frequency spectrum.

References

[1] Greco, C, Pace, P, Basagni, S, and Fortino, G. "Jamming detection at the edge of drone networks using multi-layer perceptrons and decision trees." *Applied Soft Computing* 111 (2021): 107806.

[2] Chen, W, Qiu, X, Cai, T, Dai, H.-N., Zheng, Z., and Zhang, Y. "Deep reinforcement learning for Internet of Things: A comprehensive survey." *IEEE Communications Surveys & Tutorials* 23(3) (2021): 1659–1692.

[3] Jiang, B, and Givigi, SN. "A MARL approach for finding optimal positions for VANET aerial base-stations on a sparse highway." *IEEE Access* (2021).

[4] Ismaeel, AR, Ouns, B, Moayad, A, and Azzedine, B, "Design guidelines for cooperative UAV-supported services and applications." *ACM Computing Surveys (CSUR)* 54.9 (2021): 1–35.

[5] Raza, A, Bukhari, SHR, Aadil, F, and Iqbal, Z. "An UAV-assisted VANET architecture for the intelligent transportation system in smart cities." *International Journal of Distributed Sensor Networks* 17.7 (2021): 15501477211031750.

[6] Feltus, C. "AI's contribution to ubiquitous systems and pervasive networks security-reinforcement learning vs recurrent networks." *Journal of Ubiquitous System and Pervasive Networks* 15.02 (2021): 1–9.

[7] Liu, X, Ma, O, Chen, W, Xia, Y, and Zhou, Y. "HDRS: A hybrid reputation system with dynamic update interval for detecting malicious vehicles in VANETs." *IEEE Transactions on Intelligent Transportation Systems* (2021). Doi: 10.1109/TITS.2021.3117289.

[8] Singh, S, Sulthana, R, Shewale, T, Chamola, V, Benslimane, A, and Sikdar, B. "Machine learning assisted security and privacy provisioning for edge computing: A survey." *IEEE Internet of Things Journal* 9(1) (2022): 236–260.

[9] Xiao, L, Lu, X, Xu, D, Tang, Y, Wang, L, and Zhuang, W. "UAV relay in VANETs against smart jamming with reinforcement learning." *IEEE Transactions on Vehicular Technology* 67.5 (2018): 4087–4097.

[10] Xia, Z, Wu, J, Wu, L, Chen, Y, Yang, J, and Yu, PS. "A comprehensive survey of the key technologies and challenges surrounding vehicular ad hoc networks." *ACM Transactions on Intelligent Systems and Technology (TIST)* 12.4 (2021): 1–30.

[11] Mishra, S, and Tyagi, A.K. The "Role of machine learning techniques in Internet of Things-based cloud applications." In: Pal, S, De, D, Buyya, R. (eds) *Artificial Intelligence-based Internet of Things Systems. Internet of Things (Technology, Communications and Computing)*, (2022). Springer, Cham. https://doi.org/10.1007/978-3-030-87059-1_4.

[12] Ftaimi, S, and Mazri, T. "Benchmarking study of machine learning algorithms case study: VANET network." In Ben Ahmed, M, Mellouli, S, Braganca, L, Anouar Abdelhakim, B, and Bernadetta, K.A. (eds) *Emerging Trends in ICT for Sustainable Development.* Advances in Science, Technology & Innovation. Springer, Cham, 2021. pp. 171–179.

[13] Rajesh, L, Bhoopathy Bagan, K, Thayal Sankar, P, and Suchitra, V. "Performance analysis of UAV relay in VANETs against smart jamming with Q-Learning techniques." In Tavares, JMRS, Chakrabarti, S, Bhattacharya, A, and Ghatak, S. (eds) *Emerging Technologies in Data Mining and Information Security.* Lecture Notes in Networks and Systems, 164 Springer, *Singapore*, 2021. 771–783.

[14] Wang, W, Lv, Z, Lu, X, Zhang, Y, and Xiao, L. "Distributed reinforcement learning based framework for energy-efficient UAV relay against jamming." *Intelligent and Converged Networks* 2.2 (2021): 150–162.

[15] Bangui, H, and Buhnova, B. "Recent advances in machine-learning driven intrusion detection in transportation: Survey." *Procedia Computer Science* 184 (2021): 877–886.

[16] Ouahouah, S, Bagaa, M, Prados-Garzon, J, and Taleb, T. "Deep reinforcement learning based collision avoidance in UAV environment." *IEEE Internet of Things Journal* 9.6 (2022): 4015–4030.

[17] Li, Z, Lu, Y, Li, X, Wang, Z, Qiao, W, and Liu, Y. "UAV networks against multiple maneuvering smart jamming with knowledge-based reinforcement learning." *IEEE Internet of Things Journal* 8.15 (2021): 12289–12310.

[18] Xiao, L, Ding, Y, Huang, J, Liu, S, Tang, Y, and Dai, H. "UAV anti-jamming video transmissions with QoE guarantee: A reinforcement learning-based approach." *IEEE Transactions on Communications* 69.9 (2021): 5933–5947.

[19] Xu, W, Lei, H, and Shang, J. "Joint topology construction and power adjustment for UAV networks: A deep reinforcement learning based approach." *China Communications* 18.7 (2021): 265–283.

[20] Xiao, L, Jiang, D, and Liu, S. "Reinforcement learning based communication security for unmanned aerial vehicles." *Cyber Security Meets Machine Learning*. Springer, Singapore, 2021. 57–83.

[21] Kamboj, AK, Jindal, P, and Verma, P. "Machine learning-based physical layer security: techniques, open challenges, and applications." *Wireless Networks* 27 (2021): 5351–5383.

[22] Rezwan, S, and Choi, W. "A survey on applications of reinforcement learning in flying ad-hoc networks." *Electronics* 10.4 (2021): 449.

[23] Ayaz, F, Sheng, Z, Tian, D and Guan, YL. "A blockchain-based federated learning for message dissemination in vehicular networks." in *IEEE Transactions on Vehicular Technology*, 71.2 (2022): 1927–1940.

[24] Rovira-Sugranes, A, Razi, A, Afghah, F, and Chakareski, J. "A review of AI-enabled routing protocols for UAV networks: Trends, challenges, and future outlook." *Ad Hoc Networks*, 130 (2022): 102790.
[25] Varsha, R, Manoj Nair, M, Siddharth, M. Nair, Tyagi, AT. "Deep learning based blockchain solution for preserving privacy in future vehicles." *International Journal of Hybrid Intelligent System*, 16.4 (2020): 223–236.
[26] Khan, IU, Hassan, MA, Alshehri, M, *et al.* "Monitoring system-based flying IoT in public health and sports using ant-enabled energy-aware routing." *Journal of Healthcare Engineering* 2021 (2021).
[27] Boccadoro, P, Striccoli, D, and Grieco, LA. "An extensive survey on the Internet of Drones." *Ad Hoc Networks* Volume 122, 1 November 2021: 102600.
[28] Krishna, AM and Tyagi, AK. "Intrusion detection in intelligent transportation system and its applications using blockchain technology." *International Conference on Emerging Trends in Information Technology and Engineering (ic-ETITE)* (2020): 1–8.
[29] Pakrooh, R, and Bohlooli, A. "A survey on unmanned aerial vehicles-assisted Internet of Things: A service-oriented classification." *Wireless Personal Communications* (2021): 1–35.
[30] Khan, MZ, Alhazmi, OH, Javed, MA, Ghandorh, H, and Aloufi, KS. "Reliable Internet of Things: Challenges and future trends." *Electronics* 10.19 (2021): 2377.
[31] Tyagi, AK, and Aswathy, SU. "Autonomous Intelligent Vehicles (AIV): research statements, open issues, challenges and road for future." *International Journal of Intelligent Networks*, 2 (2021): 83–102.
[32] Pirayesh, H, and Zeng, H. "Jamming attacks and anti-jamming strategies in wireless networks: A comprehensive survey." *arXiv preprint arXiv*:2101.00292 (2021).
[33] Yadav, P, Kumar, S, and Kumar, R. "A review of transmission rate over wireless fading channels: classifications, applications, and challenges." *Wireless Personal Communications* 122 (2022): 1709–1765.
[34] Li, H, Xing, X, Bi, A, and Qian, J. "Proactive flexible interval intermittent jamming for WAVE-based vehicular networks." *Wireless Communications and Mobile Computing* 2021 (2021) Article ID 8261808, https://doi.org/10.1155/2021/8261808.
[35] Chen, J, Chen, P, Wu, Q, Xu, Y, Qi, N, and Fang, T. "A game-theoretic perspective on resource management for large-scale UAV communication networks." *China Communications* 18.1 (2021): 70–87.
[36] Shahwani, H, Attique Shah, S, Ashraf, M, Akram, M, (Paul) Jeong, J and Shin, J. "A comprehensive survey on data dissemination in vehicular ad hoc networks." *Vehicular Communications* 34 (2022): 100420.
[37] Mohamed, Samir, A, Parvez, M, AlShalfan, K, Alaidy, M, Al-Hagery, M, and Othman, M. "Autonomous real-time speed-limit violation detection and reporting systems based on the Internet of Vehicles (IoV)." *Journal of Advanced Transportation* (2021): 1–15.

[38] Sharma, S, Ajay, K, Suhaib, A, and Surbhi, S, "A detailed tutorial survey on VANETs: Emerging architectures, applications, security issues, and solutions." *International Journal of Communication Systems* 34.14 (2021), https://doi.org/10.1002/dac.4905.

[39] Batth, RS, and Vashisht, S. "A survey of medium access control protocols for unmanned aerial vehicle (UAV) networks." *International Journal of Computer Networks and Applications* 8.3 (2021): 238–257.

[40] Kumari, A, Gupta, R, and Tanwar, S. "Amalgamation of blockchain and IoT for smart cities underlying 6G communication: A comprehensive review." *Computer Communications* 172 (2021), 102–118.

[41] Garg, PK. "Potentials of network-based unmanned aerial vehicles." *Cloud and IoT-Based Vehicular Ad Hoc Networks* (2021): 369–397. https://doi.org/10.1002/9781119761846.ch17.

[42] Nygard, Kendall E, Rastogi, A, Ahsan, M, and Satyal, R. "Dimensions of cybersecurity risk management." In Daimi, K. and Peoples, C. (eds), *Advances in Cybersecurity Management*. Springer, Cham, 2021. https://doi.org/10.1007/978-3-030-71381-2_17.

[43] Alzyoud, FY, Alnuaimi, AA, and Shrouf, FA. "Adaptive smart traffic accidents management system." *International Journal of Interactive Mobile Technologies* 15.14 (2021), 72–89.

[44] Gu, X, Zhang, G, Wang, M, Duan, W, Wen, M, and Ho, P.-H. "UAV-aided energy-efficient edge computing networks: Security offloading optimization." *IEEE Internet of Things Journal* 9.6 (2022): 4245–4258.

[45] Le, TD, and Kaddoum, G. "LSTM-based channel access scheme for vehicles in cognitive vehicular networks with multi-agent settings." *IEEE Transactions on Vehicular Technology* 70.9 (2021): 9132–9143.

[46] Hahn, DA, Munir, A, and Behzadan, V. "Security and privacy issues in intelligent transportation systems: Classification and challenges." *IEEE Intelligent Transportation System Magazine* 13.1 (2021): 181–196.

[47] Yu, S, Gong, X, Shi, Q, Wang, X, and Chen, X. "EC-SAGINs: Edge computing-enhanced space-air-ground integrated networks for Internet of Vehicles." *IEEE Internet of Things Journal* (2021). Doi: 10.1109/JIOT.2021.3052542.

[48] Ali, ES, Elmustafa, Hasan, M. "Machine learning technologies for secure vehicular communication on Internet of Vehicles: Recent advances and applications." *Security and Communication Networks* 2021 (2021), 1–23.

[49] Mao, B, Fengxiao, T, Yuichi, K, and Nei, K, "AI-based service management for 6G green communications." *IEEE Communications Surveys and Tutorials*, (2021): 1–34. https://arxiv.org/pdf/2101.01588v1.pdf.

[50] Lucic, MC. Efficient Infrastructure Planning Frameworks for Intelligent Transportation Systems. Diss. Stevens Institute of Technology, 2021.

[51] Sharma, D, Gupta, DS, Wani, A, and Gupta, S, and Rashid, M. "A novel approach for securing data against intrusion attacks in unmanned aerial vehicles integrated heterogeneous network using functional encryption technique." *Transactions on Emerging Telecommunications Technologies* 32 (2021): 1–32.

[52] Abbas, A, Krichen, M, Alroobaea, R. *et al.* "An opportunistic data dissemination for autonomous vehicles communication." *Soft Computing* 25 (2021): 11899–11912.
[53] Al-Absi, MA, Al-Absi, AA, Sain, M, and Lee, H. "Moving ad hoc networks: A comparative study." *Sustainability* 13.11 (2021): 6187.
[54] Liu, Xing. "Towards blockchain-based resource allocation models for cloud-edge computing in IoT applications." *Wireless Personal Communications* (2021): 1–19.
[55] Zeng, Y, Xu, X, Jin, S, and Zhang, R. "Simultaneous navigation and radio mapping for cellular-connected UAV with deep reinforcement learning." *IEEE Transactions on Wireless Communications* 20.7 (2021): 4205–4220.
[56] Yahuza, M, Idris, M, Ahmedy, I. *et al.* "Internet of drones security and privacy issues: Taxonomy and open challenges." *IEEE Access* 9 (2021): 57243–57270.
[57] Al-Sabaawi, A, Al-Dulaimi, K, Foo, E, and Alazab, M. "Addressing malware attacks on connected and autonomous vehicles: Recent techniques and challenges." *Malware Analysis Using Artificial Intelligence and Deep Learning*. Springer, Cham, 2021. 97–119.
[58] Kim, Y, and Choi, W. "Lyapunov-based energy-efficient path diversity for data transmissions in UAV networks." *IEEE Wireless Communications Letters* (2021).
[59] Dong, R, Wang, B, Cao, K, and Cheng, T. "Securing transmission for UAV swarm-enabled communication network." *IEEE Systems Journal* (2021). Doi: 10.1109/JSYST.2021.3111746.
[60] Calagna, A. Machine Learning-Driven Management of Unmanned Aerial Vehicles Networks. Diss. Politecnico di Torino, 2021.
[61] Sur, SN, and Bera, R. "Intelligent reflecting surface assisted MIMO communication system: A review." *Physical Communication* (2021): Volume 47, August 2021, 101386.
[62] Reus-Muns, G, Diddi, M, Singhal, C, Singh, H, and Chowdhury, KR. "Flying among stars: Jamming-resilient channel selection for UAVs through aerial constellations." *IEEE Transactions on Mobile Computing* (2021). Doi: 10.1109/TMC.2021.3102883.
[63] Guo, H, Zhou, X, Liu, J, and Zhang, Y. "Vehicular intelligence in 6G: Networking, communications, and computing." *Vehicular Communications* 33 (2021): 100399.
[64] Pirayesh, H, Sangdeh, PK, Zhang, S, Yan, Q, and Zeng, H. "JammingBird: Jamming-resilient communications for vehicular ad hoc networks." *2021 18th Annual IEEE International Conference on Sensing, Communication, and Networking (SECON)*. IEEE, 2021, 1–9.
[65] Zeng, Y, Qiu, M, Zhu, D, Xue, Z, Xiong, J, and Liu, M. "Deepvcm: A deep learning-based intrusion detection method in VANET." *2019 IEEE 5th International Conference on Big Data Security on Cloud (BigDataSecurity), IEEE International Conference on High Performance and Smart Computing, (HPSC) and IEEE International Conference on Intelligent Data and Security (IDS)*. IEEE, 2019, 288–293.

[66] Huang, SC, Pareek, A, Seyyedi, S, Banerjee, I, and Lungren, MP. "Fusion of medical imaging and electronic health records using deep learning: A systematic review and implementation guidelines." *NPJ Digital Medicine* 3.1 (2020): 136.
[67] Topal, OA, Gecgel, S, Eksioglu, EM, Karabulut Kurt, G. "Identification of smart jammers: Learning-based approaches using wavelet preprocessing." *Physical Communication* 39 (2020): 101029.
[68] Liu, X, Xu, Y, Jia, L, Wu, Q, and Anpalagan, A. "Anti-jamming communications using spectrum waterfall: A deep reinforcement learning approach." *IEEE Communications Letters* 22.5 (2018): 998–1001.
[69] Tyagi, AK, Rekha, G, and Sreenath, N (eds). *Opportunities and Challenges for Blockchain Technology in Autonomous Vehicles*. IGI Global, 2021. http://doi:10.4018/978-1-7998-3295-9.
[70] Tyagi, AK, Kumari, S, Fernandez, TF, and Aravindan, C. "P3 block: privacy preserved, trusted smart parking allotment for future vehicles of tomorrow." In Gervasi O. *et al.* (eds) *Computational Science and Its Applications – ICCSA 2020. ICCSA 2020. Lecture Notes in Computer Science*, 12254 (2020). Springer, Cham.

Chapter 6

Vehicular cloud computing: introduction to cloud-based vehicles

Abstract

In the past few years, cloud computing (CC) has taken the attention of the scientific community and researchers. CC has been further extended to fog and edge computing (in terms of better efficiency). It can be used in vehicles to store data at a remote server and provide efficient users/passengers. In autonomous cars, cloud-based vehicles or vehicles based on cloud computing can react to users' responses smartly and quickly. Vehicle CC (a hybrid technology of cloud and cars) can help society/governments in traffic management and road safety by instantly using vehicular resources, such as computing, storage, and the Internet, for decision-making. Hence, these services can be provided as an automated and intelligent vehicle to reduce the crowd over the road network. Also, the security and privacy of passengers/users will be protected against any cyber-attacks like a bot, worm, trojan, ransomware, etc. Hence, this chapter briefly details vehicular CC and its scope (including various interesting remarks).

Key Words: Vehicular cloud computing; Internet-of-Things-based cloud environment; Smart vehicles; Smart objects

6.1 Introduction

Cloud technology in automobiles can save our lives and make them more pleasurable. Succeeding car generations will be safer and enjoy their driving experience, thanks to cloud technology. Cloud technology in wheels is not new but evolves fleetly due to the cooperation and tie-up between automotive sedulousness and software originators. Automakers realize that embracing cloud technologies to accelerate and upgrade the driving experience since punters' exigence points to exhaustively uncombined cars/vehicles with cloud technologies. However, cloud technology is a requisite in every up-to-date car to offer comfort and security. Suppose we consider that we spend 7% of our life driving a car. The way we interact with our motorcar changed entirely in uncountable decades. Presently, cloud technology in motors is present in several ways. For representative, ultimate electric motors can switch information with remote data centers to inform the automobilist about the road and misty conditions. Cloud technology in engines has a lot of possibilities. It is uniquely suited to set up, scale, manage, and streamline features and services efficiently with a connected unit.

Automobile companies continuously develop and put their connectivity software to allow their patrons a better relationship between the cloud, the motorcar, and smartphones using infotainment as the primary interface. Our smartphone can be used as a digital key to open and start the automobile. Also, we will be capable of yielding and transporting temporary digital keys to our family and pals when we need to share the car, pre-heat or pre-cool the cabin somehow, transport route directions, or check the power stratum from any device. Of course, some imminences must be considered by the automakers when designing these types of vehicles. One of them is their vulnerability to a hacker attack. Because a bus cannot afford software crashes while driving, it is essential that the adaptability, which is necessary for the cloud performance, can respond incontinently to any system error in the automobile to find the passenger's life. Another peril might be present if the smartphone connected to our machine is stolen. In this situation, the cloud should block and track the vehicle, including the stolen smartphone. Even though cloud technology in devices is evolving fleetly, we must bide a little bit to see this promising technology in day-to-day life since all new technologies are not affordable for everyone when launched into the marketing sphere.

6.2 Related/background work

Vehicle intrusion detection was attempted long before it was recognized as a problem by scientists. In addition to aircraft that are manned and unmanned, there are also robots and self-driving machines. When a vehicle's communication, activation, and feelings are all taken into account, it is possible to detect intrusions in all three of these areas. An attack on an in-vehicle controlled area network (CAN) motor network can be inferred from its voltage profile, which has been learned by Cho and Shin [1] uncovering system. Even if the scope of an attack is limited, understanding the standard bearing of different vehicle components can help identify and locate it. Like Moore *et al.* [2], they developed an algorithm for detecting signal injection attacks in the CAN car network by monitoring command refresh rates. According to Martinelli [3], fuzzy fashions can model every day CAN communications based on actual human activity. They have developed a style based on fuzzy-rough nearest neighbor's class to distinguish between legal dispatches and those caused by a bushwhacker. The automatic-dependent surveillance-broadcast protocol is used by aircraft to broadcast their position and other situational data to other planes and ground stations for commercial air traffic control. They have been trained on this protocol. A spoofing attack on Announcements-B could compromise the safety of aircraft. This is not currently the case, but direct data requests between aircraft are possible [4]. A ground-based bushwhacker would likely have a significantly different responsibility-sensitive safety (RSS) from an airplane, so we only used RSS statistics as input features in [5] to account for this potential discrepancy. There are various ways to approach this problem. Examine the vehicle's ability to move, which is what distinguishes it as a car. According to Gwak *et al.* [6], an experimental monitoring system can detect paraphernalia malfunctions, tampering with paraphernalia, and suspicious conduct by the flight control system. Roll, pitch, yaw, and rudder/aileron control parameters are all recorded by the servo motors. The parameters of the airframe and regulators are estimated using the recursive least patios strategy. Using a set of parameter estimations for each drone's control law, the monitoring system can compare parameters between flights. Consequently, a large value for the last indicates significant parameter differences and can be used as the foundation for an anomaly-finding system. The method has been tested in the field using open-source flight simulators.

For a vehicle to operate safely, it needs to detect scents and communicate and actuate. The robustness of their olfactory systems is especially important for self-driving cars. Consequently, diplomatic channels and networks hung with false data injection attacks that can disrupt a vehicle's collision avoidance subsystem are attractive targets for attacks on these entities. Treating detection system failures as expected and using statistical anomaly detection methodologies can help identify attacks on detection systems. Small robotic vehicles, for example, can benefit from using recursive least-square defilement to remove data that exceeds a predetermined rate of change. A vehicle's collision avoidance procedures will not use an unreliable sensor if allowed to do so.

Other researchers have taken a more comprehensive approach to intrusion detection, considering the overall state of the vehicle. Denial-of-service attacks, false data injection, and various malware attacks on a robot have taken up most of their time [7]. A variety of spells and their effects on a set of cyber and physical characteristics are included in the training phase to train the system. Electronic components are surprisingly heavy, particularly those relating to networks. Physical features like battery consumption and shell vibration have improved the precision of a specific spotting style significantly. Every hour the car's network connection to its remote regulator is disrupted, the vehicle enters fail-safe mode, causing a physical sensation. By spotting this physical manifestation early, the chances of spotting a moratorium are reduced. For a drone, this study by Mitchell and Chen [8] provides a complete onboard security monitoring framework that includes GPS inputs, ground control station, detector reading, regulator yield data, flight software status, and software updates. System address is monitored to ensure the safety of the aircraft and its occupants, including aircraft oscillations around any of its axe, a departure from the flight path, abrupt changes or coherent drift of detector readings and memory leaks, real-time failures, and other unusual software addresses. Bayesian network machines are used to support probabilistic security calls. GPS signal strength transients could indicate an impending attack if barometric dimensions and beam altitude coincide, for example. A reconfigurable field-programmable gate array implementation and performance evaluation on a NASA Dragon Eye drone have yielded promising results in detecting burlesqued GPS signals and sexist commanding. Bayesian net factories can also be used for this purpose.

Classifying an attack on a robotic vehicle that uses the generic vehicle architecture is developed by Bezemski and his colleagues [28]. An expert addict specifies the norms of what is expected in an intrusion detection style known as "action specification." For example, according to Mitchell and Chen [8], some of the rules for drones include the necessity of de-arming before entering a target area, and so on. As soon as one of the rules of the specified action is broken, an attack state is triggered [11]. For each state, their state machine came up with a list of 165 safe and 4,443 unsafe commonwealths, respectively. The percentage of time a device spends in safe commonwealths can be used to determine how strictly the action's rules are being followed. They went on to demonstrate in [12] how adaptable and agile the strategy is. For example, it can reduce false negatives while simultaneously boosting false positives. When it comes to safety, this strategy necessitates that a large number of commonwealths document all safety specifications accurately for each elevation, environmental condition, and so on.

Robotic vehicles can play an important role in armies that share a physical space and follow a standard set of commercial rules, whether or not they are accessible. Because this implies that a cyber intrusion movie's drones will always follow the rules and turn right when they collide, it is a dangerous assumption to make. A distributed misbehavior finding method based on a Boolean concurrence protocol has been developed by Martini *et al.* [9] for each drone in the air to observe events. According to the terms of the agreement, each drone can use its detectors and information from its neighbors to predict what other drones will do. A drone's actions are presumed to be uncooperative if they do not match the foretold bone. Using a crew of four real drones, including a misbehaving one, the method was tested successfully. To apply these findings to larger teams with multiple errant bones, it is necessary to investigate how closely this problem is linked to the Elaborate Generals issue [14]. In the event of an attack by a driverless car on another driverless car network, what happens? Alheeti *et al.* [11–14] investigated Vehicular Ad Hoc Networks to communicate with other driverless vehicles in the same area. In this preliminary study, we used NS-2 and mobility simulation tools to assess the performance of an artificial neural network-based intrusion detection system. Standard network intrusion spotting systems typically use input features such as haul sizes and hop counts as inputs. Magnetometer and gyroscope detectors [15–17] and silver hole attack simulations [18] have been added to the work of the original authors.

Consider all of the above-mentioned detection methods. If this is the case, you will notice that they are all geared toward reducing processing load, whether through the heavy use of statistical recipes or the pre-definition of simple behavioral rules that are easy to observe. Why? Because the vehicle in which they are currently riding has a limited capacity. Because of this, they have little or no influence on cutting-edge recipes, such as those now being developed in the field of deep learning, for example. If you are driving a resource-constrained vehicle, the more powerful the spotting algorithm, the less energy it consumes, and the less engaging the result. A surfacing field called shadow robotics will be used to solve this problem. In our opinion, the bulk of the processing should be moved from the scholarly depths to a more accessible skeleton (whether a single server, cloudlet, or shadow). In other words, the offloading computation means entrusting specific computational tasks to machines that are not immediately accessible to the stoner. Most of the online forensic methods used for Android device malware and corrupted data recovery conform to the stereotype [19,20]. However, rather than relying on crowdsourcing, we practice computational offloading to make advanced scholarship resting techniques available to vehicles at a lower processing and energy cost. In both scenarios, the data is collected and aggregated on the vehicle. There is an onboard investigation to determine whether or not an attack has occurred. Unloaded vehicles use a network to send the vehicle's onboard computer data. Recent advances in deep learning algorithms have opened up a new image and natural language processing applications, such as detecting malware or devil certificates issued by well-respected certification authorities. Traditional computer networks, not cyber-physical systems like automobiles, have also made use of it. An exception, however, is the recent work of Kang *et al.* [21], which focuses on automotive assiduity and the detection of attacks on CAN vehicles [21]. However, this is only the beginning of something great. One data source is covered, and a simple and generalist deep neural network configuration is used; it does not take into account the vehicle's overall state or the temporal information of a given attack (the fact that the impact of the spell changes over time during the episode). The only way to test this theory is to use a computer simulation [22–25].

6.3 Motivation

With each new model year, automakers introduce models with an ever-increasing number of individualized convenience features. The desire for people to speak at all times has gone far beyond what was considered possible just a few years ago. Today, many people use their mobile devices to communicate or even send an email while commuting to work or traveling long distances to their destinations. Many drivers want to have fun while driving, which means they require Internet connectivity for activities like browsing the Internet, checking and responding to email, and other work that necessitates being online. Drivers need to be aware of what lies ahead in the most technologically advanced and rapidly developing countries to make the best possible decisions about which route to take during peak hours of traffic [10]. Consequently, today's goal for automakers is to make it possible for drivers to manage their connectivity while on the road. Cloud computing, IPv4, and IPv6y all have a role in transforming the future of vehicles in this format.

6.4 Cloud computing: importance, types, characteristics, and applications in trends

Cloud computing is becoming more and more important and popular by the day. Everyone's perspective on the situation is different. While "the cloud" has long been used as a metaphor for the Internet, its meaning changes when coupled with the word "computing." In the opinion of some reviewers and vendors, cloud ciphering is nothing more than an online version of mileage computing. Those who argue that everything we consume outside our firewall is cloud-based include even traditional outsourcing. Cloud computing has emerged as a viable option for increasing capacity or adding capabilities to a marquee without the need to invest in a new frame, train a new pool, or enable new software [26,27]. Pall computing, a subscription-based or pay-as-you-go service, extends information technology's (IT) living capabilities over the Internet in real time. Large and small companies offer various services based on the dark arts in the early stages of darkness computing. In addition to software as a service (SaaS) providers like Microsoft and Salesforce, there are also service-style cadre providers. A few cloud computing aggregators and integrators are beginning to emerge, but for the most part, information technology has to plug into cloud-based services on its own. InfoWorld interviewed dozens of vendors, evaluators, and I.T. customers to discover the rainbow ingredients of cloud computing. Based on those discussions, these are the various cloud computing methods that are out there.

6.4.1 Software as a service

A multitenant skeleton in the cloud can use the netizen to distribute a single user across thousands of other users. There is no upfront investment in servers or software licensing on the customer's part, so the provider's costs are lower than traditional hosting. Aside from the well-known Salesforce.com, an excellent example of SaaS in the enterprise, there are similar competitors to Workdays in H.R. apps and ERP. In addition, SaaS desktop applications like Google Apps and Zoho Office have unexpectedly risen in popularity.

6.4.2 Computing utility

On-demand access to virtual magazines and waiters is being revived by Amazon and Sun using information technology. Enterprises are currently using service computing for non-mission-critical tasks, but it could one day replace the neck of the data center. Virtual data centers can be built using App Logic and Cohesive Flexible Technologies' Elastic Waiter on Demand. Two other companies offer solutions to help I.T. implement virtual data centers from commodity waiters. An I. T. virtualized resource pool can be created using LiquidQ's correspondent capabilities and accessed via the network to sew together memory, I/O, magazine, and computing capacity.

6.4.3 Web services offered through cloud

Examples of SaaS APIs include those offered by providers of web services rather than full-blown employment, such as those offered by providers of SaaS. In addition to their core business, companies like Xignite and Strike Iron also provide business services. Google Charts and ADP payroll processing are two more examples.

6.4.4 Using platform as a service

In this instance, cloud computing provides developers with operating systems and development environments as a service. This is yet another variation of SaaS. Using the provider's resources, we create our operations that run on their servers and are delivered to our computers via the Internet. While we do not get complete freedom with these services, we get spiciness and pre-integration from the seller's design and capabilities. Cog head and Google App Engine are two of the best examples. For light development, mashup platforms like Yahoo Pipes and others burst.

6.4.5 Managed service providers

An old-fashioned cloud computing model, a managed service exposes itself to I.T. rather than the people who use it, such as an email scanning service or a play monitoring service (which Mercury, among others, provides). SecureWorks, IBM, and Verizon services, as well as matching, fall under this category. As Postini, recently purchased by Google, cloud pioneered anti-spam services. Desktop regulation services, such as those provided by Center Beam and other corporations, are also victims.

6.4.6 Integration of Internet

In general, cloud-based services have been integrated earlier than they would have otherwise. New to OpSource is the OpSource Services Bus, a SaaS product. The technology was developed by Boomi, a cloud-based integration technology company. As part of its acquisition strategy, Workday recently purchased Cape Clear, an ESB provider edging toward b-to-b integration. A decade ago, Grand Central was launched to connect service providers and provide aggregated answers to customers' questions. A more accurate term for Pall computing is typical sky computing, characterized by numerous shadows of services that I.T. customers must plug into separately at this time, with cloud devices rarely in evidence. There is only a matter of time before every company is reduced to a nodule in the shadow of virtualization and service-oriented architecture. It has been going on for a long time and has a long-term future outlook. As far as long-term trends go, shadow computing is the most difficult to dismiss.

Data storage, business applications, and entertainment industries can all benefit from cloud computing. All you need to access data stored in the cloud is a web connection (e.g., data files, images, audios, and videos). The cloud service provider offers a variety of tools for regaining lost data to ensure security. For most business applications, cloud service providers are the primary pillar. Every company today needs cloud commercial enterprise software to grow.

There is also a guarantee of round-the-clock business application availability. The entertainment industry employs this strategy. A few entertainment options that can be found through cloud computing include online video games and video conferencing. Playing video games on the Internet is becoming increasingly popular. From a remote location, you can play a variety of online video games. Some of the best cloud gaming services include Shaow, GeForce Now, Vortex, Project xCloud, and PlayStation Now. Using video conferencing apps is simple and convenient. Cloud-based video conferencing services allow us to communicate with our business partners, friends, and family members. It is possible to save money, increase efficiency, and eliminate interoperability with videoconferencing. These tools include those that aid in deployment, data integration, and disaster recovery for cloud administrators. Platforms, programmers, and infrastructures are all under the control of these devices.

6.5 IPv4 importance in previous decade and limitations in current

IPv4 is the abbreviation for Internet Protocol Version 4, or Internet Protocol (IP) Bracing technology allows us to connect our feelings to the Internet. This unique IP address, 198.164.000.001, is assigned to a device that is connected to the Internet. Transferring a packet of data across the network containing both IP addresses of the dispositions is required to pack data from one computer to another. We must realize the importance of recognizing that the Internet is currently constrained to IPv4 addresses to make the most of current methods. People and Habitude can exchange information and build relationships through these addresses (phones, tablets, computers, etc.).

The IPv4 addresses of computers that have been removed or switched to Internet Protocol version 6 (IPv6) remain inactive (they are also known as "sleeping addresses"). These require a lot of energy, even if they are not actively working, like a bear in hibernation. There are tens of thousands or even hundreds of thousands of inactive IPv4 addresses belonging to outside companies, particularly in the telecom, banking, healthcare, and Internet of Things (IoT) sectors. This is both a huge challenge and an opportunity for those who are not as passionate about building a safe Internet. The houses believe that the administration's legislation on ISPs would not affect thousands of inert IPv4 emails. The tasked yet static IPv4 addresses must be dismantled hourly by data center-predicated stress groups, or the power loss must be re-established. Allowing stockholders of IPv4 to return to available subsistence is a necessary alternative to constantly relying on our limited natural resources. GoDaddy first used precinct names and addresses in the early 2000s as a screening tool. There was nothing else they could do with them; they could return them to a neighborhood sector. IPv4 addresses that are no longer in use must be given the same treatment as IPv6 addresses. Breaking the problem can be done in various ways. One way to ensure the long-term viability of the Internet is to lease out unused IPv4 addresses. In the following paragraphs, we will discuss the various ways we can achieve our goal of sustaining the Internet.

Internet Service Providers (ISPs) play a critical role in ensuring the long-term viability of the Internet. Added to that, there are the data centers. That being said, it is foolish to ignore our own and our college's role in our lives. To show our appreciation to companies, we should allow them to rent their unused IPv4 addresses safely and securely. As a result, businesses can better identify and respond to people wasting an idle IPv4 resource. The Internet's long-term sustainability can be achieved with small steps like deciding on procedures that reward businesses for matching practices. It reduces the number of IPv6 addresses assigned and significantly reduces the number of inactive IPv4 addresses afterward. Whether we can influence the company to create a new revenue stream by leasing its IPv4 addresses would be a great challenge to creating a sustainable Internet. However, it is up to us to help speed up the decommissioning of sleeping IPv4 talks in our networks. The first step to bringing about change is when each body becomes a part of the attention devoted to reducing the gush cost in technology. An encouraging trend is reducing the gap between IPv4 addresses commonly used and those not, which is a good sign.

Current generation IPv4 has several drawbacks and limitations, including configuration issues, security concerns, infrastructure concerns, mobility concerns, and geographical boundaries. IPv4 requires both manual and automatic configuration options. Manually configuring IPv4 involves the use of the dynamic host configuration protocol (DHCP). DHCP configurations are challenging to manage because separate infrastructure control is required. Since IPv4 was published a long time ago, it is no longer expected to be secure against the threats that are being imposed today. Since the Internet protocol security (IPSec) protocol specifies the use of the net, it provides network security to IPv4. When IPSec is not built-in and implementation is optional, the problem arises. IPv4 prefixes are assigned to individual addresses so that each can become a router. Flat and hierarchical routers are now commonplace in the net, as well. However, the net spine routers contain more than 85,000 routes. IPv4 includes a mobility specification, but it is considered inefficient because of the miles it travels. That is because it has its infrastructure, and that is why.

As a result, its nodes of movement are also inefficient. Another issue that IPv4 users face is the depletion of public addresses. This is primarily due to the preliminarily employed practices for allocating elegance. To combat this problem of limited space, network address translation (NAT) has been developed. NAT divides a single public IPv4 address among multiple privately addressed computers. The United States of America is credited with expanding the Internet. As a result, the United States is concerned about how intellectual property is distributed. In reality, most of the negotiating power is concentrated in the United States.

6.6 Uses of IPv6 in cooperative intelligent transportation system/automatic vehicle identifiers (ITS/AIV): a standardization viewpoint

IPv6 is the new Internet protocol (IP) address standard that will eventually replace IPv4. All Internet-connected devices, including computers, mobile phones, home automation components, IoT sensors, and more, require an IP address. The original IP address scheme, known as IPv4, runs out of talks as the number of connected devices grows.

Protocols and address standards like IPv6 are designed to supplement and eventually replace the current IP and address standards, such as IPv4. All devices connected to the Internet, including computers, mobile phones, home automation components, IoT sensors, and more, require an IP address. The original IP address scheme known as IPv4 runs out of talks as the number of connected devices grows. Only IPv6 has been addressed in the recent development of low-power radio technologies for IP transport. When it comes to 802.15.4, IPv6 has already been defined, whereas IPv4 has not. A subnet prefix and an interface ID (IID) are required for every 128-bit IPv6 address in IPv6. A 64-bit prefix and a 64-bit IID are the most common lengths when it comes to link types. The 64-bit boundary has become the Sedulity standard in most common IPv6 performances. Among the most popular games today, you will notice that the IID length is always 64 bits. The IID bits can opt-in in various ways, depending on the use-case and deployment script. Because the 64-bit IID is so large, the development of IIDs with meaningful information has been facilitated.

Note that Link-node and non-native link-node are two types of unicast addresses in the IPv6 address framework. Non-native link addresses are used for machine discovery and configuration, and at least one is permanently set for each lump interface. Link-immigrant packets will not be promoted to other links by routers because link-immigrant lessons are not guaranteed to be unique across a larger network. Globally unique global width addresses are expected to be used across the entire breadth of the Internet. Unique Local Address (ULA) 18 are designed for use in international networks with more than one link and require a global IP address to communicate over the Internet. They cannot be used to send out Internet-based dispatches; however, ULAs are designed to feed a reasonable distance apart to avoid address collisions. To the extent that the network spans multiple links and routing hops, these addresses are truly beneficial. An address independence issue may be raised in some implementations. ULA's co-driver may provide stability, but an outside provider may limit the message's reach. Using global unique addresses, the Internet is guaranteed to remain private and accessible. When it comes to regulating and managing the Internet's address space, Internet Assigned Numbers Authority (IANA) is responsible for administering the universal pool of addresses. At the same time, Regional Internet Registries (RIRs) allocate addresses from the IANA pool to specific regions. The RIRs give address space to the exotic Internet registries (LIRs), a collection of automobile enthusiasts, businesses, and other organizations. They put themselves and their clients to the test.

IoT visionaries are concerned about the security of the billions of new smart things being created every year. As a result of the IoT, cyber-punks pose a real and immediate threat to organizations and constituents. Having the ability to steal millions of credit card numbers by manipulating an established network is extremely troubling. If they relocated to an intelligent metropolis or a nearby innovation district, the consequences could be far more devastating. When it comes to IoT security, IPv6 is a better option than IPv4—and IPSec makes it even more secure.

First, IPv6 can encrypt all of its communications. Using IPv4 is an option if installed, but this technology is not very widely used. A standard Virtual Private

Network (VPN) feature in IPv6 is encryption and integrity checking, supported by all compatible routers and operating systems. Assuming we are entering a secure bank log-in while walking into a cyber trap will make "Man in the Middle" attacks much more vulnerable if IPv6 is removed from the equation. It is also more secure in IPv6 for name resolution. SEcure Neighbor Discovery (SEND) can be used to verify that a host is who it claims to be. An ARP poisoning or other selection-based attack cannot take place as a result of this. When used in conjunction with IPv6, a higher level of trust can be established in the connections between nodes. While a bushwhacker can easily manipulate or at the very least observe traffic between two legitimate hosts using IPv4, IPv6 significantly increases the difficulty. According to various reports, more than 25 billion computers will be connected to the Internet by 2020. To put this in perspective, the same report claims that 4.9 billion biases will be eliminated by 2015. As a result of this 400% growth in just five days, IoT is expected to grow exponentially in the next 10–50 days. To understand the IoT bias caused by IPv6's trillions and trillions of new addresses, we need to look only at the number of unused lessons. Transmission control protocol/Internet protocol (TCP/IP)-based IoT productions can rest easy knowing they will always access a unique identifier.

The ability of a network-chained association to "connect" with another is a powerful feature of the IoT. IoT devices' ability to communicate with one another over IPv4 was severely hampered. One of these issues was network address translation (NAT). For groups of people with shared interests and affinities, NAT was created as a workaround. This is something we will talk about later. However, this raises security concerns for IoT products as well. IPv6 has made it possible to ship IoT productions without having to deal with NAT and firewalls. As a result, working with firewalls and NAT routers is more difficult for smaller IoT devices than larger hosts. IPv6 will allow TCP/IP-enabled IoT devices to deal with a lot more of these issues.

6.7 Benefits of IPV6 for Industry 4.0/Cloud-Based Vehicles and Society 5.0

A few car manufacturers in Europe, China, and Japan make vehicles with IPv6 as standard equipment. Because IPv6 has a nearly infinite number of IP addresses, this is why this is the case. Due to the current trend, infrastructure for future "smart" transportation networks is being built in those countries. Traffic congestion on the road, reduced gas consumption, pollution management, and the other difficulties of a tour are among the top priorities of the DOT's ITS. Tracking traffic flow down to a vehicle's precise location and speed is now possible, thanks to sensors built into today's automobiles.

IPv6's support for more unique addresses than IPv4 is a major advantage (340 trillion versus 4 billion). With IPv6, in addition to providing more IP addresses, IPv6 improves the Internet's stability and usability. IPv4 is no longer a viable option for the future because of its numerous flaws. As an example, IPv4 was never meant to be secure. Although it was originally designed for military service, it has since been adopted in public schools and examinations. We cannot rely on

IPv4 security features because they have to be retrofitted. New exercises are putting more strain on the network to keep up with the rapid advancement of technology. On their list of must-haves are improved security, emptier bandwidth, and delivery that is always on time. Adding on to IPv4 is expensive, time-consuming, and error-prone. IPv6 is the future. Network video products would not be affected by IPv6, but the efficiency of networks will improve. Keep an eye out for how people used to receive their mail back in the day. It used to be enough to know a person's name and address to shoot them at a shooting range. To keep up with the increasing number of mail recipients, a zip-code system was implemented.

Furthermore, IPv6 is simply an improved method of transmitting data. Moreover, IPv6's built-in quality of service (QoS) capabilities outperforms IPv4 in yet another area. The resource reservation protocol (RSVP), a router control protocol that speeds up the transfer of data packets from the source to the destination, makes it possible to transmit audio and video over IPv6 at a high quality and reliability. The result is high-quality audio and video.

Other IPv6 features include better IP address framing, IP motor-configurations for easier swapping of entire mass-market networks between providers, faster routing, and point-to-point encryption and connectivity using the same address across various networks. IPv6 makes it easier to frame and characterize talks. Instead of using hexadecimal memos, IPv6 uses 128-bit strings separated by colons instead of the 64-bit lines used by IPv4.

Numerous methods exist for allocating IPv6 addresses, including using the DHCPv6 server or the link-alien IPv6 address. fe80 will be the first external IPv6 address for a motor-configured link-enabled device. The suffix is now used in the extended unique identifier-64 (EUI-64) numbering scheme. Additionally, router notifications can be used to fine-tune motor-configured networks' network preferences. They will demonstrate how to set up a routable network predilection. To determine if an IPv6 address is routable, you can use the router notification system. In addition to the router and mesh prefix addresses, this motor-configured address has a EUI-64 address. The router notifications may be used to indoctrinate the network device with DHCPv6. It is possible to charge for IP configuration using Dynamic Host Configuration Protocol (DHCP) servers in IPv4 and DHCPv6 servers in IPv6 networks. Using stateless DNS or NTP is possible, but the server will not assign IPv6 addresses for network preference in stateless mode. This will necessitate the use of other systems. Aside from this, DHCPv6 stateful mode will also teach IPv6 lessons on network and server importance.

There are no broadcasts in IPv6, only multicast and unicasting like in IPv4. Broadcast and multicast are two methods for transmitting data over a computer network. Broadcasting is the practice of disseminating information to everyone with an Internet connection. Sending and receiving messages is essential for processing dispatches. The web and the hosts connected to it are slowed down due to an excessive number of communications. Broadcasts are not the best choice for transmitting network video. They are used only when necessary for protocols such as DHCP.

One message can be sent to many people at once using multicasting, on the other hand. For example, on a live tape with multiple receivers, these technologies transmit the same data to all of them. This method of transmission reduces business by delivering a single rivulet of data to a large number of people. Unicasting necessitates a carbon for each item, whereas tape rivulets only require a carbon once. Multicasting is the best option for viewing live surveillance video for a large number of developers. It is possible to assign IPv6 addresses to hosts using hostnames, and it is possible to retrieve IPv6 addresses using the IP.

6.8 Opportunities and challenges with IPv6

A significant upgrade to IPv6 from IPv4 is necessary, but the IoT is not finished yet. Companies face a difficult task in promoting IPv4 and IPv6 adoption. IPv4 will no longer be supported by Internet service providers in the future.

Note that TCP/IP is not available to all IoT devices, so they cannot all be directly connected. Some applications require fragile packet heads to maximize haul allocations in extremely small datagrams, but this is not universal. Only 20 people can be shot at once, and 18 of those are required for addressing. There is no room for user data. There should be as little information as possible to manage long-distance dispatches.

Consequently, IPv6 is not a good option due to the high cost of its components. The new IPv6-based IPv6 Internet connectivity is expected to transmit data 1000 times faster than currently possible, up to 40 terabytes per second. For new devices and appliances, IPv6 is the best option because of its large address space. By using serial numbers imprinted on community processors as IP addresses, cellular connectivity can be achieved. Advances in technology will allow drivers and passengers to communicate while maintaining their safety and security in the years to come.

In early 2008, Ford's Lincoln and Mercury models will be equipped with Microsoft's advanced Sync technology. A cell phone, smartphone, and music player can all be used without having to hold them. The Sync system uses wireless Bluetooth or stressed USB connections to connect music and communication devices, controlled by voice instructions or switches on the steering wheel. Synchronization makes it possible to use a computer or personal digital assistant to do as much as possible. Furthermore, IPv6's built-in QoS capabilities outperform IPv4 in yet another area. An IPv6 router control protocol like RSVP allows for high-quality and reliable audio and video transmission: data packets are sped along a routed path for the resultant high-quality audio and video. A total of 630,740 vehicles with more than 5 billion miles on the odometer were seized by the federal government in 2006. Using IPv6-enabled sensors, governments can track and recover stolen or misplaced vehicles.

The IPv6 Platform can improve automobile safety, reduce upkeep costs (via internal sensors), better emergency control preparedness, green stock control (for decisions on phasing out and changing automobiles), reach far-off querying, and the automobile price range is making plans.

According to Juniper Networks' director of structures engineering, Tim LeMaster, vehicular site visitors running on the IPv6 Platform have a bright future.

We may prevent accidents if vehicles have their IP addresses and can communicate with sensors installed on the roads. In the end, drivers will receive alerts from other vehicles via IPv6 technology, indicating that they should slow down. In addition, parents should monitor their teen's motors and see how fast they are going. Chatting at the pass and chatting with other drivers on the street are two possibilities being considered. Suppose you are stuck in a situation where your state-of-the-art gas injection system is malfunctioning. In that case, you may want to think about this: To get a mechanic to remotely connect via cell phone to the car's computer system, he needs to contact his car manufacturer's support network. This is all done via an IP link using a simple network management protocol. Although this feature is still in its infancy, automobile manufacturers are racing to implement it.

6.9 Conclusion

For both drivers and urban communities, automobiles are becoming increasingly important sources of computing and sensing resources. These assets can be tapped into by creating a cell vehicle platform that makes and shares numerous utilities by all vehicles on the road. Comparable in some ways but distinct from those of the Internet cloud is the mobile vehicle cloud. There are several benefits to using target vehicle cloud packages that include secure navigation and green site visitor management, and domestically relevant facts and entertainment. To achieve the goals above, we have discussed a cloud computing approach with IPv6 technology. There are many issues with IPv4 technology that need to be addressed if cloud-based vehicles are to achieve the speed and reliability they need.

References

[1] K.-T. Cho and K. G. Shin, "Viden: Attacker identification on in-vehicle networks," in Proc. 24th ACM Conf. Comput. Commun. Secur. (CCS), 2016, pp. 164–170.
[2] M. R. Moore, R. A. Bridges, F. L. Combs, M. S. Starr, and S. J. Prowell, "Modeling inter-signal arrival times for accurate detection of can bus signal injection attacks: A data-driven approach to in-vehicle intrusion detection," in Proc. 12th Annu. Conf. Cyber Inf. Secur. Res., 2017, p. 11.
[3] F. Martinelli, F. Mercaldo, V. Nardone, and A. Santone, "Car hacking identification through fuzzy logic algorithms," in Proc. IEEE Int. Conf. Fuzzy Syst. (FUZZ-IEEE), Jul. 2017, pp. 1–7.
[4] M. Strohmeier, V. Lenders, and I. Martinovic, "Intrusion detection for airborne communication using PHY-layer information," in Proc. 12th Conf. Detection Intrusions Malware Vulnerability Assessment (DIMVA), 2015, pp. 67–77.
[5] Z. Birnbaum, A. Dolgikh, V. Skormin, E. O'Brien, and D. Müller, "Unmanned aerial vehicle security using recursive parameter estimation," in Proc. Int. Conf. Unmanned Aircraft Syst. (ICUAS), 2014, pp. 692–702.

[6] C. Gwak, M. Jo, S. Kwon, H. Park, and S. Son, "Anomaly detection based on a recursive least-square filter for robust intelligent transportation systems," in Proc. Korea Inst. Commun. Sci. Summer Conf., 2015, pp. 438–440.

[7] R. Mitchell and I. Chen, "Specification-based intrusion detection for unmanned aircraft systems," in Proc. 1st ACM MobiHoc Workshop Airborne Netw. Commun., 2012, pp. 31–36.

[8] R. Mitchell and I. R. Chen, "Adaptive intrusion detection of malicious unmanned air vehicles using behavior rule specifications," *IEEE Trans. Syst., Man, Cybern., Syst.*, 44(5), 2014, 593–604.

[9] S. Martini, D. Di Baccio, F. Alarcón Romero, A. Viguria Jiménez, *et al.*, "Distributed motion misbehavior detection in teams of heterogeneous aerial robots," *Robot. Auto. Syst.*, 74, 2015, 30–39.

[10] V. Akshay Kumaran, A. K. Tyagi, and S. P. Kumar, "Blockchain Technology for Securing Internet of Vehicle: Issues and Challenges," 2022 International Conference on Computer Communication and Informatics (ICCCI), 2022, pp. 1–6, doi: 10.1109/ICCCI54379.2022.9740856.

[11] Khattab M. Ali Alheeti; Anna Gruebler; Klaus D. McDonald-Maier, "An intrusion detection system against malicious attacks on the communication network of driverless cars," in Proc. 12th Consum. Commun. Netw. Conf. (CCNC), 2015, pp. 916–921.

[12] K. M. A. Alheeti and K. McDonald-Maier, "An intelligent intrusion detection scheme for self-driving vehicles based on magnetometer sensors," in Proc. Int. Conf. Students Appl. Eng. (ICSAE), 2016, pp. 75–78.

[13] K. M. A. Alheeti, R. Al-Zaidi, J. Woods, and K. McDonald-Maier, "An intrusion detection scheme for driverless vehicles-based gyroscope sensor profiling," in Proc. IEEE Int. Conf. Consum. Electron. (ICCE), Jan. 2017, pp. 448–449.

[14] K. M. A. Alheeti, A. Gruebler, and K. McDonald-Maier, "Intelligent intrusion detection of a grey hole and rushing attacks in self-driving vehicular networks," *Computers*, 5(3), 2016, 16. https://doi.org/10.3390/computers5030016.

[15] G. Hu, W. P. Tay, and Y. Wen, "Cloud robotics: architecture, challenges and applications," in *IEEE Network*, 26(3), 2012, pp. 21–28, May-June 2012, doi: 10.1109/MNET.2012.6201212.

[16] A. Houmansadr, S. A. Zonouz, and R. Berthier, "A cloud-based intrusion detection and response system for mobile phones," in Proc. IEEE/IFIP 41st Int. Conf. Dependable Syst. Netw. Workshops, Jun. 2011, pp. 31–32.

[17] G. Portokalidis, P. Homburg, K. Anagnostakis, and H. Bos, "Paranoid Android: Versatile protection for smartphones," in Proc. 26th Annu. Comput. Secur. Appl. Conf., 2010, pp. 347–356.

[18] W. Hardy, L. Chen, S. Hou, Y. Ye, and X. Li, "DL4MD: A deep learning framework for intelligent malware detection," in Proc. Int. Conf. Data Mining (DMIN), 2016, p. 61.

[19] J. Kim, J. Kim, H. L. T. Thu, and H. Kim, "Long short term memory recurrent neural network classifier for intrusion detection," in Proc. Int. Conf. Platform Technol. Service (PlatCon), 2016, pp. 1–5.

[20] A. Y. Javaid, Q. Niyaz, W. Sun, and M. Alam, "A deep learning approach for network intrusion detection system," in Proc. 9th EAI Int. Conf. Bio-Inspired Inf. Commun. Technol. (BIONETICS), 2016, pp. 21–26.

[21] J. W. Kang and M. J. Kang, "Intrusion detection system using deep neural network for in-vehicle network security," *PLoS ONE*, 11(6), 2016, p. e0155781.

[22] T. P. Vuong, "Cyber-physical intrusion detection for robotic vehicles," Ph.D. dissertation, Dept. Comput. Inf. Syst., Univ. Greenwich, London, U.K., 2017.

[23] T. G. Barbounis, J. B. Theocharis, M. C. Alexiadis, and P. S. Dokopoulos, "Long-term wind speed and power forecasting using local recurrent neural network models," *IEEE Trans. Energy Convers.*, 21(1), 2006, 273–284.

[24] Y. Du, W. Wang, and L. Wang, "Hierarchical recurrent neural network for skeleton-based action recognition," in Proc. IEEE Conf. Comput. Vis. Pattern Recognit., Jun. 2015, pp. 1110–1118.

[25] G. Rekha, S. Malik, A. K. Tyagi, M. M. Nair, "Intrusion detection in cyber security: role of machine learning and data mining in cyber security," *Adv. Sci. Technol. Eng. Syst. J.*, 5(3), 2020, 72–81.

[26] A. K. Tyagi and N. Sreenath. 2016. "Providing trust enabled services in vehicular cloud computing," In *Proceedings of the International Conference on Informatics and Analytics (ICIA-16)*. Association for Computing Machinery, New York, NY, USA, Article 3, pp. 1–10. DOI:https://doi.org/10.1145/2980258.2980263.

[27] A. K. Tyagi, and N. Sreenath, "Providing trust enabled services in vehicular cloud computing," 2016 International Conference on Research Advances in Integrated Navigation Systems (RAINS), 2016, pp. 1–7, doi: 10.1109/RAINS.2016.7764391.

[28] A. Bezemskij, G. Loukas, D. Gan, and R. Anthony, "Detecting cyber-physical threats in an autonomous robotic vehicle using Bayesian networks," in Proc. IEEE Cyber, Phys. Social Comput. (CPSCom), Aug. 2017, pp. 1–6.

[29] G. Pearson and M. Kolodny, "U.K. MoD land open systems architecture and coalition interoperability with the U.S.," *Proc. SPIE*, 8742, 87420C, May 2013.

[30] Z. Dong, K. Kane, and L. Camp, "Detection of rogue certificates from trusted certificate authorities using deep neural networks," *Trans. Privacy Secur*, 19(2), 2016, p. 5.

Chapter 7

Advanced vehicle motion control and LTE/5G/6G for vehicular communications

Abstract

Presented in this chapter is an overview of vehicle-to-everything connections as part of the long-term evolution (LTE) project. Specifically, researchers look at how infrastructure can communicate with vehicles via broadcast/multicast 4G LTE service, as well as how vehicles can communicate with each other using the LTE side connection. Designers also go through the major vehicular services. But in the interim, there is a new platform being built called 5G, which aims to improve existing services while also opening the door to a host of new ones that require ultra-reliable, low-latency connections to be developed. This is a novel radio access technique for increasing network density while simultaneously improving coverage and accessibility on existing platforms [such as 2G, 3G, 4G, and wireless fidelity (Wi-Fi)]. What this means is that the purpose of 5G is for it to meet the communications requirements of numerous different groups. In terms of both technology and application, 5G will have the greatest impact on automobiles. In light of the deployment of 5G, experts and industry insiders are already looking forward to 6G. According to these experts, 6G will be the most important technology for information exchange and social interaction after 2030. Artificial intelligence will fuel 6G's highly autonomous closed-loop network, making up for 5G's communications, processing, and global coverage constraints (Artificial Intelligence of Things). 6G life may require vehicles to replace smartphones, and the ultimate goal of vehicle development is zero-pollution vehicles that are both incredibly safe and self-sufficient. To keep drivers and passengers safe and engaged in future automobiles, more 6G vehicular intelligence research is required. There is an increasing number of people that rely on wireless communication. Mobile has had a major impact on the younger generation. Wi-Fi protocols are being rolled out all over the world to help with this. Wireless phones, like individuals, have free will. The wireless standards for the fourth generation have now been created. A wide range of traffic efficiency and road safety applications are being developed to meet the pressing demand for smarter, greener, and safer mobility.

Keywords: Information and communication technology, Long-term evolution (LTE), Internet of Things (IoT), Artificial intelligence (AI), Massive machine-type communications (mMTC)

7.1 Introduction

When it comes to the realm of information and communication technology, experts are witnessing the birth of new software platforms and business ecosystems consistently [1]. Many of these developments are being driven by market forces, and they have the potential to have significant ramifications on how we go about our daily lives in the future. Much anticipation exists that the fifth- and sixth-generation broadband cellular networks (5G and 6G) will fundamentally alter current society and technology [2]. While the market is excited about 5G, it also has the potential to be used for consumer behavior study and influence, so it is worth investigating. The 6G standard is only relevant as a conceptual guide for the time being, as it has not yet been finalized. Asset tracking and management, intervention planning and monitoring, remote surgery, cloud-based robotics, and remote monitoring are just a few of the potential applications for 5G technology that have been proposed [3]. Other conceivable applications include smart homes and cities, smart medicine, automated passenger transportation, and self-driving automobiles, to name a few examples. Figure 7.1 describes the feature, advantages, and downside of future vehicular communication.

Several applications, such as multiple-view holograms, could need data rates that are beyond the capability of 5G. Therefore, future use cases may need 6G rather than 5G network deployment. End-users may not reap the benefits of 5G until smartphones become bottlenecks for data transfer speeds in 2020, according to industry estimates [5].

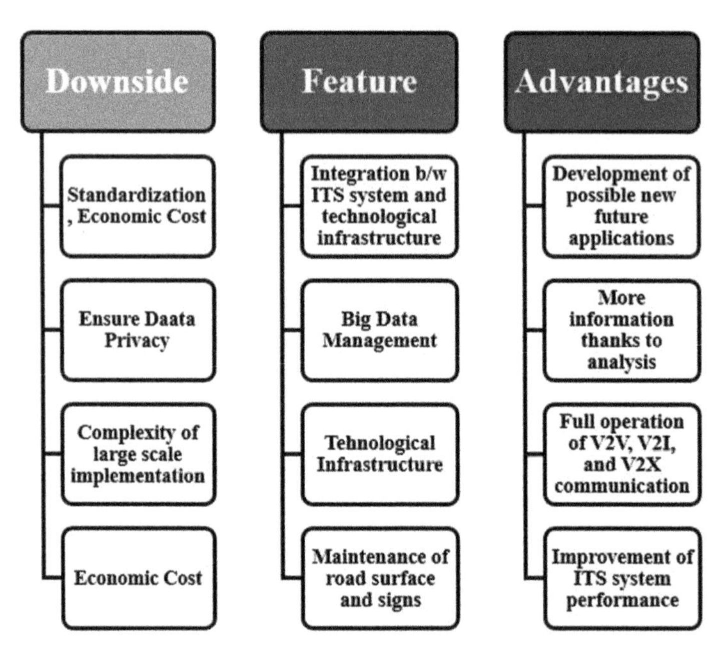

Figure 7.1 Feature, advantages, and downside of future vehicular communication [4]

As indicated by recent technological advances such as 3G and 4G long-term evolution (LTE), mobile broadband networks are poised to make even larger strides in the future decade [6]. It is critical to include both 5G and 6G applications while planning for 6G to be fully prepared. Almost everything indicates that mobile technologies will continue to be crucial for internet connectivity in the future and that 5G and 6G will enable people to adjust their behavior in a way that is more environmentally friendly in the long term [7]. A fundamental theme of Agenda 2030, the United Nations' long-term development strategy, is the need of leaving no one behind. According to this viewpoint, technological advancements should not be limited to modern metropolitan contexts. According to Figure 7.2, cars and infrastructure can exchange four different types of communications in Vehicular Ad Hoc Network (VANET) systems.

This section contains notable establishments such as nearby eating establishments, gas stations, and tourist destinations, among other things [8]. A plethora of articles discusses VANET standards, routing protocols, or a combination of these two concepts. After being identified as a vital component of the intelligent transportation system (ITS), vehicular communications and networking (VCN) have attracted significant attention since its introduction in the early 2000s. VCN must have high dependability and low latency to enable a diverse variety of applications, such as vehicle safety, entertainment, and transportation efficiency [9]. When it came to major research and standardization initiatives, the VANET was the first to make use of mobile architecture. Compared to the VANET area, the growth of

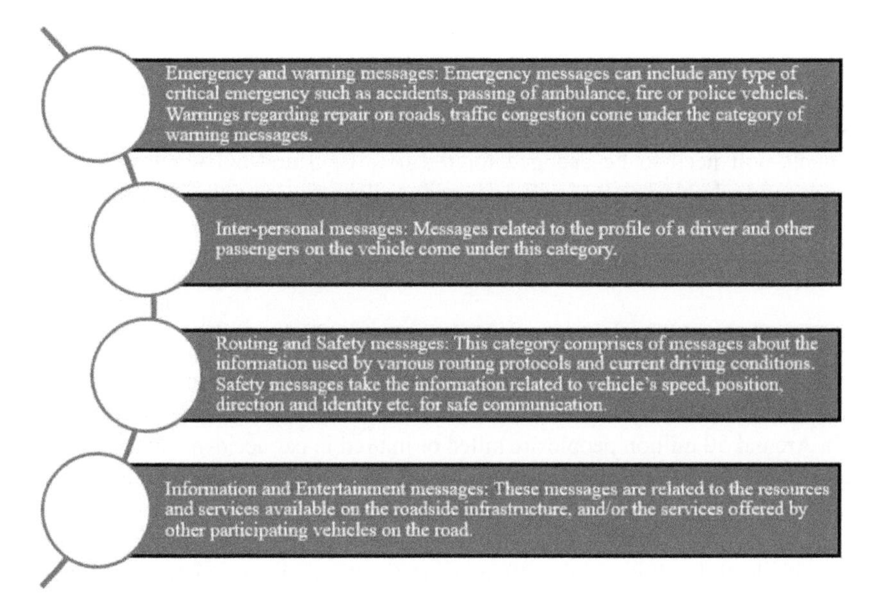

Figure 7.2 Cars and infrastructure can exchange four different types of communications in VANET systems

cellular networks, maturation of the business model, and standardization of cellular networks have all occurred far more quickly [10]. Beginning in 1983, when the first cellular system generation (1G) was developed and established as an industry standard, a whole new cellular system generation has been introduced every decade afterward. In terms of establishing and commercializing the 4G LTE network, tremendous progress has been accomplished so far in this process [11]. Beyond the fifth-generation (B5G) and sixth-generation (6G) periods, vehicle communication systems are expected to provide more complex vehicle-to-everything (V2X) services to meet the needs of an ever-growing population. To support future car applications such as autonomous and intelligent driving, vehicle communications must have extremely low latency for massive data flows and extraordinarily high dependability.

This is especially important for 5G and 6G networks [12]. As a result, the construction of a B5G and 6G vehicle communication system is quite sophisticated, and it draws a great deal of attention. It is necessary to have precise channel information and correct model channels to have effective communication systems, particularly in dynamic high-mobility situations [13]. So far, just a few research publications have been published on the topic of vehicle communication modeling. The research works described above provide good descriptions and points of view on vehicle communication channel measurements and channel modeling, and they are highly recommended. For their part, these studies only looked at vehicle communication channels operating at frequencies lower than 6 gigahertz and only looked at conventional VANET communication systems [14]. A popular study issue for B5G and 6G vehicle communications has risen in recent years due to the rapid expansion of VCN's LTE-V. However, no papers evaluating B5G and 6G vehicular channels have yet to be publicly available. The measurements and modeling of vehicle B5G and 6G frequency band channels are now being examined and analyzed. The fact that measurements and modeling of vehicle communication channels will need to be changed for the new B5G and 6G wireless frequency ranges and technologies that will be introduced shortly is particularly noteworthy. To develop better communication systems, it is necessary to fit the system to the available channel [15].

7.1.1 Safety benefits

Aiming to make driving safer while also reducing the financial damage caused by accidents, the vast majority of vehicle communication systems are designed to do both. Around 50 million people are killed or injured in car accidents each year in the United States [16]. If no preventive actions are taken, traffic-related deaths will grow to the third most common cause of death by 2020. Each year, according to the American Automobile Association, car accidents cost the United States $300 billion in lost productivity. It is a highly effective traffic management technique that may be used to automate crossings [17]. Car accidents, on the other hand, are the primary cause of deaths that occur as a result of them. Local warning systems, together with vehicle communications, can significantly reduce this occurrence. When cars arrive on the scene, they can send warning signs to oncoming traffic, and when they leave, they can inform approaching traffic of their desire to exit the roadway. Additionally, if they need to change lanes or if traffic is held up, they can alert the other driver of their intentions [18].

The rest of the chapter follows this pattern: Section 7.2 discusses the background of advanced vehicle motion control and LTE/5G/6G for vehicular communications. VoLTE/4G/5G/6G has been described in Section 7.3. Section 7.4 describes control over LTE; control over 5G is described in Section 7.5; and control over 6G is described in Section 7.6. 5G over 6G and opportunities for Society 5.0 are discussed in Section 7.7, and Section 7.8 describes the opportunities with 5G toward the future vehicle. As a final note, the chapter wraps up with a summary of the findings in Section 7.9.

7.2 Related/background work

The 1970s were a watershed decade in the history of automotive communications. Beginning in the United States and Japan, work on programs such as the electronic route guidance system and the comprehensive automobile traffic control system was underway. Inter-vehicle communication (IVC) became more popular in the late 1970s and early 1980s [19]. Before the beginning of standardization initiatives, the usage of diverse media such as lasers, infrared, and radio waves was widespread. Between 1986 and 1997, the PATH project, centered in the United States, made significant advances in automotive communications. The PROMETHEUS project in Europe, which began at the same time, was the impetus for the development of vehicle communications. VANET technology was first launched in the early 2000s as a vehicle application of mobile ad hoc networks technology, and it has since gained widespread adoption. The phrases VANET and IVC are interchangeable when referring to communications between cars that are either dependent on or independent of roadside infrastructure, contrary to what some have asserted in the past [20].

Guo *et al.* (2021): With the advent of 5G, many experts and researchers are already looking forward to 6G. According to these experts, 6G will be the most important technology for information exchange and social interaction after 2030. AI-powered 6G is a fully autonomous closed-loop network that makes up for 5G’s shortcomings in communications and processing while also providing global coverage and “AI of things.” The goal of vehicle development will be to create clean, safe, and fully autonomous automobiles, as cellphones are expected to become less important in the future of 6G living. To keep drivers and passengers safe and engaged in future automobiles, more 6G vehicular intelligence research is required. This paper has discussed potential technological advancements and applications in networking, communications, computing, and intelligence [21].

Ekram Hossain *et al.* (2021): ITSs in smart cities are one of the most promising future uses of the Internet of Things (IoT), and the Internet of Vehicles (IoV) has been acknowledged as a critical technology for their development. The sixth-generation (6G) of communications technology will see massive network infrastructures widely disseminated and the number of network nodes rise at an exponential rate, resulting in extremely high energy consumption. It has become increasingly fashionable in the post-6G age to work on green IoV for sustainable vehicle communication and networking technologies. In this article, we discuss the most important green IoV concerns from five different perspectives, including communication, computing, and traffic, as well as electric vehicle management and energy harvesting. An analysis of the literature on energy optimization is conducted for each scenario, including resource allocation and workload scheduling as well as the associated aspects that have an impact on the system's overall efficiency (e.g., resource limitation, channel state, network topology, traffic condition, etc.) [22].

Barbieri *et al.* (2021): For a while now, research on smart connected cars has been centered on how to integrate V2X networks with machine learning (ML) technology and distributed decision-making. Instead of using raw sensor data via V2X links, federated learning (FL) allows automobiles to collectively train a deep ML model by swapping model parameters. Early FL approaches made use of a parameter server (PS) as an edge device to orchestrate the learning process. For fog systems where automobiles share model parameters and sync them using consensus approaches, novel FL tools are being developed. These tools use low-latency V2X communications as an alternative to PS. According to current studies, distributed FL techniques are being investigated for improving Lidar-based road user/object categorization in this work. A novel modular, decentralized approach to FL is used to describe our consensus-driven FL (C-FL) solution, which is tailored for use with point-net-compliant deep ML architectures and Lidar point cloud processing for road actor categorization. The C-FL method is evaluated by simulating a real V2X network and exchanging point net model parameters via the collective perception service. During testing, various factors are taken into consideration, including how connected a vehicle's network is, the benefits of continuous learning overusing heterogeneous training data, and the time it takes for convergence to occur [23].

Dibaei *et al.* (2022): Thanks to recent developments in communication technology and its diverse uses, vehicular networks are already a reality. When it comes to automotive networks, cyber-security is an issue that has yet to be overcome, new attack defense mechanisms are needed. An introduction to vehicular networking communication technologies, as well as how they are being employed, is presented at the beginning of this article. In the following step, we will look into emerging technologies like machine learning and blockchain to see if they can be used to safeguard vehicle networks from cyber-attacks [24].

Mizmizi *et al.* (2021): Mobility solutions that are intelligent, autonomous, and connected are pushing the development of cutting-edge, yet adaptable, sixth-generation cellular networks. An important change in vehicle communication is anticipated with the shift to millimeter-wave or sub-terahertz communications using beams. This would necessitate accurate beam pointing to maintain the communication link, especially in environments with high mobility. An essential design component is a quick and proactive initial access (IA) algorithm for identifying the ideal beam. To create a more efficient 6G system, we investigate several IA tactics to assist in speeding up the current 5G standard. As shown by numerical simulations on realistic settings, probabilistic codebook-based beam selection outperforms the 5G standard in terms of the number of IA trials and performance comparable to position-based techniques [25].

Almohamad *et al.* (2021): In today's world, the next generation (5G scheme and beyond) of wireless communication networks is rapidly growing. Many of today's difficulties might be solved with the help of these strategies, which would ensure that we have reliable communications in all areas of our lives in the future. Fast-moving vehicles (such as trains or planes) have blind spots that make universal coverage and steady connectivity performance problematic. The Third Generation Partnership Project (3GPP) proposed an early rule for V2X networks based on LTE to enable dependable interconnection solutions for V2X. The Federal Communications Commission (FCC) endorsed the idea. This word should encompass communications among vehicles, infrastructure, and pedestrians. It could also refer to communications among vehicles and networks (V2N), infrastructure (V2I), and pedestrians (V2P). An increase in the efficiency, safety, and security of V2X communications may be feasible, as may be the availability of V2X entertainment services in general (any service of user interface exists inside a vehicle). This chapter will go through all of the previously listed topics in depth [26].

Wang *et al.* (2021): The IoV infrastructure relies heavily on cellular vehicle-to-everything (C-V2X) technology. C-V2X core technologies, network topologies, and service application scenarios have been studied by numerous standards bodies throughout the world. Advances in the IoT have upped the bar for 5G mobile broadband networks, necessitating better bandwidth, lower latency, higher concurrency, and less power. On-board units (OBUs) and roadside units (RSUs) are crucial terminal units in the IoV architecture, and this article investigates 5G and C-V2X technologies to help. Starting with an overview of present IoV architecture, it goes on to discuss IoV technology and standards evolution [27].

Noh *et al.* (2021): Evaluating the viability and effectiveness of the system on a real highway with V2I communication utilizing millimeter-wave, several approaches are being investigated by the author in an attempt to improve in-vehicle QoS while also combating the performance degradation caused by high mobility in a millimeter-wave-based system. These include a user-oriented beam switching beamforming scheme, a fast and reliable handover procedure, and mobile relaying. The suggested system design and accompanying fundamental technologies are validated on a genuine highway test site using a real-world testbed. Provide the trial's experimental results and the testbed's implementation characteristics. The validation results show that high data rate vehicular communication over a highway is both viable and cost-effective using 5G New Radio (NR)-based millimeter-wave technology [28].

Pan Hui *et al.* (2021): In the early 2030s, most people will have access to the mobile networks of the sixth generation. Internet-connected autonomous robots are becoming increasingly dense, with hundreds of gadgets per cubic meter by that time. Their only interaction is between each other and remote servers at the network's core and edge. Machine-type communications (MTCs) refer to communication between machines. To use MTC, a large number of devices must be used to collect and process multi-dimensional information simultaneously. Regardless of the scenario, wireless communication is used in both cases. Driverless autos, crewless drones, smart grid energy trading, and many more are instances of this. Compute, energy and communication will all be converted in 6G, making it possible to communicate with devices and applications simultaneously [29].

Bagheri *et al.* (2021): While 5G NR is widely expected to enable connected and cooperative self-driving, it is not well known that it will. For the time being, it is unclear how 5G NR will integrate into connected autonomous communication networks in the near future. The purpose of this post is to address the research gap on 5G NR by describing the various technology components. We will also look at the transition from the existing cellular V2X to NR-V2X. There are several essential features and functionalities that we focus on, such as communication and resource allocation on the side links, architecture flexibility, security and privacy procedures, and precise positional algorithm development at the physical layer [30].

Farah Mahdi *et al.* (2021): The statistical model for the visible light communication channel in a vehicle for dynamic systems should be explained. The proposed model takes into account the loss of the channel's inter-vehicle spacing variation and geometrical changes in inter-vehicular spacing. The research also makes use of a realistic traffic flow model that takes into account fluctuations in traffic density throughout the day. With the use of Monte Carlo simulation, we test the statistical model under a variety of transmission scenarios. The Kolmogorov–Smirnov test has been used to confirm that the statistical model and Monte Carlo simulation represent the same distribution. In light of the findings, traffic volume and vehicle spacing have an impact on the statistical distribution of the channel, causing it to fluctuate during the day. Only one reflector results in larger route loss values for the non-line-of-sight component than for the channel's line-of-sight component. Reduced inter-vehicle spacing during rush hours reduces path loss values considerably for several reflections [31].

Silva *et al.* (2021): There are numerous ways to set up a vehicle's network. Here, we will talk about what VANET is and how it is different from the IoVs. Implementing each of these paradigms satisfies the requirements of existing smart transportation systems. Background information and logical reasoning are provided to back up the author's arguments for each paradigm. There is therefore no conflict between the application of vehicular fog computing and that of vehicular cloud computing. The author discusses some of the security issues associated with these paradigms and vehicular networks, showing how they work together to address problems and overcome restrictions [32].

7.3 VoLTE vs 4G vs 5G vs 6G

To make new goods and compare them with one other easier, standards have been developed. A wide range of wireless standards is available to provide radio access for communication between vehicles, between vehicles and infrastructure, or between infrastructure and infrastructure. The main objective of these communication standards is to promote road safety, traffic efficiency, and driver and passenger comfort by allowing a variety of comfort applications. VANETs mostly use cellular access technologies such as 2G/3G/4G/5G/6G, dedicated short-range communication (DSRC), and wireless access in vehicular environment (WAVE) [33].

7.3.1 Cellular access in vehicular environment

The design of cellular access relies heavily on reusing the restricted frequency for many services. For data rates, the global system for mobiles (GSM) is the most well-known and widely used cellular system. Cellular service of the second generation (2G) was first offered in Finland in the year 1991 [34]. It is important to note that GSM makes use of two different types of transmission methods: time-division multiple access (TDMA) and frequency-division multiple access (FDMA). TDMA is the most extensively employed. However, the data rate at which multimedia data could be delivered over GPRS was insufficient. Up to 2 Mbps of data transfer speed is possible with the 2005-created universal mobile telecommunication system (UMTS). As an improved version of UMTS, 3G (high-speed downlink packet access) was created in 2008. In these two sectors, there were no GSM specifications for high-speed continuous data transfer rates, so the GSM association stepped in to fill the hole. In recent years, cellular communication standards have progressed from 3G to 4G. Mobile data transmission speeds must be increased while remaining high with 4G [35]. It was in North America in 2010 that the LTE standard was developed. Data transfer speeds, mobility, and smooth handoff are all projected to improve dramatically with the advent of 4G technology. Studies reveal that 4G does certainly fulfill these criteria, although they have not been implemented yet. Cellular technology is becoming more popular as a way to take advantage of pre-existing infrastructure. Cellular networks, on the other hand, have a problem with latency due to the existence of base stations. DSRC has solved this issue by eliminating the need for a communication base station.

7.3.2 Dedicated short-range communication

DSRC was established in 2003 by Europe and Japan to support principally communications between vehicles and between vehicles and infrastructure [36]. Europe and Japan introduced DSRC. By supporting low-overhead operations, DSRC hopes to make communication networks run more smoothly. For instance, traffic and accident information, road conditions, and warnings concerning vehicle safety and toll collection could all be included in these communications. By using DSRC, high data transmission rates can be achieved while reducing communication latency. There are no restrictions on the use of the radiofrequency spectrum in the DSRC. The FCC, on the other hand, gives it away for free to limit spectrum usage to reduce congestion and, as a result, charge for its use [37]. The FCC regulates radio stations, for example, and they must use specific channels to comply. The DSRC spectrum has seven channels. The control signal occupies one channel, while the service signals use the other six. The control channel is used to provide urgent messages and management data to make sure everything is working well. Other data is transferred over service channels that can be turned on and off while the control channel is being watched. DSRC standards used by different countries, such as the United States, Europe, and Japan, can be found in the literature [38].

7.3.3 Wireless access in vehicular environment (WAVE)

The Wi-Fi or wireless local area network can enable vehicle-to-vehicle (V2V) or vehicle-to-infrastructure (V2I) connection [39]. Wi-Fi provides a better throughput but comes at a higher cost due to the addition of wireless routers and adapters. Data interchange speeds are required to keep vehicles connected in real time. According to IEEE Standard 1609, wireless communications for vehicle-based basic components (such as RSUs and OBUs) are defined by the communication model and security measures employed. To get around the limitations of IP, WAVE applications might make use of the short message protocol (WSMP). The WAVE standard describes devices of all kinds, including stationary and mobile ones. OBU, which is a mobile device, can either be a service provider or an end customer [40].

7.4 Control over LTE

LTE has recently become more popular among car manufacturers as a communication access method. LTE is the most promising mobile broadband technology because of its high data speeds and minimal latency [41]. All cellular networks have advantages in terms of wide coverage, high penetration, and high-speed terminal support. Including automotive applications in its utilization would open up new income opportunities for telcos and service providers. People spend the most time in their cars, which are the third most popular choice after their homes and employment. LTE is well suited to fulfill the high bandwidth demands and quality of service requirements of this category of car applications, and the term "information and entertainment" (infotainment) encompasses both conventional and emerging internet applications (e.g., content download, media streaming, voice over internet protocol, web browsing, social networking, blog uploading, gaming, cloud access). No one knows yet if it will support car-specific apps that improve traffic efficiency and increase traffic security on public roads [42]. The adoption of centralized LTE design raises concerns regarding message latency since it necessitates the interchange of data across infrastructure nodes even when it is not required for safety-critical applications. Message broadcasts from several vehicles in high-traffic areas put a tremendous load on LTE bandwidth, which could compromise traditional service delivery.

7.4.1 LTE as a vehicle application support technology

LTE's application in vehicles has numerous factors to do with the concerns shown in Figure 7.3.

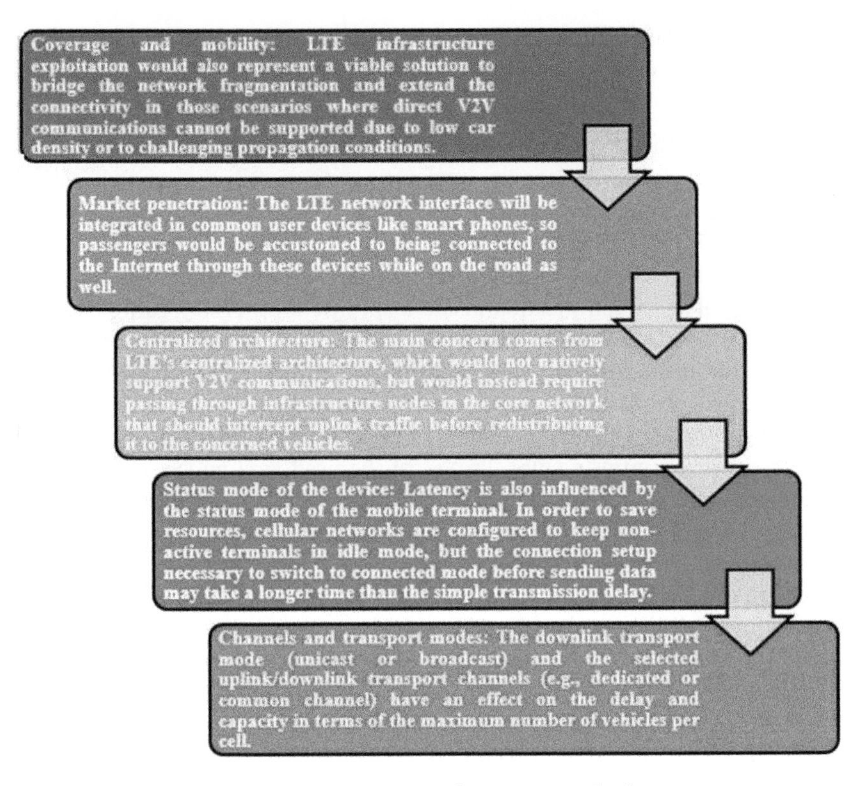

Figure 7.3 LTE's application in vehicles

7.5 Control over 5G

There are two major advantages to adopting 5G: interoperability with current 4G LTE networks and the high-performance levels required for vehicular communications [43]. Wi-Fi and unlicensed airwaves can now be used in 4G LTE networks, thanks to 4G LTE. In the future, 5G networks will be able to incorporate legacy systems like 3G, 4G, Wi-Fi, Zig-Bee, and Bluetooth, boosting their capabilities even further. Vehicles and passengers will be able to utilize this functionality to connect to the best network for safety, non-safety, and infotainment (such as content sharing). To keep up with the growing number of connected vehicles on the road, 5G would necessitate novel approaches such as spectrum sensing and direct device identification node coordination systems [44]. Direct device recognition and communication are now possible with 4G LTE without the need for infrastructure.

To enable revolutionary proximity-based services and data access, 5G would necessitate device-to-device communication over wider coverage areas. Nodes working together can also meet capacity requirements [45]. 4G LTE features intercell cooperation by design. Cloud radio access networks can help with 5G base station coordination. 5G intends to make better use of spectrum by spreading it more efficiently among its users (primary, secondary, and tertiary).

7.5.1 5G vehicular communications building blocks

Beaconing platform for proximity services for 5G-enabled communications to function, the proximity service must be available. Using geolocation data, ProSe's present purpose is to increase public awareness of devices and services that are available [46]. In the context of a 5G-enabled IoT, there are numerous potentials for chance encounters or conversation inside a specific area. A key part of the functionality of ProSe-based apps is the ability to use location data and communicate through social media networks. Instead of usual network-based location discovery (like Facebook), ProSe offers ad hoc location discovery and communication alternatives (e.g., moving vehicles on roads). Because it can discover and communicate without relying on infrastructure, ProSe can be used as a public safety communication platform. As a result of the lack of latency imposed by the core, faster data rates may be sent, resulting in greater resource efficiency [47]; to put it another way, i.e., by avoiding transmissions passing through the core network). As a result, the core network experiences less congestion. To make full use of ProSe's basic functionalities, all user equipment (UE) must be subscribed to an evolved Node B (eNB) (considering that the coverage is provided by the same eNB, although other configurations are also possible). If that is the case, then any nearby LTE-licensed UE can use the licensed spectrum to broadcast service.

The UEs that are within the service advertising coverage zones find all of the services available. To make sure that only relevant services are found, users can utilize the appropriate filters. Following service discovery by a UE, the advertised content can be retrieved across the 5G network. For safety and non-safety applications alike, 5G may make use of this form of communication in a vehicle network [48]. This is also an option if the data path may be accessed via the local eNB. The ProSe is a great place to start when it comes to 5G safety communications in vehicles. Security attack sources tracking is a key use case in the linked autonomous vehicle (AV) scenario, together with vehicular safety application communication [49]. One of kind security risks come with self-driving autos. Avast can be protected from remote assaults using encryption techniques, but finding out where the attack originated is challenging. Caching expires quickly because of the frequency with which messages are sent and the limited shelf life of the cached version of them [50]. As an alternative, the locally routed data stream can be used to identify the attacker's location. As observed in Figure 7.4, stations can be put to several uses.

7.6 Control over 6G

6G needs to be better than 5G in many areas. For starters, higher frequencies will be used in 6G than in 5G to send data at 100–1000 times the rate of 5G. With a positioning precision of 10 cm indoors and 1 m outdoors in comparison to 5G, 6G

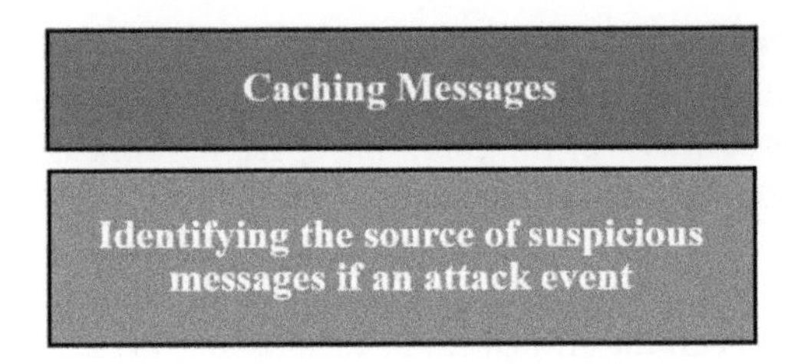

Figure 7.4 Caching messages

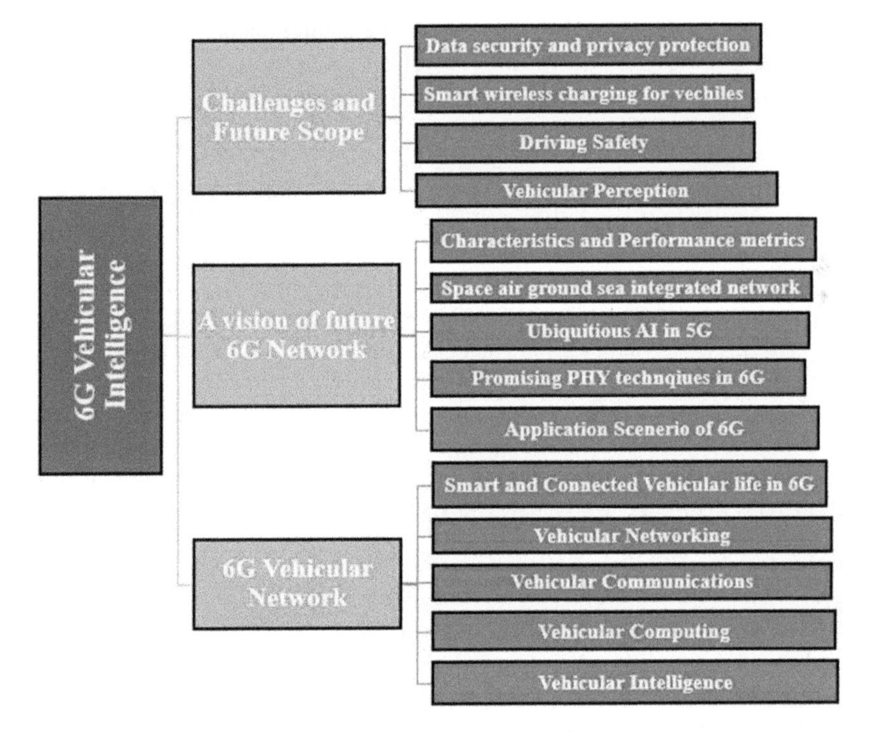

Figure 7.5 Overview of 6G [52]

is 10 times more precise [51]. The timing and phase synchronization standards for 6G are also projected to be higher than those of the 5G standard. Though 6G characteristics will be superior to those of the 5G network, it is projected that the 6G network will provide wireless access for less than 0.1 dollars per year, 1000 times less than the old 5G system. Figure 7.5 shows an overview of 6G.

7.6.1 An integrated network of land, air, and sea

The majority of experts believe that the 6G network will be a multi-modal one, connecting space, air, land, and sea [53].

7.6.1.1 Space network

Existing networks on the ground could be expanded via the space network. In terms of distance from the surface of the earth, the space network is the farthest away. Satellites in three orbital classes are used in this system: geostationary, medium, and low (LEO). There is an increasing reliance on satellite-based satellite control networks (SCNs) for communications in maritime, emergency rescue, and other fields. In light of the current difficulties in establishing IoT ecosystems in remote areas, SCNs may be able to guide worldwide IoT implementation [54]. As a result, IoT satellite solutions have emerged, and LEO satellites are better suited to these services due to their advantages over other satellite types in terms of power, propagation delay, and range. As the number of satellites increases and spacecraft become smaller, such as nanosatellites, precision and quality will improve dramatically for ground users.

7.6.1.2 Air network

The air network is designed with mobility, altitude adjustment, and intelligent placement in mind. Aircraft like an airplane, hot air balloons, planes, and high-altitude platforms (HAPs) such as an unmanned aerial vehicle (UAV) are all employed. A UAV outfitted with plenty of resources might potentially operate as a ground station because building and deploying a flying base station costs less than doing so [55]. The second benefit is that HAPs are faster and have a larger coverage area than terrestrial networks because of their lower latency. Third, low-cost UAVs are widely used in industries such as network coverage and transportation because of their excellent mobility and hovering characteristics. To make satellite services more convenient for clients, FBSs can also operate as ground to satellite communication relays. Emergency vehicles in metropolitan areas can be guided to the most direct routes by UAVs.

7.6.1.3 Ground network

As opposed to air and space networks, the 6G network's backbone is built on the ground. It is expected that the 6G ground network will be a heterogeneous, ultra-dense network made up of buildings, smart vehicles, and other ground-based infrastructure. The equipment and facilities of the ground network are all linked together [56].

7.6.1.4 Sea-based network

Future oceanographic data collection and military activities will require the support of a sea-based network. The majority of the network is made up of sea-based vehicles, such as ships, submarines, underwater gliders, and others. Due to a lack of computing capacity and a complex environment, the current sea-based network is unable to precisely place ships in real time, among other problems. The development of key underwater wireless communication techniques, such as radio-frequency, acoustic, and optical communications, will allow for high-speed data transfer between underwater devices and an increase of the sea-based network's scope.

7.6.1.5 Ubiquitous artificial intelligence (AI) in 6G

The network will be incredibly complicated, with numerous nodes and levels. New 6G networks may have a more human-centric, adaptive architecture than current ones. As the number of users and devices increases, so will the amount of data have created, making AI with data-driven characteristics more crucial on the 6G network [57]. Today, AI is emerging as a critical enabler capable of coordinating network resources at all levels, allowing it to be used in both current and next-generation systems. The use of AI can thereby help operators administer the 6G network in a fully autonomous closed-loop fashion and reduce manual labor requirements. Numerous studies have been conducted on the 6G network's structure. Computers and safety monitoring have both benefited from AI's use in addition to communications. Using each one of these applications, we will show how important AI is for the entire 6G network and how it is everywhere [58].

7.6.2 Promising physical (PHY) techniques in 6G

7.6.2.1 Ultra-massive multi-input multi-output (UM-MIMO)

It is feasible to supply more users with low-latency services while improving energy and spectrum efficiency by using more antennas on transmitters and receivers. Because of the growth in the size of MIMO, it has been renamed massive multi-input multi-output (M-MIMO or even UM-MIMO). Since researchers are considering putting MIMO technology on reconfigurable intelligent surfaces, the enormous surface area of Reconfigurable Intelligent Surfaces (RIS) could provide a suitable location for antenna deployment [59].

7.6.2.2 Non-orthogonal multiple access (NOMA)

There are numerous ways in which NOMA varies from standard orthogonal multiple access (OMA). It employs non-orthogonal transmission at both the transmitting and receiving ends, introduces interference information on purpose, and utilizes SIC technology for proper demodulation. Because of the sheer number of devices on 6G, traditional OMA can only allocate one wireless resource per user. It is possible to increase the number of connections and increase the efficiency of the spectrum by employing the NOMA technique. Improved user fairness and connection density can be achieved using a cognitive radio network that makes use of NOMA technology [60].

7.6.2.3 Directional antennas (DAs)

Omni-DAs are frequently used in mobile networks, particularly ad hoc ones, due to their simple design and low cost. There are several limitations to using these devices in a communication network, though. Signal strength and anti-interference performance can both be improved while radiated power usage is increased when these devices are used correctly. About 6G, THz is one of the most widely anticipated communication methods [61].

7.6.2.4 Orbital angular momentum (OAM)

OAM is a fundamental quantity in mechanics and electromagnetics with an infinite state space. The use of orthogonal electromagnetic waves in OAM multiplexing technology can improve spectrum efficiency while also pointing mode multiplexing technologies in a new path. OAM mode multiplexing and regular MIMO technologies can be combined to boost capacity even further. For example, rate-splitting multiple access technology, vector digital signal channel estimation software, and others have been investigated or proposed by researchers [62].

7.6.3 Application scenarios of 6G

As previously said, 6G will have a significant impact on our society and daily lives in various ways, as observed in Figure 7.6.

7.6.4 Summary of existing literature on 6G

For the time being, 6G research is mostly concerned with the future potential of the technology. Before doing anything else, they should air their views on 6G. Various 6G performance indicators have been demonstrated by comparing them to 4G and 5G to show that, in their perspective, the 6G network is a three-dimensional (3D) space, air, ground, and sea integrated network. Second, they showed a lot of interest in investigating the 6G spectrum's perspective communication technologies [63]. New technologies, as well as enhancements to present ones, are necessary for 6G communication rates and an ultra-dense communication environment. 6G networking and edge computing have also received a lot of attention due to the work being done in this area. Sixth-generation networks were considered perfect because of the privacy and security benefits of blockchain technology. Finally, it was emphasized that the 6G network will utilize AI and machine learning. Future 6G networks will be smarter, they found, because they will be using machine learning technology like AI. They were excited about the potential of future 6G network analytics.

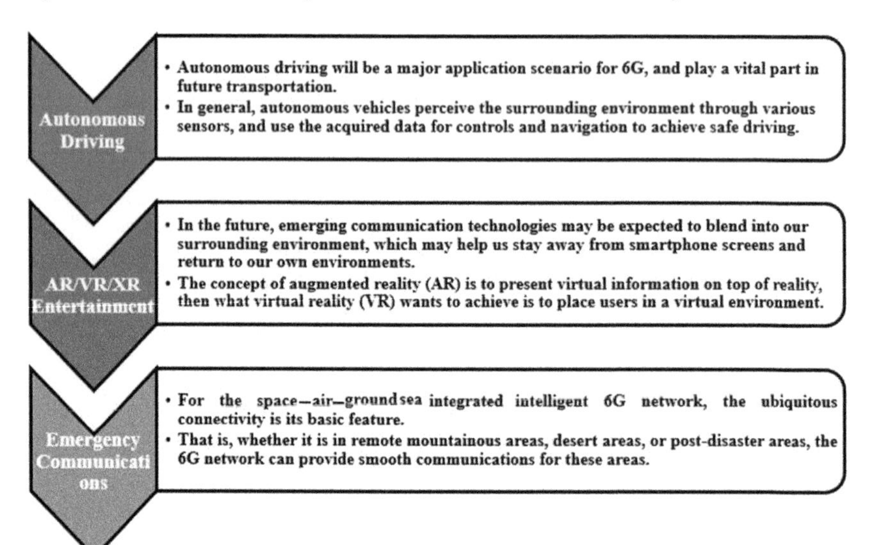

Figure 7.6 Application scenarios of 6G

7.7 5G over 6G: opportunities for Society 5.0

Higher requirements and more support will be needed shortly for new B5G and 6G applications such as self-driving and intelligent vehicles. Massive volumes of data are required for vehicular 5G and 6G connectivity [64]. Enhancing spectrum efficiency and expanding transmission bandwidth are two major approaches used by wireless systems to boost transmission speeds. Wireless system users are familiar with these strategies. The efficiency of the B5G and 6G spectrum can also be improved by a wide range of technologies. Broadband 5G and 6G communication systems are looking into using this frequency band with GHz/THz capacity to expand their available spectrum. Using THz communications in 6G networks has several clear advantages. There are significant differences in the ultra-high-frequency band's channel parameters measured in millimeter-wave/THz channels compared to those in frequency bands below 6 GHz. Large-scale MIMO may now be implemented in smaller and more portable packaging, thanks to the development of millimeter-wave and THz [65–67].

7.7.1 Vehicular channel measurements for B5G and 6G

Transmission and reception (Tx/Rx) of vehicular communication channels use low-elevation antennas. Around Tx and Rx, there are also multiple highly mobile scatterers/clusters that cause them to move rapidly. In other words, cellular channels do not have the same features as vehicular channels. Some vehicle communication channels operating below 6 GHz have been measured using one to four antennas. As a result of new technologies like UM-MIMO and millimeter-wave/THz, the 5G and 6G vehicle channels have unique characteristics. Channel measurements and channel modeling are linked and affect one another; therefore keep this in mind when performing your analysis. Channel measurements are most often used as a starting point for channel modeling and as a way to verify the final results before they are implemented. For the time being, channel modeling will assist in the development of channel measurements. Sections on 5G and 6G vehicular channel measurements will discuss millimeter-wave/THz and UM-MIMO briefly.

7.7.1.1 Millimeter-wave/THz perspective

When compared to lower frequency ranges, such as those below 6 GHz, vehicle communication channel measurements in the millimeter-wave/THz region are still in their infancy. Today, millimeter-wave/THz channel measurements make considerable use of antennas with a concentrated beam. Realistic millimeter-wave channels in portable form. Millimeter-wave channels for car wireless transmission have a moderate multi-path dispersion, according to the data. The above-mentioned data can be used to derive the fundamental small- and large-scale properties of vehicular millimeter-wave/THz communication channels, such as the lesser number of multi-paths in a cluster, increased shadowing, and higher path loss. In addition to these features, higher delay resolution is also found in vehicle millimeter-wave/THz channels. This scenario allows for the resolution of rays included within clusters.

7.7.1.2 UM-MIMO perspective

UM-MIMO channel measurements are notoriously difficult and time expensive in comparison to regular MIMO. Virtual antenna arrays are presently a regularly used testing approach for UM-MIMO in the 6 GHz band and lower, with an emphasis on revealing variations and features of channels with different antenna layouts. Aspherical wavefront and spatial non-stationarity in UM-MIMO scenarios below 6 GHz were found in measurements in typical cellular setups below this frequency range. Massive MIMO systems use a high number of antennas to cover a broad area, and as a result, these arrays are likely to be closer than the Rayleigh distance. UM-MIMO channels will appear to have a spherical wavefront due to the near-field effect. As a result, the conventional wisdom that assumes a flat wavefront is incorrect. In antenna array evolution, clusters are (re)appearing due to the incorporation of an (ultra-)large-scale antenna array in UM-MIMO. Spatial non-stationarity, that is, non-stationarity in the space domain, may result from different antennas having different collections of observable clusters.

7.7.1.3 Vehicular channel models for B5G and 6G

Depending on the modeling approach used to construct them, channel models can be classified into a plethora of different sorts. It is possible to classify channel models in various ways. For instance, they can be categorized as deterministic or stochastic. These sub-categories cross-reference one other. When it comes to VCN channel modeling, the distinction between deterministic and stochastic behavior is frequently drawn upon. Physical channels are predetermined according to models with deterministic channel behavior. Geometry-based stochastic models (GBDMs) imitate radio transmission in a specific environment using electromagnetic field theory. To define the site-specific propagation scenario, GBDMs must undertake time-consuming and thorough measurements of vehicular scenarios. In stochastic models, channel parameters are randomly represented rather than deterministically. These stochastic models do not take scattering environment geometry into account because they are NGSMs. No assumptions about scattering geometry are made while describing vehicle channels in NGSMs. When compared to NGSMs, GBSMs use simplified versions of the fundamental equations of wave propagation to generate a random distribution of effective scatterers.

7.7.1.4 3D space–time–frequency non-stationarity

Channel non-stationarity, which occurs when statistical properties of a channel change over time in space, time, and frequency, is well-known. Because of the dynamic complexity of the environment, vehicular communications channels exhibit clear time-variant characteristics. In this case, multi-path components develop and dissipate rapidly due to the high Tx, Rx, and scatterer/cluster velocities of the environment. In this scenario, the communication channels and statistics of the channels alter over time. As a result, the time domain, or the channel's temporal domain, is non-stationary. Channel temporal non-stationarity is a property peculiar to vehicle communication channels. Using an (ultra-) large-scale MIMO antenna array can cause channel characteristics to fluctuate throughout its dimensions, as demonstrated in the analysis above, resulting in spatial non-stationarity in the channel. Spatial non-stationarity is a unique property of (ultra-) large MIMO communication systems.

7.8 Opportunities with 5G toward future vehicles

Connected vehicles have long been hailed as the next generation of the automotive industry as the IoT boom spreads. Traveling and using transportation will change when vehicles can connect to our homes, other vehicles, and electronics while on the road. This year's developments will surely affect connected vehicle plans, but there are still obstacles to be solved before they become reality [68–70].

7.8.1 The importance of 5G to vehicle connectivity

In the future, the success of linked cars will depend on how well 5G works. Over-the-air (OTA) updates are critical for keeping vehicle data and systems up to date and functional, but they necessitate a fast connection between the car and a nearby base station to be effective. There is still work to be done on 5G, but one solution is to provide cars with extra flash storage so that data can be sent to a central location for processing.

7.8.2 A rocky 5G rollout exposes storage as essential

Delaying 5G will make it more difficult to link vehicles to the cloud and meet the growing need for better infrastructure. To send data offloading, linked vehicles will need to hold it until they have good enough coverage, which will cause a delay in data transfer. With no coverage or capacity, the automobile will be forced to cache data for the next year and beyond. V2X security keys, software, application, data logging, and OTA buffering are all stored on NAND flash-based devices that are currently in vehicles and will be there in the future.

7.8.3 Moving to 5G with edge computing

Manufacturers of automobiles believe anticipating and notifying you to an abrupt stop would be a significant safety feature When data has to travel a long distance to reach its destination on time, it may be delayed. If 5G and edge computing can be avoided, then manufacturers should explore them as well. Thanks to 5G, automakers and original equipment manufacturers will be able to innovate more quickly. Consequently, data will be easily available and processing workloads can be dynamically altered to better balance on-site research with cloud computing. Self-driving cars and the data computing required to provide innovative safety and infotainment services will be supported on roads full of 5G and the edge, thanks to a significant shift in computing closer to the vehicle. Making the switch to 5G and edge computing, on the other hand, may be challenging if you lack connectivity expertise. In today's highly competitive business, the time it takes to bring new features to market can mean the difference between a mediocre quarter and a successful one for many automakers.

7.8.4 Connected ambulance

One use case for 5G is a connected ambulance. Staff at the hospital have little information when an ambulance is headed to the emergency department. Assume all of the questions an ER doctor might be asked. While the ambulance is still on the road, a robotic arm might be attached in the back seat to allow the doctor to operate on the patient while the vehicle is still moving—thus triaging an important piece or scenario. Virtual reality connectivity may allow a virtual doctor to save a life.

7.8.5 Smart transportation

Vehicles will be able to see around corners before they arrive in a city with 5G connectivity. 5G networks will be implemented at all major intersections to reduce accidents due to blind areas, stop signs running, or running red lights. Other vehicles approaching the crossroads will be provided with real-time traffic data via these networks. 5G has the potential to reduce the number of accidents by handling data more quickly. Vehicles that are always aware of their location in the smart grid will be able to assist with inner-city parking, which is a major source of traffic.

7.8.6 Fleet deployment

Trucking firms are continually transporting something, whether it is commodities or supplies. Because these devices contain potentially toxic substances, a collision with one of them could result in an oil spill on the road, causing serious safety issues for drivers. Preventative maintenance of 5G might help fleets be more proactive in resolving issues. Looking for early indicators of a vehicle problem can help prevent maintenance-related accidents.

7.8.7 Autonomous VANET

Autonomous VANET parking, like last-mile delivery, might provide a contactless 5G experience as well. Instead of paying for VANET service, drivers can use self-parking cars and save time by not looking for a parking spot. As more people own cars and take control of their hygiene and health, one could argue that 5G will provide consumers with greater flexibility and assurance by laying the basis for autonomous features that will provide a seamless experience by 2020.

7.9 Conclusion

Our WM-SIMA tool was used by researchers to simulate links on vehicle channels to show how well LTE works. There are trade-offs to be made while determining which MCS is best for your communication needs. I2V communication coverage can be improved with a lower MCS but at the expense of increased transmission time. Further coverage would be required if multiple neighboring base stations were to transmit at the same time. In the future, 5G will enable a slew of new features for cars that are not possible right now. Although it increases productivity, the technology also has the potential to improve safety to a great extent. Capgemini has created a 5G lab as a service for automotive manufacturers so that they can test new 5G network virtualizations and install them in a safe and secure environment. Being able to stay connected in so many areas of our everyday lives will be exciting and energizing for drivers when we have 5G available in our vehicles. New, high-impact use cases for 5G technology will accelerate its introduction.

Decreased latency for transmitting input to users may lead to new sorts of user behavior and new strategies for influencing it. It is feasible that improving self-monitoring skills will help us exploit existing user behavior and social influence aspects in a more sophisticated way. For example, this might be highly interesting if done with completely customized digital treatments, as well as new computationally intensive virtual rehearsals of both online and offline behaviors that help behavior modification. People and society will feel the effects of these new technological advancements, but those effects will be mixed. Persuasion strategies that use subliminal or manipulative cues are likely to gain traction in the future. Technologies can quickly become invasive or dangerous if they are poorly designed. Modern and evolving information technology can influence human behavior in any scenario; hence the general people should be better informed. New technology designers should think about their societal obligations and seek to fulfill the highest ethical standards feasible when developing new technologies. For scientists to benefit from all of these developments, they must make a deliberate effort to keep on top of the game and stay ahead of the curve.

References

[1] Gui, J, Liu, Y., Deng, X. and Liu, B. "Network capacity optimization for cellular-assisted vehicular systems by online learning-based mmwave beam selection." Wireless Communications and Mobile Computing *2021* (2021). Article ID 8876186, https://doi.org/10.1155/2021/8876186.

[2] Wild, T, Braun V, and Viswanathan H. "Joint design of communication and sensing for beyond 5G and 6G systems." *IEEE Access* 9 (2021): 30845–30857.

[3] Tyagi, A. K., Rekha, G., and Sreenath, N. (eds). *Opportunities and Challenges for Blockchain Technology in Autonomous Vehicles. IGI Global,* 2021. http://doi:10.4018/978-1-7998-3295-9.

[4] Arena, F, and Pau G. "An overview of vehicular communications." *Future Internet* 11(2) (2019): 27.

[5] Li, O, He, J., Zeng, K., *et al.* "Integrated sensing and communication in 6G A prototype of high-resolution THz sensing on portable device." *2021 Joint European Conference on Networks and Communications & 6G Summit (EuCNC/6G Summit)*. IEEE, 2021, 544–549.

[6] Tang, F, Mao, B., Kato, N., and Gui, G. "Comprehensive survey on machine learning in vehicular network: technology, applications, and challenges." *IEEE Communications Surveys & Tutorials* 23(2) (2021): 2027–2057.

[7] Bera, R. "Generation, detection and analysis of sub-terahertz over the air (OTA) test bed for 6G mobile communication use cases." *Generation,*

Detection, and Processing of Terahertz Signals. Springer, Singapore, 2022, 97–122.

[8] Barakat, B, Taha, A., Samson, R., *et al.* "6G opportunities arising from the internet of things use cases: a review paper." *Future Internet* 13(6) (2021): 159.

[9] Kaja, H. Survivable and Reliable Design of Cellular and Vehicular Networks for Safety Applications. Diss. University of Missouri-Kansas City, 2021.

[10] Moser, K. "Current vision of 6G networks: exploring machine learning, holographic communications, and ubiquitous connectivity." *Communication Systems XIV* (2021): 48. https://files.ifi.uzh.ch/CSG/teaching/FS21/IFI_2021_02.pdf

[11] Mohsan, SAH, and Amjad, H. "A comprehensive survey on hybrid wireless networks: practical considerations, challenges, applications, and research directions." *Optical and Quantum Electronics* 53(9) (2021): 1–56.

[12] Nguyen, V-L, Lin, PC, Cheng, BC, Hwang, RH and Lin, YD. "Security and privacy for 6G: a survey on prospective technologies and challenges." *IEEE Communications Surveys & Tutorials* 23(4) (2021): 2384–2428.

[13] Bhat, ZA, Mushtaq, H, Mantoo, JA, Yadav, VS, Shrivastava, AK and Swati, S. "Beyond 5G: reinventing network architecture with 6G." *2021 2nd International Conference on Intelligent Engineering and Management (ICIEM)*. IEEE, 2021, 316–321.

[14] Tripathi, S, Sabu, NV, Gupta, AK, Dhillon, HS. "Millimeter-Wave And Terahertz Spectrum for 6G Wireless." 6G Mobile Wireless Networks. Computer Communications and Networks. Springer, Cham (2021). https://doi.org/10.1007/978-3-030-72777-2_6.

[15] Reebadiya, D, Rathod, T, Gupta, R, Tanwar, S and Kumar, N. "Blockchain-based secure and intelligent sensing scheme for autonomous vehicles activity tracking beyond 5G networks." *Peer-to-Peer Networking and Applications* 14(5) (2021): 2757–2774.

[16] Minokuchi, KAA, and Min, KHT. "5G advanced technologies for creating industries and co-creating solutions." NTT Docomo Technical Journal, 22(3) (2021).

[17] Li, F, Chen, W, and Shui, Y. "Analysis of non-stationarity for 5.9 GHz channel in multiple vehicle-to-vehicle scenarios." *Sensors* 21(11) (2021): 3626.

[18] Mahmood, NH, Böcker, S and Moerman, I. *et al.* "Machine type communications: key drivers and enablers towards the 6G era." *EURASIP Journal on Wireless Communications and Networking* 2021(1) (2021): 1–25.

[19] Li, Y, Huang, J, Sun, Q, Sun, T and Wang, S. "Cognitive service architecture for 6G core network." *IEEE Transactions on Industrial Informatics* 17(10) (2021): 7193–7203.

[20] Rampriya, RS, and Suganya, R. "Review on various communication mechanism for railway patrol using UAV." *2021 5th International Conference on Trends in Electronics and Informatics (ICOEI)*. IEEE, 2021, 617–625.

[21] Guo, H, Zhou, X, Liu, J and Zhang, Y. "Vehicular intelligence in 6G: networking, communications, and computing." *Vehicular Communications* 33 (2022): 100399.

[22] Wang, J, Zhu, K, and Hossain, E. "Green Internet of Vehicles (IoV) in the 6G Era: Toward Sustainable Vehicular Communications and Networking." in *IEEE Transactions on Green Communications and Networking*, 6(1) (2022): 391–423.

[23] Barbieri, L, Savazzi, S, Brambilla, M and Nicoli, M. "Decentralized federated learning for extended sensing in 6G connected vehicles." *Vehicular Communications* 33 (2022): 100396.

[24] Dibaei, M, Zheng, X, Xia, Y., *et al.* "Investigating the prospect of leveraging blockchain and machine learning to secure vehicular networks: a survey." *IEEE Transactions on Intelligent Transportation Systems* 23(2) (2022): 683–700.

[25] Mizmizi, M, Linsalata, F, Brambilla, M, *et al.* "Fastening the initial access in 5G NR sidelink for 6G V2X networks." *Vehicular Communications* 33 (2022), 100402.

[26] Almohamad, TA, Güneşer, MT, Mahmud, MN and Şeker, C. "Improving communication system for vehicle-to-everything networks by using 5G technology." *New Perspectives on Electric Vehicles. IntechOpen*, 2021. 10.5772/intechopen.99394.

[27] Wang, Y, Ning, W, Zhang, S, Yu, H, Cen, H and Wang, S. "Architecture and key terminal technologies of 5G-based internet of vehicles." *Computers & Electrical Engineering* 95 (2021): 107430.

[28] Noh, G, Kim, J, Choi, S, Lee, N, Chung, H and Kim, I. "Feasibility validation of a 5G-enabled mmwave vehicular communication system on a highway." *IEEE Access* 9 (2021): 36535–36546.

[29] Braud, T, Chatzopoulos, D, and Hui, P. "Machine type communications in 6G." *6G Mobile Wireless Networks.* Springer, Cham, 2021, 207–231.

[30] Bagheri, H, Noor-A-Rahim, M, Liu, Z, *et al.* "5G NR-V2X: toward connected and cooperative autonomous driving." *IEEE Communications Standards Magazine* 5(1) (2021): 48–54.

[31] Alsalami, FM, Ahmad, Z, Zvanovec, S, Haigh, PA, Haas, OC and Rajbhandari, S. "Statistical channel modeling of a dynamic vehicular visible light communication system." *Vehicular Communications* 29 (2021): 100339.

[32] Silva, L, Magaia, N, Sousa, B, *et al.* "Computing paradigms in emerging vehicular environments: a review." *IEEE/CAA Journal of Automatica Sinica* 8(3) (2021): 491–511.

[33] Tyagi, AK, Kumari, S, Fernandez, TF, and Aravindan, C. "P3 block: privacy preserved, trusted smart parking allotment for future vehicles of tomorrow." In Gervasi, O. *et al.* (eds), *Computational Science and Its Applications – ICCSA 2020. ICCSA 2020. Lecture Notes in Computer Science*, 12254 (2020). Springer, Cham. https://doi.org/10.1007/978-3-030-58817-5_56.

[34] Taha, A-EM. "Quality of experience in 6G networks: outlook and challenges." *Journal of Sensor and Actuator Networks* 10(1) (2021): 11.

[35] Song, Q, Zeng, Y, Xu, J and Jin, S. "A survey of prototype and experiment for UAV communications." *Science China Information Sciences* 64(4) (2021): 1–21.

[36] Wu, Q, Xu, J, Zeng, Y, *et al.* "A comprehensive overview on 5G-and-beyond networks with UAVs: from communications to sensing and intelligence." *IEEE Journal on Selected Areas in Communications* 39(10) (2021): 2912–2945.

[37] Feng, D, Lai, L, Luo, J, Zhong, Y, Zheng, C and Ying, K. "Ultra-reliable and low-latency communications: applications, opportunities, and challenges." *Science China Information Sciences* 64(2) (2021): 1–12.

[38] Zhang, R, Ning, L, Li, M, Wang, C, Li, W and Wang, Y, "Feature extraction of trajectories for mobility modeling in 5G NB-IoT networks." *Wireless Networks* (2021): 1–13.

[39] Azari, MM, Solanki, S, Chatzinotas, S, *et al.* "Evolution of Non-Terrestrial Networks From 5G to 6G: A Survey."*arXiv preprint arXiv:2107.06881* (2021).

[40] Rovira-Sugranes, A, Razi, A, Afghah, F, and Chakareski, J. "A review of AI-enabled routing protocols for UAV networks: trends, challenges, and future outlook." *Ad Hoc Networks,* 130 (2022): 102790.

[41] Wijethilaka, S, and Liyanage, M. "Survey on network slicing for Internet of Things realization in 5G networks." *IEEE Communications Surveys & Tutorials* 23(2) (2021): 957–994.

[42] Aguiar-Castillo, L, Guerra, V, Rufo, J, Rabadan, J, and Perez-Jimenez, R. "Survey on optical wireless communications-based services applied to the tourism industry: potentials and challenges." *Sensors* 21(18) (2021): 6282.

[43] Liu, Q, Liu, R, Wang, Z, Han, L, and Thompson, JS. "A V2X integrated positioning methodology in ultra-dense networks." *IEEE Internet of Things Journal* 8(23) (2021): 17014–17028.

[44] Choi, S-W, Choi, SN, Cho, D, *et al.* "Performance evaluation of MN system in highway environment." *2021 Joint European Conference on Networks and Communications & 6G Summit (EuCNC/6G Summit)*. IEEE, 2021, 449–453.

[45] Türkmen, H, Rafique, S, and Arslan, H. "Radio environment monitoring." *Design and Analysis of Wireless Communication Signals: A Laboratory-based Approach* (2021). Available at https://onlinelibrary.wiley.com/doi/abs/10.1002/9781119764441.ch13.

[46] Amakawa, S, Aslam, Z, Buckwater, J, *et al.* "White Paper on RF Enabling 6G–Opportunities and Challenges from Technology to Spectrum." (2021). Available at https://oulu.fi/6gflagship/6g-white-paper-rf-spectrum.

[47] Zeb, S, Mahmood, A, Hassan, SA, Piran, MJ, Gidlund, M, and Guizani, M. "Industrial digital twins at the Nexus of NextG wireless networks and computational intelligence: a survey." *Journal of Network and Computer Applications* (2021): 103309.

[48] Ali, ES, Hassan, MB, and Saeed, RA. "Machine learning technologies on the Internet of Vehicles." *Intelligent Technologies for Internet of Vehicles.* Springer, Cham, 2021, 225–252.

[49] Bertizzolo, L. Software-Defined Wireless Networking for 5G and Beyond: From Indoor Cells to Non-Terrestrial UAV Networks. Diss. Northeastern University, 2021.

[50] Arendt, C, Patchou, M, Böcker, S, Tiemann, J, and Wietfeld, C. “Pushing the Limits: Resilience Testing for Mission-Critical Machine-Type Communication.” In *2021 IEEE 94th Vehicular Technology Conference (VTC2021-Fall)* (2021), pp. 1–6. IEEE. Available at: https://kn.e-technik.tu-dortmund.de/.cni-bibliography/cnidoc/1626853048.pdf.

[51] Liu, F, Pan, J, Zhou, X, and Li, GY. “Atmospheric ducting effect in wireless communications: challenges and opportunities.” *Journal of Communications and Information Networks* 6(2) (2021): 101–109.

[52] Malik, UM, Javed, MA, Zeadally, S and Islam, SU. “Energy-efficient fog computing for 6G enabled massive IoT: Recent trends and future opportunities.” *IEEE Internet of Things Journal* (2021). Doi: 10.1109/JIOT.2021.3068056.

[53] Imoize, AL, Adedeji, O, Tandiya, N, and Shetty, S. “6G enabled smart infrastructure for sustainable society: opportunities, challenges, and research roadmap.” *Sensors* 21(5) (2021): 1709.

[54] Tyagi, AK, Rekha, G, and Sreenath, N. “Beyond the hype: Internet of Things concepts, security and privacy concerns.” In Satapathy, S, Raju, K, Shyamala, K, Krishna, D, and Favorskaya, M. (eds), *Advances in Decision Sciences, Image Processing, Security and Computer Vision. ICETE 2019. Learning and Analytics in Intelligent Systems*, vol 3 (2020). Springer, Cham. https://doi.org/10.1007/978-3-030-24322-7_50.

[55] Labriji, I, Meneghello, F, Cecchinato, D, *et al.* “Mobility aware and dynamic migration of MEC services for the Internet of Vehicles.” *IEEE Transactions on Network and Service Management* 18(1) (2021): 570–584.

[56] Zhou, M, Guan, Y, Hayajneh, M, Niu, K, and Abdallah, C. “Game Theory and Machine Learning in UAVs-Assisted Wireless Communication Networks: A Survey.” *arXiv preprint arXiv:2108.03495* (2021). Available at https://arxiv.org/abs/2108.03495v1.

[57] Agbotiname, LI, Ibhaze, AE, Atayero, AA, and Kavitha, KVN. “Standard propagation channel models for MIMO communication systems.” *Wireless Communications & Mobile Computing (Online)* 2021 (2021). Article ID 8838792, https://doi.org/10.1155/2021/8838792.

[58] Baltaci, A, Dinc, E, Ozger, M, Alabbasi, A, Cavdar, C and Schupke, D. “A survey of wireless networks for future aerial communications (FACOM).” *IEEE Communications Surveys & Tutorials* 23(4) (2021): 2833–2884.

[59] Ye, J, Qiao, J, Kammoun, A, and Alouini, MS “Non-terrestrial communications assisted by reconfigurable intelligent surfaces.” (2021). Available at https://arxiv.org/abs/2109.00876v1.

[60] Wu, Q, Wan, Z, Fan, Q, Fan, P, and Wang, J. “Velocity-adaptive access scheme for MEC-assisted platooning networks: access fairness via data freshness.” *IEEE Internet of Things Journal* 9(6) (2022): 4229–4244.

[61] Luong, NC, Lu, X, Hoang, DT, Niyato, D, and Kim, DI. “Radio resource management in joint radar and communication: a comprehensive survey.” *IEEE Communications Surveys & Tutorials* 23(2) (2021): 780–814.

[62] Suo, J, Zhang, W, Gong, J, Yuan, X, Brady, DJ, and Dai, Q. "Computational Imaging, and Artificial Intelligence: The Next Revolution of Mobile Vision." (2021). Available at https://arxiv.org/abs/2109.08880v1.

[63] Visconti, P, Velazquez, R, Del-Valle-Soto, C, and de Fazio, R. "FPGA based technical solutions for high throughput data processing and encryption for 5G communication: a review." *Telkomnika* 19(4) (2021): 1291–1306.

[64] Shahwani, H, Shah, SA, Ashraf, M, Akram, M, Jeong, JP, and Shin, J. "A comprehensive survey on data dissemination in vehicular ad hoc networks." *Vehicular Communications* 34 (2022): 100420.

[65] Su, Z, Hui, Y, Luan, TH, Liu, Q, and Xing, R. "Bargain game-based secure content delivery in vehicular networks." *The Next Generation Vehicular Networks, Modeling, Algorithm and Applications*. Springer, Cham, 2021, 111–129.

[66] Khan, MA, and Alkaabi N. "Rebirth of distributed AI—A review of eHealth research." *Sensors* 21(15) (2021): 4999.

[67] Božanić, M, and Sinha, S. "The quest for faster data rates: unlocking millimeter-wave and terahertz frequencies." *Mobile Communication Networks: 5G and a Vision of 6G*. Springer, Cham, 2021, 63–98.

[68] Kumari, S, Tyagi, AT, and Aswathy, SU. "The future of edge computing with blockchain technology: possibility of threats, opportunities and challenges." *Recent Trends in Blockchain for Information Systems Security and Privacy*, CRC Press, 2021.

[69] Madhav, AVS, and Tyagi, AK. The "World with future technologies (Post-COVID-19): open issues, challenges, and the road ahead." In Tyagi, AK, Abraham, A, and Kaklauskas, A. (eds). *Intelligent Interactive Multimedia Systems for e-Healthcare Applications*. 2022, Springer, Singapore. https://doi.org/10.1007/978-981-16-6542-4_22.

[70] Tyagi, AK, Fernandez, TF, Mishra, S, and Kumari, S. "Intelligent automation systems at the Core of Industry 4.0." In Abraham, A, Piuri, V, Gandhi, N, Siarry, P, Kaklauskas, A, and Madureira, A. (eds). *Intelligent Systems Design and Applications. ISDA 2020. Advances in Intelligent Systems and Computing*, vol 1351, 2021. Springer, Cham. https://doi.org/10.1007/978-3-030-71187-0_1.

Chapter 8

Computer vision in Autonomous Intelligent Vehicles (AIV)

Abstract

Autonomous intelligent vehicle (AIV) is no longer a science fiction as it has been realized in today's era of the computing world. The research and development on autonomous vehicles have been going on for decades with constructive outcomes. Autonomous vehicles have expanded their areas of applications apart from the transportation sector. The internal structure and major components of the vehicle play a crucial role in making it autonomous. After the sophisticated hardware and sensors used in them, they are required to be learned and trained to mimic the driving style of a normal human driver. For this purpose, artificial intelligence and its subsets have played a pivotal role, without which the idea of self-driving vehicles would have just remained abstract. Software tools provided by machine learning can help assist an AIV in scene comprehension, decision-making, and planning. Deep learning algorithms like long short-term memory can be implemented to make such vehicles more intelligent and take decisions just like a human brain does. Besides, if not applicable for transporting people, AIVs have been used as robotic vehicles to transport goods and carry out tasks in industries. Computer vision can immensely help AIVs and could potentially help in reaching higher levels of autonomy. Nonetheless, with all the advantages, autonomous vehicles also suffer from various issues and challenges which still remain to be improvized to achieve the far-fetched dream of fully automated driverless vehicles.

Key Words: Computer vision, Artificial intelligence, Robotics, Autonomous vehicles

8.1 Introduction

Transportation and locomotion have been an integral part of human civilization. With growing technology, we have seen advancements in ease, comfort, convenience, safety, and majorly the speed with which transportation is carried out. With an increase in facilities and demand for quicker transportation, the costs incurred have also shot up. Recently, there have been many initiatives taken by innovative companies to launch intelligent and autonomous vehicles in the market. They are not just limited to human transportation but could also be used inside factories, shopping complexes, manufacturing plants, etc. for transportation of goods.

A truly autonomous intelligent vehicle (AIV) can be defined as a vehicle that can operate on its own by making decisions by itself. They are capable of planning alternate routes and operating in highly dynamic environments. They can safely operate alongside people and work collaboratively in a fleet as well. No workspace retrofit is required for such vehicles as they can be rapidly installed. Besides, they should adopt a non-threatening operation for carrying out their tasks. A vehicle is detrimental for the people and the company's property if it is not capable of taking a sensible and safe decision. For the same purpose, they must be validated and tested thoroughly in varied scenarios before they are deemed fit for use in real-time scenarios. All vehicles have some source and a destination. A path is thus formed from the source location to the destination point. As we humans plan our journey when we go out traveling to certain places, autonomous vehicles also require that they can perform intelligent path planning. A map must be created or made available for this purpose. The map data, once fed to the vehicle, must be comprehensive and clear enough so that the vehicle does not get stuck in between. During path planning and in motion, an AIV is required to (i) have clearly defined goals, (ii) recognize forbidden sectors, (iii) check for one-way/two-way paths, (iv) check for resistance during traveling, (v) object detection, (vi) trigger events, (vii) control speeds, and (viii) mitigate traffic as far as possible. All such necessary features can contribute toward the autonomous navigation of the vehicle.

One of the practical applications of an AIV is depicted in Figure 8.1. The figure depicts the evolution of human beings with respect to carrying of goods, from ancient times till date. The second last stage shows the contemporary style of carrying goods, which involves the use of a shopping cart, a commonplace in shopping malls. The last stage is the most advanced stage for carrying good, which has completely eradicated the need for a person's evolvement. The AIV is responsible for carrying the goods and transporting them to the desired place without any human intervention.

AIVs fall under the research area of intelligent transportation system (ITS). Besides, these intelligent vehicles can offer sophisticated applications aimed at providing services to numerous transportation modes, and in relation to traffic management and fetching information on safe usage of the transport network for the users [2]. Thus, AIVs can be termed as self-controlled vehicles without any driver. Even if a

Figure 8.1 The evolution of human beings in terms of carrying goods [1]

driver is present, he/she would have to put minimum or negligible efforts in controlling the vehicle, provided that the autonomous vehicle is reliable enough. Such vehicles can also prevent accidents on roads, in contrast to the scenario when a human driver is involved. This, in turn, can guarantee the safety of the passengers.

Autonomous vehicles aimed at the transportation of people from one place to another provide convenience in numerous scenarios with an uncountable number of quality-of-life improvements. It shall give independence to the elderly and the physically disabled population. The autonomous car could transport items from one place to another, not just people! One could even think of sending their pets like a dog to a veterinary appointment. Nonetheless, the real benefit of autonomous vehicles would be the dramatic decrease of CO_2 levels in the atmosphere. As per the reports from [3], three trends have been identified, if adopted concurrently, would unleash the full potential of autonomous cars: vehicle automation, vehicle electrification, and ride-sharing. By 2050, these "three revolutions in urban transportation" could

- reduce traffic congestion (30% fewer vehicles on the road);
- cut transportation costs by 40% (in terms of vehicles, fuel, and infrastructure);
- improve walkability and livability;
- free up parking lots for other uses (schools, parks, community and centers); and
- reduce urban CO_2 emissions by 80% worldwide.

In this chapter, we shall discuss the related background work on AIVs in Section 8.2. Section 8.3 gives the internal structure of a typical AIV. Moving on, Section 8.4 provides some learning techniques for AIVs including artificial intelligence (AI) learning techniques and robotics-based AIV. Further on, Section 8.5 discusses the possibilities of computer vision contribution toward AIV development. Next, Section 8.6 explains various issues and research challenges toward AIV. Finally, Section 8.7 gives the conclusion of the chapter.

8.2 Some related works on AIV systems

The earliest idea of developing an actual autonomous vehicle arose in the 1920s. In 1925, a driverless vehicle was reported to be demonstrated when Houdina Radio Control showcased the "American Wonder," which is a remote-controlled vehicle that traveled along the Broadway curb in New York City with the help of an operator in another vehicle [4].

Later in 1986, an innovative effort led by the Navlab team at Carnegie Mellon University (CMU) in the United States and Ernst Dickmann's team at the Bundeswehr University, Munich in Germany resulted in the development of a self-driving car prototype. The Navlab car (Figure 8.2) did not rely on any dedicated infrastructure and was used as a portable laboratory for navigational research purposes [5]. The car was believed to be 98% autonomous while it drove from Washington D.C. to San Diego, CA, USA.

Further on, in order to enrich driver's comfort and safety, Mitsubishi launched the first light detection and ranging-based (LiDAR) distance control vehicle in 1995. On

Figure 8.2 The Navlab – the self-controlled car of the 1980s [6]

similar lines, Mercedes-Benz implemented an innovative adaptive cruise control mechanism incorporating radar assistance. The development of intelligent vehicles was enhanced with the advancement of GPS navigation systems and digital road maps in 2000. Notably, in 2007, the defense advanced research projects agency (DARPA) organized a race competition between autonomous cars belonging to various teams. The vehicles were required to travel a path of 96 km while obeying traffic laws, detecting and overcoming obstacles, negotiating with other vehicles on the road, and thus incorporating itself as a normal vehicle into the traffic. Most of the successful teams implemented multi-beam LiDAR technology to complete the race [7].

Another notable mention is the KITTI Vision Benchmark, which was released in 2012 [8,9]. The vehicle based on such a benchmark was good at self-driving perception tasks, including motion estimation and object recognition with a fair amount of accuracy. The robustness, speed of perception, and precision were revolutionized by the advent of deep learning, computer vision, and robot technologies. The next year, in 2013, Mercedes-Benz demonstrated the S500 Intelligent Drive. The Mercedes S500 vehicle was able to successfully cover a distance of 103 km in autonomous mode. The major proponents toward its development were the Daimler research in collaboration with the Karlsruhe Institute of Technology (KIT) [10]. The vehicle was having close-to-production sensor hardware which assisted in object detection and free-space analysis with the aid of radar and stereo vision. Additional features included a monocular vision for use in stereo vision and traffic light detection. Besides, two complementary vision algorithms – point-feature-based and lane-marketing-based algorithms – allowed the vehicle to make use of available HD maps and perform autonomous driving. However, the vehicle was focused on traveling on a single route considering the feasibility of the production hardware, complexity of the environments, and the HD maps available.

As the history of autonomous vehicles is being discussed, it is worthy to mention the six SAE levels of autonomy, proposed by the Society of Automotive

Engineers in 2014. There are five levels in the classification scheme (Figure 8.3). The levels are, namely, Level 0 (no automation), Level 1 (driver assistance), Level 2 (partial automation), Level 3 (conditional automation), Level 4 (high automation), and lastly Level 5 (full automation). The magnitude of autonomy of the vehicles increases as the levels increase, and at the same time, less intervention is required from the driver. In the same year, Tesla launched its Autopilot car with Level 2 autonomy. The Tesla driver had to monitor the system at all times, but the car provided facilities like lane keeping, braking on the highway, acceleration, and autonomous steering [11]. Even the famous ride-hailing and cab rental service company Uber made endeavors to launch a self-driving car in 2017 [12]. However, both Tesla and Uber witnessed lethal accidents in which neither the driver was paying full attention, nor the autonomous vehicle functioned as per expectation.

Table 8.1 summarizes the five levels of automation of a motor vehicle as per the SAE Standard J3016. The table takes into consideration the type of motor vehicle operation, whether the acceleration is manual or automatic, how the driving is being monitored and performed, and the system capability in terms of driving modes. It serves as the industry's most-cited reference for automated vehicle (AV) capabilities.

One of the most successful models of autonomous vehicles that have ever been created is the Waymo autonomous vehicle (Figure 8.4). It has been built through Google's self-driving efforts after completing over 1.5 million miles. Waymo offers full self-driving features for its riders with a safety driver in the back seat [14]. Apart from this, NVIDIA adopted a single convolutional neural network to demonstrate a 98% autonomous ride on an autonomous vehicle [15,29]. In 2019, Bosch and Daimler launched a fleet of autonomous cars called as San Jose pilot project, which provides customers with a shuttle service with autonomous vehicles on a limited number of routes [16,17]. Recently in 2021, Audi announced that it plans to launch the Audi Grandshpere Concept EV to the customers by 2026. It shall be equipped with Level 4 autonomous driving technology. This effort is being jointly developed by Audi and Porsche. The Grandshpere shall be a self-driving luxury "private jet" for the roads, once launched [18].

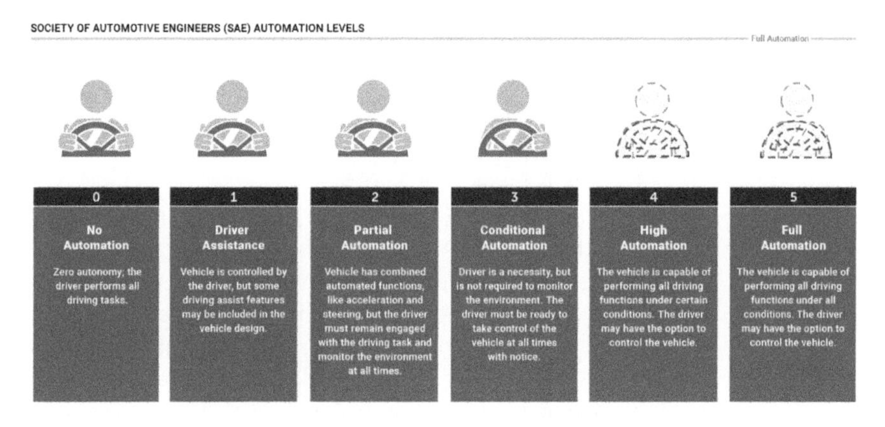

Figure 8.3 The six SAE levels of automation [28]

Table 8.1 Levels of autonomy for an AIV adapted from SAE Standard J3016 [2,13]

Levels of automation (LoA)	Type of motor vehicle operation	Steering, acceleration, deceleration	Monitoring driving environment	Fallback performance of a dynamic driving task	System capability (driving modes)
LoA 0	Not automated, traditional car	Driver	Driver	Driver	N/A
LoA 1	Driver assistance	Driver/Automatic	Driver	Driver	Some driving modes
LoA 2	Partial automation	Automatic	Driver	Driver	Some driving modes
LoA 3	Conditional automation	Automatic	Automatic	Driver	Some driving modes
LoA 4	High automation	Automatic	Automatic	Automatic	Some driving modes
LoA 5	Full automation	Automatic	Automatic	Automatic	All driving modes

Figure 8.4 The Waymo autonomous vehicle on the road

8.3 Internal structure of AIV

AIVs rely on multiple factors including complex algorithms, machine learning (ML) systems, AI, sensors, actuators, and powerful processors to execute the software.

For instance, the typical autonomous cars create and maintain a map of their ambience based on the data sensed by the sensors of the vehicle. The sensors are situated in a different part of the vehicle's body. Video cameras, mostly situated on the roof of the car, detect traffic lights, track other vehicles, read road signs, and look for pedestrians and bicyclists. The cameras use parallax from multiple images to find the distance to various objects. On the other hand, the radar sensors monitor the movement and position of nearby vehicles on the road. It is also used in measuring the distance from the car to various obstacles. LiDAR sensors detect road edges, measure distances from certain objects, and identify lane markings. This is achieved when the LiDAR sensors bounce pulses of light off the car's surroundings and the light strikes the object and returns to the sensor. The LiDAR unit can be constantly spinning and it generally uses laser beams to generate a 360° image of the car's surroundings. In addition, ultrasonic sensors in the wheels detect curbs and other vehicles when parking. Finally, a processor (the main computer) is located in the trunk of the car. The processor comes with sophisticated software installed in it.

As far as the software's part is concerned, the sophisticated software present in the system processes all the sensory input. It stores maps to compare and assess current conditions. It plots a path and sends instructions to the car's actuators, which control acceleration, braking, and steering. To ensure that traffic rules are followed and the objects are well navigated, the software uses hard-coded rules, predictive modeling, object recognition, and avoidance algorithms.

Figure 8.5 shows the major components of an autonomous car. The positioning of certain components on the car is indicative and can vary depending upon the design overlayed by the manufacturing company.

To explain the internal structure of an AIV further, we consider an autonomous vehicle developed by the center for intelligent transportation research (CITR) at the

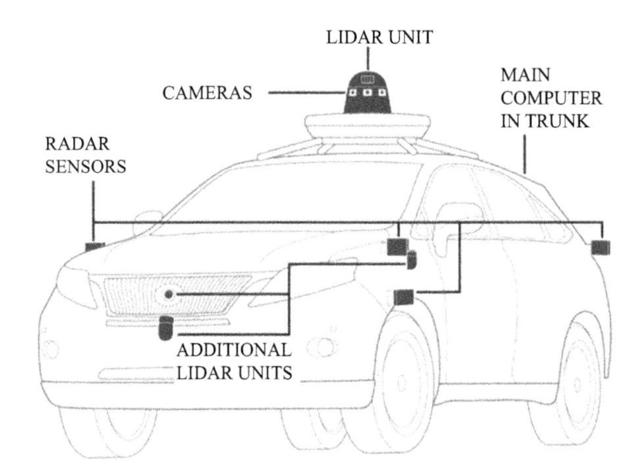

Figure 8.5 The fundamental components of an autonomous car / Source: Google / Figure drawn by Guilbert Gates

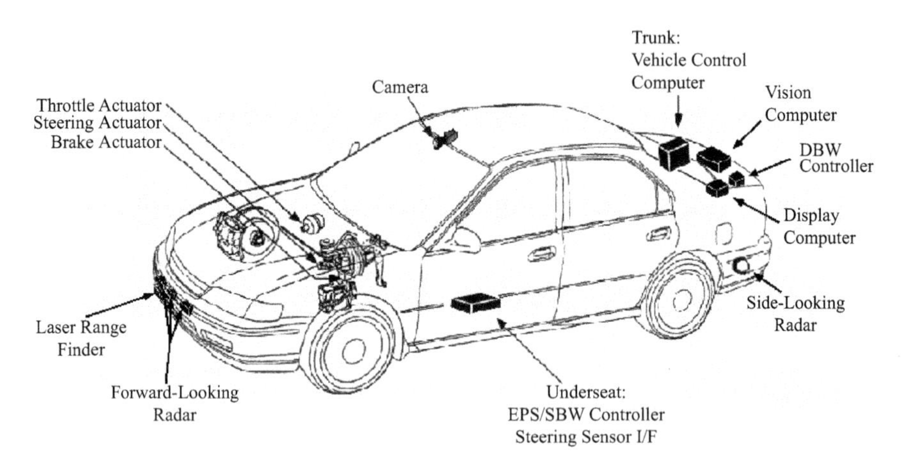

Figure 8.6 The internal structure of autonomous car built by Ohio State University, OH, USA [19,28]

Ohio State University, OH, USA, which has developed three AVs in the past. Various sensors were used, including a radar reflective stripe system and a vision-based system for lane position sensing, a radar system, and a scanning laser range finding system for the detection of objects ahead of the vehicle, and various supporting sensors including side looking radars and an angular rate gyroscope. Whenever multiple sensors were available, data fusion and fault detection were employed to maximize functionality without driver involvement. Figure 8.6 shows the physical layout of certain components and equipment used in the vehicle. Brake actuator, throttle, and steering locations are depicted. The steering electronic control units (ECUs) are mounted under the front of the driver's seat. The DBW ECU is mounted along the right wall of the trunk. The positioning of the video camera, radar radiofrequency components, and the laser range finder are indicated. The contents of the car's trunk such as the vehicle control computer, radar signal processing components, angular rate gyro, the image processing computer, graphical status display computer, radar signal processing components, and the interface electronics are also presented in Figure 8.6 [19].

8.4 Learning techniques of AIV

The credit of achieving a high level of autonomy in AIVs goes to the complex algorithms that control the overall working of the vehicle. Before the vehicle is put into a real-world scenario or tested in laboratory environments, it is trained and made to learn to tackle numerous scenarios that could otherwise take place in real-time cases.

In this section, we shall discuss AI techniques used in AIV in Section 8.4.1. Next, we discuss the role of ML in AIV in Section 8.4.2. Further, Section 8.4.3 discusses the deep learning techniques, namely, the long short-term memory (LSTM) method used in AIV. Lastly, Section 8.4.4 explains the robot-based AIVs.

8.4.1 AI-based AIV

AIVs offer the opportunity to harness the latest technology of AI to make driving decisions and mitigate human-driving-induced risks. The combination of sensor technology and AI provides a topographical representation of the road phenomenon which supports the AI to take more immediate and accurate decisions. This also reduces the wrong decisions that could be taken by human drivers otherwise, if they are fatigued, intoxicated, or are having misperception during driving [20].

It has also been observed that to achieve Level 5 automation as per SAE standard, AIVs need to adopt a hybrid intelligence system, which would enable the solution of driving tasks under undetermined conditions. According to [21], there are three main waves in the development of AI, which are listed as follows:

1. Handcrafted knowledge: In this wave, the engineers defined a set of rules that represented the available knowledge in well-defined systems. However, the developed systems were characterized by poor perception, no learning, and no abstracting abilities, but good reasoning.
2. Statistical learning: In this wave, the engineers created statistical models for specific problem domains and then trained them using big data. They even employed deep learning techniques such as CNN to create the model. The creations were characterized by goof perception and good learning, but poor abstracting and reasoning abilities.
3. Contextual adaptation: In this wave, the engineers combined the advantages of the above two methods leading to the formation of a hybrid AI system. The outlook of the current innovations was positive in every possible way.

Now, we shall discuss some significant uses of AI in autonomous vehicles.

8.4.1.1 Data processing from sensors

Throughout the operation of the vehicle, numerous sensors provide data to the vehicle's central computer. Sensors offer data on the road, other vehicles on the road, and any other obstacles that may be detected in the same way that a human can. Some of these sensors may be able to provide superior perception than the average human, but we will need smart algorithms to decipher the real-time data streams. The video cameras can also be employed to recognize the traffic signs and symbols and thus act accordingly on roads. For example, ten parameters could be taken as AI parameters for ten different traffic signs as shown in Figure 8.7.

8.4.1.2 Path planning

Path planning is essential for optimizing a vehicle's trajectory and resulting in better traffic patterns. This can help decrease traffic congestion and delays on the road. AI algorithms are well suited to the task of planning. It is a dynamic task that takes various factors into account and solves an optimization problem while following the path. "Path planning for AVs enables self-driving vehicles to determine the safest, most convenient, and economically beneficial routes from point A to point B by using previous driving experiences that assist the AI agent make far more accurate decisions in the future," according to the definition.

8.4.1.3 Path execution

After the path has been planned, the vehicle can negotiate the road conditions by identifying objects, pedestrians, bicycles, and traffic signals to arrive at the destination. The AI community is particularly interested in object detection algorithms because they enable human-like behavior. The difficulties arise when varied road and weather conditions are present. Many testing car accidents occurred because the simulation environment differed from real-world settings, and AI algorithms can behave in unpredictable ways when presented with unexpected data.

8.4.1.4 Monitoring vehicle's condition

Predictive maintenance is the most promising sort of maintenance. "Predictive maintenance" is described as "the use of monitoring and prediction modeling to

Figure 8.7 Dashboard interface with resulting AI data (ten parameters)

determine the status of a machine and to predict what is likely to fail and when it will happen." It aims to foresee future problems rather than current ones. Predictive maintenance can save a lot of time and money in this regard. For predictive maintenance, both supervised and unsupervised learning can be applied. The algorithms may make maintenance decisions based on both on-board and off-board data. For this purpose, ML classification algorithms like logistic regression, support vector machines, and random forest algorithm can be used.

8.4.1.5 Insurance data collection

Vehicle data logs can contain information about a driver's conduct, which can be useful in traffic accident investigations. These details can be utilized in claims processing. Because the safety is more predictable and assured, all of this might lead to a reduction in insurance premiums. The manufacturer will assume responsibility for fully autonomous automobiles, as the passenger will no longer be a driver. We will very certainly have some driver culpability in the semi-autonomous car. The use of smart data gathered by the vehicle's AI system will become increasingly important in proving these kinds of instances. Massive volumes of data are generated by data from all sensors. While preserving all data at all times may be impractical, saving snapshots of important data appears to be the ideal mix of acquiring evidence that may be utilized for post-analysis of a specific traffic incident. This method is comparable to how black box data is collected and examined following a crash.

8.4.2 Machine learning in AIV

ML is a technique in which a computer software learns from its previous experiences in order to enhance its performance on a given task. Supervised learning, unsupervised learning, and reinforcement learning are the three main types of ML algorithms. Unsupervised learning algorithms are based on techniques like density estimation or clustering applied to unlabeled data, whereas supervised learning algorithms are based on inductive inference where the model is often trained using labeled data to achieve classification or regression. In the RL paradigm, on the other hand, an autonomous agent learns to improve its performance at a task by interacting with its environment.

The algorithms of ML may be employed in different activities of the principle of operation of autonomous cars based on their characteristics and principles of operation. The use of supervised learning in the "Scene understanding" exercise is suitable, and the use of reinforcement learning in the "Decision-making and planning" task is appropriate. As a result, ML techniques may be used to build the whole autonomous vehicle control system. Other algorithms linked to sensor data processing should only execute duties for gathering sensory data [22].

The supervised model feeds instructions to an algorithm on how to interpret the incoming data. This is the recommended method of self-driving car learning. It enables the algorithm to assess training data using a fully labeled dataset, making supervised learning more effective in classification. Controller optimization, path planning and trajectory optimization, motion planning and dynamic path planning, development of high-level driving policies for complex navigation tasks, scenario-based policy learning for highways, intersections, merges, and splits, reward learning with inverse reinforcement learning from expert data for intent prediction for traffic actors such as pedestrians, and finally learning of policies that ensures the safety and perform risk estimation.

ML algorithms can be loosely divided into four categories: regression algorithms, pattern recognition, cluster algorithms, and decision matrix algorithms. More details on the same can be obtained from [23]. We can say that supervised learning provides the necessary environmental information for reinforcement learning.

Some of the popular ML algorithms used in AIVs are as follows [24]:

1. AdaBoost for data classification: This algorithm collects and categorizes data in order to improve vehicle learning and performance. It combines several low-performing classifiers into a single high-performing classifier to improve decision-making. The approach automatically adapts its settings to the data based on the current iteration's actual performance. Iteratively re-compute both the weights for re-weighting the data and the weights for the final aggregate. In practice, this boosting strategy is applied to simple classification trees or stumps as base learners, resulting in enhanced performance when compared to a single tree or other single base-learner classification.
2. You only look once (YOLO): Objects such as persons, trees, and automobiles are detected and grouped using this technique. It assigns unique characteristics to each class of object it groups in order for the automobile to recognize them. Identifying and classifying objects is easiest with YOLO. Convolutional neural networks are used by YOLO because they are good at comprehending spatial information. They can extract details such as edges, lines, and textures. There are 24 of these convolutional layers in YOLO. When AVs LiDAR sensors are combined with YOLO, they can maneuver through thick traffic while recognizing many objects and their unique relationships.
3. TextonBoost for object recognition: The TextonBoost algorithm works in a similar way to AdaBoost, but it takes into account data from shape, context, and appearance to improve Texton learning (micro-structures in images). It collects visual data with similar characteristics.
4. Gradient boosting: This is a technique for regression, classification, and other tasks, which produces a prediction model in the form of an ensemble of weak prediction models, typically decision trees. When a decision tree is a weak learner, thc resulting algorithm is called gradient boosted trees, which usually outperforms random forest. Gradient boosting and AdaBoost work in a similar way but each method uses different mathematical models and algorithms.
5. Histogram of oriented gradients (HOG): HOG is a feature descriptor that is frequently employed in the extraction of features from picture data. HOG makes it easier to analyze an object's location, known as a cell, to see how it changes or moves. The HOG approach is mostly used to recognize faces and

classify images. From driverless vehicles to surveillance techniques to smarter advertising, this topic offers a wide range of applications. This method is also used to recognize and classify different types of vehicles.

6. Scale-invariant feature transform (SIFT) for feature extraction: SIFT algorithms recognize and analyze items in photos. The three points of a triangular symbol, for example, are entered as features. Using those locations, an automobile can quickly detect the sign.

Let us have a look at the key components of the self-driving automobile control system in a diagram (Figure 8.8). The major activities are separated into two groups: scene comprehension and decision-making and planning. ML techniques can be used to completely accomplish these activities The control system's major responsibilities are as follows:

- *Sense*: gathering sensor data from the environment;
- *Perceive and localize:* recognize and locate objects and markers;
- *Scene representation:* understanding the environment parameters and characteristics;
- *Plan and decide:* path and motion planning, finding optimal trajectory according to the driving policy;
- *Control:* setting the necessary vehicle parameters for acceleration, deceleration, steering, and braking.

8.4.3 Deep learning in AIV

Deep learning is a subset of ML, which is in fact, a subset of AI. It is inspired by the functioning of the human brain. Deep learning algorithms analyze data with a predetermined logical framework in order to reach similar conclusions as humans. It does this by employing a multi-layered structure of algorithms known as neural networks. We may use neural networks to accomplish various tasks, such as grouping, classification, and regression. We can use neural networks to group or sort unlabeled data based on similarities between the samples. Alternatively, in the instance of classification, we may train the network on a labeled dataset in order to categorize the samples in the dataset.

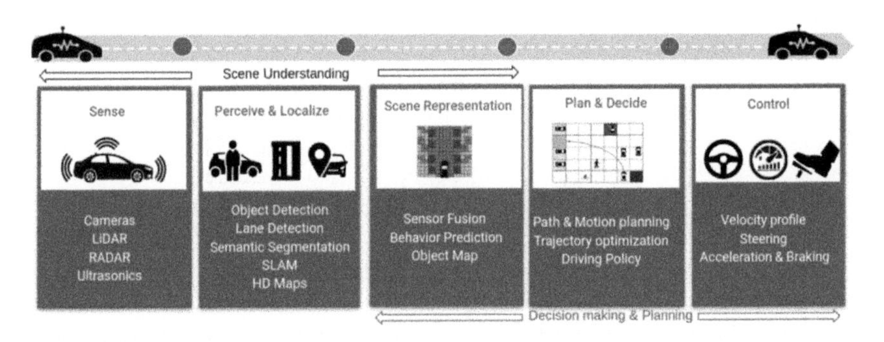

Figure 8.8 Major components and functionalities of AIVs

We shall focus upon a special type of deep learning artificial recurrent neural network known as LSTM in the context of its application in AIVs. LSTM features feedback connections, unlike typical feedforward neural networks. It can handle not just individual data points (such as pictures) but also whole data streams (such as speech or video). A cell, an input gate, an output gate, and a forget gate make up a typical LSTM unit. The three gates control the flow of information into and out of the cell, and the cell remembers values across arbitrary time periods. We shall see the implementation of an LSTM-based model for understanding the behavior of Waymo's self-driving model as proposed in [25].

Let us consider how an acceleration prediction model is built for AIV like the Waymo autonomous car using the LSTM approach. Treating 12 fundamental features as the model's input is one of the simplest ways to design the acceleration prediction model. Because the acceleration curve is a trajectory based on prior experiences, the "encoder–decoder" design developed for trajectory prediction in SS-LSTM [26] is a viable architecture for the acceleration prediction problem. The 12 fundamental features are crammed into one single input channel in our suggested basic model, whereas SS-LSTM employs various input information such as the occupancy map. These characteristics are fed into an "encoder" module, which extracts the important information from the input features and provides an intermediate output, or more accurately, a latent representation of them. After that, the intermediate result is transmitted to a "decoder" module, which decodes the representation and provides the acceleration prediction. This architecture's figure is shown in Figure 8.9.

Because only 12 features are supplied, the basic architecture is relatively simple. The data is initially encoded using the "encoder" module, which then generates an intermediate latent vector that contains the key value as well as the relationship between the elements. Then, in order to match the forecast frame numbers, this vector is stacked and sent to the "decoder" module, which generates a new representation of the original input. The final dense layer is trained to predict acceleration using the new representation. Given a ten-frame video clip, the input is a vector consisting of 12 features from those ten frames. Starting from the end of the video clip, the output is the acceleration for the next five frames. There are 128 LSTM cells in the "encoder" module and 128 LSTM cells in the "decoder" module. This network's optimizer is RMSprop, with a learning rate of 0.0001 and a dropout probability of 0.2 for each LSTM cell.

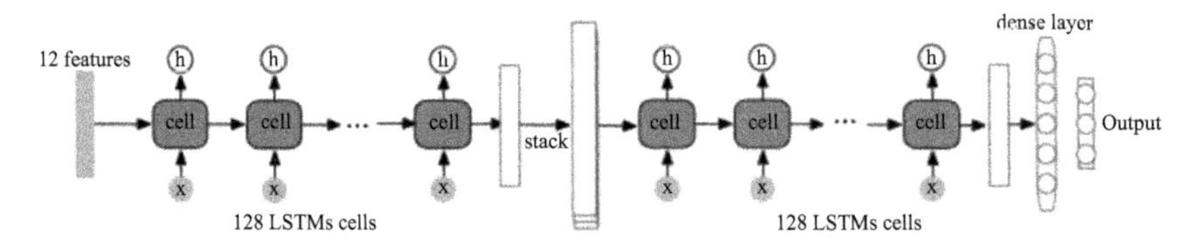

Figure 8.9 LSTM architecture for acceleration prediction model consisting of 12 features

The limitation of the information offered by the 12 input features, on the other hand, cannot be overlooked. These simple and limited numerical features make it difficult for the model to capture and assess the complex surrounding environment for the AV, and the model trained on these features may fail to provide a suitable forecast. To boost the model's capacity, more high-quality features with more vital information are required.

8.4.4 Robot-based AIV

Robotics has become an integral part of AIVs today. Robots are made up of a specific combination of components that enable them to carry out a specific mission in a specific environment. Based on the concept of modular components, algorithms and functions have been characterized as modular modules in numerous modular robot techniques for the comprehension and validation of software components like navigation, mapping, learning, and planning. AIVs based on robotics technology have been used for transporting goods and products from one location to another. They have been primarily used in manufacturing industries to automate a certain process that otherwise required human effort and inclusion. Besides, wireless robotic components could be integrated into autonomous vehicles to make them more robotic in nature.

A typical example of a robot-based AIV is Nuro R2. Jiajun Zhu and Dave Ferguson founded Nuro, an American robotics firm based in Mountain View, CA, USA. Nuro is a firm that creates self-driving delivery cars. It was the first company to get an autonomous exemption from the National Highway Traffic Safety Administration because its vehicles are designed to transport things rather than people. R2, the company's second-generation self-driving car, is designed without a steering wheel, side view mirrors, or pedals. The R2's mobility is guided by radar, thermal imaging, and 360° cameras. It also has no steering wheel, pedals, or side mirrors. The vehicle's frame is egg-shaped, making it smaller than most cars in the United States. It also features two temperature-controlled delivery containers. After the recipient enters a code, the doors open to expose the items. Figure 8.10 shows the main features of Nuro R2.

8.5 Possibilities for computer vision toward AIV

In AIVs, computer vision can contribute to the design and development of improved and next-generation vehicles that can overcome roadblocks while keeping passengers safe. Passengers can be transported to their destination without the need for human interaction.

However, driverless vehicles are still in their infancy and will take some time to be deployed on congested city streets. Because even a tiny flaw in the vehicle's design or development might result in tragic accidents and life-threatening situations.

Computer vision technology is being used by researchers and professionals to make autonomous vehicles safer for passengers and pedestrians.

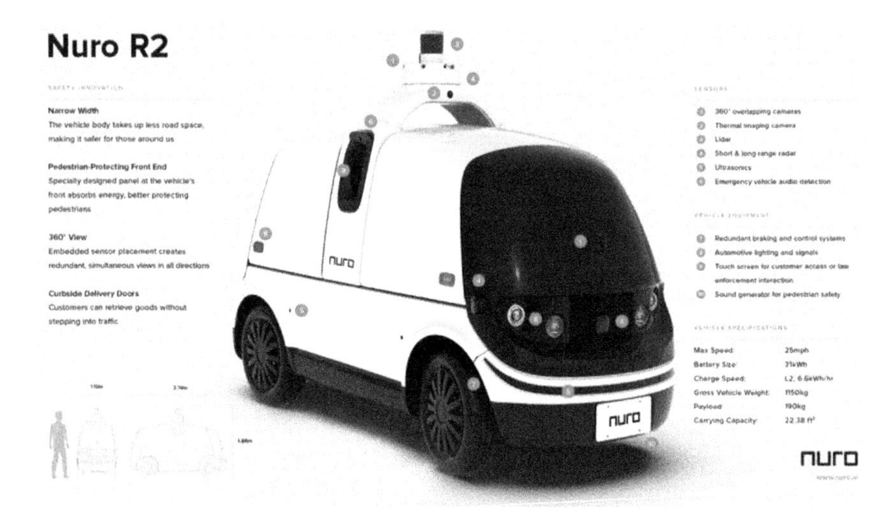

Figure 8.10 Nuro R2 – a robot-based AIV [27]

The technology of computer vision can be used in the following manner in an autonomous vehicle:

1. Classifying and detecting objects: Self-driving vehicles could use the technology to classify and recognize various objects. LiDAR sensors and cameras can be used by the vehicle, with the former using pulsed laser beams to detect distance. The information gathered can be used in conjunction with 3D maps to detect things such as traffic signals, vehicles, and pedestrians. These high-tech vehicles process data in real time and make judgments based on it. As a result of computer vision, self-driving vehicles will be able to detect impediments and avoid collisions and mishaps.
2. Computer vision-enabled low-light mode: Self-driving cars utilize different algorithms to analyze low-light photos and videos than they do for daylight images and movies. Low-light photos may be grainy, and such data may not be accurate enough for certain cars. When the computer vision identifies a low-light situation, it can automatically switch to low-light mode. LiDAR sensors, infrared cameras, and HDR sensors can all be used to get this information. These devices can be used to produce high-resolution photos and films. Using computer vision technologies, self-driving vehicles can be made intelligent, self-reliant, and dependable. However, the vehicles may face additional difficulties throughout development.
3. Gathering data for training algorithms: Using cameras and sensors, computer vision technology can collect vast amounts of data, such as position information, traffic conditions, road maintenance, and busy areas, among other things. These specific details can help self-driving cars employ situational awareness and make critical decisions as quickly as feasible. These details can be utilized to build deep learning models in the future. For example, a thousand computer vision photos of traffic signals can be used to train Deep Learning (DL) models

to recognize traffic signals while driving. It can also assist self-driving vehicles in classifying various types of objects.

4. Creating 3D maps: It will make it possible for self-driving cars to collect visual data in real time. The cameras mounted on such vehicles can capture live video and use computer vision to produce 3D maps. With these maps, autonomous vehicles can better understand their environment, recognize obstructions in their path, and choose alternate paths. Self-driving cars can use 3D maps to predict accidents and can activate airbags to protect passengers quickly. This method improves the safety and reliability of self-driving cars. As a result, technology can assist in the development of safe autonomous vehicles that avoid collisions and safeguard passengers.

Hence, computer vision can assist in the development of self-driving vehicles and make them more intelligent and decisive.

8.6 Issues and research challenges toward AIV

We can see few Level 2 autonomous vehicles like Tesla cars on roads after a constant research and development effort of several decades. Still, there are many issues and challenges associated with AIVs such as:

1. Traffic conditions: Autonomous vehicles would have to get on the road and navigate through various traffic situations. They would have to share the road with other autonomous vehicles while also dealing with a large number of humans. There are a lot of emotions involved whenever humans are involved. The flow of traffic might be carefully controlled and self-regulatory. However, there are many instances where persons may be breaking traffic laws. Unexpected circumstances may lead to the discovery of an object. Even a few centimeters per minute of mobility matters in congested areas. One cannot wait indefinitely for traffic to clear and for some sort of preconditioning to begin moving. If there are more of these automobiles on the road waiting for traffic to clear, it may result in a traffic jam.
2. Weather conditions: Another stumbling block is the weather. It could be sunny and clear, or it could be rainy and stormy. Autonomous vehicles should be able to operate in various weather situations. There is no possibility of failure or downtime.
3. Road conditions: Road conditions can be unpredictable and vary greatly from one location to the next. There are smooth and well-marked broad roadways in some cases. In other areas, the road is severely eroded, with no lane markings. Lanes are not well defined; there are potholes, mountainous, and tunnel roads with unclear exterior direction signals; and so on.
4. Accident liability: The most crucial component of autonomous vehicles is their culpability in the event of an accident. Who is responsible for self-driving automobile accidents? In the case of self-driving automobiles, the software will be the primary component that will operate the vehicle and make all critical choices. While the original concepts had a person physically behind the steering wheel, Google's subsequent designs do not include a dashboard or a

steering wheel! How is the person in the automobile intended to operate the car in the event of an adverse incident if the car has no controls like a steering wheel, a brake pedal, or an accelerator pedal? Furthermore, due to the nature of autonomous vehicles, passengers will be primarily relaxed and may not be paying close attention to road conditions. When their attention is required, it may be too late to avert the situation by the time they are required to act.

5. Radar interference: For navigation, self-driving automobiles use lasers and radar. The lasers are installed on the vehicle's roof, while the sensors are mounted on the vehicle's body. Radar works by detecting radio wave reflections from objects in the environment. When driving on the road, an automobile continuously emits RF waves, which are reflected by other cars and other objects. The distance between the car and the object is calculated by measuring the time it takes for the reflection to occur. Following that, based on the radar data, appropriate action is taken. Radar works by detecting radio wave reflections from objects in the environment. When driving on the road, an automobile continuously emits RF waves, which are reflected by other cars and other objects. The distance between the car and the object is calculated by measuring the time it takes for the reflection to occur. Following that, based on the radar data, appropriate action is taken. Will a car be able to discern between its own (reflected) signal and the signal (reflected or transmitted) from another vehicle when this technology is used for hundreds of vehicles on the road? Even if numerous RFs are available for radar, the frequency range will almost certainly be insufficient for all of the vehicles produced.
6. Social acceptability: Several high-profile accidents involving Tesla's existing AVs, as well as other automated and autonomous vehicles, have occurred. Social acceptability is a concern not only for people considering purchasing a self-driving car but also for those who share the road with it. The public must be included in decisions about self-driving vehicle introduction and adoption. Without it, we risk having this technology rejected. To defeat the latter two, we must first tackle the first three problems. Of course, there is a race to be the first business to introduce a completely autonomous vehicle. However, until we collaborate on how to make the automobile safe, offer evidence of that safety, and engage with authorities and the public to obtain a "seal of approval," these cars will remain on the test track for years.
7. Cybersecurity: Another threat to self-driving cars is that they may be comprised of software security if appropriate mechanisms are not incorporated to mitigate or avoid them. Ensuring that AIVs cannot be hacked remains a big challenge for the industry.

8.7 Conclusion

In this chapter, we discussed the definition of autonomous vehicles and their numerous applications in everyday life, especially in the transportation sector. Beginning from the earlier models of self-driving cars, we came across some names of the latest and contemporary autonomous vehicles which are commercially sold and available to avail service from. Next, an import classification scheme called Six Levels of Autonomy by SAE was introduced. The major components of an autonomous car were discussed briefly. We also saw that AI and its subsets, namely, ML and deep learning contribute significantly toward the development of autonomous vehicles. Computer vision in AIVs can show promising contribution and significance. Lastly, there is a plethora of issues and challenges associated with autonomous vehicles that need to be tackled in present times and discovered for future mitigation. Overall, autonomous intelligent vehicles are a growing industry that still requires a lot of research and development, and hence, it has become an open research community welcoming a multitude of ideas from numerous enthusiasts.

References

[1] M. Vestal, "Manufacturing with AIVs". Available at https://robotics.org/userAssets/riaUploads/file/18-MobileRobots-AnotherTypeofCollaborationAdept-MattVestal.pdf.

[2] D. Tokody, I.J. Mezei, and G. Schuster, "An overview of autonomous intelligent vehicle systems." In Jármai K., and Bolló B. (eds), *Vehicle and Automotive Engineering. Lecture Notes in Mechanical Engineering*. 2017, Springer, Cham. https://doi.org/10.1007/978-3-319-51189-4_27.

[3] "Three Revolutions in Urban Transportation – Institute for Transportation and Development Policy." https://www.itdp.org/2017/05/03/3rs-in-urban-transport/ (accessed Oct. 12, 2021).

[4] "Science: Radio Auto – TIME." https://www.content.time.com/time/subscriber/article/0,33009,720720,00.html (accessed Oct. 10, 2021).

[5] C. Thorpe, M. H. Hebert, and S. A. Shafer, "Vision and navigation for the Carnegie-Mellon Navlab," *IEEE Transactions on Pattern Analysis and Machine Intelligence*, IO, 3, 1988.

[6] "NavLab: The Self-Driving Car of the '80s." https://www.rediscoverthe80s.com/2016/11/navlab-the-selfdriving-car-of-the-80s.html (accessed Oct. 10, 2021).

[7] J. Janai, F. Güney, A. Behl, and A. Geiger, "Computer vision for autonomous vehicles," *Foundations and Trends in Computer Graphics and Vision*, 12, 1–3, 1–308, 2020, doi: 10.1561/0600000079.

[8] A. Geiger, P. Lenz, and R. Urtasun, "Are we ready for Autonomous Driving? The KITTI Vision Benchmark Suite", https://www.cvlibs.net/datasets/kitti (accessed Oct. 12, 2021).

[9] A. Geiger, P. Lenz, C. Stiller, and R. Urtasun, "Vision Meets Robotics: The KITTI dataset." vol. 32, no. 11, pp. 1231–1237, 2013, http://dx.doi.org/10.1177/0278364913491297.

[10] J. Ziegler, P. Bender, M. Schreiber, *et al.*, "Making bertha drive – an autonomous journey on a historic route," *IEEE Intelligent Transportation Systems Magazine*, 6, 2, 8–20, 2014, doi: 10.1109/MITS.2014.2306552.

[11] "Autopilot | Tesla." https://www.tesla.com/autopilot (accessed Oct. 12, 2021).

[12] "Uber Self-Driving Cars: Everything you Need to Know | TechRadar." https://www.techradar.com/news/uber-self-driving-cars (accessed Oct. 12, 2021).

[13] "SAE J3016 Automated-Driving Graphic." https://www.sae.org/news/2019/01/sae-updates-j3016-automated-driving-graphic (accessed Oct. 13, 2021).

[14] "Waymo – Waymo." https://waymo.com/ (accessed Oct. 12, 2021).
[15] M. Bojarski *et al.*, "End to End Learning for Self-Driving Cars," Apr. 2016, https://arxiv.org/abs/1604.07316v1 (accessed Oct. 12, 2021).
[16] "Metropolis in California to Become a Pilot City for Automated Driving | Daimler." https://www.daimler.com/innovation/case/autonomous/pilot-city-for-automated-driving.html (accessed Oct. 12, 2021).
[17] "California, Here We Come! | Daimler Mobility AG > Innovations > Autonomous Driving > Our Pilot Tests." https://www.daimler-mobility.com/en/innovations/autonomous-driving/pilot-project/ (accessed Oct. 12, 2021).
[18] "Audi Grandshpere, a Self-Driving Luxury 'Private Jet' for the Road, Debuts." https://auto.hindustantimes.com/auto/cars/audi-grandshpere-a-self-driving-luxury-private-jet-for-the-road-debuts-41630636774557.html (accessed Oct. 12, 2021).
[19] "OSU Autonomous Vehicles." http://www2.ece.ohio-state.edu/citr/Demo97/osu-av.html (accessed Oct. 13, 2021).
[20] M. Cunneen, M. Mullins, and F. Murphy, "Autonomous vehicles and embedded artificial intelligence: the challenges of framing machine driving decisions," *Applied Artificial Intelligence*, 33, 8, 706–731, 2019, doi: 10.1080/08839514.2019.1600301.
[21] "A DARPA Perspective on Artificial Intelligence – Artificial Intelligence Videos." https://www.artificial-intelligence.video/a-darpa-perspective-on-artificial-intelligence (accessed Oct. 13, 2021).
[22] P. J. Navarro, C. Fernández, R. Borraz, and D. Alonso, "A machine learning approach to pedestrian detection for autonomous vehicles using high-definition 3D range data," *Sensors (Switzerland)*, 17, 1, 2017, doi: 10.3390/S17010018.
[23] "Machine Learning Algorithms in Autonomous Driving – IIoT World." https://iiot-world.com/artificial-intelligence-ml/machine-learning/machine-learning-algorithms-in-autonomous-driving/ (accessed Oct. 13, 2021).
[24] "Machine learning in Autonomous Vehicles – Perelik Soft." https://pereliksoft.com/index.php/2021/07/08/machine-learning-in-autonomous-vehicles/ (accessed Oct. 14, 2021).
[25] P. Sun, H. Kretzschmar, X. Dotiwalla, *et al.*, "Scalability in perception for autonomous driving: Waymo open dataset," *Proceedings of the IEEE Computer Society Conference on Computer Vision and Pattern Recognition*, 2443–2451, 2020.
[26] H. Xue, D. Huynh, and M. Reynolds, "SS-LSTM: a hierarchical LSTM model for pedestrian trajectory prediction," *Proceedings – 2018 IEEE Winter Conference on Applications of Computer Vision, WACV 2018*, 2018, 1186–1194, 2018, doi: 10.1109/WACV.2018.00135.
[27] "Nuro – On a Mission to Beveryday Life Through Robotics. | Nuro." https://www.nuro.ai/ (accessed Oct. 14, 2021).
[28] A.K. Tyagi, and S.U. Aswathy, "Autonomous Intelligent Vehicles (AIV): Research statements, open issues, challenges and road for future," *International Journal of Intelligent Networks*, 2, 2021, 83–102.
[29] D. Agrawal, R. Bansal, T.F. Fernandez, and A.T. Tyagi, "Blockchain integrated machine learning for training autonomous cars. *Hybrid Intelligent Systems*, HIS 2021. Lecture Notes in Networks and Systems, vol 420, 2022. Springer, Cham. https://doi.org/10.1007/978-3-030-96305-7_4.

Chapter 9

Biometric and blockchain-based vehicular technology

Abstract

With new discoveries being made every day in the field of technology and new inventions rising among the society, those tasks that once required manual labors have now been hugely replaced with softwares and automation systems that simplifies the tasks with the click of a button. With a huge population across the world and with emerging trends in the industrial sector, the human economy has not only observed huge changes in growth and development but also rising sophisticated demands to meet every day needs of the common man. With India becoming a trillion economy, a huge rise in population and growth in various sectors, have led to drastic changes in the environment. Today, travel has become a necessity for many people in the workforce. With public transports, hugely meeting the demands of a large population, people today prefer to use their own forms of transport for commuting. This practice no only increases the amount of carbon-footprint and consumption of resources such as fuel but also requires a huge demand in infrastructure to store and park vehicles. According to a recent survey made by World Health Organization (WHO), lack of infrastructure for parking the vehicles has resulted in increased waiting time, fuel consumption and high amounts of air pollution. With parking systems mostly being man-handled, today with the help of technologies, we have been able to automate this process of allotting a parking space for vehicles with the help of software and apps. This system gains access to the user information and allots a parking space manually by analyzing user data and the vehicle requirements. But the question of security of user-data is one question that remains unanswered. Since user data is vulnerable to the breach of security attacks and manhandling, we need a solution that can not only satisfy the storage requirements but also meet the security requirements. One such solution that we will be considering is a hybrid technology of blockchain and biometrics that can provide two layers of protection to the user level data, offering storage in the cloud at the same time. Blockchain revolves around distributed computing and architecture that takes care of integrity and authentication of data., whereas biometrics plays the role of authenticating and analyzing the user information for the purpose of verification and control of data. Deploying and studying the use of this hybrid technology can help in providing a secure environment for user data in vehicle transportation systems.

Key Words: Blockchain, Biometric, Internet-of-Things', ITS – intelligent transportation systems, V2V – vehicle to vehicle, V2I – vehicle to infrastructure, VANET – Vehicular Ad Hoc Networks

9.1 Introduction

Today, with rapid innovation in the field of automotive and mechanics, Technology has aided in the growth of smart transportation systems such as electric cars and self-driving cars that work on the basis of algorithms that drive them. These transportation systems are not only smart because of their working mechanisms but also due to a variety of factors such as physical and locomotive design, but also due to their interconnection with other vehicles in terms of proximity and roadside infrastructure. Such interconnection of networks that links proximity and infrastructure is termed as VANET [1]. Recently, one such automated parking systems involve the use of IoT and wireless communication networks to assist humans in finding a vacant space for parking lots. In such a system, a set of IoT devices are installed in parking lots and with the help of sensors, the status of parking slots is detected by these sensors, and the status is collected and stored in cloud-based system. Since the system involves the use of an application software to keep the driver informed, the data service provider collects, analyses, and stores the information from collected from the application user, using a system Application Programming Interface (API), and using the status of parking space from the cloud storage, the user is informed whether a parking lot is vacant in the nearby space. Though this system offers great flexibility in terms of user convenience, there arises concerns regarding the privacy and security of user data. In the case of IoT, since data is communicated through a series of networks, there is a high risk of data transmission leakage and mutation, wherein the service provider can gain access to vulnerable information of the user, which is a cause of security concerns [2]. Though a variety of security protocols are discussed, the level of security that it provides seems to still not have advanced along with recent trends when it comes to security standards and hazards. Through this chapter work, we shall try to propose a system that integrates data privacy, storage, and security using advanced technologies.

We shall study the use of two mechanisms that ensure high-level security of data, namely Blockchain and Biometrics which can be integrated with the existing smart parking system. The Blockchain mechanism collects and stores user data in the form of a block and secures the data using a hashing algorithm using a data structure. Blockchain is a distributed system that does not rely on the use of a third-party system. The user data is stored in the form of a block, where each block is interconnected to one and another through links. The trust between each block in a blockchain network is created through mathematical methods and cryptographic techniques instead of a central authority. Talking about intelligent vehicles, there exist two kinds of data sharing scenarios, namely vehicle to vehicle or vehicle to vehicle (V2V) data sharing and V2I (vehicle to infrastructure) data sharing that require a secure protocol for the regulation of safety standards and hazards [3].

In that data sharing network of intelligent vehicles (IV), security is at a major risk during V2V or V2I communication [4]. Such a network demands the use of a biometric system, for transmitting reliable data, over the data sharing platform. The blockchain network has a cloud storage and manages the blockchain protocol in the system, from where the data becomes accessible to the system users. Combining this mechanism of the blockchain network with a secure biometric, a unique crypto id with biometric enabled feature along with transactions that are self-executing, over the cloud networks [4]. Through these two mechanisms, we shall try implementing a mechanism that ensures robustness by providing a secure data environment, or enabling transactions and transport of data over a secure V2V network.

The organization of this chapter work is done as follows: In Section 9.2 , we shall discuss the previous works and the reason behind implementing Blockchain and Biometrics in intelligent vehicular systems. Section 9.3 shall provide insights about what is biometric technology, Section 9.4 shall cover about what is biometric based vehicular technology, Section 9.5 shall provide insights about blockchain technology, Section 9.6 shall discuss about blockchain based vehicular technology, Section 9.7 shall discuss about integrating biometric and blockchain technologies in vehicular technology and in the final section, we shall discuss about the various issues that we shall face with regard to biometric and blockchain-enabled vehicles.

9.2 Related works

Most of the ITS used today are majorly based out of IoT and cloud storage networks that involve data flow through channels, which at times may result in third party service providers gain unauthorized access to data. This can increase the vulnerabilities of various cyber-security attacks that can lead to manipulation and misuse of the user's personal data records. The VANET's which are a major part of ITS today pose various risks associated with security and data breaches, due to which it is unfit for large scale use [5]. Many scientists who are currently researching on ITS are working on other ways that can enhance sustainability and data protection in the long run. Through this work that will be carried out, we shall try to propose a mechanism that shall ensure secure transmission of data using double level of encryption and cryptographic techniques that not only provides data storage, but as an alternative to cloud, it shall offer two levels of security through access and authentication of secure data using blockchain–biometric mechanism, ensuring data communication and access to service providers in a doubly secured way. Since most of the methods leverage the use of Intelligent Vehicle Biometric Credits (IVBC), our method involves the use of Vein and Iris-based recognition to authenticate user access by the service provider.

Out of all the biometric-based approaches, the Iris recognition method is one of the most prominent methods, where the muscular folds found in the iris of human beings are unique for each individual and offer a strong sense of security. The Iris pattern in each individual can be tracked to be unique, and be matched and validated with pattern recognition techniques. A unique characteristic behind such recognition system is that unlike the normal recognition techniques that use fingerprint and other exposed areas of the body, which can be used without user authentication, iris pattern securely resides in the eye, which makes it a unique form of a verification technique.

A Vehicular Ad Hoc Network (VANET) allows cars equipped with communication, sensing, and networking capabilities to link with other vehicles (V2V) or roadside infrastructure (V2I) for data or information sharing. Accident reporting, infotainment, aberrant vehicle behavior warning, smart parking, congestion warning, and advertising are some of the uses it can provide. Despite the phenomenal growth of vehicle use among users and drivers, a considerable portion of onboard capacity remains chronically underutilized. The cloud environment is preferred to construct the vehicular network for maximum exploitation of idle applications and boosting vehicle capacity, dominating the appearance of Vehicular Cloud Computing (VCC). VCC is a combination that has an impressive encounter on roadside and traffic authority by totally handling idle vehicular utilizations, like storage and execution for opinion selection. By coordinating idle onboard resources, a collection of vehicles in a parking area or on a highway can make a cloud environment to combine information, technical data, and make agreements for developing the aspect of facility for both passengers and drivers. The vehicular cloud (VC) provides a profitable method to facilitate the promotions of surplus utilizations of car or vehicles, which are correlated and operated so as to offer maximum benefit to the users.

Commonly, a VC is temporary and dynamic because of vehicle mobile applications. The temporary VC is a significant subsidiary for the conventional cloud (CC) for developing storage and other capacities for clients. It is forecasted that vehicular cloud computing (VCC) is capable to develop different vehicular services and applications such as enhanced riding and driving actions, vehicular crowd-sensing, downloading video stream, and road traffic controlling. Because of the remarkable increase in the number of smartphone users, it can provide a crucial interface between networks and drivers. For example, a smartphone equipped with biosensors can gather information about the physiological condition of the driver of a vehicle, and the assembled information can be forwarded to VCC. In case of danger or accident, the warning bell can be launched to the smartphone for alarm sound. With association of smartphones, applications supported by VCC become ascendable, enhanceable, and reasonable for implementation.

9.3 Biometric blockchain-based technology

Several security features have been enabled in vehicles but biometric is the most reliable and secure feature to secure vehicle against any cyber-attack. This biometric-based vehicular technology can be found in further subsections.

9.3.1 Biometric-based vehicular technology

Biometrics leverages the use of those traits that are unique and distinct for every human being [6]. The most commonly used biometric technique today is that of fingerprint and palm-based verification. Such a technology uses the geometry of the palm and other dimensional characteristics such as width and length of the fingers to verify the identity of the user [7,8]. Authentication and key agreement is a key method deployed in biometric systems to verify the identity of the users and the network service providers for providing secure communication channels [9]. In the first stages of verification, a biometric sample from the user is collected and analyzed and is placed on a smart card reader for reading the fingerprint data [10]. If this data is exposed, it tends to expose the identity of users, resulting in data leakage [11]. As a next step, different techniques of biometric verification are applied to verify the extent of security that is being offered [12–14]. Some literature surveys have been focusing widely on the security of intelligent transportation systems in the recent years [15–19].

In an ITS, the possible threats encountered within are three aspects. The first aspect is that of the vehicle, where the vehicle is built up of three components electronic control units like sensors, networks, and a gateway to facilitate communications such Bluetooth or Wi-Fi. The second aspect looks through how physical devices are configured to a vehicle and the third discusses about the medium used by the vehicle to communicate with the external environment to detect any input. Several authentication processes have been proposed for securing vehicular networks. Suggestions were made by several researchers for combining several authentication schemes to exploit key-chains [12,20]. This process is carried out at the Data Security Sharing and Storage Based on a Consortium Blockchain (DSSCB) by the Preselected Nodes (PSNs) with recording rights. Of order to allow collaborative management in new blocks, the data block requires all participants to verify together. As a result, to tackle the problem of distributed storage consistency, an efficient distributed consensus is required. Cooperative control is essential in non-centralized control, according to neighbor-based distributed control [13], for all PSNs to reach consensus under limited information exchange and in dynamic interactions. To provide consistent accounting for the bitcoin network, the blockchain employs a Proof of Work (PoW) technique, which is heavily reliant on node power.

9.3.2 *What is biometrics?*

Biometric devices have revolutionized the ways in which users gain authenticated access to a system. Biometrics involves the use of those traits that are unique with respect to each and every individual such as fingerprints and facial features, to provide authorized personal authentication to the users. This method of authentication is now the most preferable and secure means, which can provide ease of authentication compared to complex passwords and pins. The use of biological traits in biometrics becomes difficult to clone, copy, and share among third-party applications. Apart from physical aspects in human beings, biometrics also leverages the use of non-physical characteristics these days such as voice and signatures.

Note that with huge growing trends in the field of technology (in current smart era), these two concepts, i.e., biometric and blockchain provide two levels of security/ high level of security. In biometric, there are other patterns that are unique to humans such as retinal and hand patterns. And Blockchain keep this unique biometric information in decentralized and secure manner in blocks.

9.3.3 How does biometric work?

First, the type of biometric that is supposed to be used is decided on the basis of the solution. Depending upon the type of solution to be deployed in the scenario, such as facial or fingerprint based, the image is captured as an input, followed by which it is processed and extracted on to a template. This template is now mounted on to a repository in which all the biometrics collected are stored and read using a smart card reader. Now the chosen biometric is scanned live and extracted onto a template. Now this biometric is then processed and extracted onto a final template. The scanned biometric is now matched with the one that is extracted on to the template [14].

9.4 Blockchain technology

Blockchain technology is a distributed ledger technology (DLT) which is the most common technology for today's smart era/applications to make their computer systems/communication networks more secure. This technology is immutable, trustable about cost- consuming. More details about blockchain can be found in further subsections.

9.4.1 What is blockchain technology?

Blockchain is a distributed repository that consists of data in the form of records that are held together by interconnecting nodes [15]. A block generally contains hash value of the previous block, payload, signature, and time stamp of the parent block. The hash value of the previous block makes the current block resistant to any changes that are to be made. It is a distributed repository that cannot be forged through cryptographic mechanisms. Blockchain systems can be classified into three types: Public, private, and consortium blockchain [16]. Public as the name suggests is that network in which all data records inside the network are visible and accessible by the public, for example, Bitcoin. In a private blockchain network, it is a network that is controlled by a centralized repository, wherein a central admin only has the responsibility to access the records in the network. A consortium blockchain can be managed and controlled by several other repositories. In a consortium blockchain, only those repositories that are granted permissions shall be able to access the records in the system. A consensus mechanism is a method of fault tolerance used in a blockchain network to ensure agreement among various other nodes in a network. This mechanism ensures that all other internodes are synchronized, with each other and have a mutual agreement on transactions, which are authorized among the blockchain network. This mechanism is the crux of ensuring privacy among the blockchain network [21–26].

9.4.2 Blockchain-based vehicular systems

Blockchain has played a major role when it comes to decentralized management of data in vehicular networks [17]. In such a system, important data related to vehicles such as stored files, digests, tokens, and data to check the integrity is stored in the blockchain network, which provides fair judgments for the storage devices that are present within the vehicular network [18]. It is assumed that the vehicles within the network interact with one another through V2V communication, such that the vehicles can connect with each other through the Internet more and more effectively. We need a blockchain network specifically designed to monitor, secure data, and not the ones used for cryptocurrency. A single blockchain network is balanced and supervised autonomously to record the details of the vehicles through networks. All vehicles track their respective locations using beacon messages. A location certificate is used as a proof that is used to specify the proximity of one vehicle from the other. This location certificate helps in detecting the event messages of a vehicle in a particular geographical area. The events here are local and the event messages are confined to vehicles in a particular location.

A 7-layer conceptual blockchain model has been proposed specially in a Vehicular Ad Hoc Network (VANET). The presence of blockchain technology enables the smart contract system in the ad hoc network and allows the usage of multiple softwares, including modularity functionalities such as traffic regulations, licenses, and road traffic statistics. The role played by the use of blockchain technology in this case is to facilitate the integration of these functionalities with enhanced trust and security [13]. The main advantage behind deploying blockchain is that it can provide peer to peer (P2P) communication in a way that the user data is secure, allowing data sharing and facilitating multiple communications across different nodes in the IV network. The blockchain technique's distributed consensus is the system's heart and soul, and it is critical to the system's smooth operation. Because VANETs and blockchain have similar node distribution characteristics, combining the two can solve some of the issues that plague VANETs. When all dispersed network nodes update their ledgers and make a consistent statement in a copy of the ledger, a consensus is obtained. The vehicle first delivers data to the RSU's record pool, which it subsequently packs into blocks when a certain amount of time has passed. Before the data block can be written in the digital ledger, a distributed consensus must be created. Such process has been carried out in the Data Security Sharing and Storage Based on a Consortium Blockchain (DSSCB) by the preselected nodes (PSNs) with recording rights [12,20]. To verify/allow collaborative management in new blocks in a blockchain network, the data block requires all participants to verify together. As a result, to tackle the problem of distributed storage consistency, an efficient distributed consensus is required. Note that Cooperative control is essential in noncentralized control, i.e., for all preselected nodes (PSNs) to reach consensus under limited information exchange or for dynamic interactions. To provide consistent accounting for the Blockchain network, the Blockchain concept implement a Proof of Work (PoW) technique, which is heavily reliant on node power.

There are currently a number of anti-malware filters that use pattern-matching algorithms to detect questionable files. The viral patterns are stored and updated on a central server in these pattern-matching techniques. These centralized countermeasures, on the other hand, are vulnerable to malicious attackers. The blockchain technology has the potential to assist secure dispersed networks. A recent study proposed BitAV, an unique anti-malware environment in which users can distribute infection patterns on blockchain to improve the fault tolerance of BitAV. BitAV can increase scanning speed while also increasing fault reliability. As a result, in order to fully comprehend the privacy implications of the blockchain's decentralized architecture, we must first determine whether the privacy benefits derived from decentralized coordination outweigh the privacy costs derived from the disclosure of metadata that may reveal personal information [13]. Another recent study presents two scenarios involving shopper identity. Third-party trackers acquire information about user purchases for advertising and analytics on most shopping websites.

Secure data provenance is critical for data accountability, forensics, and privacy because cloud data provenance is metadata that captures the history of the creation and actions conducted on a cloud data item. A recent essay proposes leveraging blockchain technology to create a decentralized and reliable cloud data provenance that blockchain technology enhances the privacy and availability of blockchain-based data provenance by delivering tamper-proof records, maintaining data openness and accountability in the cloud, and helping to improve the performance of blockchain-based data provenance [13]. ProvChain is an architecture that collects and verifies cloud data provenance by embedding provenance data into blockchain transactions, which was created and implemented by the researchers mentioned above. ProvChain contains three key phases: (i) collection of provenance data, (ii) storage of provenance data, and (iii) verification of provenance data.

According to the scope of the network, blockchain networks can be classified as public chain, private chain, or consortium blockchain. The public chain is an open blockchain in which anyone may participate, with no identity identification or permission settings. The chain's transactions are entirely open and transparent. Every user in the chain has access to the entire account book. Bitcoin and Ethereum are two popular public chain platforms. Only permitted network users can join the consortium blockchain, which is usually maintained by many institutions or organizations and has a quicker processing speed than the public chain.

The block header and block body make up the block of the consortium blockchain suggested in this study. The block header contains the following information: block ID, block size, previous block hash, Merkle tree root hash value, timestamp, and digital signature. To ensure that the transaction sheet has not been updated, the Merkle tree's root hash value is the SHA256 hash value of the transaction content. The digital signature is the block producer's signature, indicating that the block has passed verification

9.5 Integration of biometric and blockchain in vehicular technology

Despite the increased use of blockchain offering reliability, there still seems to arise a sense of security issues in the system. Given the complex architecture of the blockchain network, there occurs a series of drawbacks when it comes to sharing data, considering the potential vulnerabilities of data attacks of the user. Hence, more amount of research and analysis were required to provide an extra layer of security, when it comes to securing the user data. Hence, a decision was made to integrate, biometric technology into the existing blockchain systems in order to bring out the best security for the system, scalability, and privacy. Hence, biometric blockchain was altogether brought as a new framework, to ensure better security among V2V networks. The function of biometrics in the system would be to scale off ledgers and generate unique identities with respect to each and every use accessing the system. Biometrics as a service is used to offer an extra level platform to enable large scale and secure data transfer across the network. Such a system can offer numerous amounts of benefits through which we can keep a track of data sharing in the vehicular network between a service provider and the system user and then access of data shall be restricted to only few users. A personal credit-based system can be assigned to and hence, an extra layer of security shall be added to facilitate data sharing.

A car parking service provider oversees each smart parking system. Every parking place is linked to a blockchain-based, secure smart car parking system. A local duplicate of the ledger is kept in each parking place. The data collected by the parking sensor is the first transaction in a smart parking system. As a parking area is coupled with an IoT device such as a parking sensor, automobile parking availability can be generated as a transaction. To generate the transaction, each automobile parking service provider has a smart contract. An IoT device generates a transaction when the status of a parking space is changed from "empty" to "engaged."

9.5.1 Biometric blockchain-based IVs

Biometric blockchain comprises of blockchains and its interconnected ledgers that are linked together with the help of cryptography, in which each block is assigned an individual and a unique signature. Each block is treated as an individual transaction, in which the data records of the vehicular network are stored. The overall system is comprised as several layers as follows:

(a) Physical layers: These layers mainly comprise of physical devices IoT devices, sensors, cloud storage and other devices that are a part of the vehicular network.
(b) Data link layer: In this layer, the records of the vehicular networks are processed and encrypted with the help of appropriate data structures and cryptographic hashing techniques. Here, the keys required for hashing are generated by the double SHA 256 algorithm.

(c) Network layer: This layer is usually responsible for the transmission of data in the form of packets over a peer-to-peer network. It validates the communication between happening between two nodes in a network and validates the transmission of data over the network.
(d) Handshake layer: Also known as consensus layer of the blockchain network. It has an authority over the transmission of data, taking place across the network and also enables build integrity over a secure communication environment in the network. The Proof of Driving (PoD) algorithm is used for validating the secure communications among the network of intelligent vehicles.
(e) The biometric layer: This layer is assigned a unique crypto based data, which is allotted to all the vehicles in the IV network. The vehicle with higher number of credits shall be the favorable one in the network. The vehicle with the maximum IVBC credits leads in the network.
(f) Presentation layer: This layer is responsible for aggregating all the scripts, contracts, and algorithms that are present in the IV network.
(g) Service layer: Consists of the use cases of the software present in the IV network.

Hence, biometric and blockchain based solutions will be the necessity in the near future for society 5.0 and industry 4.0.

9.6 Issues concerned with biometric-based Blockchain solutions

Though biometric-based blockchain offers multiple levels of security than a simple cloud-based storage system, it does experience sensitivity in certain factors. Biometrics offers an added layer of security in the blockchain network. While BBC is being deployed at a large scale, it may require access to certain personal information of the user which will be highly sensitive to exposure. Hence, such a system would require additional support and assistance with a mechanism that sought to protect its privacy, such as a user software agreement or a standard license. Solutions proposed to tackle this issue include the use of end–end encryption for biometrics. These encrypted biometrics could form a miniature version of a digital signature [19,24]. Though BBC's offer multiple levels of security, these security checks may be time consuming.

9.6.1 Security

Though blockchains follow end-to-end encryption using hashing mechanisms and encryption techniques, the weaker links present in the blockchain may be a prime cause of concern. Since every consecutive node is linked together in a blockchain network, gaining access to one particular node in a blockchain may risk exposing the consecutive nodes in the network. With such an issue in picture, it may be possible to expose the data and make it vulnerable in the hands of third-party attackers.

9.6.2 Scalability

The greater the number of nodes in the blockchain network, the greater will be the vulnerability of the entire network. Hence, blockchain is not suitable in the case of a large number of records. In a blockchain network, for every record, there is a clone of a data that is generated as a backup. Hence, when it comes to handling large amounts of data, the greater will be the number of copies that will be generated in the network, due to which there is a high chance of redundancy occurring.

9.7 Conclusion

Through this chapter work, we have tried analyzing the scope of combining blockchain and biometrics and studying its advantages. We have also analyzed as to how merging these two technologies can provide an extra layer of security in communication networks and can be useful in ensuring privacy, integrity, and security in an IV network. The system shall collect data, encrypt, and generate a hash value for the same and store it in the form of records in a block. The biometric plays the role of providing user verification by analyzing the physical input delivered by the user and providing end-to-end encryption in a peer-to-peer blockchain network. Though there does exist certain pitfall when it comes to the ease of access to data such as fingerprint and physical verification, Iris verification in biometrics is seen to be offering a higher level of security to retain privacy in the system.

References

[1] A Shrestha, R Bajracharya, AP Shrestha, and SY Nam. A new type of blockchain for secure message exchange in vanet. *Digital Communications and Networks*, 6, 2, pp. 177–186, 2020.

[2] K Yang, K Zhang, J Ren, and X Shen, "Security and privacy in mobile crowd sourcing networks: challenges and opportunities," *IEEE Communications Magazine*, 53, 8, 75–81, 2015.

[3] R Jiang, T Li, D Crookes, W Meng, and C Rosenberger. (Eds.), Deep Biometrics. Unsupervised and Semi-Supervised Learning, 2020. Available at https://link.springer.com/book/10.1007/978-3-030-32583-1#about.

[4] C Wang, Q Wang, K Ren, and W Lou, "*Privacy-preserving public auditing for data storage security in cloud computing," in Proceedings of the IEEE INFOCOM*, San Diego, CA, 2010, 1–9.

[5] L Zhu, C Chen, X Wang, and AO Lim, "SMSS: symmetric masquerade security scheme for VANETs," in Proceedings of the 10th International Symposium on Autonomous Decentralized Systems, March 2011, pp. 617–622.

[6] P-SL Chih-Yu Hsu, K-K Tseng, and Y. Li, Palm Personal Identification for Vehicular Security with a Mobile Device, IJVT, MDPI, Volume 2013, Article ID901524, https://doi.org/10.1155/2013/901524.

[7] Wei Xiong, Kar-Ann Toh, Wei-Yun Yau, Xudong Jiang, "Model-guided deformable hand shape recognition without positioning aids," *Pattern Recognition*, 38, 10, pp. 1651–1664, 2005.

[8] CC Han, HL Cheng, CL Lin, and KC Fan, "Personal authentication using palm-print features," *Pattern Recognition*, 36, 2, 371–381, 2011

[9] Q Jiang, N Zhang, J Ni, J Ma, X Ma, and K-KR Choo, "Unified biometric privacy preserving three-factor authentication and key agreement for cloud-assisted autonomous vehicles." *IEEE Transactions on Vehicular Technology*, 1–1, 2020. doi:10.1109/tvt.2020.2971254

[10] X Li, J Niu, J Ma, W Wang, and CL Liu, "Cryptanalysis and improvement of a biometrics-based remote user authentication scheme using smart cards," *Journal of Network and Computer Applications*, 34, 1, 73–79, 2011.

[11] BS Abhilasha, S Anna, and M Shimon, "Privacy preserving multifactor authentication with biometrics," *Journal of Computer Security*, 15, 5, 529–560, 2007.

[12] B Groza and S Murvay, "Efficient protocols for secure broadcast in controller area networks," *IEEE Transactions on Industrial Informatics*, 9, 4, 2034–2042, 2013.

[13] B Leiding, P Memarmoshrefi, and D Hogrefe, "Self-managed and blockchain-based vehicular adhoc networks," in Proceedings of the 2016 ACM International Joint Conference on Pervasive and Ubiquitous Computing: Adjunct (UbiComp '16). New York, NY: ACM, 2016, pp. 137–140.

[14] S Liu and M Silverman, "A practical guide to biometric security technology." *IT Professional*, 3, 1, 27–32, 2001. doi:10.1109/6294.899930

[15] Y Rahulamathavan, RC Phan, M Rajarajan, S Misra, and A Kondoz, "Privacy-preserving blockchain based IoT ecosystem using attribute-based encryption," in *2017 IEEE International Conference on Advanced Networks and Telecommunications Systems (ANTS)*, 1–6, 2017, doi: 10.1109/ANTS. 2017. 8384164.

[16] WB Wang, DT Hoang, PZ Hu, *et al.*, "A survey on consensus mechanisms and mining strategy management in blockchain networks", *IEEE Access,* 7, 22328–22370, 2019.

[17] Z Yang, K. Yang, L. Lei, K. Zheng, and VCM Leung, "Blockchain-based decentralized trust management in vehicular networks." *IEEE Internet of Things Journal*, 1–1, 2018. doi:10.1109/jiot.2018.283614.

[18] C Cai, X Yuan, and C Wang, "Hardening distributed and encrypted keyword search via blockchain," in Proceedings of the IEEE 2017 Symposium on Privacy-Aware Computing (PAC), Washington, August 2017, pp. 119–128.

[19] CC Han, HL Cheng, CL Lin, and KC Fan, "Personal authentication using palm-print features," *Pattern Recognition*, 36, 2, 371–381, 2003.

[20] X. Zhang and X. Chen, "Data Security Sharing and Storage Based on a Consortium Blockchain in a Vehicular Ad-hoc Network," in IEEE Access, vol. 7, pp. 58241–58254, 2019, doi: 10.1109/ACCESS.2018.2890736.

[21] D Wu, J Yan, H Wang, D Wu, and R Wang, "Social attribute aware incentive mechanisms for video distribution in device-to-device communications," *IEEE Transaction on Multimedia*, 19, 8, 1908–1920, 2017.

[22] Y Li, Q Luo, J Liu, H Guo, and N Kato, "TSP security in intelligent and connected vehicles: challenges and solutions," *IEEE Wireless Communications*, 26, 3, 125–131, 2019.

[23] Q Luo, Y Cao, J Liu and A Bensilmane, "Localization and navigation in autonomous driving: threats and countermeasures," *IEEE Wireless Communications*, 26, 4, 38–45, 2019.

[24] AK Tyagi, SU Aswathy, G Aghila, and N Sreenath, "AARIN: affordable, accurate, reliable and innovative mechanism to protect a medical cyber-physical system using blockchain technology," *IJIN*, 2, 175–183, 2021.

[25] A. K. Tyagi, D. Agarwal and N. Sreenath, "SecVT: Securing the Vehicles of Tomorrow using Blockchain Technology," 2022 International Conference on Computer Communication and Informatics (ICCCI), 2022, pp. 1–6.

[26] A. K. V, A. K. Tyagi and S. P. Kumar, "Blockchain Technology for Securing Internet of Vehicle: Issues and Challenges," 2022 International Conference on Computer Communication and Informatics (ICCCI), 2022, pp. 1–6.

Chapter 10

Internet of Things-based vehicular technology (IoV): introduction and its scope

Abstract

Several changes have occurred in the automobile business over the past decade. Internet of Things (IoT) devices has been employed in a wide range of applications. As automobiles become increasingly automated and sophisticated due to smart devices being used in their technology, however, the term "intelligent automation" refers to automatically responding to user inquiries by taking necessary action to decrease or avoid any casualties in the event of an accident. Here we present a comprehensive discussion of IoT-based vehicles or the importance of IoT-based vehicular technology. Researchers can use this chapter to gain new insights into IoT-based vehicles. Aside from the broad application of machine learning and deep learning, IoT-based vehicles are typically equipped with these concepts so that they can respond quickly to any user queries or increase their efficiency.

Key Words: Internet of Things-enabled vehicles, Autonomous vehicle, Intelligent vehicles, Internet of vehicles

10.1 Introduction

In recent years, the Internet of Things (IoT) has been a hot topic in academic and artificial environments. It is vital to alter "things" because of changes in "things" that can link through a network and progress to gain knowledge from diverse holdings. Internet of Everything (IoE) is a term used on the IoT to refer to all things that can communicate across a network. IoT has made it feasible for items to speak with one other and exchange information between them. This specificity enables us to address a variety of needs in each industry. The Internet of Vehicles (IoV) communicates between vehicles in intelligent transportation systems [1]. To ensure road safety, it enhances market operations and services. Our taxonomy for IoV includes services, plays, and framings. In either case, we look at the concerns and obstacles that plants face in the womb. We illuminated spaces and services because of the needs of drivers and malfunctioning vehicles and because we were concerned about the quality of the performances. A summary of the current condition of IoVs in terms of frames, services, and operations is provided in this document. It is possible to get started right away with the answers to problems in commerce advice in megacities from this perspective. The findings of this study support the creation of intelligent megacities and their supervision.

It provides an overview of IoV terminology and a survey and literature evaluation on the topic. In addition, the middle layer is the data processing layer, which combines data from various sources. Edge nodes act as both data sources and data centers, analyzing data from a variety of sources. One of the components of distributed computing. Computing and data processing can be done on-site at sensors and devices, allowing for better response times and lower costs. To meet the needs of various services and applications, data processing, and collection are necessary. Monitoring events are used to gather data in this tier. When it comes to processing and calculations, fog nodes can be employed that. Data transmissions are reduced, bandwidth is saved, and resources are more easily managed using these nodes. The cloud layer delivers a variety of services and applications and applications and services at the highest level. Processing and customized services are provided between devices and the cloud tiers and cover scalable service complexity. Many services and applications are available for intelligent transportation's functional requirements in the application layer, such as notifying drivers based on data analysis, monitoring vehicle running status, mishaps, and other clinical warning administrations concerning driver status and energy.

In our digital age, technical advancements are going toward more intelligent and linked products. This has ramifications across all sectors, and the automotive industry is no exception. With the help of sensors and communication techniques, stylish vehicles are gradually becoming more and more autonomous [2]. The web of Things is a new technology that's growing in every part of our life. The IoV [3] inspires an entirely new transportation and conveyance networks industry. IoT and the improvements in wireless and mobile communication have made the IoV a promising paradigm shift in the realm of wireless and mobile communication. As a newer version of the classic Vehicle Ad Hoc Network (VANET), it offers short-term communication between various road transport entities such as motor vehicles and pedestrians and roadways, parking lots, and city infrastructure [4].

How will the web of vehicles work?

To create a social network with intelligent objects like players, the IoV uses vehicle-to-vehicle (V2V) interconnectivity as well as V2R (vehicle-to-road), V2H (vehicle-to-human), and V2S (vehicle-to-sensor). This leads to the existence of the SIoV, which is essentially a transportation instance of the SIoT. The IoT is a complex network in which two or more entities can communicate throughout time using various technologies, such as a navigation system, mobile communication, and device networks to exchange information and instruction [5]. Modern cities' infrastructure is dotted with various sensors, which collect and process data and communicate powerfully among all the various entities that populate them. Vehicles are radio-controlled for a short length of time to support this knowledge.

10.2 Related/background work

Surveys on the IoV and connected technologies are discussed in this section. Transport unintended networks is one of the many related technologies (VANET). These unwanted mobile networks, referred to as VANETs, are used in automated transportation systems. According to [6] the architecture of ad hoc transport networks is examined, considering various features and events. The challenges and needs are categorized, and the most frequent simulation tools are introduced. This document does not include any evaluations or parameters. During the writing of this article, a thorough Quality of Services (QoS) assessment was carried out, and categorization measurements were linked with the review of applications and services.

Furthermore, there are no metrics for implementing VANET in an IoT environment. Cars, infrastructure, ad hoc, and service domains are separated in [7]. It is thus possible to categorize applications and services based on the QoS requirements for safety and non-safety purposes. These issues were discovered and addressed in the assessment of all protocols, standards, and procedures.

Various IoV and VANET attacks are discussed in [3] by authors. They divide the attacks into active and passive attacks and use the Open Systems Interconnection (OSI) layer model for cluster attacks on networks to characterize them. Attacks on applications and knowledge can both be classified. There are, however, no methods for detecting and responding to attacks within the systems. In the IoT's upper layer of knowledge communication, security is examined in [4]. In this work, various approaches to addressing user authentication and privacy concerns are challenged to build trust and safety. There is no unified approach to dealing with security issues at an entirely different level. These publications deal with a narrow subset of a much broader topic. If you read our work, you will notice that we try to reason architectures more thoroughly and several other areas. Alternate studies focus on a specific site of studies, such as protocols and networks, communications, and routing algorithms. From the perspective of communication, structural elements in [5] are categorized and examined as a single system. The first step was to go over all of the past work. Then, a generalized stratified design is given.

It is given a layered architecture like the IoT reference design, but the qualitative alternatives needed for IoV are not considered. The researchers classify various routing protocols and technologies according to their benefits and drawbacks. Alert generation, vehicle maintenance, and communication services all fall under the category of routing applications. Classifying applications, methods, and algorithms allow us to divide them into distinct groups for easier management. In topology-based approaches, distributed routing tables are actively changed proactively. Responding to demand and load on network routes is enabled by reactive routing systems.

In contrast, geographic routes are enforced using destination positioning information, and hybrid courses combine geographic and topology-based methods. Agglomeration protocols classify nodes in a network. In network communications, they use a cluster header for knowledge and route management. Agglomeration and control of knowledge on related nodes are both possible with knowledge fusion. A thorough examination is given to each category of the route. In this context, no research has been done on evaluating the feasibility and constraints of misusing transit traces from different settings with radically varied node densities. A look at V2X and V2R/V2I communication is provided in [8]. V2X applications are categorized to facilitate their use and are designed with them in mind. As part of this investigation, researchers look at how security affects V2V and wayside unit communication and service. To improve safety, several methods have been identified and assessed, including protocols and standards of communication. To lower the costs of more significant layers of security, physical layer techniques are discussed in this study.

Improved environmental needs of VANETs supported transportation network safety needs in specific articles. Ref. [9] investigates the social networks of automobiles. Vehicles and social networks are used to prevent hold-ups and avoid road accidents in VSN. Vehicles and social networks (also called Vehicular Social Network) are used to prevent hold-ups and avoid road accidents in Vehicular based Environment (VBE). Vehicle location and social information (such as traffic data) are used for services and applications to gather user knowledge. It offers voters an entirely new set of services and strongly emphasizes the human qualities crucial to transportation infrastructure. Situations and applications in these networks are created to support safety requirements and anomaly detection in traffic.

Vehicular Social Network (VSN) offers data-driven apps and location services. Because of the limited resources and information available in these networks, messages must be sent reliably. There is no priority for message delivery, and communication security and trust are unresolved issues for users' and social groups' collective knowledge. This issue must be addressed by expanding the capabilities of information technology and social network methods. Because of the variability in node quality, managing VANET networks can be a problem. The communication layer was designed to accommodate a slew of options. Named knowledge networking (NDN) in transport networks is the subject of research in [8]. This technology aids in content caching and enhances data sharing and resource management by selecting and replacing caching strategies. The authors of transport NDN cover design and cache management approaches.

It was discussed the difficulties of adopting this strategy in communication networks. Caching-related benefits were analyzed and classified by application and evaluated to support performance, a method of transmitting and receiving knowledge packets. They must be algorithmically introduced into the nodes. The communication and network layer is prescribed in [10–13]. They do not have to think about what comes after them. For the sake of this essay, we are going to focus on the underlying layers and basic settings of the system's supported services and applications. We have established a taxonomic classification that all levels of the design process can understand.

Many intelligent transportation systems and traffic applications require high-level discourse information and access to information at any given time. Nonfunctional needs were identified as a source of the difficulty. One of the most pressing issues in many apps and services is user access security. As a result, several of these publications categorize attacks and recommend various responses to each type. Using this method, it is possible to alter the implementation of security measures. Quality and changing specifications necessitate a security solution; The performance of nonfunctional requirements for network topologies is one of the many issues facing applications and services. Protocols and applications for transportation networks are examined using [14]. Security checks and security threats are considered on a separate layer. As a group, security challenges, nonfunctional requirements, and their trade-offs are all handled. Analysis measurements are not included in the information provided. Intelligent transportation systems can begin with the usage of information like this. The style and implementation of transport networks are also covered in depth. In this work, we attempted to rationalize the documents supporting how they are enforced and cumulatively implemented. Applied mathematics charts supporting the factors and measurements were extracted after simultaneously analyzing the implementation and simulation methods. The network of connected automobiles was intended to be used for a slew of different purposes. A wide range of studies have been conducted on the classification and style of applications: The framework, communication, and nonfunctional factors are examined in [15], which provides a wide variety of applications used for traffic control services.

VANET cloud design was to include a taxonomy. This taxonomy studies transport clouds and cloud environments for transport networks. On the other hand, this taxonomy is dependent on technology and is not comprehensive enough to classify all applications. This is the subject of several research projects. Atmospheric, system and application-related services and applications are examined in [16]. Safety, traffic control, and enjoyment are all linked to these applications. They provide a set of metrics for measuring performance. Our work analyses and classifies papers based on the IoV technology-supported devices and sensors in our study. IoV architectures, services, and applications have not been systematically surveyed; the classes were selected to meet domain-specific needs. In this post, we have attempted to conduct classifications and debates that meet the general transportation atmosphere demands of the readership. IoV taxonomy is extensive. Therefore, we analyze architectures, services, and applications in great detail. In the section on services, we tend to focus on the most critical aspects of service quality. Finally, we provide an application of mathematics to the evaluation criteria. In this article, we summarize the current state of the applying and repair layer. It is possible to hide open questions and future research by other scholars.

10.3 Working of IoV technology

An intelligent object social network is created using vehicle to vehicle (V2V) interconnection and vehicle to the road (V2R) interconnectivity to create a social network with competent object participants. As a result, the social Internet for vehicles (SIoV) was born. A transport instance of the SIoT can be described as such. The IoT is a complex network in which two or more entities can communicate over time using various technologies, such as a navigation system, mobile communication, and sensing element networks to exchange data and issue instructions. Sensors installed in vehicles, suitable terminals, and platforms spread across the infrastructure of modern cities and collect data, which is processed and communicated securely among all parties. The fact that vehicles can be remotely controlled over some time lends credence to this theory [17].

10.4 The architecture of the IoV

"How heterogeneous networks actuate in IoV system with completely different functions and jobs sorted and designed in network stratified architecture" has been the subject of numerous studies. The sixth approach is derived from combining the three, five, and seven network stratified architectures discussed previously. All the sensors and devices that gather information about the vehicle, the driver's position, and events around it are part of the customer layer; they are responsible for transmitting the information to the second layer (connection). All heterogeneous networks (V2V, V2P, and V2I) can communicate through the affiliation layer, sending the information to the third layer (cloud). IoV system functions are processed and computed in the cloud layer to meet the needs of the applications [18]. Conveyor, location, and cloud layers were combined to create a three-layered design, with the work of each layer being identical, i.e., shopper, location, and cloud layers all performing the same functions. Even though short-range communication is frequently used for long-distance communication by connecting so many vehicles and infrastructure networks via neighboring vehicles, it has become a bottleneck [19].

A three-tiered architecture (area network, network management, and D2D applications) was also proposed by Gandotra *et al.* for less expensive communications than devices to devices (D2D), without causing any data to be transmitted through the network base station (BS). In the first layer, all devices are painted and communicate with each other (area network). D2D information is gathered and sent to the core network (D2D applications layer) to produce the chosen application, such as public safety and security services, etc., in network management [8]. However, Kaiwartya *et al.* proposed a design with five layers (perception, coordination, AI, application, and business) on the other hand. As the vehicle moves through the space, sensors gather information about its speed, direction, and position, affecting the driver's attitude. The perception layer also collects information about traffic conditions and the weather outside the vehicle and sends it to the coordination layer. The coordination layer's job is to ensure that data gathered from the perception layer is transferred for AI processing firmly and in a unified structure. The latter is collected from entirely different heterogeneous networks like WAVE, Wireless-Fidelity (WiFi), 4G/LTE, and 5G. There is an AI layer that is represented by a virtual cloud substructure. Because of this, it is to blame for addressing "the data or knowledge of data" from the coordination layer and using victimization decision-making algorithms to analyze this information. An evaluation of knowledge received is used to manage several cloud services. The application layer is responsible for creating the applications tested by the AI layer before being made available to the public [10]. The higher layer could be a business model that analyses usage information of the promising applications by employing multiple techniques (such as use case diagrams and graphs, differentiation tables, and flowcharts.) to determine how much money is needed to manage the applications.

Contrary to popular belief, however, the seven-layered design was proposed by Juan Contreras Castillo and colleagues (user interaction, acquisition, pre-processing, communication, management, business, and security). This layer is responsible for creating an intelligent user interface for passengers, which monitors the outside environment to gather traffic, route conditions, and car parking information. This case was supported by having the interface in the vehicle track the driver's actions and then display the most important activity within the interface. The interface's best feature is that it reduces driver errors, making driving more secure. This layer's responsibility is to decide how to collect and transmit various types of data and knowledge from inside and outside of vehicles, using the most appropriate network technologies (short-range technology for within the car, e.g., Bluetooth or ZigBee, and for outdoor transmission long-range technology, e.g., WiFi or ultra-wideband). To prevent the dissemination of unrelated information, the pre-processing layer analyses the collected data and filters it. When it comes to deciding how devices can communicate, it's crucial that we know which network to select from a variety of heterogeneous networks to send the gathered data. fuzzy algorithms and other intelligent technologies work on various parameters, such as QoS, privacy, and security data [19]. The management layer's primary role is to manage the information exchanged between the various components of the IoV system to improve the system's performance. The business layer should focus on process and data analysis, which can be done using multiple statistical and critical tools, such as flowsheets, with different cloud computing substructures. The final layer (Security layer) interacts with all other layers to ensure that all layers function and knowledge transfer securely.

According to another study, the Universal IoV design (UIoV) for good cities comprises seven-layered (identification layer, physical objects; inter-intra device; communication layer; cloud services; transmission & massive information computation; application) components. The identification layer is in charge of naming and addressing all of the system's objects that are clear to all users. All information from other things is sent to the inter-intra devices layer for further processing by object layer. The inter-intra devices layer may be a new and distinct UIoV layer that works in conjunction with the communication layer to produce entomb communications between all system actors like V2I, V2P, V2V, V2R, V2S, and V2D. The cloud layer's responsibility is to ensure the quantifiability of IoT applications and services across all computing environments, including software packages, hardware infrastructure, and process platforms. Information pre-processing, massive information computation, and intelligent transport sub-layers could make up a new and distinct layer in the UIoV called the transmission and large information layer. The transmission and large information layers are responsible for all computation in the overall UIoV layers. In the final layer of the UIoV, the application layer, the higher layer, provides relevant data to the lower layer. It defines several protocols for sending the message.

It may be challenging to design a universal network that incorporates multiple heterogeneous networks in a single design. Identifying and effectively clustering similar functionalities and representative components of heterogeneous networks as a layer is necessary. Many of the superimposed design style goals are focused on optimizing layers and improving the differentiation between layers. Beyond the layering of functions and components, the network characteristics of the heterogeneous design are taken into account. The features of the network include interoperability, accountability, modularity, and measurement. Cars should be able to connect to various networks and devices via the IoV architecture. Therefore, an open and flexible superimposed design in terms of technology adaptability is even more appropriate for the procedure. Incorporating a plug-and-play interface, service-oriented architecture (SOA), and robust web integration are all essential goals of the

IoV design style. A primary effort is being made toward the superimposed branch of knowledge style of IoV because investigations into IoV are still at an early stage, with the majority of actions coming from industry and research [20–22].

10.4.1 Perception layer

Several different types of sensors and actuators are used to define the framework's first layer of design. These devices include vehicles, remote sensing units, smartphones, and other personal electronic devices. The layer's first task is to gather information about the vehicle, the traffic environment, and other devices. More than 25 billion records are included in this enormous collection: speed and direction of travel; acceleration and position; engine condition; travel documents; on-road vehicle density; climate; multimedia and picture records of people [23]. Magnetic attraction transformation and secure transmission of perceived knowledge to the coordination layer are also responsibilities. For the most part, the layer's main issues are related to the efficient collection, classification, and storage in terms of value and energy.

10.4.2 Coordination layer

Virtual universal network coordination modules define the second design layer for WAVE, WiFi, 4G/LTE, and satellite networks. The information gleaned from lower layers is passed on to the artificial intelligence layer for further processing. Since there are not any established standards or protocols for interoperability between different types of networks, this layer's ability to provide a reliable network property is compromised. At its core, this layer is responsible for transforming heterogeneous networks' data into a single structure that can be processed by any network that receives it.

10.4.3 Virtual cloud infrastructure layer

The virtual cloud infrastructure serves as a defining feature of the final design layer. It is IoVs brain, in charge of storing, processing, and analyzing the data it receives from lower layers and higher cognitive processes based on the evaluation. Vehicular cloud computing (VCC), Big Data analysis (BDA), and an expert system are major operational components of this information management center. Service management is also a significant concern in IoV, where exclusive and dedicated services are required by suitable applications that are also handled by this layer due to the large number of services offered in cloud environments [24].

10.4.4 Application layer

A good application ranges from traffic safety and efficiency to multimedia "based on primarily based" picture and net-based utility applications in the fourth layer of the design. The coating is responsible for delivering high-quality services to end-users supported by the AI layer's intelligent and demanding analysis of processed data. VANETs' application layer was also depicted in its design to ensure safety and potency, but good industrial-purpose apps were not expected. One of the primary responsibilities of the AI layer is finding and combining the best services for end-users, which is one of those services. Both the business layer and the user application layer benefit from this knowledge. For this reason, IoV has evolved into a world network for reliable conveyance communication as the range of good applications has been realized. The layer's suitable applications are the driving force behind the research and development efforts in IoV.

10.4.5 Business layer

The operational management module of IoV is the fifth layer of the design. Foresight methods for developing business models supported by usage knowledge and applied mathematics analysis of information are the layer's primary responsibility. Tools like flowcharts, comparison tables, and use cases are essential parts of the layer of analysis. The other duties of the layer embrace higher cognitive processes related to economic investment and usage of resources, pricing of use of applications, overall budget preparation for operation and management, and mixture knowledge management.

Existing protocols are grouped into a single layer for a protocol stack at the top of the pile. The protocol stack has been designed to meet the functional requirements of every layer known in the architecture. WAVE, C2C, CALM, IoT-A, IoT6, and HyDRA are among the protocols used in VANETs and IoT projects. There are three planes in the protocol stack that include security, operation, and management. The appropriate protocols for different layers and planes of IoV design have been identified by efficiently managing the most valuable requirements using existing protocols.

10.5 Characteristics of IoV over intelligent transport system (ITS) and autonomous intelligent vehicle (AIV)

Aside from cutting costs and optimizing knowledge center resources, server virtualization can shine in I/O performance and its role in facilitating more economical application execution of virtualized servers. A revolution in PCI bus virtualization has introduced the PCI-one SIG's root I/O virtualization (SR-IOV). Multiple virtual machines (VMs) can share PCI hardware resources thanks to SR-IOV, which adds additional definitions to the PCI express (PCIe) specification to allow for this sharing.

System designers for IoV benefit greatly from virtualization, which provides several advantages over intelligent transport system (ITS) and autonomous intelligent vehicle (AIV), can be listed here as:

- It allows for many virtual machines to be run on a single server, which reduces the need for hardware and the associated costs of space and power.
- It provides the ability to start or stop and add or remove servers multiple times, increasing flexibility and allowing for quantifiable outcomes.

- Allows you to run completely different operating systems on a single host machine, which further reduces your dependence on dedicated equipment.
- As SR-IOV-enabled hardware improves I/O performance and makes it easier to implement in PCIe devices, planning systems that are natively engineered on this hardware are also the most cost-effective [25].

Consideration has been given to the functionalities and representations of the layers in a five-tiered design for IoV. Design considerations include operational and security planes in a protocol stack for stratified design. A network model for the IoT is depicted by identifying three network components: the cloud, the association, and the customer. Cloud computing, heterogeneous network association, and IoV customer applications are examined. Playacting a qualitative comparison between IoV and VANETs highlights the benefits of realizing IoV. IoV's design and development challenges and problems are discussed. IoV's long-term aspects are depicted. One of the IoV's most important features is its "cloud," which symbolizes the brain. The primary cloud services include a wide range of services related to intelligent computing and process. Cloud infrastructure is used to host the services. "Connection" is the second component of IoV, which provides access to cloud-based intelligent computing and process services. It is possible to identify an association using a spread of seven wireless access technologies due to various wireless access technologies. The other transport communications of IoV form a unique association.

10.6 Benefits of IoV

IoV's third component, "client applications," makes use of a variety of different types of connections. The needs of each service provider application may differ from those of other service providers. For a service provider, a wireless access technology's characteristics describe their expected services. As a result, vehicular applications place a high priority on wireless access technologies. IoV components and their roles are well-illustrated in the following sections.

- *Safety*: The automotive industry can take advantage of a wide range of IoT hardware and software solutions to make their vehicles more aware of their surroundings and capable of operating on their own. As an illustration, if the vehicle is equipped to communicate with the adequate town infrastructure, it will be able to obtain up-to-date information on traffic, accidents, and other road hazards, as well as the current weather. The driver can alter his route to avoid traffic and possible accidents thanks to the information [26]. To do so, the vehicle will gather information about the road around it, including curbs and speed bumps, along with other vehicles, making it much easier to park the vehicle and creating a more dominant position for the vehicle.
- *Efficiency*: As a result of the gathered information, we will benefit from using a connected car package. The package is designed to help us avoid traffic jams, accidents, busy streets, or roads currently under construction, thus saving us time on our journeys. It is easier to manage large fleets of vehicles with

connected automobile package solutions [27]. As an example, a manager can track the location of the vehicle's victimization GPS and analyze its condition and schedule maintenance, enhancing the vehicles' power.

- *Cost-effectiveness*: There is a cost associated with implementing IoT infrastructure on the roads and the connected vehicles themselves [28]. However, we will still reap the benefits of reduced traffic jams in terms of the amount of time and fuel we save. In addition, a well-developed infrastructure makes it easier for people to share a vehicle.
- *Environment*: There will be fewer cars on the road, which means less pollution due to fuel emissions resulting from more people carpooling. Electric cars are also becoming more common, and this trend is expected to continue [29].
- *Accessibility*: Disabled people now have more options for getting around, thanks to the advancements in automotive technology. When it came to transportation, they needed more help than simply relying on public transport. Those who are disabled will be able to drive independently in the future thanks to connected self-driving cars, which will make use of advanced technology.
- *Data*: Many information is gathered by sensors on connected vehicles and given to the driver or used by the vehicle itself to respond autonomously to various situations. However, manufacturers and firms that provide high-quality automobile development services can also analyze the data to improve user expertise.

10.7 Applications of IoV

Modern cities' infrastructure is dotted with various sensors, which collect and process data then communicate to all parties involved. Based on this information, vehicles are steered by time.

A revolutionary technology

Currently, only a tiny portion of the research conducted on the IoV is available to the general public. However, many organizations like Huawei and Google are already conducting trials, but the IoT is still in its infancy and will take some time to fully mature. The design of this network still has some issues to overcome. No new technology can be completely secure, as we have already learned from the past. In its early stages, the IoT is still vulnerable to security threats. Such revolutionary automotive technology could hurt human life or cause economic losses due to damages if something goes wrong. Net of a car is ready to change our driving experience once any security concerns have been addressed. To make our roads as safe as possible, we need good town infrastructure and IoT. What are the specifics? We are about to find out. Here are a few more specific examples of how IoV can transform the experience of operating a motor vehicle:

Safe driving

In addition to what has already been said, IoV technology ensures a safe driving experience. Systems that use sensors to detect a possible collision and alert the driver are called cooperative collision avoidance systems (CCAs). Messages about the vehicle's performance or warnings about potential emergencies are sent out regularly. Traffic jams, dangerous road conditions, or accidents can all trigger an emergency message with the help of the most recent connected roadway technology.

Convenience service

The IoV allows for remote access to a vehicle. Small door lock and sick a purloined vehicle or "Find My Vehicle" are handy if we leave it in a large parking lot. In addition, transportation agencies can significantly benefit from this technology. It improves traffic, transit, and parking knowledge, making it easier to manage transportation systems to reduce congestion. Advanced docudrama systems also fall under IoV's remit. Connected cars will offer online, in-vehicle entertainment options that can change streaming music, media, navigation, or other information via the dashboard [26].

Traffic and crash response

Urban congestion management, transportation and logistics, and urban traffic are areas where IoV has a significant impact. Emergency services can receive crash data and the location of the vehicle automatically from new, connected cars. This could save lives by speeding up the response time to emergencies.

More applications

There are many other uses for the net of cars that are not limited to traffic systems and safe navigation. These include crash interference and electronic toll collection. Using IoV, we will see that fashionable cars can try to do everything independently while ensuring the driver's safety. The company's goal is to meet the increasing demands of customers and provide them with cutting-edge driving skills.

IoV turns vehicles in our assistants

One possible application of the IoT is the IoV. It was explicitly designed with the automotive industry in mind. V2V and pedestrian-to-vehicle communication via the IoT is becoming more and more of a standard feature in cities across the world.

IoT-based pure electric vehicles

When it comes to the design of electric vehicles, the IoT is becoming a more popular platform to replace older, non-smart wireless technologies. The introduction of integrated electric vehicles would solve all of the issues with current vehicles while also protecting the environment. Sensors can be placed throughout the body of an electric vehicle at a fraction of the cost of a traditional sensor. Consequently, the introduction of electric vehicles with IoT-based technology that monitors the battery life of electric vehicles is emphasized in this text as being necessary and essential. Things Speak, a web-based monitoring system, has been used to keep tabs on all-electric vehicles since they are all connected to the Internet (day-by-day). Once an effective boosting formula is integrated with the objective operation, these online results can be visualized in MATLAB. Visual analysis and performance results show that using IoT-based technology improves the projected technique for electric vehicles. It has also been found that the cost of implementation is reduced, and the electric vehicle's capacity is increased by 74.43% when sensors are continuously monitored.

10.8 Security, privacy, and trust issues and integration challenges on the IoV

Many attacks, such as timing, transition, Trojan horse, botnet, and so on, have been mitigated on VANET in the previous decade. The following attacks are explained as:

- *Sybil attack*: It is an extreme attack on VANET in that the assailant perniciously claims or takes completely different characters and uses these personalities to disrupt the VANET's utility by dispersing phony feelings.
- *Impersonation attack*: There are many ways in which a malicious user could impersonate a legitimate user to gain an advantage in the system. This type of attack is frequently referred to as an impersonation attack.
- *Denial of Service* (*DoS*): There are several types of DoS (denial of service), including DDoS (distributed denial of service) attacks. Due to a DoS attack, attackers cannot communicate with each other, and vehicles cannot obtain information, such as street position and transportation, which could lead to catastrophic consequences. During a DDoS attack, nodes can send out an attack from multiple locations, making it more challenging to discover a victim. Node's fuel in a DDoS attack may be expected to harm vehicles and RSUs, which are a significant component of VANETs.
- *Non-repudiation*: The legal concept of information non-repudiation, which alludes to the foundation of knowledge and trustworthiness, is used in information security.
- *Replay attacks*: This exploits the network's conditions by storing messages to use them later when they become invalid and untrue.
- Assailant pretends to be a user device and controls all of the system's functions. For example, an eavesdropping attack occurs when an unapproved party intercepts or modifies data transmitted between two electronic devices.
- *Auxiliary attack*: when the attacker collaborates with the aggregators to compromise privacy. For example, an attack by the man in the middle (MITM). The man-in-the-middle strategy is that the aggressor or engineer captures a link between two frameworks. It is a dangerous attack because the perpetrator acts as if they were the original sender. The recipient will be fooled into thinking they have received a simple message because the offender has the initial correspondence.
- As well as several additional attacks like timing attack, transition attack, etc.

In the last, some other interesting work like issues, attacks, etc., towards future vehicles or Internet of Vehicles can be found in [30–32].

10.9 Conclusion

As a result, the IoV framework stores a great deal of individual information like supply position, goal, and so on, making security and protection a minor undertaking for them. IoV's success has sparked a variety of new approaches to security and safety in the field, recommending a wide range of approaches to these issues and their solutions. This is an issue that must be addressed, given the rapid advancement of automobiles and the prevalence of electronic data transfer. There are numerous threats to IoV security and protection, including DoS attacks, replay and forgery attacks, and an outsider association. Customers and structures must be productive and contribute equally to maintain consumer singularity in the face of these challenges. The documented attacks have prompted us to come up with a few plans in this work. This work can serve as a model for a solution to these problems if it grows in scope.

References

[1] Kabalci, Y.; Kabalci, E.; Padmanaban, S.; Holm-Nielsen, J.B.; Blaabjerg, F. Internet of things applications as energy internet in smart grids and smart environments. *Electronics* 2019, 8, 972.

[2] Farmanbar, M.; Parham, K.; Arild, O.; Rong, C. A widespread review of smart grids towards intelligent cities. *Energies* 2019, 12, 4484.

[3] Yao, L.; Chen, Y.Q.; Lim, W.H. Internet of things for electric vehicle: an improved decentralized charging scheme. In Proceedings of the 2015 IEEE International Conference on Data Science and Data Intensive Systems, Sydney, Australia, 11–13 December 2015; pp. 651–658.

[4] Benedetto, M.; Ortenzi, F.; Lidozzi, A.; Solero, L. Design and implementation of reduced grid impact charging station for public transportation applications. *World Electr. Veh. J.* 2021, 12, 28.

[5] Sousa, R.A.; Melendez, A.A.N.; Monteiro, V.; Afonso, J.L.; Ferreira, J.C.; Afonso, J.A. Development of an IoT system with intelligent charging current control for electric vehicles. In Proceedings of the IECON 2018, 44th Annual Conference of the IEEE Industrial Electronics Society, Washington, DC, USA, 21–23 October 2018; pp. 4662–4667.

[6] Savari, G.F.; Krishnasamy, V.; Sathik, J.; Ali, Z.M. Abdel Aleem SHE. Internet of Things based real-time electric vehicle load forecasting and charging station recommendation. *I.S.A. Trans.* 2020, 97, 431–447.

[7] Gao, D.; Zhang, Y.; Li, X. The internet of things for electric vehicles: wide area charging-swap information perception, transmission, and application. *Adv. Mater. Res.* 2013, 608, 1560–1565. [CrossRef]

[8] Asaad, M.; Ahmad, F.; Alam, M.S.; Rafat, Y. IoT enabled electric vehicle's battery monitoring system. *EAI SGIOT* 2017, 8. [CrossRef]

[9] Helmy, M.; Wahab, A.; Imanina, N.; *et al.* IoT-based battery monitoring system for electric vehicle. *Int. J. Eng. Technol.* 2018, 7, 505–510.

[10] Divyapriya, S.; Amudha, A.; Vijayakumar, R. Design and implementation of grid connected solar/wind/diesel generator powered charging station for electric vehicles with vehicle to grid technology using IoT. *Curr. Signal Transduct. Ther.* 2018, 13, 59–67.

[11] Muralikrishnan, P.; Kalaivani, M.; College, K.R. I.O.T. based electric vehicle charging station using Arduino Uno. *Int. J. Sci. Technol.* 2020, 29, 4101–4106.

[12] Ayob, A.; Mahmood, W.; Mohamed, W.M.F.; *et al.* Review on electric vehicle, battery charger, charging station and standards. Res. *J. Appl. Sci. Eng. Technol.* 2014, 7, 364–372.
[13] Motlagh, N.H.; Mohammadrezaei, M.; Hunt, J.; Zakeri, B. Internet of things (IoT) and the energy sector. *Energies* 2020, 13, 494.
[14] Kong, P; Karagiannidis, G. K. Charging schemes for plug-in hybrid electric vehicles in smart grid: a survey, in *IEEE Access*, 2016, 4, 6846–6875.
[15] Phadtare, K.S. A review on IoT based electric vehicle charging and parking system. *Int. J. Eng. Res.* 2020, 9, 831–835.
[16] Vaidya, B.; Mouftah, H.T. IoT applications and services for connected and autonomous electric vehicles. *Arab. J. Sci. Eng.* 2020, 45, 2559–2569.
[17] Florea, B.C.; Taralunga, D.D. Blockchain IoT for smart electric vehicles battery management. *Sustainability* 2020, 12, 3984.
[18] Elakya, R.; Seth, J.; Ashritha, P.; Namith, R. Smart parking system using IoT. *Int. J. Eng. Adv. Technol.* 2019, 9, 6091–6095.
[19] Issrani, D.; Bhattacharjee, S. Smart parking system based on internet of things: a review. In Proceedings of the 2018 Fourth International Conference on Computing Communication Control and Automation (ICCUBEA), Pune, India, 16–18 August 2018; pp. 10281–10285.
[20] Rupanr, S.; Doshi, N. A review of smart parking using internet of things (IoT). *Procedia Comput. Sci.* 2019, 160, 706–711.
[21] Mukadam, Z.; Logeswaran, R. A cloud-based smart parking system based on IoT technologies. *J. Crit. Rev.* 2020, 7, 105–109.
[22] Rajbhandari, S.; Thareja, B.; Deep, V.; Mehrotra, D. IoT based smart parking system. In Proceedings of the 2016 International Conference on Internet of Things and Applications (IOTA), Pune, India, 22–24 January 2018; pp. 121–136.
[23] Chandran, M.; Mahrom, N.F.; Sabapathy, T.; *et al.* An IoT based smart parking system. *J. Phys. Conf. Ser.* 2019, 1339, 266–270.
[24] Sivaraman, P.; Sharmeela, C. IoT-based battery management system for hybrid electric vehicle. *Artif. Intell. Tech. Electr. Hybrid Electr. Veh.* 2020, 1–16.
[25] Lekshmi, M.; Mayur, P.; Manjunatha, B.; Kavya, U.; Anil Kumar, J.H. IoT based smart car parking with wireless charging feature for electric car. Int. Res. *J. Eng. Technol.* 2020, 7, 2188–2191.
[26] Girish, B.G.; Gowda, A.D.; Amreen, H.; Singh, K.M.A. IoT based security system for smart vehicle. *Int. Res. J. Eng. Technol.* 2018, 5, 2869–2874.
[27] Sun, Y.; Jin, K.; Guo, Z.; Zhang, C.; Wang, H. Research on intelligent guidance optimal path of shared car charging in the IoT environment. *Wirel. Commun. Mob. Comput.* 2020, 2020, 1–13.
[28] Bajaj, R.K.; Rao, M.; Agrawal, H. Internet of Things (IoT) In the smart automotive sector: a review. *J. Comput. Eng.* 2018, 36–44.
[29] Sun, S.; Zhang, J.; Bi, J.; Wang, Y.; Moghaddam, M.H.Y. A machine learning method for predicting driving range of battery electric vehicles. *J. Adv. Transp.* 2019, 2019.

[30] Mohan Krishna, A.; Tyagi, A.T.; Prasad, S.V.A.V. "Preserving privacy in future vehicles of tomorrow", *JCR*. 2020, 7(19), 6675–6684.

[31] Tyagi, A. K., Agarwal, D. and Sreenath, N. SecVT: securing the vehicles of Tomorrow using Blockchain Ttechnology. 2022 International Conference on Computer Communication and Informatics (ICCCI), 2022, pp. 1–6, doi: 10.1109/ICCCI54379.2022.9740965.

[32] Akshay Kumaran, V.; Tyagi, A. K.; Kumar, S. P. Blockchain technology for securing Internet of vehicle: issues and challenges. 2022 International Conference on Computer Communication and Informatics (ICCCI), 2022, pp. 1–6, doi: 10.1109/ICCCI54379.2022.9740856.

Chapter 11

Intelligent transportation system: Introduction and its scope in the near future

Abstract

It is the purpose of this chapter to discuss the introduction of the smart transportation system (STS) and its working practices, as well as to categorize the STS and its advantages. In addition to its chapter, the literary evaluation examines the motivation and sensitivity of the organization. The automotive and academic sectors have also taken an interest in cyber-attacks on STSs, problems, and problems relating to intelligent transportation systems (ITSs) (including security and privacy issues), future research opportunities in STSs (including ITS Architecture), and security potential for conflict (which includes cyber-attacks on STSs. In addition, the open nature of the STS, which functions as a wireless communication technology, poses a threat to security and privacy in the workplace. Additionally, the security and privacy mechanisms for the ITS are classified, and the vulnerabilities of these systems are highlighted. ITSs are a complicated field at the beginning of their development, requiring sophisticated data models, dynamic behavior, and strict time restrictions. It is a difficult undertaking that is dependent on the safety and efficacy of public transportation systems. One of the most essential aspects in the development of an ITS is the implementation of basic architecture standards as well as unique security requirements for each system.

Key Words: Mobile ad hoc networks (MANETs), Intelligent transportation system (ITS), Traffic management center (TMC), Information technology (IT), Operational technology (OT), Vehicular Ad Hoc Networks (VANETs)

11.1 Introduction

An intelligent transportation system (ITS) is a sophisticated application that delivers new services to better inform users in many forms of transport and traffic management and ensures the safe, core, and intelligent use of transport networks [1]. ITS can enhance transport efficiency and safety, for example, road transport, traffic management, mobility, etc., in several scenarios. Worldwide, ITS technology is used to improve road capacity and minimize travel time [2]. The architecture of a smart city that converts cities into a digital civilization streamlines people's lives on all fronts. Intelligent transport system. In any city, mobility is an important problem; people use a transport system for their movement throughout the city, whether it is for school, school, business, or other purposes. The smart transport system can save time and make the city smarter [3]. The ITS attempts to achieve efficiency of traffic by minimizing traffic problems. It improves traffic information, local convenience, operational data in real-time, sitting access, etc. to minimize travel time and promote safety and comfort for passengers. The usage of ITS is well recognized and is now employed in many countries [4]. The town of Glasgow is one such example. In cities, the intelligent system of transport provides regular daily passengers with public bus information, timings, access to seats, current bus status, time to reach certain destinations, next to bus location, and the density of bus passengers [5]. Glasgow City Council Iain Langlands, GIS, and Data Manager explain that urban bus operators have sensors in their cars. "If the bus travels early to the next bus stop, the bus will slow down quickly and slow down somewhat in red light rather than scheduling the bus." The system has been so intelligently developed those passengers and even drivers do not know of the delay, as there is virtually little wait [6].

To enhance the efficiency or safety of surface transport networks, ITS employs computer, communication, and sensor technologies. Most industrialized nations such as the United States, Canada, Australia, and the European Union embraced the ITS initiative. There are also several barriers to doing the same thing—social, political, and economic—while transit capacity is rising. The demand to convey products and passengers quickly and securely is increasing [7]. All this encouraged the development of new transit options. ITS combines established communication, controls, electronics, computer hardware, and software technologies to improve the operation of the surface system. ITS aims to convert surface transport into an efficient, fully integrated, fully accessible, customer-centric, and cost-effective system to ensure the fast and safe movement of products and persons [8]. ITS offers drivers realistic, reliable, and relevant knowledge for the first time to tackle this challenge. Problems such as traffic congestion; poor transportation efficiency, and endangered environments may be managed by new and sophisticated management technology created in recent years comprising computer technology, electronics and communications, road, and transport management approaches [9]. Figure 11.1 shows a flow diagram of data in an ITS.

How does the ITS work?

The traffic management center (TMC) is the most important ITS unit. It is primarily a transportation authorities-managed technology system. All information is collected and analyzed in real-time or local vehicle information for future operations and traffic control. The operations of the well-organized and skilled TMC focus on automatic data collecting with proper location information rather than data processing to deliver accurate information to travelers as shown in Figure 11.2 [11].

As urbanization develops fast, so does the number of automobiles on the road. A combination of the two in return puts considerable pressure on cities to maintain improved transport facilities so that the city continues to operate problem-free [12]. The only solution is to build a smart transit system. ITS is a win–win scenario for both city and people managing, offering citizen safety and comfort,

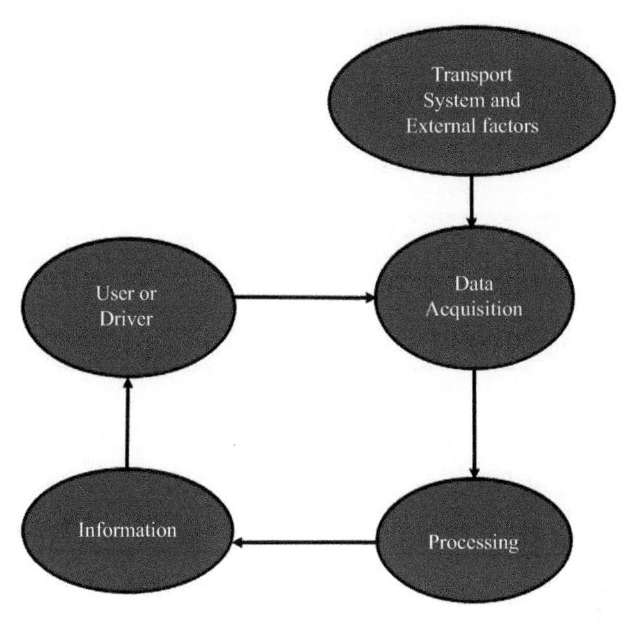

Figure 11.1 Data flow chart in an ITS [10]

Data collection	Data Transmission	Data Analysis	Traveler Information
•For strategic planning, accurate, broad, and rapid data collection, including real–time observation, is necessary. Through different hardware devices, the data are collected here, which constitute the basis of subsequent ITS activities. This comprises automated vehicle identification system, GPS automatic locators for vehicles, sensors, cameras, etc.	•Rapid and real–time information transmission is essential to ITS competencies so that this element of ITS involves passing data collected from the area to TMC and then transferring the processed TMC data back to travelers. Traffic notifications are delivered to travelers via the internet, SMS, or onboard car electronics.	•Data collected and received by TMC are handled in several ways. These procedures include error correction, data purification, data synthesis, and logical adaptive analysis. Inconsistencies in data are detected and rectified using expert software. Then the data is altered and collected for analysis.	•Travel advisory systems (TAS) are used to inform transportation users. Traveler Information: It offers real–time information such as travel time, driving speed, delay, road accidents, route modification, diversions, working conditions, etc.

Figure 11.2 ITS work

easy-to-maintain municipal managers, and surveillance. Traffic management was an issue as the first wheels were mounted on the first cart. The modern world demands mobility. ITS, an intelligent transportation system, sometimes leverages communications and IT to tackle this congestion and other traffic control issues. In several aspects of transport, ITS represents a major transformation [13]. ITS is a

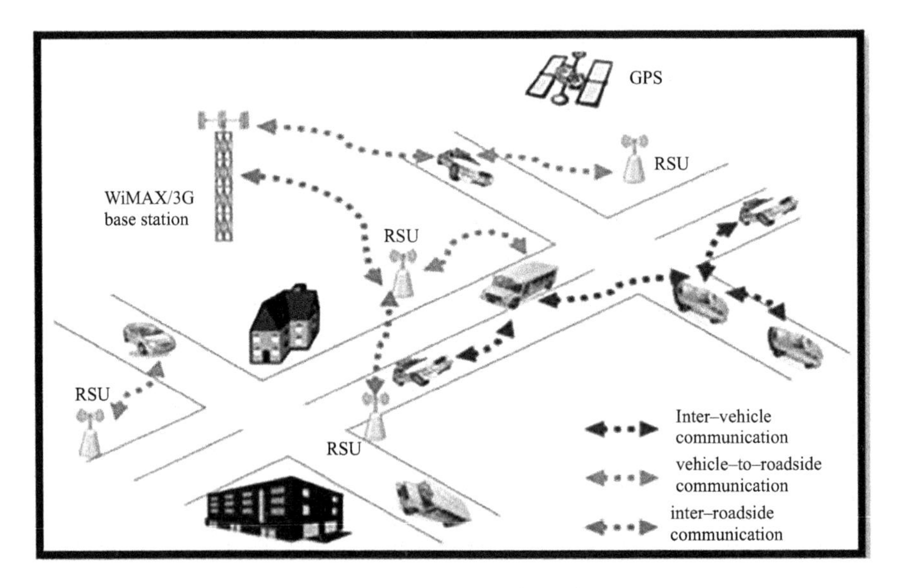

Figure 11.3 Example of an ITS [14]

global project to enhance surface transport system efficiency and efficiency with advanced IT and sensor technologies. ITS is a sophisticated technological system designed to assure efficiency and allow safe movement of the Electronic Toll Collection System (ETC) and Figure 11.3 shows an example of an ITS.

11.1.1 Commercial vehicle ITS operations

The economic lifeblood depends on the capacity of the transport system to sell products securely and effectively—from food, food, and computers to hazardous substances. In commercial vehicle operations, the ITS is being used to improve safety and regulatory processes for commercial vehicles, and to enhance efficiency via continuous appraisal and adjustment of operations in both the motor carrier and state-run engines sectors. ITS is changing how the federal and state authorities deal in this way with the motor carrier industry [15]. For commercial vehicle activities, the following applies.

11.1.2 Safety information exchange

The interchange of security information comprises the electronic communication of safety information and support for the credentials of carriers, vehicles, and drivers participating in commercial vehicles' operations. This information is used by the enforcement community and related agencies and agencies to make better-educated decisions based on past safety performance information [16]. Systems for the exchange of safety information make it easier:

- Enhanced access to carriers, vehicles & drivers, and credential information's with the collection of automated credential information.

- Enhanced access.
- Proactive updates on a carrier, vehicle, and driver safety and credentials/ encourage efforts that detect and encourage insecure operators to improve their performance.

11.1.3 ITS benefits

The benefits of ITS will increase transport system benefits and efficiency, avoiding the need for much more expensive construction of the physical infrastructure. This expectation must be misled with the illusion that the reinstallation of ITS may integrate new infrastructure expansion in India [17]. A major transportation development plan in India may enhance the number of system beneficiaries, significantly improve transport safety and, under some conditions, decrease the scale of infrastructure growth. Below is a list of the advantages of ITS efforts as shown in Figure 11.4.

11.1.4 Intelligent transport technology

Automotive navigation, traffic control, changeable signage containers, automated cameras, and platform detection, as well as application monitoring systems such as CCTV safety systems, automatic detection, and/or stopped vehicle systems, are all

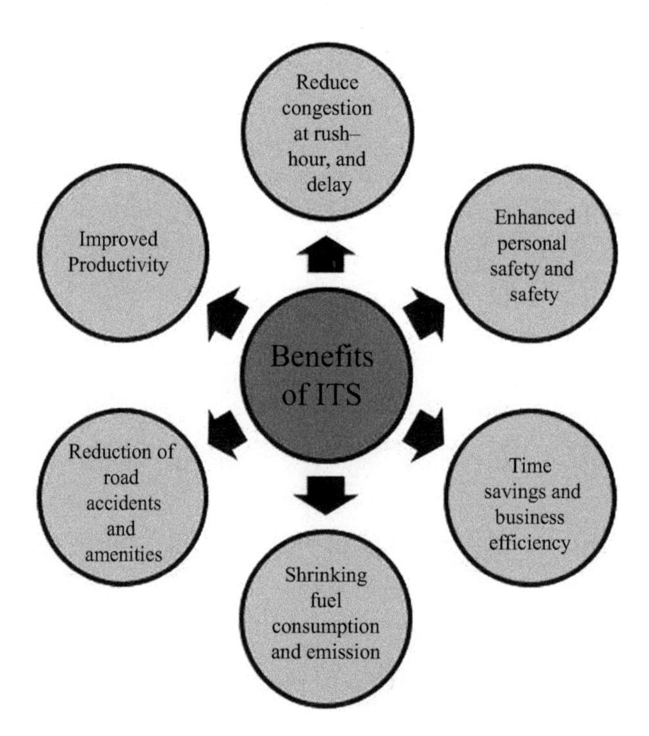

Figure 11.4 Key benefits of ITS

included in smart transport technology. Smart transport technology also includes autonomous vehicles. Aside from that, predictive techniques are developed, and comparisons with past data are carried out [18].

11.1.5 Wi-Fi communication

Wireless communication technology for smart transport systems has been proposed in many ways. For short- and long-term ITS, radio modem communication between UHF and VHF bands is generally utilized. In principle, the range of such protocols may be increased by employing mobile ad hoc networks or mesh networks. Long-range communications infrastructure networks like 5G are used. However, contrary to short-term protocols, these approaches are well established for long-term communication and they need significant and costly construction of infrastructure [19].

11.1.6 The technology of computers

Recent advances have led to fewer and more powerful computer processors for automobiles in automotive electronics. In the early 2000s, two or one hundred networked logical system controllers were incorporated into a standard automobile [20]. Currently, CPU modules have lower and lower costs of hardware memory and operating systems. Now, a more elaborate software program, incorporating model-driven process control, artificial information, and ubiquitous computing, can work with new embedded system platforms. Artificial intelligence [21] is arguably the most important for intelligent transport systems.

11.1.7 ITS applications

Applications for ITS is divided into four areas as shown in Figure 11.5.

The rest of the chapter is arranged into the following: Section 11.2 describes the ITS literature survey. Section 11.3 describes ITS evolution and Section 11.4 discusses ITS motivation. Section 11.5 describes vulnerabilities of ITS and

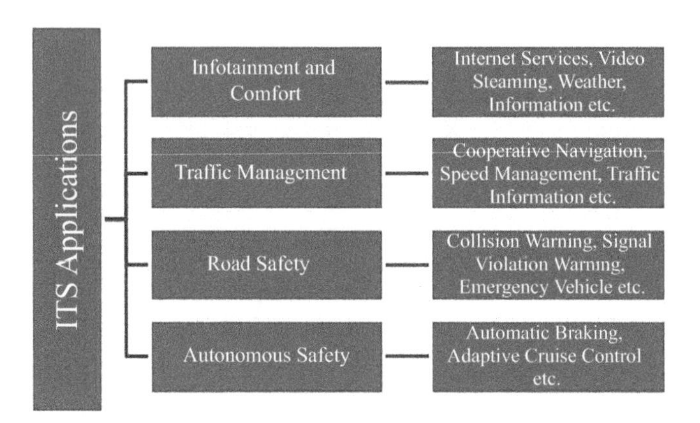

Figure 11.5 Classification of ITS application [22]

Section 11.6 describes ITS cyberattacks, Section 11.7 describes ITS architectures and security challenges and Section 11.8 describes intelligent security in IoT. Section 11.9 describes blockchain-based ITS and Section 11.10 describes future research directions in ITS, whereas Section 11.11 describes the conclusion over ITS and a summary of the chapter.

11.2 Literature survey

ITS is a sophisticated application for providing users with new and more coordinated and intelligent transport networking services in connection with a wide range of transport and traffic management modalities. Some of these technologies include camera applications to enforce traffic rules or signs, which permit speed limit modifications in the event of an accident to be recorded based on the call-up conditions. ITS is used globally to boost traffic and minimize travel times. The movement from rural to urban regions has evolved unevenly in emerging nations. Many parts of the developing world were urbanized without significant urban engineering and development. A small percentage of the population can afford automobiles, yet there is a considerable increase in congestion inside these multimodal transport networks. They also generate huge air pollution, severe dangers to safety, and promote social inequities. A multimodal walking, biking, motorcycling, bus, and railway system can sustain a high population density. Other parts of the rising world, like China, India, and Brazil, remain primarily rural but are quickly expanding and urbanizing.

A motorized infrastructure is being developed here with the motorization of the people. The grand disparity in the rich shows that only a portion of the public can drive, thus the motorized transport system for the affluent cross the dense multimodal transport system of the poor [23].

Joe Zhu *et al.* (2021): Stress the need for data-driven solutions owing to the volume and availability of ITS data. In the academic and industrial sectors of ITS, the usage of enormous data methods has become more essential. The ITS only uses signals, object recognition, traffic flow projections, time schedules, planning of routes, vehicles, and road safety. This research gives a bibliography, a thorough assessment of the ITS application, and an analysis of the most common huge data ITS models. This study examines the deployment of large-scale ITS data approaches that emphasize various aspects and integrate applications and models [24].

Chen *et al.* (2021): Identify the smart transportation system (STS) as essential to public transport and safety and other issues. Traffic detection is an important part of the ITS. ITS offers smart advice to minimize congestion and pollution based on the collection of information on urban road flows in real-time. Typically, cloud computing is utilized to identify traffic in an ITS. All collected pictures are transferred from the edge of the network to the cloud computing center. However, increasing traffic monitoring has generated considerable storage, connectivity, and processing difficulties for typical cloud-based transport solutions [25].

Joanne Mun-Yee *et al.* (2021): The trustworthy traffic redirection system defines traffic simulation as an accurate model necessary for traffic management.

On the other hand, there remains a lack of data since it eventually leads to errors in the congestion forecast, which contributes to a lower recirculation efficiency. The lack of realistic simulations of traffic also prevents the creation of an efficient traffic management system. This article provides three solutions for these issues: (i) a simulation for live transport, (ii) a pheromone forecast and a re-route network, and (iii) a missing data management approach based on weighted, historical data (WEMDI). The re-routing function of Google Maps was utilized to assess the model of traffic simulation [26].

Nencioni *et al.* (2021): Describe the use of the city in smart technologies and sensor data collection. Intelligent cities use public funds to improve the quality of their services by highlighting comfort, maintenance, and sustainability. Wireless mobile communications in its fifth generation (5G) offer a new kind of communication that links everyone and everything. 5G will have significant economic and social implications as it provides the backbone for critical communication in many smart city applications. The STS is one of the numerous smart city applications that 5G technology can allow. This article seeks to explore the effect and implications of 5G on ITS from different perspectives. The paper examines the technological foundation and economic benefits of 5G and the influence of important vertical firms as an intelligent city, including energy, healthcare, production, entertainment, vehicles, and public transit [27].

Bagga *et al.* (2021): A network spread that allows networked vehicles and ad hoc automobiles to connect with other internet networks in real-time may be created by configuring an internet vehicle as a network spread. Communication takes place between a large number of entities through open channels in general (e.g., vehicles, pedestrians, fleet management systems, and road infrastructure). An opponent can attack open communication by dropping, modifying, inserting fake (or malicious) messages, or deleting transit information. This can result in replays, impersonation attacks, man-in-the-center attacks, underprivileged attacks, and other forms of aggressive behavior from the opponent. In addition to security, secrecy, and traceability, the authentication mechanism must also include two other key characteristics [28].

Raza *et al.* (2021): The specific subject of ad hoc automobile networks investigation throughout the last decade is described. It was used efficiently in building smart cities for smart transit and entertainment systems. The major issues impeding wide usage are intermediate links, expensive routing costs, strict communication design, unscalable networks, and a massive packet crash. When utilized in metropolitan settings, these problems are significantly more serious. These barriers might be removed through the integration of small unmanned aircraft with ad hoc automobile networks. In this study, we provide an ad hoc communication architecture for automobiles that use unmanned aircraft to fly over the region in use and provide the underlying communication services. Ad hoc aviation support network benefits include line-of-sight communication, load balance, flexibility, and economic deployment. A case study of a car collision evaluates the performance of the model proposed. Results show that the use of unmanaged aircraft provides rapid and reliable emergency signal delivery to adjacent cars, thus allowing the adoption of further damage prevention safety measures [29].

Khan *et al.* (2021): The worldwide shift in future transportation networks is being explored to expand vehicle Internet technology for a variety of reasons and to establish a significant demand for autonomous road transportation. This new environment may be characterized by the presence of unmanned aerial vehicles (UAVs), which can provide a robust, adaptive, and business-friendly connectivity and network infrastructure as well as a computer system platform. When dealing with non-homogeneous and non-static data flows, the intelligent transmission system (ITS) for UAVs can provide a low-cost communication alternative that can increase security and efficiency. By contrast, the directional antenna provides more network coverage, reuse area, and bandwidth than the omnidirectional antenna. In addition, the MEC provides high bandwidth services and ultra-low latency services to meet the rising need for latency-sensitive automation applications, such as the analysis of vehicle video data, autonomous driving, and smart navigation, among other things [30].

Chandrappa *et al.* (2021): Explain that the proposed system correctly counts and classifies the car colors for the observed cars for real-time video input. The recommended approach for reducing the dynamic shadows and extracting vehicle regions to identify moving vehicles is KNN, a binary mask, and morphs, such as erosion and dilution. Furthermore, a unique ID has been provided for identified automobiles to prevent repetitive numbers of cars from being seen as they cross the counting area. In addition, the K-nearest neighbor machine learning classification was used to determine the color of cars to analyze the similarity between training and tested vehicles. The vehicle color is decided by the greatest RGB intensity [31].

Pinho *et al.* (2021): The antenna synthesis technique utilized in a STS is described in detail in this study (ITS). Intelligent mobility is made possible by wireless communication between roadside devices (RSUs) and onboard devices (OBUs). The planned RSU antennas must, however, cover a broader variety of high-speed and high-data-throughput vehicles than previously anticipated. RSU antennas emit in a certain pattern, which is determined by the emission patterns of the antennas. A key difficulty in a multi-route network is road interference, which negatively affects the reliability of single-to-interference and noisy communication (SINR). As a result of the large SINR distribution coverage area, this study provides an interference-conscious antenna synthesis approach that is also used to optimize several planar modulations. The proposed antenna range increases the reliable range of connections for misaligned cars in the center of the track while also increasing the SINR dispersion throughout the whole track surface. Resulting from the communication reliability tests, it is evident that the antenna array provided surpasses conventional designs [32].

Sysoev *et al.* (2021) describe the recommendation for the approach to traffic regulations for smart transportation networks. The technique focuses on enhancing the performance of the transport system, movement speed (or time). Control means changes in flow rates of traffic for certain components of the system (graphical edges), e.g., by monitoring traffic signals operations and changing sections capacity, e.g., via reversing lanes [33].

Misra *et al.* (2021): Demonstrate the capacity of end-users to use Safe-aaS to develop customized security solutions for their needs. The safety service provider's

(SSP) service area is now restricted due to the nature of the work. However, many SSPs working together might accomplish the distance that end users are seeking in a single connection. Moving from one SSP service area to the ultimate customer service of another SSP service area results in damage. None of the handoff solutions offers end-users the ability to customize their safety settings. An approach for leveraging road transportation as a Safe-aaS application scenario in the 5G intelligent transportation system is proposed by the author [34].

Sharma *et al.* (2021) described intelligent transport systems as a mix of ICT with traffic monitoring and management applications (ITS). The growing number of road vehicles in metropolitan areas makes automated traffic management technologies necessary. Smart transport systems can make choices through vehicle recognition, categorization, and analysis. This article employs the modified Regional Convolution Neural Network (RCNN) to give an automatic vehicle recognition video analysis technique. Vehicle identification in a particular frame is evaluated using traffic data gathered by CCTV cameras on highways. A Google network is pre-trained for extracting features. The regionally based Convolution Neural Network uses this characteristic to identify vehicles. The automobiles have a likelihood score created by the object's intersection. The recommended network integrates information from previous levels to reduce losses and enhance the accuracy of vehicle detection. Apart from the automobile identification technology, vehicle numbers, and behavioral analysis are included. Vehicle counting data may be utilized in smart cities for traffic congestion. The calculation of vehicle speed is part of the behavioral analysis. The statistics on speed are important for the enforcement of traffic in smart cities [35].

Gautam *et al.* (2021) provide examples of cities that place a considerable emphasis on smart technology to regulate traffic flow. According to different aspects, it is utilized for traffic control and management, as well as for the maintenance of traffic routes. The VANET ad hoc network is the most common method of constructing an ITS as well as managing and forwarding traffic swiftly over time. The topic of media transmission is now being researched by several different persons. VANET is the correspondence screening industry that is seeing the greatest growth. When data is spread over a large number of vehicles, the correspondence push paradigm is used in the majority of VANET applications. Comprehensive textual analysis is required for the wide range of VANET applications and the many connection protocols that are used in these applications [36].

11.3 Evolution of ITS

This section identifies and describes [37–39]:

- Car2Car consortium.
- EU Commission Action Plan on ITS applications.
- US Connected Vehicles Programme;
- Australian Gatekeeper.

11.3.1 Car2Car consortium

Car2Car is a non-profit, European automaker-led association, sponsored by equipment suppliers, research institutions, and so on. Car2Car focuses primarily on C-ITS, which uses wireless communication technologies to relay information among telematics systems in motor vehicles and on-road infrastructure. The word vehicle-to-vehicle communication is used as a vehicle-to-infrastructure communication for communication between car telematics and communication between telematics and infrastructure. The partnership with Car2Car thus focuses on a small number of ITS applications that can exchange information directly: collision prevention, risk prevention, etc.

11.3.2 Europe (EU) Commission Action Plan on ITS applications

The goal is to develop interoperable and seamless ITS services that enable the Member States to select systems for investments. The Directive drives the ITS Action Plan of the European Commission, which focuses on Figure 11.6. Support for law enforcement officers is a fundamental aim of the ITS Action Plan. The eCall application and protected parking spots are intimately linked to the activities of law enforcement officials. eCall is an emergency call generated by the activation of in-vehicle sensors by car occupants manually or automatically.

11.3.3 US Connected Vehicles Programme

The United States Department of Transport (DOT), which is committed to establishing a framework for safe, interoperable V2V and V2I wireless communications

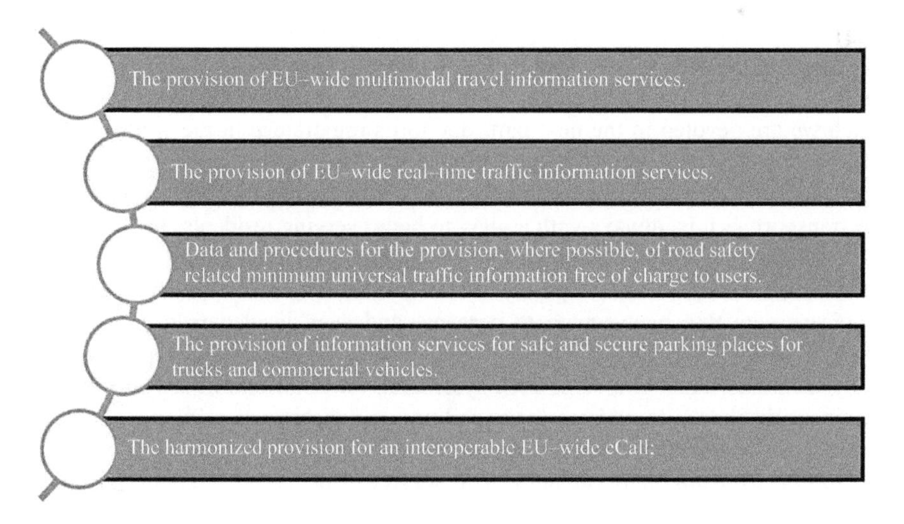

Figure 11.6 European Commission ITS Action Plan

such as the Car2car project and under a distinct regulatory and standardizing framework, is a multimodal initiative. The standard sets out three messages [SAE 10]:

(i) The main communication for data transmission between automobiles and between vehicles and infrastructure is a basic security message (BSM). Although the BSM is primarily created for security reasons, data may also be utilized in the message for other connected automotive applications.
(ii) In many traveler information applications, a roadside alerting message (RSA) is used to warn mobile users of surrounding emergency activities. In that regard, a notice is received from either the onboard public safety units (OBUs) or the traffic information infrastructure by mobile users.

11.3.4 Australian Gatekeeper

The chapter aims to explore the prospects of ITS technology for increasing police enforcement efficiency and efficiency in Australia. The study identifies several technologies and compares present technological methods with potential new approaches to technologies. Wireless communication technologies such as the DSRC can be used, for example, to evaluate vehicle speed on the road. This article contains a spectrum of ITS technologies, including camera, image processing, analytical instruments, etc. Australia has not created a distinct ITS program like US Connected Vehicles and is likely to leverage other geopolitical areas for the usage of the technology and standards. The Australian government's security sector in which the Australian Gatekeeper Framework is an important component is a particular method.

11.4 Motivation to ITS

The fast development of society poses numerous problems in terms of mobility, reflecting the essential adjustments at various levels that motivate this effort. The transport network evolves every day to be more intelligently equipped, a subject to which we are devoted in the next part, due to its importance in the world in brief [40]. In the face of possibilities and new difficulties, it will enable digitization, which is also true in different areas of society and industry, and also in the mobility and transport sectors, always with a view to better serving residents. In addition to all of these advancements, research on ITS and the safety concerns surrounding it also has to be complemented urgently. Security in many cases is a method for some to misuse data that may equate to personal and even legal damage due to the growing number of assaults we have witnessed. Therefore, it is necessary to understand the vulnerabilities of such systems in the contribution provided in this document that seeks to address the mobility components and difficulties arising, where safety is one of the major considerations.[41].

Public transport management: The service is designed to encourage people to utilize public transport. Effective automation, public transport planning, and administration can achieve this by analyzing data on different routes in real-time. The information enables the operators to know and provide vehicle planning and shipment with a quick response to deviations, delays, or other situations. It also contributes to protecting the security of public transit providers.

Knowledge of the Route: If passengers have previous information regarding the route that is most suited for their voyage, their journey in a new zone is easy and convenient. The journey can be facilitated through real-time traffic information, transit systems, abrupt curvatures, stop signs, road conditions, manual work, and other route information. The motorist can get all this information via his computer, smartphone, or phone network before traveling.

Vehicle security and control: The service provides safety support through vehicle monitoring and vehicle control information to vehicle operators. Drivers can assess their driving abilities, road conditions, and car performance. You may be aware of any frontal or rear-end incidents when changing lanes or when tracking other automobiles on crossroads. Even sophisticated car sensors can assist drivers in poor visibility by capturing pictures of the surroundings due to bad weather or night vision. The purpose is to reduce accidents or accidents by reporting the imminent collisions to drivers and emergency operators.

Electronic scheduling: These charts can allow travelers to learn about arrival and departure times, delays, transfers, and links to transport and bus stops. The information lets travelers make informed choices or modifications they choose to make within the last minute of their journey.

11.5 Vulnerabilities of ITS

An ITS is comprised of transport infrastructure and communication technologies such as road networks, transport, and transport systems aimed at decreasing traffic congestion and environmental effect. It allows automobile operators to decide informedly. Consequently, in contrast to operational technologies like, traffic signal systems and traffic management systems utilize more and more IT. The ITS technology and system combination call for a wide variety of technological and operational solutions. Recent safety events include ransomware adversaries. This is a malicious program that encrypts computer information and requests money for the decryption key and data recovery. Attacks are being concentrated on energy grids in the US and other important radar facilities as attack targets [42] as shown in Figure 11.7.

These assumptions are wrong and cause vulnerable ITS to be designed and deployed. The cyber-security framework, which is developed through a collaboration between the government and business, uses standard language and provides processes for establishing a new cyber-security program [43]. Cybersecurity Framework for critical infrastructure advance as shown in Figure 11.8.

Cybersecurity framework recommendations guide the adoption of the transportation system framework. This article discusses how NIST CSF principles are used to decrease the cyber risk of major transportation infrastructure, such as ITS.

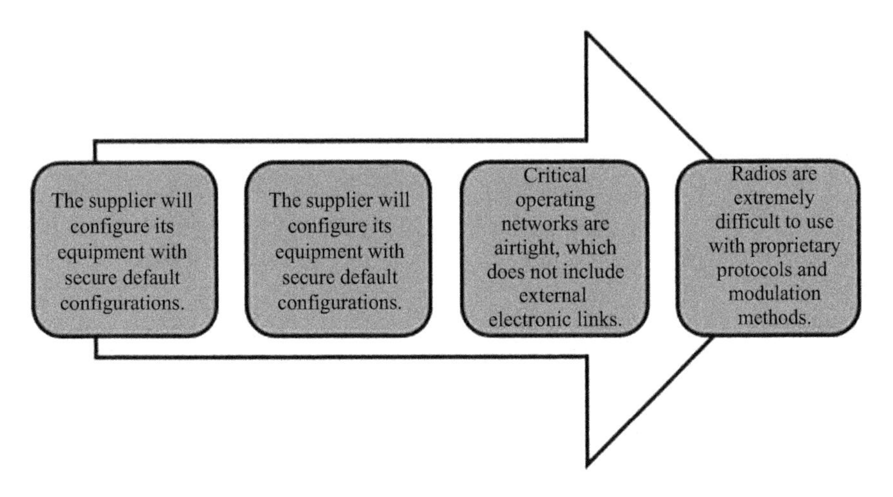

Figure 11.7 Radar facilities

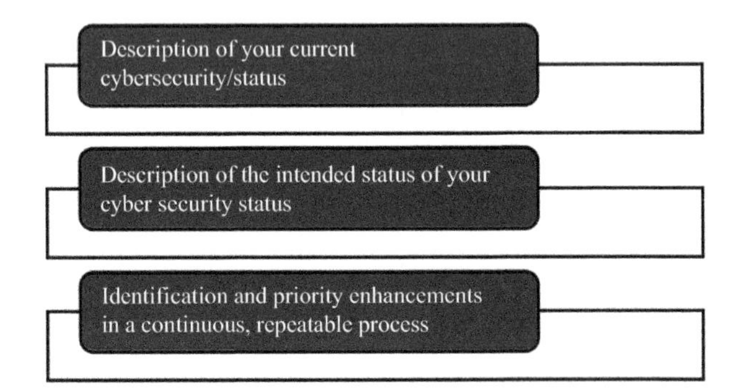

Figure 11.8 Cybersecurity framework

11.6 ITS cyberattacks

The diversity of ITS makes categorizing and identifying cyber threats more difficult [44]. ITS-specific attacks connected to architectural layers afterward as shown in Figure 11.9.

11.6.1 Vehicular Ad Hoc Networks (VANET)

It is typical for cyber attackers to use the man-in-the-center technique, which intercepts and transports the two communicators to a new state of communication. In the instance of a man-in-the-middle assault on a VANET's physical and data connections, the attacker believes the nodes are within a specific range. If the assaulting nodes are not available, either the signal must be muted or the position data must be changed; nevertheless, the attacking party may be identified in any of the two nodes.

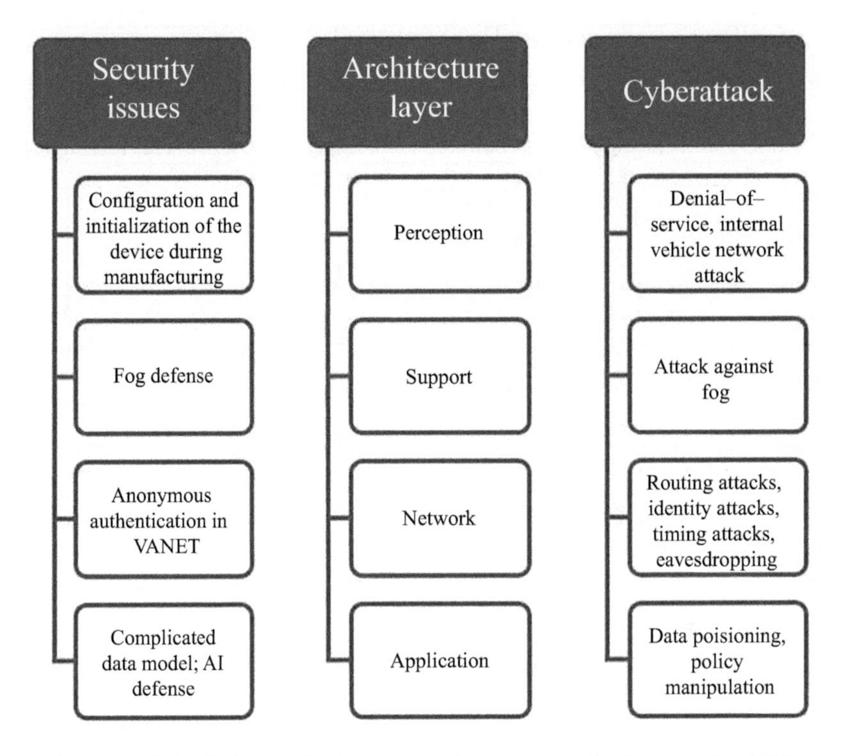

Figure 11.9 ITS security issues, architecture, and cyberattacks [45]

11.6.2 Routing attacks

The degree of VANET physical and data connection indicates one-hop communication. Multi-hop communication provides ways for routing. Routing assaults are examples of a VANET protocol routing violation in which a hostile node hinders data from achieving its final objective. A black hole attack is a routing technique in which the malicious node eliminates all of the packets to be sent quietly. Another form of routing attack is a grey hole attack, which only happens on selected packets.

11.6.3 Timing attacks

Timing attacks lead to communication delays and hence interrupt the operation of real-time applications. For instance, an emergency message is delivered to the following car in a cooperative adaptive cruise control system to avert a collision. If the assailant can cause a delay, the braking system reflexes are delayed despite the right acceptance, and the accident is not averted.

11.6.4 Spoofing

The attacker transmits misleading data in the event of a spoofing attack to produce an incorrect system reaction. For instance, the attacking party provides faulty GPS coordinates to hinder navigation system operations [46].

11.6.5 Denial of Service (DoS) attacks

DoS is a common cyber-attack that affects the availability of system components. It is especially dangerous in the event of certain key safety features. Sybil is a popular VANET DoS attack, in which a malicious vehicle broadcasts several identities and flashes on the network. This disrupts the regular exchange of information.

11.6.6 Vehicle internal network attack

Because most internal vehicle networks are intended to address threats while automobiles are not linked. For instance, the attacker may easily access the internal network via the controller area network CAN) protocol and therefore control the airbag management system.

11.6.7 Attack identity

Identity confidentiality ITS may apply to the driver, passenger, pedestrian, and so on privacy. An attacker may attempt to extract personal data, location, behaviors, habits information. The attacking party can get information about how nicknames in VANET are issued and so monitor the whereabouts of a vehicle [47].

11.7 ITS architecture and security challenges

The ITS may be seen as an IoT subtype and can thus be built using analogous approaches and concepts. It may be utilized in ITS as well. The proposed architecture consists of four layers responsible for the various IoT features [48–50]. Implementing this notion in ITS plays a more accurate role at each step.

11.7.1 ITS architecture

The ITS *perception layer* includes cell phones, sensors of cars, and infrastructure devices for users. Security concerns of the perception layer are many, as in many cases, it is not designed for connected vehicles, setup and activation of devices during manufacture and internal vehicle network architecture as shown in Figure 11.10 [51].

The network layer is a sophisticated wired and wireless alloy. Authentication must be anonymous since personal data must be safeguarded. The limited variety of nodes and timing restrictions create more problems. The standard establishes requirements for the prevention of duplication, integrity, secrecy, and the use of pseudonyms. It sets forth the requirements but does not give particular advice. The 5G concept concentrates on the service. Slicing Security-as-a-Service or SSaaS offers operators different and adjustable security packages, including encryption processes, encryption settings, blacklist and whitelist configuration choices, authentication procedures, insole strengths, etc.

The information is processed in the Fog or the cloud on *the support layer* depending on your time, space, and security needs. Fog-based systems face more safety issues as an emerging technology, as it is harder to safeguard the operating

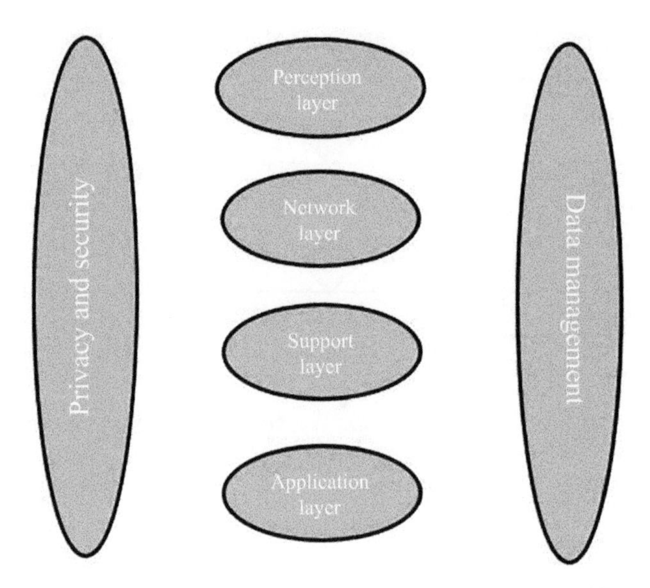

Figure 11.10 Internet of things (IoT) architecture [52]

environments of Fog-dispersed systems than a centralized cloud. Cloud computing and fog computing cannot integrate the security and privacy safeguards directly because of their properties such as mobility, heterogeneity, and large redistribution.

In the information, warning, and even activation of a particular car system, *the application layer* represents the ultimate user contact (in the case of unmanned vehicles). Before the user reaches several places, the data acquired in the sensor layer can be processed. In the vehicle itself, the safety and time limits can be applied locally on roadside units (RSUs), in Fog or Cloud, depending on the calculations of data semantics. The ITS data fulfill all of the big data criteria that require artificial intelligence. It must be carefully considered in key security systems like ITS as it is very vulnerable to many cyber-attacks. Figure 11.11 shows architecture layer with security issues and cyberattacks.

11.7.2 ITS security and privacy challenges

It is necessary to research both privacy and security since they are the most difficult challenges in information technology to solve simultaneously. It is critical to do background checks on every ITS member. Figure 11.12 demonstrates how each light must be evaluated in terms of the security and privacy considerations connected with information technology systems (IT systems) (ITS). Location and the actual identity of the ITS user are protected by the use of ITS-S (vehicle) privacy [53]. False identities are employed to fulfill the demands of security and privacy in a secure environment. Wireless, high-speed mobility, dynamic topology, sparse, and congested settings are all features of information and communication technology (IT) systems. Separated from other information and communication

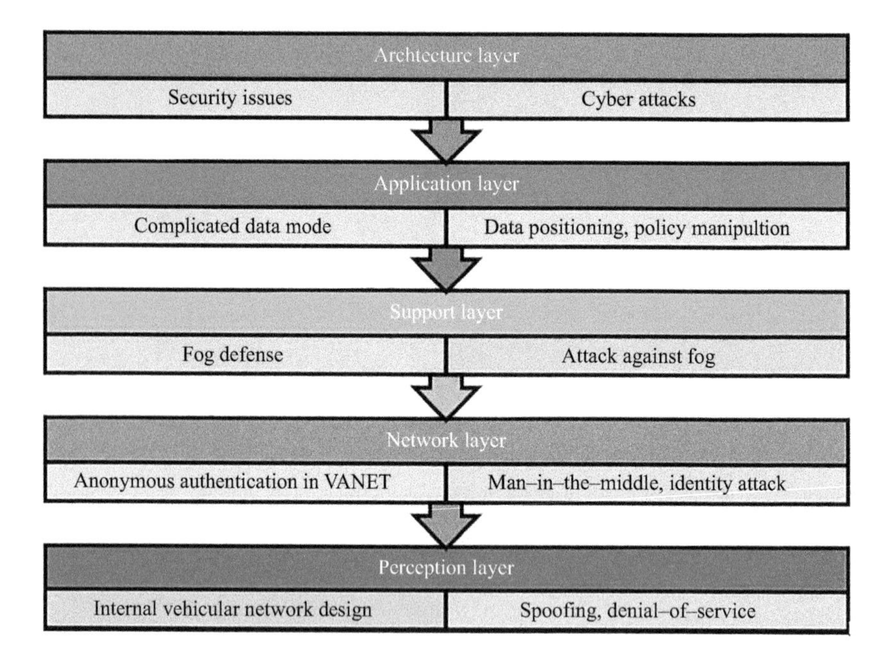

Figure 11.11 Architecture layer with security issues and cyberattacks

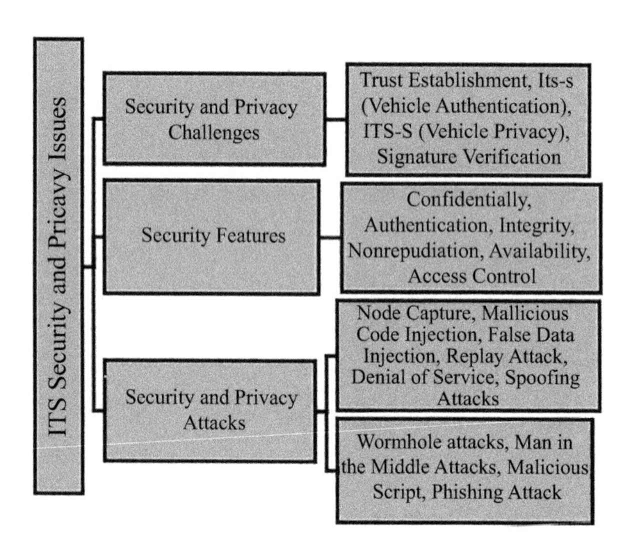

Figure 11.12 ITS privacy and security challenges [55]

technology-based networks, ITS is more vulnerable to attack than other networks. Autonomous cars maliciously broadcasting or altering original communications from a legitimate ITS user to alert other vehicles of traffic congestion are known as malicious automobiles (or malicious automobiles). As a result, the ITS aim will not be achieved, and honest ITS users may be deceived as a result of this failure. An enemy may be misled or awakened to the facts of the ITS user's predicament, depending on the circumstances. Therefore, it is necessary to put in place mechanisms to ensure that communications are authentic and complete before they are sent. [54].

Confidentiality: This function guarantees access to data for authorized ITS users only. Unauthorized users cannot spy or impede the data. Confidentiality is a major service for the ITS, registration, and acquisition of pseudonyms. It is therefore vital to guarantee that ITS is secret. Advanced cryptographic approaches should thus be taken into account.

Availability: This functionality in ITS ensures that servers and data are always available to ITS clients, therefore reducing downtime. In the realm of information technology, real-time services are essential. An investigation on the availability of resources is required. Improved security measures should be considered during the deployment of ITS to ensure its availability.

Identification: This function is capable of ensuring that illegal ITS-S (vehicles) are not connected to ITS. It is tough to identify each ITS-S. Therefore, appropriate systems for vehicle identification must be designed and developed.

Authentication: Authentication can make sure received messages are authentic. Authentication may, however, be done in ITS without disclosing the vehicle's actual identification. This feature requires efficient methods.

Privacy: An anonymity function in ITS may only be handled by servers that have been granted permission. Privacy is meant to encrypt data, whereas confidentiality is intended to ensure that messages/data do not disclose the identity of the person who sent them. The privacy afforded by a legal vehicle is the most significant element of ITS. To prevent an honest ITS user from being tracked, the design and implementation of security and privacy protections are critical.

Trust: When people have confidence in one another, they may ensure that the previously described safe and anonymous communication features are implemented in stages between different entities. In information and communications technology, the trust function may be divided into two categories: applications and entities. It is important to create dependable approaches to increase public confidence in ITS services.

11.8 Intelligent security in IoT

An intelligent and proactive protective posture is required with the complexity of ITS. Comprehensive ITS cyber safety strategy and efficient use in various areas of security systems [56–58]. Many publications provide strategies for characterizing the future overall look of ITS, however, few actual findings can be implemented by ITS.

11.8.1 Artificial Information

With the development of IoT intrusion detection systems (IDS), the rising safety risk and complexity of tasks are increasingly referring to AI. Due to the demand for flexible solutions and the need for a complete approach to a rapidly evolving system, AI will be able to play a role in future ITS cyber security.

11.8.2 Machine learning

Machine learning (ML) for cybersecurity systems is the most often used subset of AI. Its downside is that it is sensitive and the training data set needs to be carefully selected in the training phase. If noise is introduced, it might damage the whole system (envision attacks, poisoning attacks). A comprehensive classification with proactive approaches must be developed. Due to this discomfort, ML techniques are often used as an auxiliary mechanism.

11.8.3 Ontology

Ontology is a viable approach for addressing various challenges, especially unstructured data. Ontology is a growing topic in the area of IoT security. It provides a common vocabulary, describing the practical security components of accessing and sharing data for producers, consumers, and intermediaries. It aims to provide crucial information on the provision, access, and processing of data as well as on and the sources of regulations and certifications.

11.9 Blockchain-based ITS

Blockchain, which records previous transactions and provides a flawless system, can decrease well-planned opportunistic behavior. The economic atmosphere of inexpensive transaction costs that are vital to economic efficiency and progress is created by reliable news. Smart contracts and IT can provide a superior trust-building mechanism that reduces uncertainty in the behavioral transaction [59]. The distributed and decentralized accounting systems decrease intermediate expenses. Blockchain technology allows governments and audit agencies the right to access blockchain business accounts increases corporate profit and fiscal control and cuts tax evasion. However, blockchain integration with Internet technology may overcome this issue effectively and make transportation accountability of blockchain practical [60].

11.9.1 Intelligent blockchain transportation: social aspect

Blockchain is a distributed directory and database that incorporates mathematics, encoding, the Internet, computer programming, and other fields into its construction. It has decentralized and traceable characteristics, as well as shared maintenance, transparency, and openness, and it allows for the marking of the entire process. Blockchain and the IoT can automatically upload and update data on the fly, and smart contracts can enhance transaction speed in real-time.

11.9.1.1 Characteristics of blockchain have a social impact

Blockchain is a widely used directory and database that includes mathematics, cryptography, the Internet, computer programming, and other fields. There are no central servers involved, and the system is processable and traceable. It also enables comprehensive process labeling, maintenance, transparency, and openness. These characteristics help to maintain the long-term stability of the blockchain. With the rapid construction of intelligent contracts, the blockchain and things technology Internet can automatically upload and update information on a timetable, allowing transaction speeds to be significantly increased.

11.9.1.2 Problem management

A complete signature system is provided by blockchain technology, which may be used to link channels under the chain, construct a secure, large-scale payment system, and improve the performance of blockchain systems. Blockchain technology is being utilized for the recording of traffic data on the internet, as well as the broadcast of important personal information and online drivers' credit ratings, to ensure that those responsible for road accidents are held accountable. It is also possible for government officials to monitor and analyze Internet traffic data to assist them in collecting revenue and taxation from citizens.

11.9.1.3 High-level systems design

The system's high-level architecture, which includes the premiums and penalties system as well as the loan assessment system, establishes a blockchain basis for the whole intelligent traffic index structure and serves a vital purpose inside the system. In the context of blockchain and genuine information, the payment system refers to virtual money benefits for those who participate actively in blockchain and genuine information. Using blockchain technology for data integrity and non-repudiation, credit evaluation analyses customer confidence through service confidence assessment, compartmental trust, and trust in one's ability to perform one's job.

11.9.2 Intelligent transport blockchain: environmental aspect

Excessive exhaust emissions from automobiles are the most significant source of pollution in urban transportation. Methods for reducing urban pollution through transportation include traffic optimization, speed control, rapid acceleration, and minimal idling, as well as the encouragement and the use of air and fuel resistance lines for large trucks and other heavy equipment. Because of the popularity of transactions between electric vehicles and charging stations, the number of electric vehicles on the road has grown in the vehicle-to-grid (V2G) environment. The cost of public transportation has increased as a result of the expansion of public transportation infrastructure. Apart from that, a delicate urban greening system based on information supplied by the Blockchain Transport Consortium helps to support urban management.

11.9.3 Intelligent blockchain transport: economic aspect

The four criteria are as follows: transaction costs, administrative expenses, building costs, and spending on infrastructure finance. Transaction costs are the most expensive. Because of its defined method, blockchain can reduce well-designed opportunistic behavior, keep records of past transactions in perpetuity, and store records of future transactions in perpetuity. Reliable ledgers can provide an economic environment with minimal transaction costs, which is critical for the efficiency and profitability of an organization's operations. Smart contracts and information technology (IT) have the potential to establish a stronger trust mechanism in transactions, reducing behavioral uncertainty. The use of distributed accounting technology and blockchain decentralization helps to minimize the need for middlemen. All of these variables contribute to a reduction in transaction costs for both governments and companies. It is possible to minimize the cost of monitoring, monitored, and authorized money flows to prevent fraud and enhance fund flow supervision through the use of auto-downloading and traceability systems.

11.10 Future research directions in ITS

There is no comprehensive and complete research. The present research effort can be improved. The literature study given thus far has implemented ITS with various characteristics and each isolated [61]. It is important to build an ITS by integrating all these elements in a single system that addresses the needs of various users. Safety, precision, and cost-effectiveness might be regarded factors of importance in the development of intelligent ITS.

In addition, the following problems IoT faces might be considered [62] as shown in Figure 11.13. By optimizing the usage of IoT technologies, a better world may be reached by tackling open issues. Most individuals rely on public transit. Therefore, the degree of comfort should be the same as that of the private vehicle, i.e., the quantity of waiting time should be reduced and the public car seat should be guaranteed. The tracker sensors can monitor the vehicle, the seat sensors can track the number of people by calculating the crowd forecast. Gateways and mediators can be used to fuse different sensors and actuators that lead to data storage through signals or communicators at a given place. This leads to a better approach to a smart transport system.

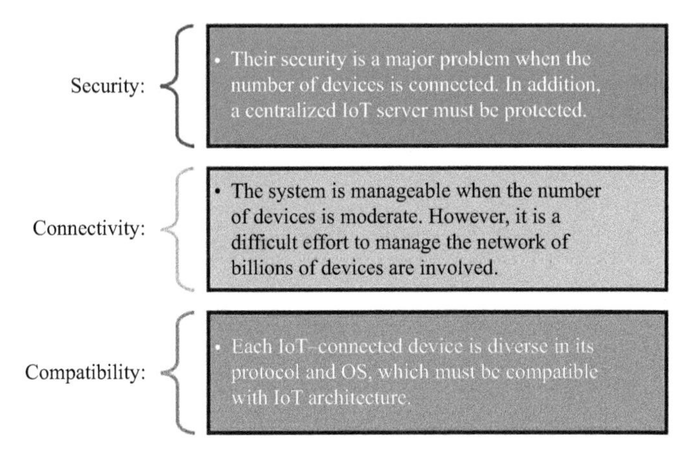

Figure 11.13 Problems IoT faces

11.10.1 Analysis of public views and attitudes from online sources

In addition to physical data acquired from various sensors, data sources such as the state and transportation of a city, public views, and internet sources may also be analyzed, in addition to physical data (e.g. social networks). Future ITSs should utilize numerous data streams to monitor and control systems. To extract relevant and meaningful information from social network data sources, it is advised that a natural language processing (NLP) algorithm and contemporary semantic data analytic frameworks be used (e.g., public comments on Twitter). The NLP algorithm should be able to detect social activities and/or public commentary that could contribute to future transportation problems (such as football congestion), as well as public opinion and understanding of existing and future transportation systems (such as public transportation systems) [63].

11.10.2 Modeling of the CSP traffic network

It is necessary to build a CSP model to better integrate CSP data with other multi-source data development to ease data connection and fusion. In the future, it will be necessary to investigate a hierarchical model of traffic networks for the digital reconstruction of CSP regions that are connected to physical, semantic, logical, and perceptual networks. When expressing network connections, the interlayered network connection (e.g., on the cyber, social, and physical levels) can be utilized in conjunction with cognitive computing and/or probabilistic models to get the desired result. Statistics and NLP can be used to investigate the cross-domain data association rule. The space–time association rule, for example, maybe described as the relationship between Bluetooth intensity and traffic volume, as well as energy consumption in buildings and peak flows. Because of a large amount of traffic information available, it is necessary to develop a hierarchical traffic network model to determine and describe the information types and amounts that are appropriate for executing the individual services effectively in terms of both temporal and geographic resolution for each service [64].

11.10.3 Connected environment flow models

With the increasing popularity of VACS, future ITSs will be utilized in conjunction with a mix of CAVs and RHVs in connected environments. Because the behavior and features of the CAV are significantly different from those of the RHV, it is necessary to understand the flow characteristics of these mixed vehicle parameters for use in ITS. It is necessary to develop a more detailed vehicle model including microscopic and macroscopic fluxes. With CAV-related characteristics, new CF models at the nanoscale level will be studied and developed (e.g., unreliable vehicular communications, communication delay, platooning driving protocols, a penetration rate of CAVs, etc.). Accordingly, this CF model may be used to the connection architecture of the ITS system [65].

11.11 Conclusion

All the systems must readily transmit information among each other, communicating both at the end of the engine operator and the different States/Federal Engine Authorities in a common way for the overall efficiency of the commercial vehicle operations. ITS is one of the most important concerns for the future in emerging countries like India and is very important for society. Different transport infrastructures must be built with increasing mobility in this Internet era, to carry people and stuff easily and conveniently to both urban and interurban areas. Smart transport systems are the focus of the age of safer, more competitive, and more cohesive intelligent mobility. A huge number of cities, which are mostly inspired by cultural, social, and environmental concerns, are now followed by intelligent cities. Increasingly, the rise of ICTs has fostered progress and creative solutions that enable cities to improve their operations automatically and automatically. This notion is supported by two major reasons: citizens themselves and technology. The confluence of these two characteristics can result in a variety of social benefits, including the maximization of resources and the considerable enhancement and enrichment of individuals. When it comes to managing business information in real-time, an ITS demands an equally sophisticated infrastructure. Intelligent mobility serves as the foundation for ITS diverse remote sensing, sophisticated analysis, integrated scheduling, and other capabilities. As a result, the smart equation is unambiguous: there can be no smart city without smart mobility, and there can be no smart mobility without ITS. In the field of information technology, trust management models are used to establish trust and provide safe communication between devices.

References

[1] Rajkumar, SC, and LJ Deborah. "An improved public transportation system for effective usage of vehicles in an intelligent transportation system." *International Journal of Communication Systems* 34(13) (2021): e4910.

[2] Jiang, L, Xia, Y., Wang, L., *et al.* "Hyperfine-resolution mapping of on-road vehicle emissions with comprehensive traffic monitoring and an intelligent transportation system." *Atmospheric Chemistry and Physics Discussions* 22(12) (2021): 16985–17002.

[3] Gangwani, D, and P Gangwani. "Applications of machine learning and artificial intelligence in intelligent transportation system: a review." *Applications of Artificial Intelligence and Machine Learning* (2021): 203–216.

[4] Tyagi, AK, Aswathy, SU, Aghila, G, and Sreenath, N. "AARIN: Affordable, Accurate, Reliable and INnovative Mechanism to Protect a Medical Cyber-Physical System using Blockchain Technology." *IJIN*, 2 2021: 175–183.

[5] Mohandu, A, and M Kubendiran. "Survey on big data techniques in intelligent transportation system (ITS)." *Materials Today: Proceedings* (2021).

[6] Suthanthira, VN, Radhika, K, Maheshwari, M, Suresh, P and Meenakshi, T. "IoT-based intelligent transportation system for safety." *Cloud and IoT-Based Vehicular Ad Hoc Networks* (2021): 47–65.

[7] Sumit, D, and R S Chhillar. "Analysis of intelligent transportation system for smart cities in current framework." *Design Engineering* (2021): 10339–10349.

[8] Zhao, L and Y Jia. "Intelligent transportation system for sustainable environment in smart cities." *The International Journal of Electrical Engineering & Education* (2021). Available at https://journals.sagepub.com/doi/abs/10.1177/0020720920983503.

[9] Krishnendhu, SP, and P Mohandas. "Intelligent transportation system: the applicability of reinforcement learning algorithms and models." *Advances in Computing and Network Communications* (2021): 557–572. https://doi.org/10.1007/978-981-33-6977-1_41.

[10] Stephenson, S. Automotive Applications of High Precision GNSS, 2016.

[11] Kumar, R, R Khanna, and S Kumar. "Vehicular middleware and heuristic approach for the intelligent transportation system of smart cities." In *Cognitive Computing for Human-Robot Interaction.* Academic Press, London, 2021. 163–175.

[12] Manogaran, G, and TN Nguyen. "Displacement-aware service endowment scheme for improving intelligent transportation systems data exchange." *IEEE Transactions on Intelligent Transportation Systems* (2021). Doi: 10.1109/TITS.2021.3078753.

[13] Yuqing, W and G Xing. "Design of intelligent transportation system based on a genetic algorithm and distributed computing." *International Journal of Engineering Intelligent Systems* 29(2) (2021).

[14] Vanni, R, SJ Luz, G Mapp, andE Moreira. Ontology-Driven Reputation Model for VANET, 2016.

[15] Priyanka, E. Bhaskaran, C Maheswari, and S Thangavel. "A smart-integrated IoT module for intelligent transportation in the oil industry." *International Journal of Numerical Modelling: Electronic Networks, Devices and Fields* 34(3) (2021): e2731.

[16] Bhatia, V, Jaglan, V, S Kumawat, V Siwach, and H Sehrawat. "Intelligent transportation system applications: a traffic management perspective." *Intelligent Sustainable Systems.* Springer, Singapore, 2022. pp. 419–433.

[17] Belhadi, A, Y Djenouri, G Srivastava, and JCW Lin, "SS-ITS: secure scalable intelligent transportation systems." *The Journal of Supercomputing* 77(7) (2021): 7253–7269.

[18] Bagula, A., K Kyamakya, and JC Chedjou. "Intelligent transportation-related complex systems and sensors." (2021).

[19] Lamssaggad, A, N Benamar, AS Hafid, and M Msahli, "A survey on the current security landscape of intelligent transportation systems." *IEEE Access* 9 (2021): 9180–9208.

[20] Cheng, D, C Li, and N Qiu. "The application prospects of NB-IoT in intelligent transportation." *2021 4th International Conference on Advanced Electronic Materials, Computers and Software Engineering (AEMCSE).* IEEE, 2021, 176–1179. Doi: 10.1109/AEMCSE51986.2021.00240.

[21] Poon, STF. "Designing for urban mobility: the role of digital media applications in increasing efficiency of intelligent transportation management system." *Smart Cities: A Data Analytics Perspective.* Springer, Cham, 2021. pp. 181–195.

[22] Hamida, EH B Noura, and W Znaidi. "Security of cooperative intelligent transport systems: standards, threats analysis, and cryptographic countermeasures." *Electronics* 4(3) (2015): 380–423.

[23] Walzer, N, R Phillips, and R Blair, eds. *50 Years of Community Development Vol II: A History of its Evolution and Application in North America.* 2. Routledge, London, 2021.

[24] Kaffash, S, AT Nguyen, and J Zhu. "Big data algorithms and applications in intelligent transportation system: a review and bibliometric analysis." *International Journal of Production Economics* 231 (2021): 107868.

[25] Chen, C, B Liu B, S Wan, P Qiao, and Q Pei. "An edge traffic flow detection scheme based on deep learning in an intelligent transportation system." *IEEE Transactions on Intelligent Transportation Systems* 22(3) (2020): 1840–1852.
[26] Chan, RKC, JM-Y Lim, and R Parthiban. "A neural network approach for traffic prediction and routing with missing data imputation for the intelligent transportation system." *Expert Systems with Applications* 171 (2021): 114573.
[27] Gohar, A and G Nencioni. "The role of 5G technologies in a smart city: the case for intelligent transportation system." *Sustainability* 13(9) (2021): 5188.
[28] Bagga, P, AK Das, M Wazid, JJ Rodrigues, KKR Choo, and Y Park. "On the design of mutual authentication and key agreement protocol on the internet of vehicles-enabled intelligent transportation system." *IEEE Transactions on Vehicular Technology* 70(2) (2021): 1736–1751.
[29] Raza, A, SHR Bukhari, F Aadil, and Z Iqbal. "An UAV-assisted VANET architecture for an intelligent transportation system in smart cities." *International Journal of Distributed Sensor Networks* 17(7) (2021): 15501477211031750.
[30] Khan, MA, TA Cheema, I Ullah, *et al.* "A dual-mode medium access control mechanism for UAV-enabled intelligent transportation system." *Mobile Information Systems* 2021 (2021). Article ID 5578490, https://doi.org/10.1155/2021/5578490.
[31] Kavitha, N, and DN Chandrappa. "A robust multiple moving vehicles tracking for the intelligent transportation system." *Advances in Machine Learning and Computational Intelligence.* Springer, Singapore, 2021. pp. 97–111.
[32] Sharma, A, A Prajapati, and P Pinho. "Interference-aware antenna synthesis for enhanced coverage in intelligent transportation system." *IEEE Transactions on Vehicular Technology* 70(8) (2021): 7803–7811.
[33] Galkin, A and A Sysoev. "Controlling traffic flows in an intelligent transportation system." In *Society 5.0: Cyberspace for Advanced Human-Centered Society.* Springer, Cham, 2021. pp. 91–101.
[34] Roy, C and S Misra. "Safe-Passé: dynamic handoff scheme for provisioning Safety-as-a-Service in 5G-enabled intelligent transportation system." *IEEE Transactions on Intelligent Transportation Systems* 22(8) (2021): 5415–5425.
[35] Sharma, P, A Singh, KK Singh, *et al.* "Vehicle identification using modified region-based convolution network for the intelligent transportation system." *Multimedia Tools and Applications* (2021), https://doi.org/10.1007/s11042-020-10366-x.
[36] Gautam, V. "Analysis and application of vehicular ad hoc network as an intelligent transportation system." In *Mobile Radio Communications and 5G Networks.* Springer, Singapore, 2021. pp. 1–17.
[37] Sharma, V, L Kumar, and S Sergeyev. "Recent developments and challenges in intelligent transportation systems (ITS)—a survey." In *Intelligent Computing and Communication Systems.* Springer, Singapore, 2021. pp. 37–44.
[38] Agrawal, D, R Bansal, TF Fernandez, and AK Tyagi. "Blockchain integrated machine learning for training autonomous cars." In *Hybrid Intelligent Systems. HIS 2021. Lecture Notes in Networks and Systems*, 420 (2022). Springer, Cham. https://doi.org/10.1007/978-3-030-96305-7_4.
[39] Al-Khrisat, W, N Hazim, and MR Hassan. "Improving traffic incident management using intelligent transportation systems, a case of Amman

City." *Turkish Journal of Computer and Mathematics Education* 12(12) (2021): 4343–4352.

[40] Kopeć, PK. Feasibility study of intelligent transport on the example of a production plant. Diss. Instytut Elektrotechniki Teoretycznej i Systemów Informacyjno-Pomiarowych, 2021.

[41] Dewi, NK and AS Putra. "Law enforcement in smart transportation systems on highway." *International Conference on Education of Suryakancana (IConnects Proceedings)*, 2021. Available at https://jurnal.unsur.ac.id/cp/article/view/1367.

[42] Hassan, T., T Fath-Allah, M Elhabiby, A Awad, and M El-Tokhey, M. "Detection of GNSS no-line of sight signals using LiDAR sensors for intelligent transportation systems." *Survey Review* (2021): 1–9.

[43] Baloni, R, S Shanthi, V Chandra Sekar, and V Shakthi Priyan. "Secured intelligent cooperative communication in vehicular networks—a comprehensive review." In Reddy, VS, Prasad, VK, Wang, J, and Reddy, KTV. (eds) *Soft Computing and Signal Processing. Advances in Intelligent Systems and Computing,* vol 1340, 2022. Springer, Singapore. https://doi.org/10.1007/978-981-16-1249-7_30.

[44] Amini, M, and S Sayyadi. "A comparative study of the legal nature of companies providing intelligent transportation services (case study: uber and snapp companies)." *Science and Technology Policy Letters* (2021). Available at: http://stpl.ristip.sharif.ir/article_22407.html?lang=en.

[45] Mecheva, T, and N Kakanakov. "Cybersecurity in intelligent transportation systems." *Computers* 9(4) (2020): 83.

[46] Li, X, J Tan, A Liu, P Vijayakumar, N Kumar, and M Alazab. "A novel UAV-enabled data collection scheme for intelligent transportation system through UAV speed control." *IEEE Transactions on Intelligent Transportation Systems* 22(4) (2020): 2100–2110.

[47] Sirohi, D, N Kumar, and PS Rana. "Convolutional neural networks for 5G-enabled intelligent transportation system: a systematic review." *Computer Communications* 153 (2020): 459–498.

[48] Kumar, N, SS Rahman, and N Dhakad. "Fuzzy inference enabled deep reinforcement learning-based traffic light control for the intelligent transportation system." *IEEE Transactions on Intelligent Transportation Systems* 22(8) (2021): 4919–4928.

[49] Yang, C, M Zha, W Wang, K Liu, and C Xiang. "Efficient energy management strategy for hybrid electric vehicles/plug-in hybrid electric vehicles: review and recent advances under intelligent transportation system." *IET Intelligent Transport Systems* 14.7 (2020): 702–711.

[50] Yang, Z, J Peng, L Wu, *et al.* "Speed-guided intelligent transportation system helps achieve low-carbon and green traffic: evidence from real-world measurements." *Journal of Cleaner Production* 268 (2020): 122230.

[51] Tyagi, AK, and N Meghna Manoj. "Internet of Everything (IoE) and Internet of Things (IoTs): threat analyses, possible opportunities for future." *Journal of Information Assurance & Security (JIAS)*. 15(4) 2020.

[52] Cui, L; G Xie, Y Qu, L Gao andY Yang. "Security, and privacy in smart cities: challenges and opportunities." *IEEE Access* 6 (2018): 46134–46145.

[53] Anand, B, V Barsaiyan, M Senapati, and P Rajalakshmi, P. "Real-time lidar point cloud compression and transmission for the intelligent transportation system." *2019 IEEE 89th Vehicular Technology Conference (VTC2019-Spring)*. IEEE, 2019, 1–5.

[54] Surnin, O, A Ivaschenko, P Sitnikov, A Suprun, A Stolbova, and O Golovnin. "Urban public transport digital planning based on an intelligent transportation system." *2019 25th Conference of Open Innovations Association (FRUCT)*. IEEE, 2019, 292–298.

[55] Ali, QE, N Ahmad, AH Malik, G Ali, and WU Rehman. "Issues, challenges, and research opportunities in intelligent transport system for security and privacy." *Applied Sciences* 8(10) (2018): 1964.

[56] Song, M and Z Hou. "Model analysis of traffic emergency dispatching in intelligent transportation system under cloud computing." *International Journal of Internet Protocol Technology* 13(1) (2020): 46–54.

[57] Krishna, AM and AK Tyagi. "Intrusion detection in the intelligent transportation system and its applications using blockchain technology." *2020 International Conference on Emerging Trends in Information Technology and Engineering (ic-ETITE)*. IEEE, 2020, 1–8, doi: 10.1109/ic-ETITE47903.2020.332.

[58] Tu Peng, Xu Yang, Zi Xu and Yu Liang, "Constructing an environmentally friendly low-carbon-emission intelligent transportation system based on big data and machine learning methods." *Sustainability* 12(19) (2020): 8118.

[59] Ning, Z, S Sun, X Wang, *et al.* "Blockchain-enabled intelligent transportation systems: a distributed crowdsensing framework." *IEEE Transactions on Mobile Computing* (2021).

[60] Zhao, J, C He, C Peng, and X Zhang. "Blockchain for effective renewable energy management in the intelligent transportation system." *Journal of Interconnection Networks* (2021): 2141009.

[61] Lei, A, H Cruickshank, Y Cao, P Asuquo, CPA Ogah, and Z Sun. "Blockchain-based dynamic key management for heterogeneous intelligent transportation systems." *IEEE Internet of Things Journal* 4(6) (2017): 1832–1843.

[62] Zhu, L, FR Yu, Y Wang, B Ning, and T Tang. "Big data analytics in intelligent transportation systems: a survey." *IEEE Transactions on Intelligent Transportation Systems* 20(1) (2018): 383–398.

[63] Camacho, F, C Cárdenas, and D Muñoz. "Emerging technologies and research challenges for intelligent transportation systems: 5G, HetNets, and SDN." *International Journal on Interactive Design and Manufacturing (IJIDeM)* 12(1) (2018): 327–335.

[64] Kalloniatis, C, D Kavroudakis, A Polidoropoulou, and S Gritzalis. "Designing privacy-aware intelligent transport systems: a roadmap for identifying the major privacy concepts." *Research Anthology on Privatizing and Securing Data. IGI Global*, 2021. 589–609.

[65] Drop, N, and D Garlińska. "Evaluation of intelligent transport systems used in urban agglomerations and intercity roads by professional truck drivers." *Sustainability* 13(5) (2021): 2935.

Chapter 12

Autonomous Vehicles (AVs): localization, navigation and tracking/tracing

Abstract

In the recent years, vehicular technology has been improved to a tremendous level. From driver to driverless, we have converted our vehicle in today era. For example, Tesla and other company is working on providing driverless cars. Here, driverless means autonomous, i.e., ability to react in every situation. To reduce number of accidents, reduce emission of gases, and improving passenger's satisfaction level or completely securing user's information against today's new types of cyber-attacks, etc. Such features need to enabled or inbuilt with vehicles. This chapter discusses about autonomous vehicles (AVs) and also discusses about localization, navigation or tracking or traces mechanisms for the same vehicles. This chapter gives clear description that "How Autonomous Vehicles (AV) can change the future and the public transportation system".

Keywords: Autonomous vehicles, Navigation in vehicle, Global positioning system, Hybrid vehicles

12.1 Introduction to autonomous vehicles

Autonomous vehicles (AVs) are quickly establishing themselves as a new type of infrastructure. This technology has piqued the interest of automakers, electronics manufacturers, and information technology (IT) service providers, and academic research has played a vital role in the development of their prototype systems. Despite this trend, autonomous cars are not arranged in a systematic way. Third-party vendors cannot easily test new AV components because commercial vehicles conceal their in-vehicle system interface from consumers. Furthermore, sensors are not all the same. Some vehicles may merely have cameras, while others may have cameras, laser scanners, GPS receivers, and milli-wave radars. Aside from hardware issues, there are also software issues to consider. Building an AV platform from the ground up is inefficient, especially for prototypes, due to its vast scale. Although open-source software libraries are recommended for this purpose, they have yet to be integrated into AV development. Algorithms for scene recognition, path planning, and vehicle control demand diverse skills and expertise, and they

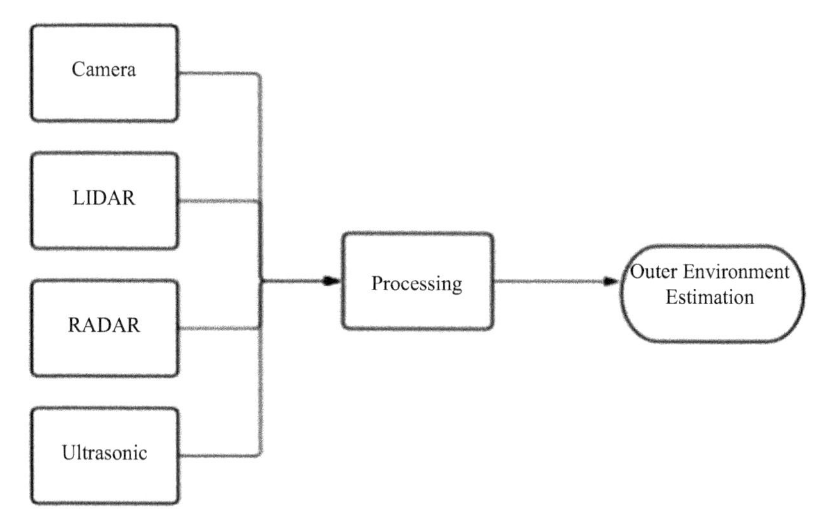

Figure 12.1 Conventional environment estimation mechanism for AVs

frequently necessitate a large amount of engineering effort. Finally, in order to drive on a public road, a fundamental dataset, such as a map for localization, must be provided. AVs are made up of a variety of technologies. The development of their platform necessitates diverse collaboration.

The vehicle and technology sectors have achieved major advances in recent years, bringing computerization to a function that has been entirely a human activity for almost a century: driving. ACC and parking assist technologies, which allow cars to manoeuvre themselves into parking spaces, are becoming more common in new cars. Some businesses have gone even further, developing nearly completely AVs that can navigate highways and urban environments with little to no human intervention. Assuming that these technologies are viable and reach a mass market, AVs have the potential to revolutionize transportation. Figure 12.1 depicts the conventional environment estimation mechanism for AVs.

12.2 Related/background work

New developments in communication and mechanical technology have had an impact on our daily lives, and transportation is no exception. These improvements have given rise to the prospect of AV innovation, which aims to reduce collisions, energy consumption, pollution, and obstruction while also increasing transportation availability. Despite the fact that the idea of autonomous cars has been known for a long time, high costs have stymied large-scale development [1]. All things considered, there has lately been a speed increase in inventive work initiatives to bring the concept of AV to fruition. For example, the Google vehicle's approach brought AVs to the forefront [2,3]. Furthermore, the automotive industry invests around

€77 billion on R&D globally in order to maintain development and remain competitive [4,5].

Because of the rapid improvement of communication technology and the need to account for the developing populace in developed nations, AVs may have become a requirement and a vital business perspective [6]. When it comes to confronting fresh ideas and innovations such as informal communities, PDAs, and AVs, a few scholars have repeatedly warned that the transportation scene is rapidly shifting [7–9]. Uber is a concept that is clearing metropolitan neighborhoods to the point that taxi companies are attempting to hold business and remain competitive. Manyika et al. [10] recalls automobile computerization as one of the top ten troublesome breakthroughs of the future.

Because of the fierce competition among vehicle manufacturers, the year 2020 has been designated as a landmark year for introducing commercial AVs to the general market [1,11,12]. Perhaps the growth of broad sections of the AV market will be the center of the present century. Based on the organization and acceptance of previous brilliant vehicle technologies (for example, programmed transmission and hybrid electric drive) [13], it is anticipated that AVs would account for roughly 50% of vehicle deals, 30% of cars, and 40% of all vehicle travel by 2040.

The AV is associated with a variety of favorable cultural consequences, for example, a more secure vehicle framework, a cheaper cost of transportation, and enabling a smidgeon of adaptability to the non-walking and disabled, as well as those in lower pay families. It is estimated that the direct cultural value created will be between 0.2 and 1.9 trillion dollars per year by 2025 [10]. Such certain impacts are the driving forces behind the emergence of AV innovation, transforming it into a viable business model sooner rather than later and then some.

Some agree that AVs should be viewed as a multidisciplinary invention with a broad-calculated primary point. Maddox *et al.* [14] depicted the AV in a graphic with two additional pieces to ensure a good operating AV worldview: "Associated" and "Large Data." As a result, the terms "Associated" or "Associated Vehicle" refer to the advancements that ensure communication between all contributing experts or partners, including walkers, specialists, and vehicles, as well as infrastructure.

The first attempt at autonomous cars dates back to the mid-1920s [15] and gained momentum in the 1980s when researchers discovered how to create robotized parkway frameworks [16,17]. This made it possible for semi-autonomous and autonomous cars to be linked to the expressway foundation. From 1980 to 2000, the majority of AV pioneer pilots were produced in Germany and the United States [18,19].

AVs owe a great deal to the defense sector's substantial research on unmanned equipment, known as (DARPA) the United States Defense Advanced Research Projects Agency [20]. Google's self-driving car generated a lot of attention for AV and attracted a lot of talent from many fields. Google's autonomous fleets covered over one million kilometers in July 2015, with only 14 minor traffic incidents on public highways reported. However, in all situations, the AV was not at fault; rather, it was either being driven manually or the other driver was at fault [21].

AVs often operate on a three-stage configuration known as "sense-plan-act," which is the basis for many automated frameworks [22,23]. Understanding the complex and dynamic driving environment is a significant challenge for AVs [1,24]. To that aim, the AVs are furnished with a variety of sensors, cameras, radars, and other devices that gather basic information and data from the surrounding environment. These data would then be used as input for programming, which would recommend appropriate methods such as speed increase, path change, and exceeding.

It involves when AVs will be found in the street organization, and not if. Accordingly, they are a significant piece of transportation arranging which requires the advancement of the suitable models. The last phase of transportation displaying is the recreation of vehicle developments in the street network which is known as traffic task. With that in mind, there are two significant recreations: miniature and full scale. In miniature recreation, nitty gritty developments and practices of the singular vehicles are considered in any investigation. The early and essential elements of mechanization (levels 1 and 2) have incited a few researchers to incorporate AV innovations into miniature recreations [25]. As verified over, the connectedness of AVs has been examined in signal control approaches which brought about critical decreases in delay [46–48]. Different investigations will generally expand the range of the current powerful traffic task models to some way or another incorporate AVs [26]. Regardless of current computational advances and current demonstrating information, the size of miniature reenactment is restricted to a part of a city and not the whole city.

The intricacy of the AVs' route lies in the way that the AVs should share street space with non-AVs, bringing about blended traffic designs. Consider briefly two kinds of vehicles: non-AVs and AVs. The non-AVs keep up with their childish conduct since there is no influence to drive them to look for the non-briefest way. Meanwhile, we have the capacity to authorize a vehicle route to anticipate the AVs. The test currently is to be seen as a model for a blended traffic design that comprises both SO and UE traffic designs. In such rush hour gridlock designs, an armada of AVs are associated and they agreeably track down their course (vehicle route), while others are childish drivers who just look for the most limited potential ways (vehicle steering) [27].

12.3 What are AV and automated vehicle?

Through its ability to sense its surroundings, an AV, also known as a driverless car, is able to operate itself and execute critical activities without the need for human interaction.

A completely automated driving system (ADS) is used in an AV to allow it to respond to external conditions that a human driver would handle. There are six various levels of automation, and as the levels progress, the driverless car's autonomy in terms of operation control grows. The car has no control over its functioning at level 0, and the human driver is in charge of all driving.

The vehicle's ADAS (advanced driver assistance system) can do level 1 tasks. In some circumstances, the ADAS can handle steering, acceleration, and braking at level 2, but the human driver must maintain undivided attention to the driving environment during the voyage while also doing the other responsibilities. In some circumstances, the advanced driving system (ADS) can undertake all aspects of the driving duty at level 3, but the human driver must be able to regain control when the ADS requests it. In the remaining cases, the human driver performs the required tasks.

At level 4, the vehicle's ADS can handle all driving responsibilities on its own in situations when human attention isn't required. Finally, level 5 entails complete automation, in which the vehicle's ADS is capable of performing all functions in all conditions and no human driver assistance is necessary. The use of 5G technology will enable full automation by allowing vehicles to communicate not just with one another, but also with traffic lights, signage, and even the roadways itself. Adaptive cruise control, or ACC, is one of the features of vehicle technology employed in driverless vehicles. This device can automatically alter the vehicle's speed to keep a safe space between it and the vehicles in front of it. This function is based on data acquired from the vehicle's sensors and allows the car to undertake duties such as braking when it detects any cars ahead. This data is then processed, and the relevant instructions are delivered to the vehicle's actuators, which control the vehicle's responsive actions including steering, acceleration, and braking. Traffic lights and other non-vehicular activity can be detected by highly automated vehicles with totally automated speed control.

12.4 Why is AV technology important now?

When it was impossible to imagine self-driving automobiles on the road, the public authority-built streets, traffic lights, and signage. While drivers are aware of traffic lights and signage that indicate when they should or should not drive their car, AVs require different alerts such as radio or mobile network signals. Various self-driving organizations are currently investigating collaborating with keen logical businesses to enhance the innovation of both the autos and the framework. Indeed, many industry leaders agree that intelligent streets are essential for the final fate of AVs. The ongoing advancement of automobile technology seeks to provide even higher safety benefits and ADSs that, one day, will be able to handle the entire chore of driving when we do not want to or cannot do it ourselves. Fully autonomous automobiles and trucks that drive themselves rather than humans will become a reality. Several new vehicles on the market today include technology that assists drivers in avoiding drifting into adjacent lanes or making unsafe lane changes. It uses a use a combination of hardware and software to help automobiles recognize risks and warn the driver to avoid an accident. Ford's independent vehicle research department is looking at how emergent innovation from clever foundation might provide further data to the AV before it even appears at a convergence.

While many cities have various types of traffic signals – flat, vertical, posted at the crossing point rather than in the intersection – AI calculations combine amazing framework with AVs, enabling the deployment of these vehicles in new places to happen much faster. While some self-driving companies do not rely on keen foundation as of now, the AV industry is realizing that it's a critical piece of the puzzle when it comes to speeding up the capacity to convey additional AV courses in various cities and nations. Furthermore, transportation frameworks cannot achieve security without a knowledgeable base. The need for innovative innovation on the streets is growing. Pete Buttigieg, the United States Secretary of Transportation, has even emphasized the importance of constructing a stronger framework as well as the innovation for self-driving vehicles. The integration of AI and video analysis in smart streets will aid in the growth of autonomous car innovation. These advancements will also change the security, comfort, and execution of autonomous automobiles, as well as the overall transportation biological system. Independent cars are becoming another component of the system. Auto producers, hardware designers, and IT management. This idea has piqued the interest of vendors.

Furthermore, scholarly research has contributed significantly to the development of their model frameworks. One remarkable work, for example, was provided through Carnegie Mellon University. Regardless of this tendency, autonomous cars exist are not methodically synchronized Taking into account. Business vehicles safeguard their in-vehicle data. Outside sellers may only test new sections of the framework interface from clients with great effort. Cars that are self-sufficient Sensors are also used. They are not indistinguishable. Independent cars, regardless of equipment, should handle programming difficulties. Because an independent vehicle stage is huge in size, developing it from scratch is inefficient, especially for models. Open-source programming libraries are popular for this reason, although they have yet to be integrated to support independent cars. The planning and execution of computations for scene recognition, route planning, and vehicle control also need interdisciplinary talents and knowledge, frequently resulting in a key design effort. Finally, a fundamental dataset, such as a limitation guide, should be provided to drive on a public street.

12.5 The promise and perils of AV technology

AVs have the potential, at least in theory, to completely transform urban development as we know it, with a revolution in ground transportation that, if regulations allow, could dramatically alter the landscape of cities around the world, with massive economic, social, spatial, and mobility implications. In terms of the benefits that driverless technologies may be able to give to the human race, there are high hopes. Enhanced transportation safety, significantly fewer traffic accidents and fatalities, reduced traffic congestion, increased road capacity, environmental benefits in terms of fewer carbon emissions and noise nuisance, relief from driving and navigation duties and thus more occupant flexibility, no driving restrictions (everyone will be able to "drive"), and transformation of the current car ownership regime are just a few of the potential benefits.

However, there is another, murkier side to this coin: despite its potential benefits, vehicle automation may provide its own set of issues. User resistance to giving up driving control, loss of situational awareness, loss of driving skills, privacy concerns, increased vulnerability to software and hardware flaws and hacking, a new human–machine ethics paradigm as a result of automated risk allocation in collisions, liability for damage concerns, the need for an entirely new legislative framework, loss of driving-oriented jobs, susceptibility of the car's navigation system to various types of malware.

Although transportation can help nations prosper, it is necessarily associated with negative externalities such as pollution, accidents, and human deaths. A great number of studies have been conducted to estimate these costs in terms of human-driven vehicles. These expenses are distinct from direct expenses like gasoline, car maintenance, vehicle registration, and licensing, or public transportation tickets. The externality cost is a hidden cost placed on the entire society; it covers costs such as traffic congestion, accidents, environmental degradation, and security. In general, AV technology is thought to have the ability to significantly reduce (if not eliminate) many of these currently present negative externalities. According to one estimate, these external costs could be as high as the fuel price imposed on the entire society, including low-income people who rely primarily on public transportation. Additional benefits of AVs include increased accessibility and mobility, as well as improved land use. Although there may be considerable disadvantages to using AVs, it is usually assumed that these disadvantages are offset by the benefits. The numbers for traffic accidents in the United States in 2010 are shocking: 32,999 people were killed, 3.9 million people were injured, and 24 million vehicles were damaged, costing 277 billion dollars in both real and intangible damages. This financial burden has a cascading effect, affecting productivity, medical expenditures, legal and court fees, workplace losses, emergency service fees, traffic congestion, insurance administration fees, and property damage. The adoption of modern technologies including airbags, anti-lock brakes, electronic stability control, 8 head-protection side airbags, and front collision alerts has contributed considerably to a decreased trend in the number of crashes in the United States. These are features that will be adopted in AV technology. In particular, some studies estimate the reduction of crashes could be as high as one-third if all vehicles are equipped with adaptive headlights, forward collision, warnings, lane departure warnings, and blind spot assistance. Human error is blamed for more than ninety percent of crashes. Therefore, AVs should be able to prevent an appreciable number of these crashes, in turn eliminating the vast majority of all traffic delays.

12.6 Effects of AV technology on safety and crashes

The advantages of computerized cars in terms of health are critical. The capacity of mechanized vehicles to save lives and reduce wounds is founded on a fundamental and tragic trust. Mechanized cars may be able to eliminate human error from the collision situation, which will help to ensure drivers and passengers, as well as bicycles and pedestrians. Automated cars may provide further monetary and cultural benefits. By eliminating the vast majority of motor vehicle accidents, these costs might be eliminated. Streets brimming with self-driving cars might also help to ease traffic flow and reduce jams. With automated automobiles, the time and money spent driving may be put to better use. While the full cultural benefits of mechanized vehicles and their driver assistance elements are difficult to project, the ground breaking capability of mechanized vehicles and their driver assistance elements can also be perceived by investigating socioeconomics and the networks these advances may help to support. In many regions, the ability to drive is more important than the ability to do business or live independently. Millions more people might benefit from this type of opportunity if robotized cars are used. According to one study, automated cars might offer up new economic opportunities for almost 2 million people with disabilities. Mechanized vehicles and driver assistance technologies (recalling those now in use on the roads) have the potential to lessen collisions, prevent injuries, and save lives. Human error or choice is responsible for 94% of all actual engine vehicle crashes. Vehicles that are completely automated that can see more and react faster than human drivers might drastically reduce missteps, following accidents, and their costs. Vehicles are evaluated by the groups that create them. Organizations should agree to Federal Motor Vehicle Safety Standards and ensure that their vehicle is free of hazards. Many businesses nowadays are attempting advanced computerized vehicles to ensure that they perform as intended, but much effort has to be done to ensure their protected operation before they are made openly available.

Maybe, it is assumed that AVs are the same as different vehicles in vehicle directing, that is, the narrow minded (or non-helpful) traffic design. Associated vehicles achieve basic ongoing traffic information (like travel time and episode reports) which would then be able to be utilized in a helpful rush hour gridlock style. In doing so, AV information would then be able to be ordered and handled in rush hour gridlock observing focuses that thus will suggest the most suitable courses. That is, every vehicle will presently in a real sense know about the courses and objections of different vehicles. Accordingly, a more complex, effective, and informed vehicle directing framework is accomplished. We allude to this as a vehicle route.

12.7 Effect of AV technologies on mobility for those unable to drive

The attitudes of 211 blind persons in the UK, as well as the factors that influence their desire to travel in AVs, are being investigated every day. The data is analyzed using a semi-automated structural topic modeling approach after participants answered an open-ended question about their sentiments about AVs. The exercise yields four "topics": (1) "hope" for future independence and freedom of travel provided by AVs for the blind, (2) skepticism that AVs would ever be designed to satisfy the needs of the blind, (3) safety concerns, and (4) AV affordability.

A number of factors, including a participant's desire for independence, comorbidity, locus of control, and level of generalized anxiety, were thought to influence the four mediating issues. Three of the mediating variables, namely, hope for future independence, safety concerns, and affordability, had a substantial impact on desire to travel in an AV. Skepticism of AVs had no effect on the results. There are a number of ramifications for AV design and the generation of public information messaging promoting AVs. Public awareness efforts should emphasize the freedom of movement that AVs will allow for those who are blind, as well as promises about safety. AVs have "tremendous mobility possibilities for visually impaired individuals." The benefits of AVs for blind persons are reported in a lot of the grey literature on the subject. For example, RNIB stated that distance will no longer be a hindrance to blind people's ability to travel, altering their lives. Blind persons should be able to engage more completely in society, eliminate social exclusion, get easier access to education and training, and improve their overall quality of life by using AVs. However, there have been some questions raised about the capacity of blind persons to interact with self-driving cars.

Though the prospect of self-driving cars is appealing on many levels, buyers must overcome the cognitive dissonance of riding side-by-side with an empty driver's seat and a steering wheel that moves by itself. In contrast, Big Tech is still coping with America's unequal geographical and legislative terrain, as well as the huge ethical and technological obstacles of operating AVs alongside impulsive and unpredictable human drivers in legacy vehicles. Despite the fact that we have some advanced technologies on the road, even the most advanced self-driving car will just sit behind a moving truck that is parked in the road until the human driver instructs it to move around it. Despite the fact that technology levels are not yet at a point where fully AVs are feasible, the advancements we are witnessing are ripe for a much-needed discussion on how this technology will be implemented once it is fully perfected. Our culture is reliant on transportation, and it would be naive to believe that self-driving automobiles will not alter our way of life. As we begin to consider the consequences, it is critical that we start with the individuals who will be most affected by the application of this technology: persons with disabilities. The Ruderman Family Foundation and Securing America's Future Energy (SAFE) collaborated with the Ruderman White Paper on Self-Driving Car Technologies: The Impact on People with Disabilities. We are delighted to share our findings with you and to continue this important dialogue about how to ensure that 20% of our population benefits from this breakthrough technology.

12.8 Energy and emissions implications of AVs

The sophistication of vehicle robotization options varies. On the one hand, the driver is in command, but the driving experience is enhanced by features such as accident avoidance, path and side assistance, stopping, and maintaining safe driving distances. A significant number of these components are now in use in cars. On the other end of the spectrum, the vehicle is capable of performing all driving functions under all situations. Vehicles that are completely computerized are being developed and tested in a variety of economic sectors in the United States, as well as globally. Vehicle automation has the potential to significantly alter transportation, with huge implications for energy and the environment. There is a significant vulnerability in terms of the impact of robotization on movement interest and vehicle productivity. Emotional productivity gains from computerization might bring down fuel expenses, thereby reducing the intensity of option-powered cars. Furthermore, these motions may have both beneficial and bad natural consequences. Some robotization circumstances resulted in illogical conclusions. For example, if evident degrees of efficiency enhancement force forth alternative fuel cars to be crucial. The MARKAL (MARKet ALlocation) model is used in this research, which is an energy framework improvement model that simulates the growth of energy innovation and fuel mix across many years. The spotlight is on light-obligation vehicle (LDV) and heavy-duty vehicle (HDV) trip requests, vehicle innovation decisions, and fills for this investigation. One advantage of using this model over a transportation-only approach is that cooperation with non-transportation domains may be gradually reestablished. This is especially important in this case since vehicle mechanization has the potential to cause progressive changes in transportation, making static speculations about the stock and pricing of electricity, oil-based commodities, and other fuel sources difficult. Scientists investigated the consequences of vehicle mechanization on energy consumption and natural outcomes. Changes in energy and emissions from vehicle mechanization have been investigated within the broader context of what have been dubbed the "Three Revolutions": computerized vehicles (AVs), shared vehicles, and electric vehicles (EVs).

With vehicle outflow decreasing a main worldwide need, prodding strategy endeavors to restrict air contamination's consequences for general well-being and alleviate the cataclysmic damages of worldwide environmental change, robotized driving has the ability to lessen these discharges through efficiencies in driving rates and aversion of gridlock. It has for some time been realized that key "measures poisons," a side-effect of fuel burning, detrimentally affect human wellbeing, with air contamination the study of disease transmission inspecting the connected effects on respiratory and cardiovascular sickness from:

- Carbon monoxide (CO),
- Nitrogen dioxide (NO_2),
- Lead (Pb),
- Sulfur dioxide (SO_2),
- Ozone (O_3),
- Particulate matter,
- Carbon dioxide (CO_2).

Transportation right now represents huge impacts and effects on ecological well-being—driven by transportation designs and related elements, like private vehicle proprietorship, eco-friendliness, and traffic thickness—and understanding and surveying the changing effects of independent transportation on natural wellbeing will be a basic worry of worldwide significance. Academic regard for ecological well-being effects of independent vehicles so far has inferred that these effects—and regardless of whether they are positive or negative—rely on a scope of elements, like the qualities of independent vehicles, the transportation organization's capacity to adjust to these vehicles, and purchaser inclination as it identifies with private responsibility for vehicles. Intensifying these natural wellbeing hurts, independent vehicle consequences for the ozone depleting substance emanations driving worldwide environmental change are hazy, for certain examinations projecting a diminished ecological effect while others project that an increment in vehicle-miles voyaged could almost twofold such discharges (Wadud et al., 2016). The twin ecological well-being chances from both air contamination and environmental change are integral to the SDGs, with targets zeroed in on "air quality"; "dangerous synthetic compounds" noticeable all around, water, and soil; and "relief and variation to environment changes," with states looking for by 2030 to "twofold the worldwide pace of progress in energy proficiency" (UN, 2015). Long haul vulnerability stays over which mechanized vehicle advances will arise and what sorts of fills they will utilize, leaving general well-being researchers with a scope of potential other options and results identified with the effect of computerized vehicle reception on natural wellbeing. Evaluating the scope of mechanical choices, their attainability, and the potential effects on natural wellbeing are vital stages for general well-being researchers to propel an examination plan around here and to guarantee that ecological wellbeing impacts are considered as advances advance and robotization reception starts.

12.9 Costs and disadvantages for implementing AIV in today's scenario

Autonomous driving will cut fuel consumption for long-distance trucks since platooning may be created on the road, resulting in less wind resistance for the vehicles inside the platoon. Furthermore, smoother driving in general can save petroleum. We predict that fuel consumption for long-distance trucks can be lowered by 10%, but not for other types of trucks because they normally cannot form platoons. Furthermore, we infer that AVs will not be implemented until they are at least as safe as MDVs, and in the example calculation, we make the modest assumption that accidents will be decreased by 10%. It is difficult to say how much capital expenses will differ between AVs and MDVs in 2025 and 2040. According to estimates, a self-driving car will cost $10,000 more when it has a 10% market share, but only approximately $3,000 more in the long term when it has 90% market share. According to Volvo reps, their self-driving car will cost around 120,000 SEK more than the same MDV when it launches in 2021, but it will eventually come down to the same price. We do not make any assumptions about the capital cost differential between AVs and MDVs. Instead, new research shows how much the capital cost (which includes vehicle technology development, higher capital expenses due to higher production prices, and changed maintenance costs for AVs) can rise without outweighing the benefits of AVs.

Additional costs linked to trucks and cars may occur when AVs begin to operate in normal traffic. Some may be transitory, such as development expenses for new technology, while others, such as increased production costs for more modern cars and upkeep, may be permanent. In the long run, however, vehicle production may become less expensive since they can be constructed without any driving aids, and they may become more productive as truck driver space is converted to cargo space. Because of learning by doing and economies of scale in the manufacturing industry, vehicle capital costs may gradually fall. The costs of creating the essential car technology, as well as any higher vehicle production costs, will be factored into the price that consumers pay for autonomous automobiles. As previously stated, we investigate how much capital expenses can rise without totally cancelling out the quantifiable and appraised net benefits. In the immediate term, greater investment in digital infrastructure may be required, and specific lanes may be required as long as traffic is mixed between AVs and MDVs. The costs of developing new technology could be substantial. Infrastructure investments, on the other hand, may decline in the long run. Private cars can make better use of road and parking capacity, using less room. As a result, there may be fewer or narrower streets and lanes, as well as less demand for parking in high-value locations. Vehicles parked in other areas, on the other hand, may cause extra traffic. Commuters' cost per kilometer would decrease and their welfare would grow when AVs were introduced, as would travel distance and city size. As a result, land rentals would rise in central locations while falling in outlying places. The scale of the effects on infrastructure and land usage, as well as whether the net effects will add to benefits or costs, are both unknown.

The ability to travel in an autonomous automobile can assist older persons, people with disabilities, and people who do not have a driver's license. Based on 2009 travel trends and assumptions, they anticipate a 14% increase in traffic as the top limit. People without a driving license could increase traffic by a maximum of 9% and disabled people by up to 2.6% according to their study. AVs may also promote car sharing, both between existing car owners and between new groups. When vast amounts of data must be processed for required communication, the consequences on people's perceived safety and privacy concerns are not evaluated and valued in this study. People may believe that vehicles without a driver pose a greater risk and that their privacy will be compromised, at least at first. On the other hand, driver errors are no longer an issue. Software, legal challenges, and safety were all concerns. It is likely that the new technology will not be widely adopted until it achieves a level of safety comparable to that of today's automobiles.

12.10 Safety applications in VANETs/ITS

AV innovations give finely tuned speed increase slowing down moves constantly while continually and enthusiastically observing the encompassing traffic climate. Along these lines, AVs can journey at higher rates while keeping more limited separations (lower types of progress). Semi-AVs outfitted with ACC have as of now shown such a promising capacity. Lower degrees of progress by a line of AVs would not think twice about, and consequently, we are probably going to see a platooning of AVs. Therefore, the throughput of the streets (or limit) will essentially expand—a few examinations have been assessed by up to multiple times. In certain examinations, the way that AVs are associated has been taken advantage of in signal control which has brought about considerably less deferral at signals or proportionately higher street limit.

The trend setting innovations of AVs should facilitate traffic dissemination and bringing down of movement costs which thus may actuate extra travel interest. Such an interest can be viewed as both a danger and a chance. The danger emerges from the way that the extra travel request might deteriorate gridlock. The extra interest is the consequence of extra venture (AVs) infused into the vehicle framework. Such worries are becoming genuine. For example, specialists at Delft University in the Netherlands have encouraged the Dutch government to go to lengths (e.g., travel request the executives) to control the development of movement and resulting externalities of the looming AV advances.

On the off chance that one expects to keep up with requests at similar levels as preceding the development of AVs, then, at that point, there is an authentic chance to take advantage of the initiated request through blockage evaluating. The estimating can be set to the level at which the incited request scatters. Clog evaluating is presently a functioning space of examination. The significant degree of correspondence advancements among AVs can enormously smooth out any kind of evaluating plans, for example, distance-based charging and dynamic valuing plans.

AVs discharge drivers from participating in the physical and mental activities related with driving, permitting them to use this time on other useful exercises in transit. Subsequently, AVs further diminish the chance expense of movement as far as the saved worth of time relating to off-wheel exercises.

AVs might prolongedly affect the land-use design. The worth of land increases relatively with its vicinity to the focal city where open positions exist in numerous ventures like banking, monetary business sectors, and numerous other help regions. Nearness is shown by transportation. The approach of vehicles in the start of the twentieth century brought about the development of rural areas. The connection among AVs and land use is both confounded and some way or another dumbfounding. In one situation, the presentation of AVs could stimulate a pattern toward considerably more scattered and low-thickness land-use designs encompassing metropolitan locales. All in all, AVs might bring about the further development of rural areas and may even drive further into exurban regions. In a completely different situation, AV innovation hinders the intense requirement for parking spots. Parking spots in the core of urban communities can be opened up for other utilization. Consequently, AVs could wind up invigorating metropolitan development in focal regions, adding to the thickness of CBDs. Note that stopping offices consume a major piece of room in CBDs. Shoup assessed that the absolute region devoted to parking spots is normal, comparable to around 31% of locale regions.

In rundown, the drawn-out assumption with the reception of level 4 AVs is that one would probably see denser metropolitan centers, more structures, and less parking spots. Simultaneously, AVs could prompt significantly more prominent scattering of low-thickness improvement in metropolitan periphery regions given the capacity of proprietors to take part in different exercises while vehicles pilot themselves. Underdeveloped nations battle with an absence of a transportation foundation, like streets, extensions, and public vehicles, which is hindering their monetary turn of events. Reception of AVs by these agricultural nations might save them the expenses related with extending capital-escalated foundation. A comparable worldview was seen when non-industrial nations jumped over to cell phone innovation which excluded them from costly landline foundation.

It involves when AVs will be found in the street organization, and not if. Along these lines, they are a significant piece of transportation arranging which requires the improvement of the fitting models. The last phase of transportation displaying is the reproduction of vehicle developments in the street network which is known as traffic task. With that in mind, there are two significant re-enactments: miniature and large scale. In miniature recreation, itemized developments and practices of the singular vehicles are considered in any examination. The early and fundamental highlights of mechanization (levels 1 and 2) have incited a few researchers to incorporate AV innovations into miniature reproductions. As indicated over, the connectedness of AVs has been explored in signal control arrangements which brought about critical decreases in delay. Different investigations will generally broaden the scope of the current powerful traffic task models to some way or another incorporate AVs. Regardless of current computational advancements and current demonstrating information, the size of miniature recreation is restricted to a piece of a city and not the whole city.

12.11 Traffic condition sensing application for AIV

The word AVs only make sense if the vehicles can sense traffic conditions around them so that calculated decisions can be taken. Among the various requirements for this task, two main ones are obstacle detection and track detection. To avoid accidents and collisions, the car should be able to recognize impediments for smooth and effective operation. It should also be able to figure out how far away the barrier is from the car. Similarly, track detection is critical since the AV must stay on a predetermined path and stay within the yellow lines on both sides of the road. This chapter describes the technologies and advancements in obstacle detection and track detection in autonomous cars/vehicles that have been presented in the literature so far. The sensor-based technique is the most frequent and widely utilized approach for both obstacle identification and track tracking. A variety of sensors and related technologies have lately been discovered and put into use. Acoustic, radar, laser/LiDAR, optical sensors, and sensor fusion are the most prevalent sensors. It also addresses their benefits and drawbacks. There are two types of sensors: cooperative and non-cooperative. There are two types of sensors: cooperative sensors and non-cooperative sensors. The camera-based approach is the second most popular method for recognizing the track and obstacles in autonomous cars. Although some academics consider it a sub-category of sensor-based approaches, it has been presented as a separate category due to the diversity and wide range of camera-based detection systems. Three types of camera-based detection systems have been identified: based on knowledge; based on stereo vision; and based on motion.

For obstacle identification and tracking, there are a variety of deep learning algorithms available. One such strategy is presented in [28], which employs numerous sources of local patterns and depth information to produce reliable car and pedestrian detection, recognition, and tracking on the road. The same paper explores the use of deep learning for obstacle detection and categorization in high-speed autonomous driving. The employment of a monocular camera to simulate human obstacle detection and avoidance behavior on UAVs [29] is a rather uncommon and new approach. Although it can be regarded as a subcategory of camera-based approaches, it has been presented as a separate category due to its uniqueness in comparison to other camera-based approaches. Similarly, employing radar and visual information, Ref. [30] presents a bionic vision-inspired technique. A mechatronics system with a PID controller that predicts and controls the vehicle heading angle in order to follow the lane or avoid obstacles is another solution mentioned in the literature [31]. Some research [32,33] show how to employ a laser scanner/rangefinder to execute obstacle detection and road following in an outdoor setting. In situations such as diverse weather conditions (e.g., sun, rain, and fog), and different road appearances, this technique surpasses commonly used camera-based vision algorithms.

12.12 Opportunities and challenges for future towards AVs

Despite the fact that transportation is a means of cultivating the flourishing of social systems, it is inextricably linked to negative externalities such as pollution, errors, and human disasters. There are several studies evaluating these costs in terms of human-driven vehicles. These expenditures differ from direct expenses such as the cost of gasoline, car maintenance, vehicle enrolment, and permits, or public transportation tickets. The externality cost is a hidden cost imposed on society as a whole; it includes costs such as congestion, disasters, and climate degradation, as well as security. As a general rule, AV innovation appears to have the potential to significantly reduce (if not eliminate) a substantial number of these present negative externalities. In one measure, these extra costs can be nearly as great as the gasoline cost imposed on society as a whole, including low-wage workers who rely solely on open vehicles. AVs can also provide other benefits such as increased availability and adaptability, as well as significantly increased area utilization. Despite the fact that there may be important barriers associated with AVs, it is widely agreed that these costs are well compensated by the advantages.

The expenditure issue is having an ever-increasing impact on efficiency, clinical expenses, legal and judicial charges, work environment disasters, crisis management costs, the clog issue, protection organization expenses, and property damage. The acceptance of technological developments, for example, airbags, electronically monitored slowing mechanisms, electronic security control, head-insurance side air packs, and forward impact alarms, has resulted in a decrease in the number of accidents in the United States. These are some of the highlights that will be addressed in AV innovation. Specifically, some studies estimate that if all vehicles are outfitted with versatile headlights, forward collision admonitions, path take-off warnings, and vulnerable side assistance that are associated with level 0 or level 1 vehicle mechanization, the number of accidents could be reduced by as much as 33%. Human error is blamed for a large number of mishaps. As a result, AVs should be able to prevent a significant proportion of these incidents, eliminating the vast majority of traffic delays.

There are three primary elements identified with AVs that effect blockage strongly and occasionally in the other direction: (i) reduced traffic delay due to a decrease in vehicle collisions; (ii) enhancing vehicle throughput; and (iii) changes in the total vehicle-kilometer-voyaged (VKT). A drop in vehicle accidents is projected to result in fewer deferrals and, as a result, greater dependability of the vehicle framework. The changes in VKT as a result of the AV method are yet unclear; however, some experts believe that VKT will increase in the long run (known as the "bounce back impact"). They speculate on a variety of factors, including increased VKT due to self-filling and self-stopping, increased usage of AVs by individuals unable to drive, an increased number of trips (both empty and involved), a shift away from public transportation, and longer drives. They speculate on a variety of factors, including increased VKT as a result of self-filling and self-stopping, increased usage of AVs by individuals unable to drive, an increased number of trips (both empty and involved), a shift away from public transportation, and longer drives. As a result, it is fully evident that AV innovation will soon favorably effect gridlock reduction until it arouses extra attention, which may add more weight to an already congested organization. The overall effect of the AV on congestion does not appear to be investigated at this time.

12.13 Conclusion

Vehicles are becoming an increasingly important source of computing and sensing assets for both drivers and metropolitan populations. In this work, we have gone over the core concepts of AVs, as well as its implications for the environment, the disabled, and the economy. Furthermore, one of the fundamental objectives of an AV is to detect an impediment and maintain a track of the lane and route. Various studies and researches that are currently available in the literature have been discussed in this chapter, with a focus on the obstacle detection and track detection capabilities of AVs. The approach and application of a certain technique have been categorized and summarized for a wide range of techniques described in the literature. This chapter clarifies and categorizes the various strategies for obstacle detection and tracking, as well as the research that has been conducted on these techniques. We have also briefly discussed the future opportunities and challenges presented by AVs in the real-world.

References

[1] Fagnant DJ, Kockelman K (2015) Preparing a nation for autonomous vehicles: opportunities, barriers and policy recommendations. *Transp Res Part A* 77:167–181.

[2] Guizzo E (2011) How google's self-driving car works. IEEE Spectrum Online, October 18. Available at https://spectrum.ieee.org/how-google-self-driving-car-works.

[3] Markoff J (2010) Google cars drive themselves, in Traffic. *New York Times 9.*

[4] ACEA (2015) The automobile industry pocket guide. *European Automobile Manufacturers Association.*

[5] Nieuwenhuijsen J (2015) Diffusion of Automated Vehicles: A Quantitative Method to Model the Diffusion of Automated Vehicles with System Dynamics. Delft University of Technology, TU Delft.

[6] Hong D, Kimmel S, Boehling R, et al. (2008) Development of a semi-autonomous vehicle operable by the visually-impaired. In: *IEEE International Conference on Multisensor Fusion and Integration for Intelligent Systems*, 2008. MFI, pp. 539–544.

[7] Anderson JM, Nidhi K, Stanley KD, Sorensen P, Samaras C, Oluwatola OA (2014) *Autonomous Vehicle Technology: A Guide for Policymakers.* Rand Corporation, Santa Monica, CA.

[8] Folsom TC (2011) Social ramifications of autonomous urban land vehicles. In: IEEE International Symposium on Technology and Society.

[9] Piao J, McDonald M (2008) Advanced driver assistance systems from autonomous to cooperative approach. *Transp Rev* 28:659–684.

[10] Manyika J, Chui M, Bughin J, Dobbs R, Bisson P, Marrs A (2013) Disruptive technologies: advances that will transform life, business, and the global economy. *McKinsey Global Institute New York, New York, NY..*

[11] Petit J, Shladover SE (2015) Potential cyberattacks on automated vehicles. *IEEE Trans Intell Transp Syst* 16:546–556.

[12] Knight W (2013) Driverless cars are further away than you think. *MIT Technology Review*. Available at: https://www.technologyreview.com/2013/10/22/175716/driverless-cars-are-further-away-than-you-think/.

[13] Litman T (2015) Autonomous vehicle implementation predictions. *Victoria Transport Policy Institute* 28.

[14] Maddox J, Sweatman P, Sayer J (2015) Intelligent vehicles? infrastructure to address transportation problems–a strategic approach. In: 24th international technical conference on the enhanced safety of vehicles (ESV). Available on: https://www-esv.nhtsa.dot.gov/Proceedings/24/isv7/main.htm.

[15] Schoettle B and Sivak M. A survey of public opinion about connected vehicles in the US, the UK, and Australia, 2014 International Conference on Connected Vehicles and Expo (ICCVE), 2014, 687–692, doi: 10.1109/ICCVE.2014.7297637.

[16] Kan Z, Qiang Z, Haojun Y, Long Z, Lu H, Chatzimisios P (2015) Reliable and efficient autonomous driving: the need for heterogeneous vehicular networks. *Commun Mag IEEE* 53:72–79

[17] Patriksson P (1994) The traffic assignment problem: models and methods, VSP BV, The Netherlands. *Facsimile reproduction published in 2014 by Dover Publications, Inc.*, Mineola.

[18] Sheffi Y (1985) *Urban Transportation Networks: Equilibrium Analysis with Mathematical Programming Methods*. Prentice-Hall Inc., Englewood Cliffs, NJ.

[19] Fenton RE, Mayhan RJ (1991) Automated highway studies at the ohio state university-an overview. *IEEE Trans Vehicular Technol* 40:100–113
[20] Ioannou P (2013) *Automated Highway Systems*. Springer Science & Business Media, New York, NY.
[21] Lantos B (2010) *Nonlinear Control of Vehicles and Robots*. Springer Science & Business Media, New York, NY.
[22] Blasch EP, Lakhotia A, Seetharaman G (2006) Unmanned vehicles come of age: the DARPA grand challenge. *Computer* 39:26–29.
[23] Behere S, Törngren M (2015) A functional architecture for autonomous driving. In *Proceedings of the First International Workshop on Automotive Software Architecture*. ACM, London, pp. 3–10.
[24] DiClemente J, Mogos S, Wang R (2014) Autonomous Car Policy Report. Available at https://cmu.edu/epp/people/faculty/course-reports/Autonomous%20Car%20Final%20Report.pdf.
[25] Siciliano B, Khatib O (2008) *Springer Handbook of Robotics*. Springer Science & Business Media, Berlin.
[26] Farhadi A, Endres I, Hoiem D, Forsyth D (2009) Describing objects by their attributes. In *IEEE Conference on Computer Vision and Pattern Recognition, 2009*. CVPR, New York, NY, pp. 1778–1785.
[27] Savasturk D, Froehlich B, Schneider N, Enzweiler M, Franke U (2015) A comparison study on vehicle detection in far infrared and regular images. In *IEEE 18th International Conference on Intelligent Transportation Systems (ITSC)*, IEEE, New York, NY, pp. 1595–1600.
[28] Iqbal A (2020) Obstacle detection and track detection in autonomous cars. In *Autonomous Vehicle and Smart Traffic*. IntechOpen, London.
[29] Al-Kaff A, Meng Q, Martin D, Escalera A, Armingol JM (2016) Monocular vision-based obstacle detection/avoidance for unmanned aerial vehicles. In *IEEE Intelligent Vehicles Symposium (IV)*, IEEE, Gothenburg, pp. 92–97.
[30] Wang X, Xu L, Sun H, Xin J, Zheng N (2014) Bionic vision inspired on-road obstacle detection and tracking using radar and visual information. In *17th International IEEE Conference on Intelligent Transportation Systems (ITSC)*. IEEE, Qindao, pp. 39–44.
[31] Al-Zaher TSA, Bayoumy AM, Sharaf AM, El-din YHH (2012) Lane tracking and obstacle avoidance for autonomous ground vehicles. In 9th France-Japan & 7th Europe-Asia Congress on Mechatronics (MECATRONICS)/13th International Workshop on Research and Education in Mechatronics (REM). IEEE, Paris, pp. 264–271.
[32] Xu Z, Zhuang Y, Chen H (2006) Obstacle detection and road following using laser scanner. In: *6th World Congress on Intelligent Control and Automation*. IEEE, Dalian, pp. 8630–8634.
[33] Wang X, Li H, Liu B (2016) Object tracking and state estimation in outdoor scenes based on 3D laser scanner. In *IEEE International Conference on Signal and Image Processing (ICSIP)*. IEEE, Beijing, pp. 607–611.

Chapter 13

Realistic vehicular mobility/channel models for vehicular communications for autonomous intelligent vehicles (AIVs)

Abstract

Among the characteristics of vehicular communication include a dynamic environment, high mobility, and the comparatively modest antenna heights of the communicating entities (vehicles and roadside units). Because of these properties, vehicle propagation and channel modeling are particularly difficult to accomplish. With an emphasis on the applicability of vehicle propagation and channel utilization models for evaluating protocols and applications, this chapter analyses vehicle propagation and channel utilization models. Understanding the transmission methods and execution methodology adopted by each model is the first step toward categorizing them properly. Models can be categorized in several different ways. One method is through the use of channel characteristics that have been implemented. Another technique is to assess how feasible certain options are (e.g., geographical data input). In addition to modeling unusual surroundings (such as tunnels, overpasses, and parking lots), other topics covered include novel forms of communication vehicles that have not previously been examined in vehicular channel modeling, etc. (e.g., scooters, public vehicles). The incorporation of information and communication technologies into road transportation systems has tremendous potential. The construction of V2V (vehicle-to-vehicle) communication lines is required to accomplish this vehicle to infrastructure (V2I). To facilitate this form of connection, the IEEE 802.11p and LTE-V (long-term evolution for vehicle communications) protocols have been proposed as two choices. It is the purpose of this chapter to study the physical and medium access control layers about the features of the communication channel between vehicles. It begins by discussing the most significant influences on V2V and V2I channels, as well as some of their most notable characteristics, before moving on to other topics. In addition to offering instances of roadway conditions, the literature also includes a description of the channel parameters, which is particularly useful. Illustrative following that, modeling methodologies are evaluated, and two of the most commonly used ones are described in greater detail.

Keywords: Vehicle-to-vehicle (V2V), intelligent transportation systems (ITSs), single carrier frequency division multiple access (SC-FDMA), wide sense stationarity (WSS), line-of-sight (LOS)

13.1 Introduction

Vehicular communication is influenced by a variety of vehicle combinations (V2V, V2I, V2P, and so on), as well as static and mobile objects (such as people) and other factors. Among the many advancements in wireless communication technology, vehicle channel modeling stands out for its originality [1]. When these characteristics are combined, you get difficult-to-model propagation conditions. When operating in a typical urban context, signal statistics on both the small and large scales fluctuate fast due to the dynamic environment, low antenna heights, and high levels of vehicle mobility. As we can see from the propagation characteristics, the signal interacts with the built-up environment as it travels from the transmitter to the receiver during its transmission. Because rays bounce off of buildings and other obstructions in cities, the total quantity of rays that arrive at the receiver increases significantly [2]. For autos, the cloud computing message confirmation authentication process shown in Figure 13.1 is depicted.

As a result of the high density of goods as well as the high mobility of communication vehicles and their surroundings, documenting the features of vehicular channels is far from a straightforward task. Despite this, many existing mobile channel models are not well-suited for vehicular systems, owing to the above-mentioned particular features of vehicle channels. The signal propagation path will be variable depending on how far apart the transmitter and reception antennas are located from one another [4]. Compared to cellular systems, the operation frequency and communication distance for vehicle communications are significantly

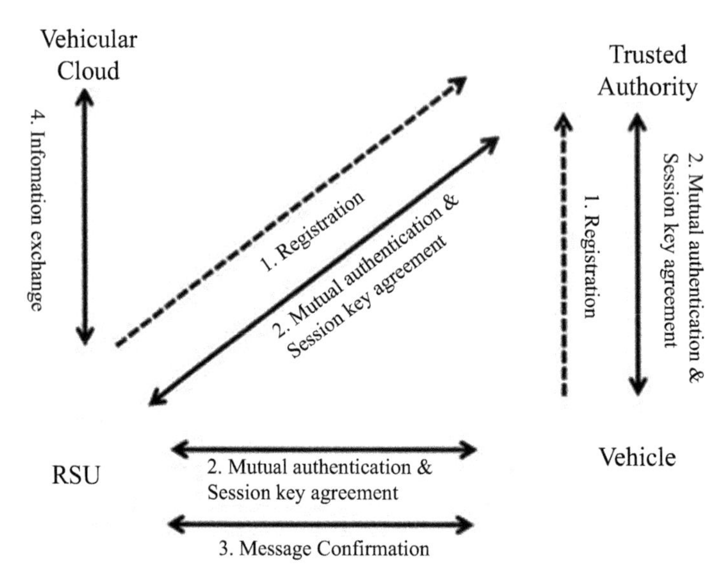

Figure 13.1 Authentication protocol for message confirmation in vehicular cloud computing [3]

different. A large number of studies on V2V channel models have been conducted, and the results are now publicly available. V2V channel measurement study in many situations, for example. This is a major difficulty in the field of V2V channels. The classification of V2V channels is done using the implementation technique, and the advantages and disadvantages of each strategy are explored. V2V and V2I propagation channels have a variety of properties, and the differences in these characteristics have a significant impact on the design of the vehicular wireless system under consideration. V2V channel measurements and model construction, including categorization of models according to implementation approach, are part of the job [5]. The process of creating V2V channel models, as well as the process of setting up a V2V measurement, are also described. The usability of vehicle channel model usability is, as a result, particularly crucial to our company's operations. It is the equivalent of questioning whether current models are capable of accurately simulating large-scale vehicular communication networks. An appropriate channel model must first be developed to adequately analyze vehicular protocols and applications before they can be put into practice. As a result, we provide suggestions for selecting a suitable channel model based on the protocol/application being assessed, the geographical information that is readily available, and the time constraints for simulation execution. On the right-hand side of Figure 13.2, you can see the current status of vehicular channel modeling, with specific emphasis being devoted to the model on display.

Intelligent transportation systems (ITS) are being developed as a result of the desire of authorities and vehicle manufacturers in applying information and

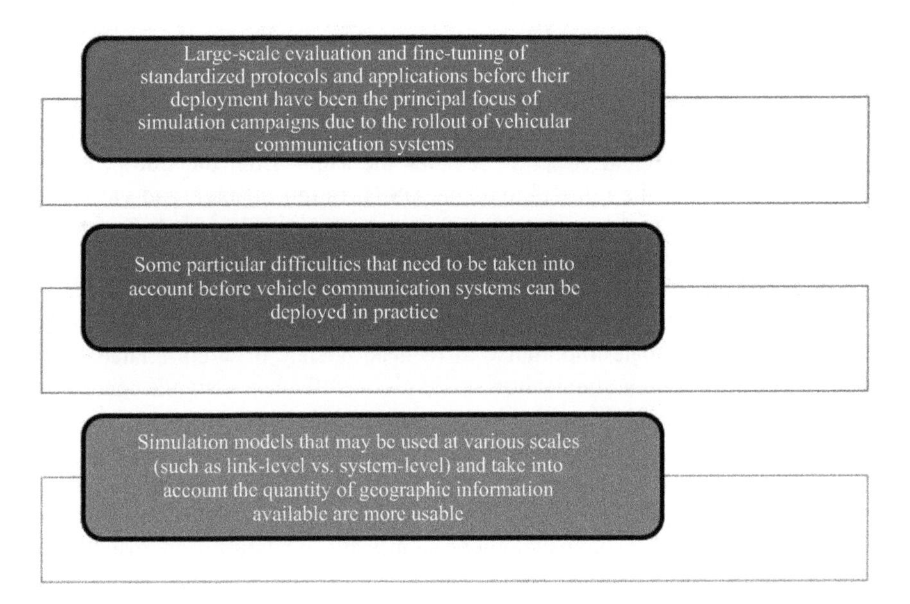

Figure 13.2 Vehicular channel modeling state of the art

communications technologies (ICTs) to this environment to reduce traffic accidents, CO2 emissions, and travel times (ITS). In-vehicle information systems (ITS) allow vehicles to communicate with one another, alerting them to safety-related occurrences and advising them on how to improve traffic flow, for example [6]. ITS according to data issued by the European Commission (EC) and the United States Department of Transportation (USDOT), information technology (IT) solutions that promote safety are particularly significant (DOT). Infotainment encompasses ITS applications such as map consulting and Internet access, which are grouped as a single idea. As a result of the rising number of vehicle sensors, the ITS application category "sensors on wheels" has developed. Data collected from vehicle sensors can be pooled and communicated to traffic control centers to aid in the management of traffic flow and congestion. The development of mobile ad hoc networks is required for the transmission of ITS data (VANETs). Virtual private networks (VANETs) allow you to communicate with other devices as long as you have an Internet connection (V2P). The term "vehicle-to-everything (V2X) communications" refers to systems that connect vehicles to other devices. To be employed for safety-related applications, V2X technology must be able to operate reliably in continuously changing surroundings while maintaining high transmission speeds and low latency [7].

The use of DSRC and cellular V2X have been proposed as potential solutions to this problem (C-V2X). The Department of Transportation (DoT) coined the term DSRC to denote a short- and medium-range communication system (SMRCS) that was specifically designed for ITS and assigned a frequency band. The IEEE 802.11p standard is used to regulate the physical and medium access control (MAC) layers of WAVE, which stands for wireless access in vehicular environments. DSRC controls the physical and medium access control (MAC) layers of WAVE. The consequence is the employment of OFDM and a MAC protocol that is based on contention. According to EU legislation, DSRC technology is referred to as ITS-G5. As with WAVE, there are minor variations between the more advanced levels, but the fundamental structure remains the same. Both the European and American systems will be referred to as the DSRC in this chapter, and vice versa [8]. The C-V2X technology was developed as part of the third Generation Partnership Project, which was launched in 2009 (3GPP).

V2I communication is carried out using cellular networks, but V2V communication is carried out over a dedicated link, which eliminates the requirement for cellular infrastructure. The importance of reduced latency in safety-related applications and car-to-car communication in locations without a network connection cannot be overstated, and this technology is essential. It is necessary to use single carrier frequency division multiple access (SC-FDMA) for data transmission uplink when transmitting data over long distances. With the fifth generation of cellular systems, more services, such as C-V2X communications, will be made available to users (5G). DSRC systems have undergone significant standardization, installation, and testing in the 10 years since they were first introduced [9]. DSRC devices are currently available for purchase on the business market. If C-V2X is compared to other wireless technologies, it has various advantages, including the ability to better match the characteristics of the channel, larger bandwidth, and the availability of an existing infrastructure (which in the case of the DSRC should be likely defrayed by public administrations). Even though commercial devices are still in the early stages of research, this technology is the most potential medium- and long-term technology due to the current expansion of cellular networks in the direction of 5G technology. The ability to communicate quickly and adapt to changes in the environment are two of the most significant qualities of V2X communications (except in cellular V2I transmissions). As previously mentioned, the mobility of communication endpoints and primary scatterers exacerbates the first of these characteristics, which is shared by all mobile cellular radio channels. Consequently, as a result of the low antenna height, the majority of propagation occurs in the horizontal plane, increasing any impacts caused by the surroundings of the transmitter and receiver. Due to the rapid change in environmental conditions, the broad sense stationarity (WSS) assumption is no longer valid. A common occurrence is the existence of correlations between path parameters, which goes against the traditional belief regarding uncorrelated scatterers (US). V2X channels have several properties, and the most important factors that influence them are discussed in this chapter [10].

The rest of the chapter is divided into 10 sections where Section 13.2 describes related work /literature of recent research works by different authors whereas Section 13.3 describes specific considerations for vehicle channels. Section 13.4 describes the classification and description of the vehicular channel model, Section 13.5 describes realistic and efficient vehicular channel modeling, and V2V channel models are described in Section 13.6. Modeling of wireless links described in Section 13.7 and 13.8 describe the modeling of wireless signal propagation. Section 13.9 describes vehicular wireless technologies and Section 13.10 describe decentralized congestion control technique for AIV and at last Section 13.11 describe the summary of the chapter.

13.2 Related/background work

It is possible to gain a better understanding of car travel on the road, interactions between vehicles (such as altering speed based on traffic circumstances), and enforcement of road restrictions through the use of mobility models, which are capable of precisely that [11]. Using networking models that contain routing protocol details, as well as medium contention and upper-layer protocols, it is possible to produce a more realistic data transport. One of the most important components of a successful VANET protocol and application evaluation is the precise modeling of the signal propagation environment. The physical features of the (simulated) received signal have a direct impact on upper-layer protocols such as medium access, routing, and transport, as well as application-relevant elements (e.g., throughput, message delay, etc.). The inclusion of static and mobile objects (such as buildings and plants) around communication vehicles is required to effectively depict VANET channels (other vehicles on the road). A channel model's scale can vary depending on how much information it contains [12]. Small, medium, and large channel models are possible. It is also possible to use both deterministic and stochastic models, depending on how the data is calculated. Geometrical information relevant to a site is used to categorize models as geometrical or non-geometrical, depending on their shape.

Kiran Jot *et al.* (2021): Robots are already being employed in a wide variety of industries and professions as a result of recent advancements in science and technological research. This necessitates increased human–robot connection, as historical experience has demonstrated that robots are regarded as friends rather than machines or tools by the general public. Human–computer interaction (HCI) research can assist us in gaining a better understanding of and utilizing computer-based technologies in our daily lives. Interaction between humans and computers human–robot interaction (HRI) is distinct from other disciplines of research in robotics because HRI builds on robot HCI techniques by incorporating autonomy, intimacy, and decision-making capabilities. Standardization and the development of standards are required due to the sheer volume of human–robot interactions. This will ensure that robotic technology is used in an appropriate and principled manner [13].

Carou *et al.* (2021): The "Industry 4.0" movement, which is a collection of emerging technologies, has the potential to drastically transform industrial operations. The concept and relevance of these new enabling technologies have been extensively debated and written about in the past several years. To accomplish this, the authors of this book researched additive manufacturing, autonomous and collaborative robotics, big data and artificial intelligence, cloud computing and the Internet of Things, as well as augmented reality and virtual reality simulations, among other topics. Therefore, the chapter's purpose is to present these technologies while also highlighting some notable examples from academia and industry to help readers understand them better [14].

Mounir *et al.* (2021): When big public meetings such as strikes, protests, parades, or other huge public gatherings take place in densely populated areas, a wide range of security concerns develop. Security forces are frequently sought at events to ensure the safety of all people in attendance. As a result of having restricted access to modern technologies, security employees may soon find themselves overworked. Fortunately, intelligent monitoring systems are being embraced and deployed by smart cities all over the world, which is excellent news. When working with big groups of people, complex crowd analysis techniques, as well as intelligent monitoring systems, are required as well. This overview examines several different crowd analysis research. Crowd analysis is commonly divided into two categories: statistics and crowd behavior analysis. It is possible to characterize the patterns of mobility and activities observed in a crowded scene if the level of service (LoS) for that scene is known. Crowd behavior analysis can be used to determine the LoS. Finding abnormalities in huge groups of people is a popular topic right now. As long as there is not a clear understanding of what constitutes an abnormality, anything relating to crowd analysis can be considered unusual. The focus of our review is on crowd analysis subfields that have been under-explored or have been addressed infrequently [15].

Austin *et al.* (2021): Many people now use wearable devices regularly to track their daily steps, calories burnt, levels of exercise, heart rate, and sleep patterns, among other things. In recent years, the use of wearable technology has increased considerably. This technology is still in its early stages of development, both in terms of research and commercialization. To put it another way, according to the Pew Research Center, the number of people who routinely use a wristwatch or activity tracker has increased significantly in recent years. ECG and PPG sensors, for example, have been more widely available in recent years as a result of the price reductions that have occurred in these devices. In the history of software development, there has never been a better opportunity to create software that will be widely utilized than today. Physiologically adaptive systems, which are still in the early stages of development, are becoming increasingly popular in research on human–robot and HCI. As part of this project, we intend to develop a framework for developing physiologically aware scenarios, as well as software and tools to make the process more efficient [16].

Emmanouil *et al.* (2021): More than just what museums provide on their websites and in publicly accessible databases, we need to figure out how to make the vast collection of exhibitions more accessible and explorable. This is crucial now, more than ever, given the current political climate. It is proposed in this project to develop the Invisible Museum, a user-centric platform that enables users to create interactive and immersive virtual 3D/VR displays utilizing a single collaborative authoring environment for 3D and VR content. Incorporating feedback from museum curators and end-users, the platform was created utilizing a human-centered design approach. It was put into production. The Europeana Data Model and the CIDOC-CRM standards for content representation both provide ontology bindings for text data in their respective formats. Deep learning technologies that are state-of-the-art are being used to assist curators in their efforts. It is possible to create and depict narratives on the platform, which can then be used to guide storytelling experiences and relate artifacts to their socio-historical environments. This is quite crucial. Important contributions are made to the disciplines of dynamic virtual exhibitions generated by users, personalized suggestions and tours, 3D/VR visualization in web-based technologies, and immersive navigation and interaction. The following disciplines benefit from the contributions: Every one of them is critical [17].

Kishan *et al.* (2021): When it comes to creating interactive hardware and software projects, Arduino, an open-source electronics platform, has emerged as the go-to solution. Arduino code written in the development environment can be used to control plugins such as inputs, sensors, lights, and displays, among other things. It is a good idea to begin using Arduino prototypes as a starting point. Due to the increasing popularity of the Arduino platform, the practice of Arduino prototyping is becoming increasingly widespread. When it comes to Arduino prototyping, those who are not programmers but have an interest in the subject will find it challenging. As the public's interest in this topic develops, the demand for publicly available knowledge will rise as well. When it comes to Arduino prototyping, this chapter presents a thorough study and comparison of the available literature. It is estimated that over 130 comparable studies have been carried out in the last 15 years, including the current year (2015–20), with all of them having been peer-reviewed and published. There was a three-step process that took a significant amount of time and effort to choose these studies [18].

Gilmore *et al.* (2021): This discrepancy between the rate at which digitalization is required in control rooms in socio-technical sectors such as the railway industry and the rate at which digitization is required is now being exploited. Techniques for designing user-oriented control rooms benefit considerably from quantitative human factors research procedures, such as situation awareness studies, which have been demonstrated to be vital. In the first section, which is based on literature analysis, the author investigates how different aspects of control room design interact with one another and demonstrates how these interactions can be modified to allow for a more user-friendly control room layout. Also included is a case study to see whether or not incorporating quantitative human factors methodologies into control-room design has any positive effects on the design. The case study will demonstrate the implementation of a limited number of quantitative human factors research methods in a practical setting. During the simulation with an existing and novel interface for train traffic controllers, quantitative data was collected to assess various metrics through eye-tracking, interval pop-up queries, and a series of self-report questionnaires based on an evaluation of various quantitative situation awareness measurement techniques. Comparing eye-tracking measures to self-report and probing question measures, it appears that eye-tracking measures are more accurate in capturing situation awareness. A successful quantitative human factors research approach is demonstrated in control room design, and it is recommended that this approach be used more frequently. Based on the findings of this scenario, it is advised that future research investigate the relationships between quantitative methodologies such as eye-tracking measurements and probing query measures that evaluate dynamic situation awareness while investigating control room design [19].

Muhibul Haque *et al.* (2021): When it comes to data storage and transfer efficiency, Spintronic (also known as spin electronics or spin transport electronics) takes advantage of the two fundamental degrees of freedom that electrons possess—spin-state and charge-state—to improve the device's operational speed. As a result, it looks to be a viable technique for dealing with the vast majority of the issues that traditional electron-based systems are currently experiencing. To integrate semiconductor microelectronics with magnetic components, this unique method makes use of a hybrid semiconductor–magnetism integrated circuit to achieve high integration efficiency (MIC). Semiconductor microelectronic devices used to process only the electronic charge of binary numbers in the past, but this has changed (0 and 1). Transistor sizes can only be lowered to a few nanometers in size at this point, and those will be approximately 1 nm. A consequence of the quantum effect is that the creation of gadgets and other operating procedures have become more challenging. As a result, alternative future technologies, such as spintronics, will be required for the transportation and storage of information in the future. This article is the first to take a comprehensive look at the field of spintronics in one place. The topics of fundamentals, materials, applications, and barriers in current research paths are all covered. For the sake of getting started, let us go over some of the most essential concepts in the spintronics field. Following that, we will go into the different types, characteristics, and other aspects of spintronic materials. Following that, the fundamental principles of building and operating spintronic devices and circuits are thoroughly discussed. This highlights the multiple uses, as well as the advantages and disadvantages of utilizing spintronic technology, among other things. Finally, but certainly not least, the current hurdles, future research areas, and prospects for spintronic technology are discussed. The charge and spin of electrons serve as the foundation for this new paradigm in electrical and magnetic devices. In the search for novel materials, modern engineering and technological advancements have given us reason to believe that this will be an extraordinarily optimistic technology in the future [20].

Kenneth Sebastián *et al.* (2021) refer to making required and timely adjustments to one's situation or one's behavior to "unfuck." A greater time has never existed to analyze and critique criminal justice systems with colonial origins than now. The future must be considered to better understand interpersonal harm, racial social control, and government violence. To do so, we must acknowledge the overlapping theoretical boundaries of critical criminology and Latino studies that have arisen from colonial legacies. By identifying particular areas where researchers can intervene to strengthen criminological inquiry, emerging Latino criminology strives to center the periphery while simultaneously moving away from the field's multiple colonial and white supremacist traditions. It is consistent with both classic and innovative methods to the study of crime and criminal justice scholarship, as well as with other approaches to criminal justice scholarship [21].

Sergey *et al.* (2021): Holographic subsurface radar (HSR) is not being used too much right now, which is unfortunate. The ground-penetrating radar (GPR) community has seen significant attenuation of electromagnetic waves in most materials over time, which they attribute to a historical perspective. Consequently, time-varying gain on the continuous wave (CW) HSR signal cannot be applied, and hence a suitable effective penetration depth on the CW HSR signal cannot be obtained. HSR has several advantages over ISR due to the way it employs CW, including the potential to amplify later (i.e., deeper) arrivals in a lossy medium. The most significant advantage of this technology is the ability to perform shallow subsurface imaging at a resolution that is not possible with ISR. HSR system design is simplified in comparison to ISR system design due to the employment of low-tech transmission and receiving antennas in the HSR system. In this study, which also covers the fundamentals of radar hologram reconstruction theory, an optical analogy is employed to investigate radar hologram reconstruction strategies. In this article, we will discuss the history, technologies, and applications of the "RASCAN" holographic subsurface radar type, which is a sort of holographic subsurface radar. Humanitarian demining, construction inspection, surveys of historic architecture and artworks, non-destructive testing of dielectric aerospace materials, security applications, paleontology, detection of damage caused by wood-boring insects, and other applications are among the subsurface imaging and remote sensing applications being considered [22].

13.3 Specific considerations for vehicular channels

13.3.1 Environments

Radio propagation is highly influenced by the environment in which it takes place. When it comes to the transmission of vehicle communications, a wide variety of circumstances come into play. Structures such as buildings, automobiles (both stationary and mobile), and various forms of plants are examples of this. Radiation propagation is significantly affected by the amount, size, and density of many item kinds in the environment [22]. Sorting out different types of items is simple; however, describing the environment that they generate is a completely different store. Generally speaking, there are three types of environments where car communication takes place: highways, suburban areas, and rural regions. In addition to the speed and density of automobile traffic, other factors such as the existence and location of roadside objects have a substantial impact on signal propagation as well. Open spaces such as motorways and communities with low-rise structures are common in metropolitan areas, and they may be developed similarly to a typical suburban setting in the United States. As a result, you should be wary of any environmental classification you come across online [23]. In the case of propagation models that are designed for a single environment, there is no guarantee that they would function correctly in another environment of the same "class." Therefore. When it comes to propagation models, the most appropriate way is to employ precise measurements of things in their surroundings to develop them. Figure 13.3 describes specific considerations for vehicular channels.

13.3.2 Link types

It is necessary to distinguish between different connection types since their propagation characteristics differ significantly from one another and the propagation environment. By comparison, a permanent base station (or access point) that can be raised if necessary is used for V2I channels, rather than the transmitter and reception antennas being mounted on top of the vehicle as in V2V channels. V2P communication networks will aid in the development of traffic safety software for vulnerable road users. Reflected waves behave differently from diffracted waves and dispersed waves because the transmitter and receiving antennas move, cast shadows, and differ in height from one another [24].

13.3.3 Vehicle types

There are significant differences in the size and motion dynamics of various types of vehicles, such as personal automobiles and pickup trucks, business vans and tractors, lorries and scooters, and public vehicles. As a result, vehicle-specific models cannot easily be converted to function on other vehicles of the same make and model year. Even if a vehicle is not serving as a transmitter or receiver, the characteristics of the vehicle can have an impact on propagation modeling. The presence of large trucks can interfere with the transmission line of sight (LOS)

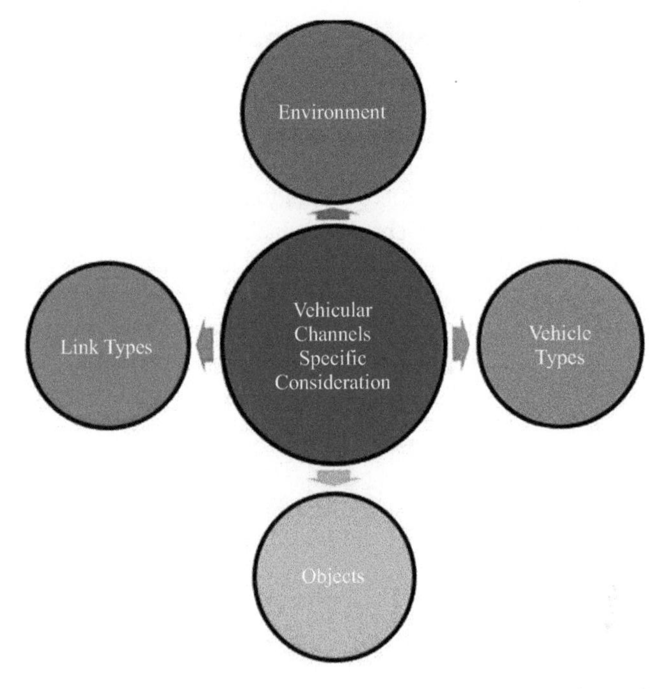

Figure 13.3 Specific considerations for vehicular channels

between the transmitter and the receiver, resulting in an additional 20 dB of attenuation. Personal autos contribute an additional 10 decibels of attenuation [25].

13.3.4 Objects

In-vehicle propagation conditions, regardless of the link type, a plethora of elements might have an impact on the signal propagation path. The degree to which an object has an impact is determined by the type of object, the type of link, and the surrounding environment. Because the majority of communication between transmitting and receiving vehicles on highways takes place on the road surface, automobiles on the road are required when simulating vehicular channels in highway environments, and they must be present in the simulation. However, in cities with a two-dimensional architecture, cars talking with one another tend to be distributed throughout the metropolis. Because both mobile and static objects can be sources of reflected light and diffraction aberrations when modeling vehicle channels, it is necessary to consider these factors while simulating vehicle channels. When it comes to modeling V2V propagation channels, several measurement efforts have demonstrated the importance of the loss of signal (LOS) scenario. Independent of propagation conditions, there are considerable RMS delay spread discrepancies between non-LOS and LOS channels (e.g., highway or urban settings). As the number of reflections and diffractions in a signal grows, so does the likelihood of multipath effects and signal attenuation [26].

13.4 Classification and description of vehicular channel models

This section contains in-depth information on vehicle propagation and channel modeling, as well as some current changes to the subject matter covered. The ability of a model to realistically reflect a variety of situations is a significant factor in determining whether or not it should be included in this category [27]. The models in this part have been evaluated on real-world data, and they are all available for download. We categorize the models according to the propagation mechanism, the implementation approach, and the channel characteristics that were used in their implementation. The classification of vehicle channel models is illustrated in Figure 13.4.

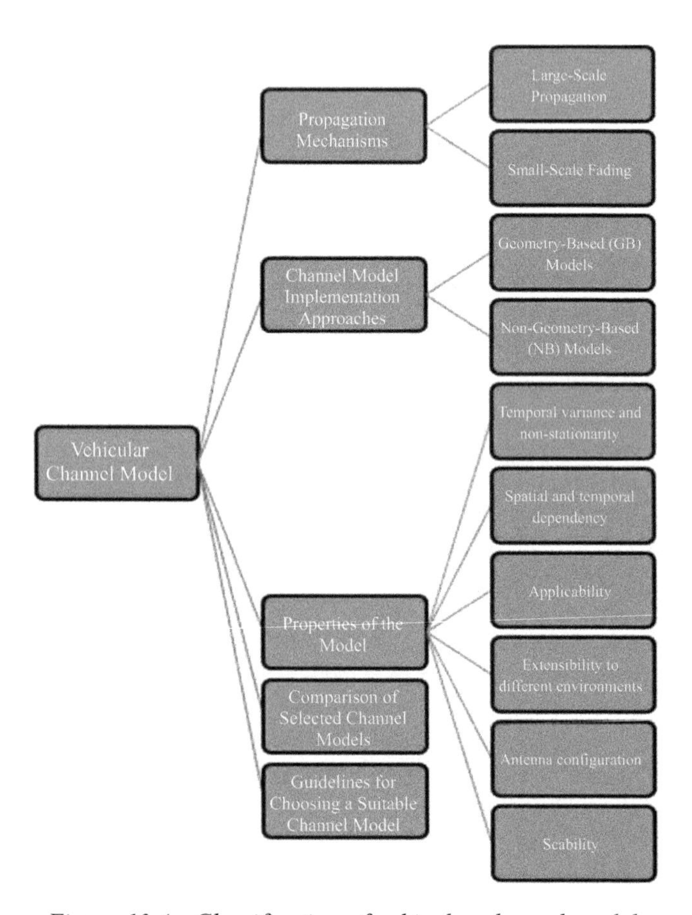

Figure 13.4 Classification of vehicular channel models

13.4.1 Propagation mechanisms

Additionally, vehicle channels may face severe Doppler shift and frequency-selective fading owing to moving and stationary objects in addition to path loss that varies with a place (e.g., different settings) and time (e.g., different times of day). Because it is impossible to model all of these different characteristics at the same time, piecemeal modeling has been the most often used technique for a long time [28].

13.4.1.1 Large-scale propagation

It is the log-distance path loss model that is most often employed for large-scale vehicular channel propagation, and it is empirical data that are utilized to estimate the route loss exponent from this model. In conjunction with suburban channel data, the dual-slope log-distance model was used to calculate the distance between two points. It is the same concept in a wide range of contexts, such as highways, rural, and highway scenarios, urban intersection scenarios, and parking garage scenarios. There are a variety of additional large-scale models that are employed. When it comes to LOS and non-LOS conditions, the geometry-based efficient propagation model for V2V Communication (GEMV^2) employs path loss models (such as two-ray ground reflection model and long-distance path loss), while large-scale propagation effects are addressed through the use of ray-tracing techniques [29]. The classification of vehicle channel models is illustrated in Figure 13.4.

13.4.1.2 Small-scale fading

Multipath propagation and Doppler effects from vehicle and object movement are responsible for small-scale signal modifications in addition to large-scale propagation. Weibull, Nakagami, and Gaussian distributions are used to represent small-scale fading; the parameters for these distributions are determined from measurement data, like that used in large-scale propagation models. In part, this is because the standard deviation of the Gaussian distribution used to explain small-scale fading varies depending on the number of cars and other items present in a given region. In a variety of situations, ray-tracing techniques have been employed to estimate small-scale fading. It is worth noting that empirical measurements have demonstrated that the presence of a LOS path has a significant impact on the propagation characteristics of vehicle communication. For this reason, large- and small-scale propagation qualities for LOS and non-LOS networks are typically examined in distinct studies. Recently, models with properties that allow them to distinguish between different types of links have been developed, and this is a significant step forward [30].

13.4.2 Channel model implementation approaches

Models can be classified into the following categories depending on how they are implemented and whether or not geographic information is available [31].

13.4.2.1 Geometry-based (GB) models:

Here, we will discuss few geometry based models as:

- *Ray-tracing models*—When it comes to vehicular channel modeling, geometry-based deterministic (GBD) models are the most widely used type of model. An in-depth description of the propagation environment must be provided to reliably compute channel statistics using ray-tracing methods, which are currently unavailable. Data interpolation and interface modeling using ray tracing, which is more scalable, are two applications (RADII). The average attenuation for each ROI may therefore be computed using RADII's preprocessing ray-tracing techniques, and the attenuation between ROIs can be computed using an off-line interpolation technique. By employing a lookup table instead of channel statistics in the simulations, it is no longer necessary to reproduce the channel statistics in the simulations.

- *Simplified geometry-based models*—When it comes to vehicular channel modeling, GBD models are the most widely used type of model. An in-depth description of the propagation environment must be provided to reliably compute channel statistics using ray-tracing methods, which are currently unavailable. Data interpolation and interface modeling using ray tracing, which is more scalable, are two applications RAy-tracing Data Interpolation and Interfacing (RADII). The average attenuation for each ROI may therefore be computed using RADII's preprocessing ray-tracing techniques, and the attenuation between Region of Interests (ROIs) can be computed using an off-line interpolation technique. By employing a lookup table instead of channel statistics in the simulations, it is no longer necessary to reproduce the channel statistics in the simulations.

When designing this model, considerations for long-range (LOS), moving object discrete components, static object discrete components, and diffuse dispersion are all taken into consideration. The parameters of the model were developed using data collected on highways and in suburban areas. To distinguish between links that are in the LOS and links that are not, it is necessary to use a LOS indicator (using this model). To foresee changes in conditions, it is possible to estimate transition probabilities based on the distributions of loss and non-loss components recorded in various situations. Specifically, the one that adds essential information about road crossings into the channel modeling process the data that was utilized to construct the model comes from measurements taken by the study's authors at various junctions.

13.4.2.2 Non-geometry-based models

All natural-language models (NG models) follow a similar recipe: they analyze the channel characteristics in a given environment and, as a result of that analysis, make adjustments to the path loss, shadowing, and small-scale fading parameters. On occasion, the NG model's delay line is used to tap into a power source. This model contains a large number of taps, each with a different delay and Doppler spectral type, to replicate signals received from many propagation pathways at the same time. Based on an extensive measuring campaign conducted in urban, suburban, and highway settings with two different degrees of traffic density, a Time-domain linear (TDL) model was developed for each of the three types of settings (high and low). The persistence process, which accounts for the finite "lifetime" of the propagation channels, is included in the model to characterize the nonstationary features of the system. It takes the use of the Markov chain approach to do this. To accommodate for an unexpected rise in the number of LOS components, the model should include birth/death operations as well as death/birth operations.

13.4.3 Properties of the model

Given that the primary focus of this chapter is on the model's usefulness for protocol and application evaluations, we have included the most important characteristics of the model in this section [32].

13.4.3.1 Spatial and temporal dependency

While small-scale fading models take into consideration propagation parameters (such as reflections and scattering) for signal attenuation that varies over time and space, statistics show that the two variables are inextricably linked. The static and dynamic characteristics of the physical cosmos are responsible for this spatial and temporal dependency. This results in multiple communication lines being affected at the same time by environmental factors such as obstructing objects, ambient noise/interference, and others. Because of geographical correlation, these connections are comparable. However, as a result of vehicle movement and fluctuating traffic intensity, signals get weaker with time (i.e., temporal dependency). When building a channel model, it is possible to take into account the effects of space and time.

13.4.3.2 Temporal variance and non-stationarity

The considerable non-stationarity of vehicle channels has been discovered through measurements. As a result, if the channel broadcasts fast-moving automobiles, the channel statistics may also fluctuate accordingly. It is possible for objects in the model that are both static and moving to cause the LOS component to appear and disappear at random. The capacity of vehicle channel models to generate non-stationarity was a deciding factor in the adoption of these models.

13.4.3.3 Extensibility to different environments

We distinguish between calibrated channel models and those that can represent effects beyond those acquired at specific locations in terms of their applicability elsewhere by extracting relevant characteristics from measurements taken at a certain number of sites. The fact that these models are based on measurements means that they are not reliable in places with a diverse variety of unique characteristics. When compared to models that rely purely on measurements, models that incorporate geometry-specific information about the simulated region can shed insight on a variety of other aspects of the simulation.

13.4.3.4 Applicability

The primary goal of vehicular channel models is to evaluate their ability to take into account application-specific situations to promote the realistic development of vehicular and ITS-related applications in the field of transportation engineering and technology. Instead of examining broad highway scenarios, we constructed channel models for a variety of highway applications, such as merging lane conditions, traffic congestion situations, and driving through traffic jams, rather than wide highway scenarios. Identifying feasible applications based on propagation circumstances can be accomplished in some cases; however, channel characterization is required in others (e.g., pre-crash and post-crash warning). It is possible to apply channel models to a variety of use-cases in addition to the ones for which they were originally established.

13.4.3.5 Antenna configuration

If small-scale fading is to be integrated, channel models must be flexible enough to accommodate alternate antenna layouts that take advantage of and mitigate the negative impacts of small-scale fading. We updated the model's documentation to include information on the several antenna combinations it supports (e.g., multiple-input multiple-output antenna configuration).

13.4.3.6 Scalability

Models are classified based on their efficacies and channel characteristics, among other factors. The capacity of a model to scale is influenced by its efficiency. As the requirement for speedy evaluations of automotive applications grows, channel models must be capable of enabling large-scale simulations. This is especially true for embedded systems. The complexity of the approaches used to construct channel statistics has a substantial impact on the model's ability to predict channel statistics. Although ray-tracing models are more accurate, they are also more constrained in terms of scalability than other models. Based on their qualitative scalability, the channel models are evaluated.

13.4.4 Comparison of selected channel models

The log-normal shadow fading model with log-distance path loss was one of the models considered. The received power values recorded during the measurement campaign in Porto are compared. The path loss parameters of the log-distance model are designed in such a way that they approximate as nearly as possible the values derived from the measurement data. The location of the other vehicles is unknown at this moment, and we have no way of knowing. The following clarification is provided: as a result, the GEMV2 model is unable to use their placements; instead, simulated locations are being used, which reduces the accuracy of non-LOS link estimation. The most notable finding of the study is that when measurements for a specific environment are not available, NG models produce results that are discordant with one another. If the route loss exponent is greater than 100 m, for example, the dual-slope Cheng model is inappropriate for such a long route. As demonstrated by the log-distance model, when the parameters of an estimate are obtained from measurements conducted at a specific site, the estimate is more accurate. In the case of people who do not have access to geographic information, GBD models, such as the GEMV2 model, are a preferable option [33].

13.4.5 Guidelines for choosing a suitable channel model

Because they do not know the environment, stochastic models have the advantages of simplicity and scalability, but their accuracy suffers as a result. In GB models, the trade-off between scalability and accuracy is a trade-off that varies substantially amongst models. To function properly, ray-tracing models require improved computing power as well as an understanding of the propagation medium. Given its large capacity and ability to deliver channel information that is specific to a certain region, it is not dependent on the user's location. A favorable accuracy/scalability tradeoff can be achieved by using simplified GB models that consider real-world item placements. Compared to standard ray-tracing models, models like these are substantially more scalable while still maintaining the same level of accuracy and usability. For a model to be applicable for a certain application or protocol evaluation, the processing power of the model and the availability of relevant data must be considered (either geographical or measurements). It is feasible to use a variety of models for system-wide performance evaluation purely based on the data collected (e.g., total packet delivery ratio, average end-to-end delay, etc.). The use of a dynamic link transition and small-scale variation model should be considered if an application demands network topology statistics (such as how many adjacent vehicles there are) or data that are dependent on location (e.g., the packet delivery rate or end-to-end delay in an area with rapid channel fluctuations). Applications that require the transmission of time-sensitive information regarding a specific incident, such as safety-critical applications, benefit tremendously from models based on GBD [34].

13.5 Toward realistic and efficient vehicular channel modeling

This section examines the current trends in vehicular channel modeling, as well as the need for models that can be used in large-scale simulations of the vehicular network, as well as the limitations of present models. Simulating vehicle channels provides a realistic protocol and application evaluation alternative, which we examine in detail. Finally, we discuss a variety of unsolved difficulties in propagation and channel simulation that have yet to be addressed. This may be seen in Figure 13.5, where the channel modeling has been included.

13.5.1 Efficient models for realistic large-scale simulation

To efficiently evaluate applications before they are deployed in the real world, realistic channel models for large-scale simulations are becoming increasingly important as the deployment phase of ITS systems in major markets approaches. Virtual-to-virtual (V2V) and virtual-to-internet (V2I) communication line simulators, such as the NS-35, use simple statistical models (such as free space and log-distance path loss) that are applied uniformly to all communication settings, no matter how unique they are. So, vehicle channels such as quick changes in latency and Doppler spreads, among other important metrics, are not well represented, nor are other key parameters taken into consideration [36]. When it comes to link-level modeling, simple models performed badly, especially in more complicated models, according to the results. Using geometrical propagation models that can distinguish

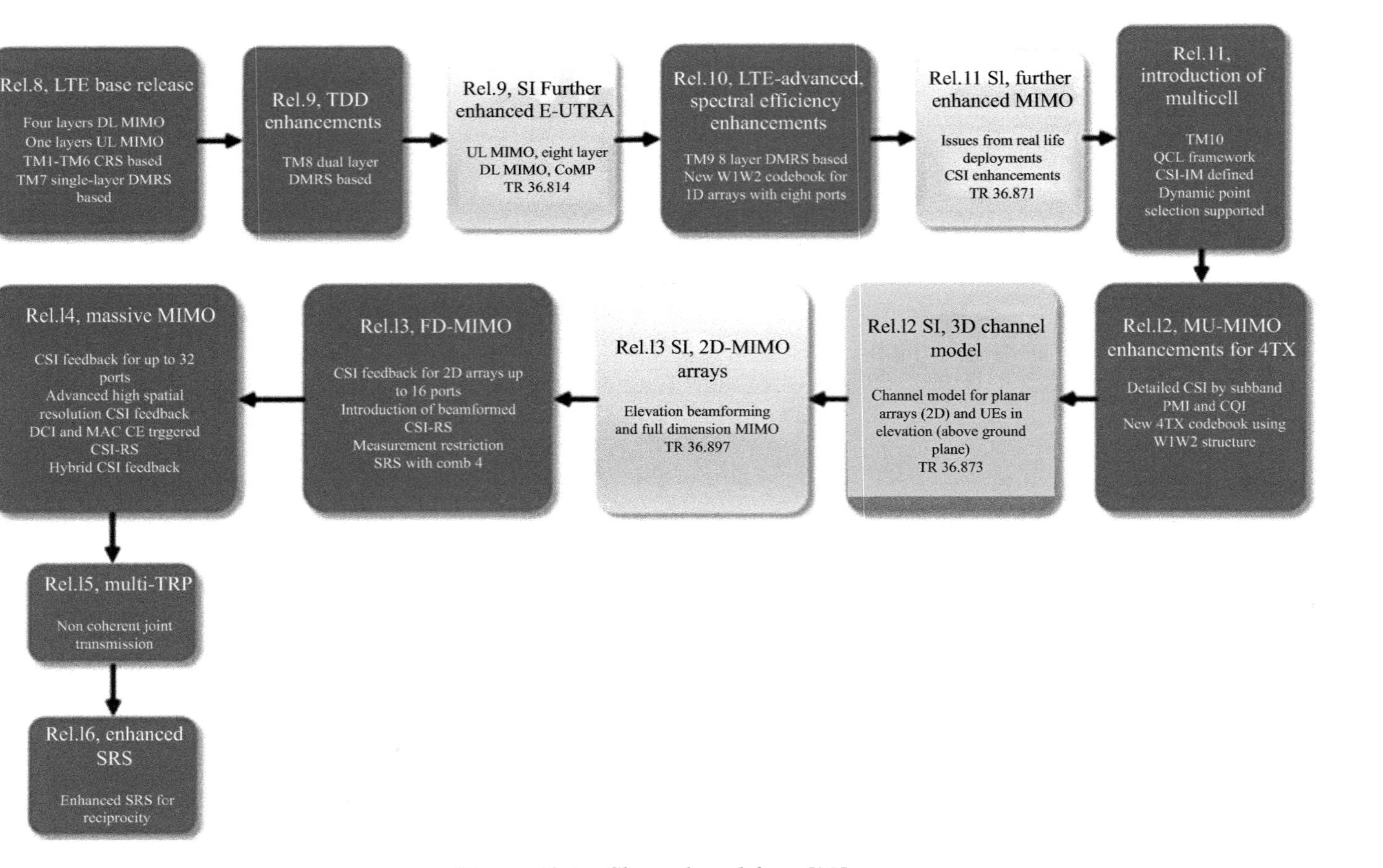

Figure 13.5 Channel modeling [35]

between different LOS conditions and environments, together with small-scale channel models that can provide enough delay and Doppler data for each representative environment, should be the initial step. This is the correct course of action to adopt moving ahead. The incorporation of realistic protocol and application evaluations into large-scale network simulators should be a high goal. The GEMV2 channel model, which can be used to city-wide vehicular networks with thousands of communication vehicles, can be used to simulate signal propagation in several scenarios (such as highways, rural areas, cities, and difficult crossings). There are no compatibility issues with SUMO data, not with building and foliage outlines, nor with Open Street Map (OSM) data [37]. Furthermore, it investigates variables linked to networking such as delivery rates, effective transmission range (effective transmission range), and neighborhood size (neighborhood size) in addition to propagation data in the GEMV2 model.

13.5.2 Vehicular channel emulations

Conducting testing in realistic situations and using realistic equipment to determine the features of wireless vehicular channels is the most realistic technique of determining these qualities. This method cannot be employed for large-scale studies due to the high expense and low reproducibility of the procedure (e.g., involving tens or hundreds of vehicles). Channel simulation, on the other hand, can ensure high levels of repeatability, configuration, and management. Simulation of a channel Simulating a genuine television channel is difficult because it necessitates the creation of a highly complex simulation environment that takes into account all components of the real system and makes certain assumptions about how things work in reality [38]. When genuine communication systems are combined with parts of simulation environments, a realistic testbed is created that maintains replication and configuration while providing a realistic testbed for researchers. Channel emulation occupies the space between channel research and simulation. For example, the Carnegie Mellon University (CMU) Wireless Emulator uses genuine devices to create signals, which are then used to simulate accurate signal propagation models before being returned to the real devices for further processing. Using real-world measurements, it was demonstrated that the emulator worked realistically. It was decided to develop and implement a vehicle channel emulation for the NS-3's small-scale statistics model for vehicular channels. In terms of frame reception rates, the findings obtained from the simulator and the CMU Wireless Emulator were found to be extremely similar to one another.

13.5.3 Open research issues

13.5.3.1 Channel models for different vehicle types

When it comes to vehicle channel measurements and modeling, personal automobiles have received the most attention. Even though the dimensions and road characteristics of various vehicle types (such as commercial vans and lorries, as well as scooters and public vehicles) vary greatly, studies on these vehicles are rare. The mobility provided by automobiles differs significantly from that provided by scooters and motorbikes [39]. In part because of their smaller size and lack of a roof on which to put antennas, scooters' mobility suggests that their propagation parameters may differ significantly from those of automobiles. In recent studies, it has been discovered that commercial trucks and vans have channel propagation parameters that differ from those of personal cars. In the end, we had varied communication ranges and packet error rates to work with. To better understand the channel characteristics of vehicles other than personal automobiles, more research will be required in the future.

13.5.3.2 Under-explored environments

However, even though vehicle communications can take place everywhere, the vast majority of signal propagation tests are carried out in common sites such as motorways and tunnels, along with parking garages and roundabouts. The transmission of V2V signals across parking garages, tunnels, and bridges has been the subject of investigation in recent years. Even in contexts with a diverse range of application use cases, more measurements and modeling studies are required.

13.5.3.3 Vehicle-to-X channels

The differences between V2I and V2P communications are so significant that there has been no investigation into the propagation characteristics of other types of channels than V2V transmissions. V2I measurement projects, as well as dedicated V2I channel models, are both in short supply these days. The Doppler-delay characteristics of V2I channels can be captured using a time-domain linear model (TDL). In urban, suburban, and highway areas, data-driven models are employed to make decisions. V2I systems, which consist of a fixed base station and mobile equipment that communicates with it, are similar to today's cellular networks in part because of their usage of mobile equipment. However, there is one exception to this rule: on highways, roadside units (RSUs) will be installed alongside the route at a height that is substantially lower than that of cellular base stations to provide coverage. This is only applicable to the installation of static (infrastructure) nodes. In urban areas, living near major junctions is the best bet for a comfortable lifestyle. According to the findings of another study, V2I communications in metropolitan areas can be unpredictable due to the presence of both stationary and mobile items, resulting in a dynamic channel that is both spatially and temporally unpredictable.

13.5.3.4 Vehicle-to-X and 5G

In recent years, research on impending 5G cellular networks has begun to pay greater attention to ITS-related applications, and it is reasonable to predict that efforts on channel modeling for V2X and 5G systems will increasingly merge in the future. When the demand for delay tightens for various 5G scenarios, the given solution, for example, eliminates the principal impediment to usage in a vehicle context, which is a significant benefit (i.e., lack of low-latency guarantees). With 5G, you may begin enabling V2X systems and predicting the channel usage of the future. When applied to highly mobile terminals, 5G's D2D concept has significant parallels to V2X communications. As a result, modeling D2D channels can benefit from existing V2V channel modeling work and vice versa.

13.6 V2V channel models

Two-channel modeling methodologies are employed in V2V communication, as well as in many other types of communication systems. Physical and electrical magnitudes, as well as the issue geometry, are the primary focus of the analysis at the lowest levels of the hierarchy [40]. Generally, raytracing techniques are used to create the illusion of depth. Geometry-based stochastic models (GBSM) and geometry-based dynamical models (GBDM) are two types of geometry-based stochastic models. According to this definition, GBSM models have parameter values that can be anticipated, whereas DEM models have parameter values that can be described statistically. Non-geometrical stochastic models, on the other hand, are concerned with high-level behavioral characteristics that are intrinsically statistical, as opposed to geometrical stochastic models (GSM). They are sometimes referred to as tapped-delay lines, which refers to the fact that their initial representation is based on an echoes-based architecture (TDLs). To summarize, all strategies result in the realization of channel response through the use of echo-based expressions [41]. This must be kept in mind at all times. V2V channels can be described in a variety of ways, the most common of which is as follows: Although they share many properties with traditional cellular mobile channels, there are a few important differences. As previously stated, the WSSUS model is a solid starting point, albeit there are some concerns regarding its application in this particular case, which is discussed further below. For starters, only small-scale V2V channels can be maintained in a stationary state (during very short periods). Even in mobile channels, the model must be complex to account for non-stationary behavior. To do so, numerous setting classes such as "urban," "suburban," and so on must be generated (with a lesser degree of mobility). Furthermore, in V2V transmission, the different transmission channels used by the signal are frequently highly correlated with one another. In contrast, while GBDM is typically accurate when compared to measured channels, it comes at a high cost in terms of processing power [42]. GEMV2 makes use of the GBDM simulation program to create realistic scenarios (in a variety of contexts, including cities, suburbs, and highways). The V2V channel's schematic is depicted in Figure 13.6.

The use of this program enables the definition of time-varying conditions in the literature. It is only possible to estimate the total strength of the signal. In part, because they define some typical scenarios over which statistical assumptions are made to simplify geometry description elements (such as the deployment of scatterers or reflectors), GBSM computation effort is less than GBDM, although it may still be excessive for many applications. The channel response is obtained by the use of ray-tracing techniques. As a result, it is more difficult to use measurement findings for validation purposes [44]. When the mobility of the transmitter, receiver, and scatterer is taken into consideration, both geometrical models are willing

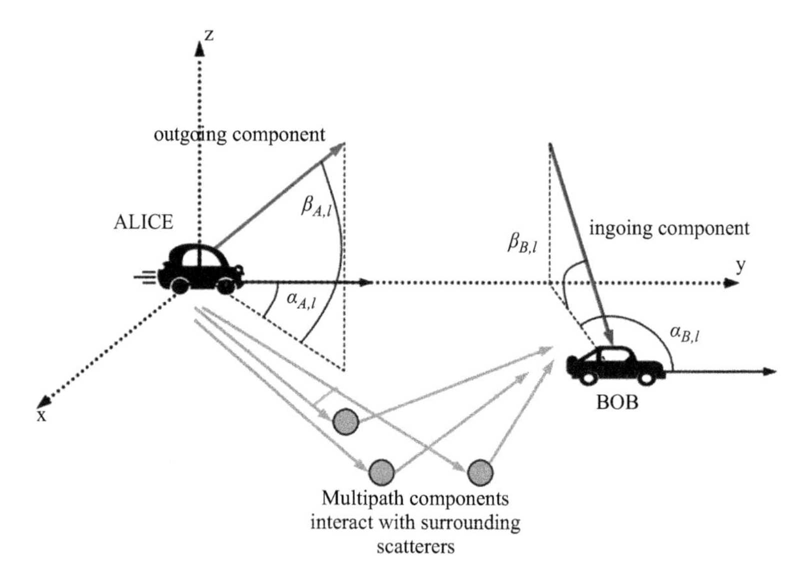

Figure 13.6 V2V channel model [43]

to accommodate non-stationarity. A few models are as simple to construct as the NGSM, and their behavioral properties (such as the number of echoes, their weights, and delays, for example) are determined by statistical processing of channel estimations from measurement campaigns, which is a relatively new technique. Non-stationarity, on the other hand, will need additional effort during the development phase. It is necessary to replicate the time fluctuation of the Ricean distribution parameter K-factor for a given set of cases, and a bimodal Gaussian distribution is employed to do so. In a word, the methods described here are two distinct approaches to NGSM and GBSM modeling [45].

13.7 Modeling of wireless links

Even though wireless or mobile hosts are becoming increasingly popular, fixed-line connections will continue to constitute a significant portion of the entire Internet infrastructure for the foreseeable future. Wireless local area networks (WLANs) and cellular connectivity are becoming increasingly common methods of connecting to the Internet. In Japan, for example, about 40 million people utilize mobile data services to access the internet through their phones. There are three basic types of wireless connectivity: LANs, wide-area cellular networks (WANs), and satellite networks [46]. LANs are the most common type of wireless connectivity. This chapter discusses wireless connection simulation models for unicast transport protocols over various wireless channels, as well as their implementation. Wireless sensors and ad hoc networks are not taken into consideration because they are considered to be a separate modeling domains. It has been established that wireless networks have a substantial impact on the performance of transport protocols from end to end. The majority of packet losses are attributed to Internet congestion, which serves as the foundation for today's congested network technologies. Packet losses on wireless networks, which are caused by corruption rather than congestion, are in direct opposition to this assumption. Forward error correction (FEC) and local retransmission are two techniques that are used to deal with corruption at the link layer in several link technologies. These techniques, on the other hand, may come with their own set of issues to contend with [47].

Wireless networks with high levels of bandwidth and latency unpredictability may have a severe impact on the performance of transport protocol implementations (including those transport protocols that consider increased delay as an indication of congestion). Consequently, TCP congestion control may punish connections that use cellular or satellite links, while rewarding flows with a high response time round-trip time (RTT). Interleaving and FEC, both of which are intended to lower bit error rates, are substantial contributors to latency. Large queues at the link level contribute to excessive latency for a variety of link technologies due to their different characteristics. When it comes to wired and wireless communication, using the same set of protocols has several advantages. As a result, anyone who has access to a local wireless network can connect to the rest of the Internet. In other words, any future transport protocols should consider this while building new ones in the future. In contrast to cable connections, wireless communications must have a minimum impact on the transport protocols that they communicate over. There has been a significant amount of research and development into wireless network transport protocols. It is tough to compare the outcomes of different studies when there are so many distinct model alternatives to choose from [48].

For those who are unfamiliar with wireless links, this chapter can be useful in describing how they affect transport protocols and how wireless link models are utilized in research. As a result of our efforts, the ns-2 network will be significantly more secure for everyone. Assumptions concerning wireless networks can have a significant impact on how different modes of transportation are evaluated. Consider the situation of an unexpected delay resulting in a fictional TCP timeout as an illustration. According to one proposal, link-level error recovery could be responsible for this type of fluctuation in delay. False TCP timeouts on GSM radio links with poor radio reception are extremely rare, as demonstrated by measurement tests [49]. The inter-packet arrival jitter pattern was more modest although there were some major spikes in delay variation. This tiny jitter raises the TCP retransmit timer to avoid false timeouts and to avoid data recovery from being delayed. It has been demonstrated through measurement and simulation that during handovers, a more noticeable pattern of delayed jitter is produced, which has the potential to cause false timeouts The wireless link characteristics of the mixed-wired/wireless network is depicted in Figure 13.7.

13.7.1 Essential aspects of models

Consider the topic of wireless communication, which will be discussed in greater depth later in the book. Furthermore, in addition to the many different types of links

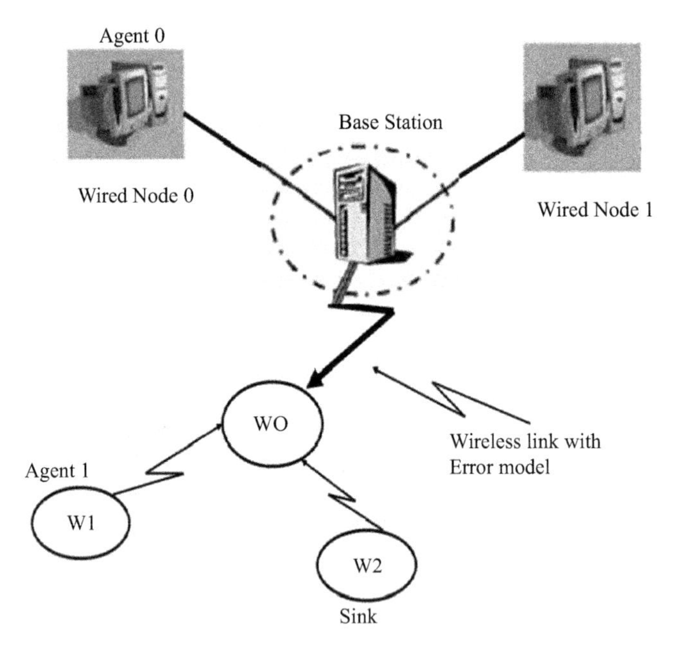

Figure 13.7 Mixed wired/wireless network model wireless link characteristics [50]

and their topologies, traffic models and performance metrics are covered in great detail. A full discussion of various types of links is provided in this section.

13.7.1.1 Types of wireless links

This chapter goes into great detail about wireless connectivity, including cellular, WLAN, and satellite connections.

Cellular: Cellular technologies such as General Packet Radio Service (GPRS) and CDMA2000 are the most frequently utilized in the world today, however, universal mobile telecommunications system (UMTS) may one day replace them (UMTS). Because of the challenging radio propagation conditions that cellular networks must contend with, methods such as FEC and link-layer retransmissions must be employed. The time it takes to get channel access may be prolonged due to the long round-trip times of the link. Every packet may necessitate the provision of a separate channel. In addition to dealing with bandwidth limits, transportation systems must also deal with battery power saving [51]. The use of wireless LANs for local area networks is becoming more common. Instead of competing for available capacity in the uplink and downlink directions, WiFi networks fight for available capacity in both the uplink and downlink directions. The range of coverage given by a single base station ranges from a few tens of meters to several hundred meters. In 802.11b, link error management is intrinsically linked to the MAC address. For each data frame, a maximum of three retransmission attempts is allowed before the frame is considered lost. Packet fragmentation can improve the efficiency of error recovery; however, it is not a popular practice in the industry.

Satellite: Satellite connections often have high latency and a large amount of bandwidth available. With just one geostationary satellite, it is feasible to cover a whole continent. Satellite connectivity offered by constellations of mobile low earth orbit (LEO) satellites such as Iridium or Global Star is yet another sort of satellite connectivity. When a satellite's signal is lost, the internet connection is instantly switched to another satellite's signal. The TCP study is an excellent illustration of this. During a handover, everything becomes disorganized [52].

13.7.1.2 Topologies

The quantity and location of wireless networks along the route can have an impact on how mobile a person is. Because of the increased latencies and error loss rates that occur when many wireless networks are used, congestion control and loss recovery become more difficult to achieve. Other design topologies that employ a single wireless link or many wireless links at each end or in the middle are also viable alternatives. Wireless links are frequently used as the final point of connectivity between the host and the network. The only way for smartphone and tablet owners to connect to the internet is through a wireless connection. In most cases, a second wireless connection between a mobile phone and a laptop is not available. A combination of infrared and Bluetooth connections is employed to achieve this purpose. Other than increasing the amount of time it takes to transfer data, these links frequently have minimal effect on transfer strategies. When the signal is strong and the distance between the two points is small, these linkages are widely employed [53]. Infrared or Bluetooth links that are subjected to WLAN interference or that are out of alignment cause transmission mechanisms to be compromised. Because of interference caused by a large number of wireless links on the same path, packet losses may occur. It is common for wireless networks to employ flow control to avoid wasting available capacity. At the moment, finding Wi-Fi connections on a route that extends to both ends is difficult to come by. This is due in part to the fact that connecting to a mobile host is a potentially dangerous proposition. A stationary network server can provide two transport connections for applications using two mobile hosts, such as instant messaging, when two mobile hosts are involved. In the future, phone calls may be done utilizing VoIP technology. In the future, two wireless lines, one on each end, will be the standard configuration.

When using VoIP, the transport protocol UDP is utilized to simulate a circuit-switched link, which is more realistic. Using a wireless link at either end of a VoIP call that is delivered as best-effort traffic over the public Internet may be relevant for the development of transport protocols in the future. Satellite connections are frequently used by Internet service providers (ISPs) in remote regions such as Hawaii to connect. VSAT satellite links are also used in Africa and other parts of the world to connect directly to the United States or Europe. Gas stations, for example, are increasingly reliant on VSAT services for credit card verification as the market in the United States continues to grow. It is possible to use satellite lines for residential access even if they have ten times less capacity and statistical multiplexing than an ISP. Because of inter-satellite forwarding, it is feasible for constellations to have several satellite links; nevertheless, these are rarely crowded and can be considered as a single link as a result of this. WLANs are typically used as a last-resort connection for users, and this is true for the vast majority of them. Sometimes point-to-point wireless local area network connections (WLANs) are necessary when joining two local area networks together. There is no difference between direct WLAN and LAN connections when it comes to speed and dependability. User-directed WLANs or LANs connected to WLANs are both feasible configurations for the model.

13.7.2 Wireless traffic

The outcomes of the simulation can be significantly influenced by the traffic model used. In particular, concluding performance from one-way mass TCP transfers is completely pointless. When other wireless networks are accessible in the path, optimizing TCP in one method may not be successful. For example, web browsing is a well-known bidirectional use that occurs frequently. The amount of data transmitted and received by mobile users is frequently bigger in the downlink direction. Consequently, it is usual for people to believe that the receiver is located close to the wireless link as a result. To provide widespread Internet access, wireless networks are preferred over fixed lines because they do not have the same limits as fixed lines. According to our calculations, consumers are less likely to use peer-to-peer file sharing because cellular connections are slow and expensive. As a result, mobile users may find themselves connecting to servers that have been specifically configured for the delivery of mobile services. Because of the limitations of small communication devices' displays, the content of the "wired" Internet cannot be displayed in a readable manner on these devices. This is the only logical conclusion that can be drawn. Unlike most servers on the wired Internet, these servers can execute customized applications such as the wireless application protocol, which is not commonly seen on the wired Internet (WAP). When comparing HTTP traffic versus WAP traffic, the difference is staggering. To compare alternative cellular technology approaches, it was decided to create a wireless traffic simulation model. Cellular networks, in conjunction with machine-to-machine control interfaces, can be utilized for particular data exchanges. According to the European Telecommunications Standards Institute (ETSI), particular assessment standards for GPRS have been developed, and models have been developed to predict the creation of e-mail (FUNET), fleet management, and train traffic. Overall, the amount of cellular data traffic has been under-measured, although this is starting to change. Due to user privacy concerns, taking measurements is prohibitively expensive, in part because of the high cost of hardware. Studies of modem dial-up links are frequently used to predict cellular link traffic because modern cellular links are extremely comparable to modem dial-up links in terms of speed and reliability [54].

13.7.3 Performance metrics

When assessing wireless transport protocols, the first thing we look at is the normal performance indicators. Among the most essential considerations are fairness, throughput, latency, and dynamic response time. When it comes to wireless communications, several indicators are crucial. It is crucial to have a high level of output since it defines how much of the information delivered is useful. The efficiency with which the transport protocol utilizes the battery power of the mobile terminal is affected by the throughput of the protocol. A high-throughput network also means that the radio spectrum is being used efficiently and that there is less interference to other users while talking about it. The fact that mobile customers are usually charged based on the amount of data they send across a wireless network means that higher throughput translates into lower expenses [55].

13.8 Modeling of wireless signal propagation

Accurate radio channel models will be required for this task, which can be constructed either using ray-tracing simulator software or by conducting extensive field observations with steerable antennas and channel sounder equipment. A measurement campaign's timetable and resources required for the equipment (e.g., directional antennas) are just two of the various factors to take into consideration when planning the campaign (e.g., vector network analyzers, channel sounders). Despite this, a great deal of study has been done on millimeter wave propagation, and it has spanned a wide range of frequencies across the spectrum. All of the effects studied in these experiments, both major and little (such as the route loss exponent, maximum coverage duration and outage time, and the penetration/reflection loss angle), were taken into consideration in these studies [56]. We chose the second alternative and employed a software ray-tracing simulator as a professional tool to create the final product (Wireless InSite). Because of the flexibility of the results produced by a ray tracer, they may be applied to a wide range of scenarios that are strikingly similar to the simulated one, regardless of whether the simulation takes place indoors or outside. If you generate or import (into the ray-tracer) an incredibly exact description of everything in the application settings, you can be confident that the simulation environment closely reflects reality. Customers can readily adapt our channel propagation models (for certain city streets) to other urban scenarios by simply changing the channel propagation parameters.

All of our conclusions will be applicable if the current circumstance is identical to the one, we faced in the previous situation. Even if the new terrain has significant differences from the old, the simulation approaches are still viable; nevertheless, the findings will be slightly different from the previous ones [57]. To save time and money, a new layout can be imported into the simulation environment or even developed from the ground up, allowing for significant time and cost savings. Even though measurements are required to verify the proposed channel model, these simulations can be followed by measurements to acquire an early understanding of the challenges to constructing a specific use-case scenario's millimeter-wave channel model. There is a wave network. A. When considering radio channel propagation, the three most important effects to consider are path loss, shadowing loss, and fade loss. Path loss is the most significant of the three effects to consider. The first two are essential for understanding the radio channel's large-scale propagation model on a larger scale. The micro-scale propagation model constitutes a significant portion of the third chapter. Increased transmitter-to-receiver distance (Tx) results in a decrease in the strength of the route loss signal. Whereas, if the route loss is increasing, the exponent n indicates how quickly this is occurring [58].

To account for shadowing losses induced by obstacles and scattering objects that absorb the transmitted signal, large-scale propagation models must be developed. By conducting data collection campaigns, it is possible to calculate statistics for both the input and output values. If reflections, diffractions, and scattering are not present, radio waves will arrive at the receiver from a variety of directions and with variable propagation times, depending on the frequency. The absence of LOS between the transmitter and receiver signifies that there is no communication between them. At the receiving end, their unpredictable amplitudes, phases, and angles-of-arrival combine to distort or fade the incoming signal (AoAs). Other factors, like Doppler dispersion caused by the mobility and speed of the transmitting (Tx) and receiving (Rx) devices, influence the small-scale propagation channel idea (represented by cars, people, and other moving objects). Rather than random frequency modulation, Doppler shifts induce temporal dispersion, which is caused by multipath propagation delays. Every one of these has a little but significant impact and each one is equally important. The link between signal attributes (bandwidth and symbol period) and channel parameters (Doppler spread/coherence time and delay spread/coherence bandwidth) determines the sort of fading that a signal undergoes on a mobile radio channel. Using this technique in conjunction with flat or frequency-selective fading, rapid or slow fading, multipath delay spread fading, and Doppler-spread fading, four different fading styles can be achieved [59]. To perform this inquiry effectively, it is being carried out in two stages. In the first stage, the path loss and shadowing models for the large-scale channel propagation model are constructed and tested. In the second study, a small-scale channel propagation model is employed to investigate multipath delay spread using a single channel. Figure 13.8 depicts the signal propagation mechanism that a wireless network employs to communicate.

13.9 Vehicular wireless technologies

All autonomous vehicles must have real-time wireless connections to communicate with their passengers, as well as a variety of sensors and other equipment on the road. Keeping people safe and mobility are all connected. You should have a better understanding of how autonomous vehicles will communicate with one another in the future if you have finished reading this article. *Radiofrequency (RF) spectrum is used by all audio-visual communication systems. The RF spectrum is a component of the electromagnetic spectrum that corresponds to radio frequencies [61].

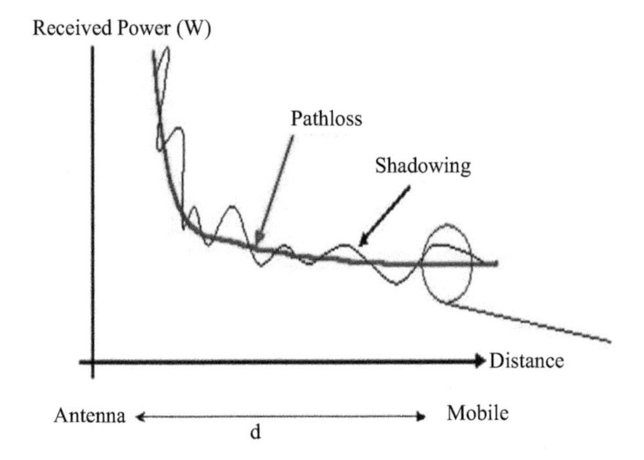

Figure 13.8 Signal propagation in a wireless network [60]

13.9.1 Commercial wireless services

Vehicle telematics applications take advantage of commercial wireless voice and data connection services to communicate with and monitor their vehicles. The data can be accessed either through an internal data connection or through a smartphone application. At the moment, the majority of connected-car platforms rely on drivers connecting their apps to the network via their smartphones. Instead of automotive radios, new cell phones can now increase the speed, capacity, and efficiency of a vehicle. They also contribute to the overall safety and comfort of the driver and his or her passengers. Additionally, the new technology would make it possible to get real-time traffic and navigational information. To boost network capacity and speed, new digital signal processing techniques and modulations are utilized in 4G, the most recent wireless data transfer standard, which is currently in use. As a result of technological improvements in automobiles, today's vehicles may include conveniences such as a Wi-Fi hotspot, Internet radio, and other online services, among other things. Another example of this trend is the development of upgraded navigation systems that deliver real-time street-level visual images while driving on the highway.

13.9.2 Dedicated short-range communications (DSRC)

Autonomous vehicle to vehicle (V2V) communication is accomplished through the use of DSRC, which is a short-range wireless service. With DSRC's assistance, safety-related applications such as avoiding collisions at crossings can be made more convenient. Additionally, non-safety communications and vehicle diagnostics are carried out using DSRC. Some people are concerned about the absence of recent applications received by the DSRC. In addition to licensed usage of this spectrum, unlicensed applications of this spectrum are also a topic of discussion, which is understandable.

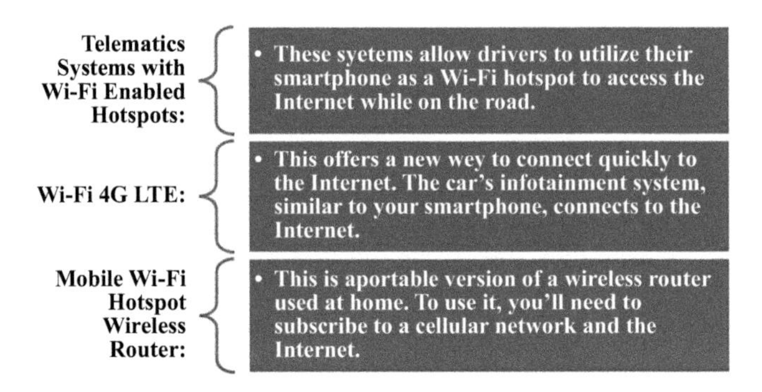

Figure 13.9 Wi-Fi classifications

13.9.2.1 Wi-Fi

Electronic devices can communicate data wirelessly over computer networks utilizing Wi-Fi, which includes high-speed Internet connections, thanks to the widespread adoption of the technology known as WLAN (wireless infrared). When it comes to providing Wi-Fi in their vehicles, automakers have a few different options to consider. This is an excellent technique if you want to give a smooth online experience to a large number of users at once [62]. The process is illustrated in Figure 13.9.

13.9.2.2 Bluetooth

Bluetooth, a short-wavelength radio transmission technology, allows mobile and stationary devices to transfer data over short distances utilizing short-wavelength radio waves. You can use it to create a private network that is extremely secure. Mobile phones that are connected to the Bluetooth technology in a car can make hands-free calls. Along with increasing the range of wireless phones in cars, this technology can be used to improve communication within the vehicle, either by enabling V2I conversations or by sending data that can be re-sent to a wireless phone over a Bluetooth connection.

13.10 Decentralized congestion control techniques for autonomous intelligent vehicle (AIV)

About safety-related software, the usage of beacons and event-driven messaging are essential. Event-driven messages are sent in response to events such as an EEBL or a PCS, rather than the more regular beacon messages that would otherwise be sent. The sensing capability of the CSMA/channel CA aids in the prevention of message collisions. Congestion arises when the channel is overcrowded because channel sensing is unable to prevent message collisions. Consequently, the PRR declines considerably, having a severe impact on the overall performance of the application. CCH should be in charge of handling beacons and events. If a beacon issues a message, it takes precedence over a message issued by an event. The need for this is dictated by the unpredictable nature of vehicle events, and as a result, a portion of the channel must be reserved for communications that are deemed to be of crucial importance for safety. If no action is taken, traffic congestion will worsen, making communication difficult, if not impossible, in some areas. It is necessary, as a result, to implement traffic flow management measures. Congestion control is important. Strategies for congested channel management should be implemented to reserve a portion of the channel capacity for messages prompted by events. Using congestion control algorithms, vehicle communication parameters such as transmit power, message rate, data rate, and carrier sensing threshold can be adjusted to reduce channel load as much as possible. Alternatively, as stated in the previous chapter, the communication parameters that are employed can have an impact on the reliability of the apps. It is, therefore, necessary to develop algorithms for congested area prevention and performance optimization. As an alternative, decentralized or centralized traffic management algorithms might be used. An RSU or other single coordinator is responsible for managing the channel load when using a centralized approach. These parameters are broadcast to every vehicle within range. Each vehicle shares a channel must separately alter its communication characteristics to reduce congestion when decentralized congestion control is used to reduce congestion. Congestion control can be control via a decentralized strategy or a non-centralized strategy. DSRC and ITS-G5 do not necessitate the establishment of a centralized infrastructure. DCC technology is used by both the ETSI and the SAE standards organizations. In contrast, a decentralized procedure has additional disadvantages, such as the inability to ensure equity, which is a significant disadvantage. Fairness in this context refers to all automobiles having the same amount of time to use the channel at the same time. It is measured by the amount of time a vehicle spends transmitting beacon messages over a certain channel. Unfair channel use time can hurt the performance of an application.

13.10.1 Effect of direct cable connection (DCC) tuning parameters

DCC algorithms allow for the adjustment of message and data speeds, transmit power, and carrier sensing threshold, all of which help to reduce congested networks. Channel load and application performance are both affected by the DCC tuning options that are selected. You should be aware that several different DCC algorithms make use of contention windows to choose from. Reduce message collisions by fine-tuning the contention window, but be prepared for additional latency as a result of this. The performance of the PRR is unaffected. As a result, we did not consider this during our investigation.

13.10.1.1 Channel load

When congestion control is not used, both the message-rate and data-rate DCC channels are fully loaded. This is referred to as the "no-DCC" case. It demonstrates the amount of channel load present without the use of DCC procedures. Because of the message-rate characteristic of DCC, the channel load can be reduced by sending out fewer beacon messages overall. Rather than employing data-rate DCC, broadcasting messages at higher data rates reduces channel load, resulting in shorter beacon message delivery times than would otherwise be the case. To vary the number of vehicles sharing a channel, it is essential to adjust the transmission power and carrier sensing threshold DCC to the transmission range and sensing range of the vehicle. If data from another vehicle is received by a vehicle, the sensing range of the vehicle will be occupied. When a message is being conveyed, the transmission range describes the area around a vehicle where other vehicles can detect that a channel is busy and hence avoid using it. Carrier sensing thresholds and transmit power DCC algorithms, on the other hand, have little effect on channel load because of how closely vehicles are spaced out in the surrounding area, which is why DCC algorithms are used. The sensor range of the centered vehicle for the DCC carrier detection threshold or the DCC transmission power Ra can be visualized by the use of an arrow. If this is the case, the central vehicle will attempt to lessen channel load by decreasing its sensing or transmission range from Ra1 to Ra2, as appropriate. Consequently, in case 1, since Ra2 has fewer autos than Ra1, lowering Ra lowers the channel load. Even though Ra2 was reduced, it did not affect channel load because the number of cars in both scenarios is the same. By transmitting beacon messages more frequently, it is possible to enhance the data rate. Message rate, transmit power, and carrier sensing threshold are all important parameters. Using DCC algorithms, DCC algorithms are more capable of managing a given channel load than DCC algorithms that do not use DCC algorithms.

13.10.1.2 Application performance

The reliability and awareness range of the app can be altered by altering the communication settings used by the app. When congestion hurts application reliability, message-rate-based DCC techniques can be used to lower the message rate below the minimum needed message rate. The thresholds for transmission power and carrier detection the transmission and sensing range constraints imposed by DCC algorithms may be incompatible with an application's desire for a greater awareness range. The application's awareness range requirements may be compromised as a result of the short communication range given by data-rate DCC approaches. There should be a compromise struck between the desire of the DCC algorithms to reduce congestion and the requirement of the application for dependability (awareness range).

13.10.2 Design goals of DCC algorithms

DCC algorithms should keep networks free of congestion while yet delivering predictable performance to applications. These design objectives must be met in a variety of traffic environments, including urban, highway, and rural settings. DCC algorithms must be scalable if they are to be effective in meeting design objectives in large vehicle quantities. As a result, altering communication parameters to

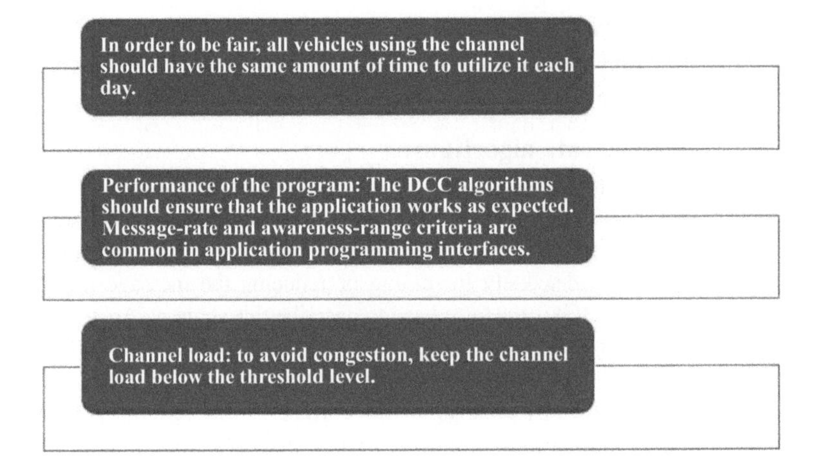

Figure 13.10 Design goals of DCC algorithms

alleviate congestion may harm the performance of your application. Given the above explanation, we propose the following DCC algorithm design objectives, as shown in Figure 13.10.

13.10.3 Metrics to assess DCC performance

While keeping in mind the DCC design objectives, this section presents metrics for measuring the channel load, application performance, and fairness performance of DCC algorithms. The dependability and awareness range of applications are used to evaluate their overall performance. Based on these criteria, the thesis evaluates and contrasts different DCC methods. In the observing zone, many metrics are measured and analyzed. Channel busy percentage (CBP) is a metric used by the SAE and ETSI standardization bodies to determine how busy each channel is throughout a given period. We maintain track of the daily customs and border protection inspections that cars in the research area are subjected to. As a result, it is a reliable indicator of how much traffic is there on any particular day of the week. The measurements are taken on a discrete basis every. A DCC-based algorithm should be employed to avoid exceeding the load threshold.

13.10.4 Overview of proposed DCC algorithms

DCC algorithms, for example, can be used to fine-tune the transmission rate and transmit power of a message. Using DCC techniques, you can accomplish this with any or all of the parameters. The transmit power and message rate of DCC algorithms created by standards bodies are the primary concerns of the algorithms. There are DCC methods that use data rate and carrier sensing thresholds. In this section, you will find an overview of the many DCC algorithms that have been presented by academics and standards-setting organizations.

13.10.4.1 DCC proposed by academics

Only a few congestion control systems employ data rate adaptation or carrier sensing thresholds, whereas the vast majority alter the message rate or transmit power.

13.10.4.2 Message-rate algorithms

It is critical to match the message rate with the number of beacon signals that each vehicle transmits per second to achieve maximum efficiency. The message rate can be easily adjusted if the observed channel load exceeds the channel load threshold, which can be accomplished by discretely increasing or reducing the message rate. Binary message-rate control (BMC) is the term used to describe this strategy. An approach for controlling the binary message rate that makes use of additive increase multiplication and division (AIMD) is presented. Because of shadowing, it is possible that the channel load measurements recorded by each car are not correct. To ensure fairness for AIMD, PULSAR employs global channel load information acquired from neighbor vehicle channel load exchanges to ensure fairness. PULSAR does not consider the requirements of the application. This work contributes to the expanding body of evidence demonstrating that message-rate distribution should be equalized. LIMERIC, on the other hand, does not take into consideration the requirements of the application.

13.10.4.3 Transmit power algorithms

The transmission range of the beacon's message can be restricted. The lookup table and channel load data are utilized to incrementally increase and decrease the transmission power. However, algorithms may encounter challenges with fairness on the other side of the coin. To account for variations in channel load, the transmit power is updated when the Stateful Utilization-based Power Adaptation (SUPRA) algorithm is used to calculate the transmit power. According to the authors' theoretical and numerical calculations, they have been successful in achieving a fair distribution of transmit power. The needs of the application, on the other hand, are not taken into consideration. As part of the optimization process to alleviate congestion, it is necessary to limit the maximum channel load while also maximizing transmit power and maintaining fairness. The inclusion of additional information in beacon messages by algorithms, such as the location of two-hop surrounding autos and the vehicle's histogram of vehicle density, results in a larger transmission channel load. When a vehicle's power-to-weight ratio is low, the transmission power is increased or decreased accordingly (PRR). The PRR is calculated based on the sequence numbers of the messages that have been received. If the average PRR for all sender autos exceeds a certain level, the program lowers the transmit power (see Figure 13.1). An increase in transmit power is triggered by a drop in the PRR. Despite these efforts, there can be no assurance that traffic on the channel will remain below the threshold.

13.10.4.4 Combined message-rate and transmit power algorithms

The transmission power and the message rate can both be controlled at the same time. You may modify the transmit power without affecting the message rate, which is a convenient feature. The tracking error of a vehicle is based on the difference between the vehicle's actual location and the projected position of other cars in its immediate vicinity.

The adaptive message rate for a vehicle is calculated using this estimate, which may be found here. When the channel load changes, the transmission power changes like that of SUPRA. The power of the transmission is regulated with the help of the PRR to fit the channel load and the density of the vehicles on the channel. Prepare a distributed solution to address these two subproblems: one dealing with message rate management under fixed transmit powers and another dealing with power control under fixed message rates. Optimization problems have been formulated in this way: fairness and applicability are not taken into consideration when developing these algorithms. Transmission power and message rate can be adjusted to meet the needs of the application using the algorithms that have been proposed. Based on the observed channel load, the application's minimum message rate need, and the channel-load threshold, these algorithms adjust the message rate. Depending on the communication range, a straightforward lookup table indicates how much transmit power is required. The transmission power is determined by the received power as well as the transmission power information transmitted in the beacon signals that the vehicle is receiving. The amount of channel load increases with each transmission of power exchange data. As a result, none of these algorithms can be guaranteed to be fair because they all aim to reduce congestion while also taking into account the app's minimal message rate and communication range expectations.

13.10.4.5 Carrier sensing threshold algorithms

Using a sensor range limiter, you may manage the channel load. The delay, channel load, and carrier detection threshold all decrease as the message size decreases. In contrast, the PRR becomes worse the more you are away from the source of the radiation. It is a carrier sensing threshold adjustment technique that alters the carrier sensing threshold depending on how long a user must wait before a connection can be established at the medium access layer. Based on the density of the vehicles, the carrier detecting threshold is modified. The channel load may still exceed the threshold if something goes wrong, even if you are employing one of these strategies.

13.10.4.6 Data-rate control algorithms

To control the channel load, adjust the data rate at which the beacon message is transmitted. In research, it has been demonstrated that high data rates can minimize channel traffic without compromising application performance. Based on the channel load that is being detected, a change to the data rate is made. Standards for application, such as fairness, are, on the other hand, completely ignored.

13.11 Conclusion

This information is utilized to alert other vehicles as well as foresee and avert dangerous situations through the employment of vehicular communication-based safety systems, which are based on the transfer of messages between cars and trucks. With DCC methods, we were able to obtain an acceptable application performance despite the high vehicle density in our test environment. Because of channel congestion, applications perform badly when there are a large number of vehicles on the road at the same time. To decrease congestion as a result of increased vehicle density, DCC algorithms optimize the utilization of the channel. Numerous hours have been spent researching the effect of vehicle density on the performance of the application. A significant deal of detail is provided in this chapter on how to model wireless networks to evaluate transport protocols. When it came to delaying variation, bandwidth-on-demand, and inter-system handover, there was a comparison between the simulation results and the actual measurement data. There are various simulation scenarios for this scenario contained in ns-2, which may be useful in examining the performance of the wireless network transport protocol in wireless networks.

Future research will most likely focus on the explicit communication that occurs between connection layers and transport protocols in greater detail. A greater amount of research is required to determine the impact of link-layer processes on protocols and applications running above the link layer. It is the purpose of this chapter to give an in-depth examination of the most recent channel model approaches and measurement campaigns in vehicle-related situations, with a particular focus on automotive communication technologies. Despite the substantial progress that has been made in this field, there are still several critical problems that must be overcome before V2X communication can be widely used and implemented. The development of micro-mobility models, massive MIMO systems, spherical wavefronts, and channel non-stationarity for automotive propagation is still in its early stages. When the simplified models are used in system simulations, it will be possible to make more accurate assessments of beam tracking and gain a better understanding of the vehicular communication link. A greater amount of data is required to extract relevant information from the vehicle's wireless communication channel, despite the numerous measurement activities. When communicating at millimeter-wave frequencies, it becomes more difficult to locate vehicle measurement projects to participate in. Even though vehicular communication in the framework of ITSs and smart cities has the potential to reach high levels of organization, optimization, and reduction of hazards and accidents, it is still considered a relatively new technology in the field. Therefore, researchers in this subject are now working on modeling vehicle communications channels and several significant hurdles that must be overcome in this area within the next several years, as stated in the research objectives presented earlier in the chapter. The primary focus of this research is on the investigation of V2V channel characteristics and how they affect communication systems. Consequently, the first half of this chapter is devoted to the description and modeling of channel behavior. In this section, we will address high mobility scenarios in terms of their physical properties that influence channel behavior, as well as how they have been modeled in the literature thus far. The characteristics of V2V channels are demonstrated by the use of an example developed by a channel simulation tool. Aside from that, it conducts studies on essential behavioral traits in several geographic contexts such as the urban, suburban, rural, and highway environments. Having examined the differences and similarities between current V2V communication technologies such as DSRC and LTE-V, this chapter focuses on the physical layer and data link layer challenges that must be addressed first. The relationship between the received signal and noise and interference is examined in realistic mobility scenarios, and specific MAC approaches are discussed in detail. OFDM modulation is described in detail, with a parametrization and an overview.

References

[1] Liu, B, C Han, X Liu, and W Li. "Vehicle artificial intelligence system based on intelligent image analysis and 5G network." *International Journal of Wireless Information Networks* (2021): 1–17.

[2] Rahman, MH, M Abdel-Aty, and Yina W. "A multi-vehicle communication system to assess the safety and mobility of connected and automated vehicles." *Transportation research part C: emerging technologies* 124 (2021): 102887.

[3] Lee, J, S Yu, M Kim, Y Park, S Lee, and B Chung. "Secure key agreement and authentication protocol for message confirmation in vehicular cloud computing." *Applied Sciences* 10(18) (2020): 6268.

[4] Chaudhry, A. *A Framework for Modeling Advanced Persistent Threats in Intelligent Transportation Systems*. Diss. 2021.

[5] Cao, Q, G Ren, D Li, H Li and J Ma. "Map matching for sparse automatic vehicle identification data." *IEEE Transactions on Intelligent Transportation Systems* (2021). Doi: 10.1109/TITS.2021.3058123.

[6] Lucet, E, A Micaelli, and F-X Russotto. "Accurate autonomous navigation strategy dedicated to the storage of buses in a bus center." *Robotics and Autonomous Systems* 136 (2021): 103706.

[7] Wu, J, B Kulcsár, and X Qu. "A modular, adaptive, and autonomous transit system (MAATS): a in-motion transfer strategy and performance evaluation in urban grid transit networks." *Transportation Research Part A: Policy and Practice*151 (2021): 81–98.

[8] Liu, Y, Q Chang, M Peng, T Dang, and W Xiong. "Virtual reality streaming in blockchain enabled fog radio access networks." *IEEE Internet of Things Journal* (2021). Doi: 10.1109/JIOT.2021.3111115.

[9] Pavani, K., and P. Sriramya. "Novel vehicle detection in real-time road traffic density using haar cascade comparing with KNN algorithm based on accuracy and time mean speed." *Revista Geintec-Gestao Inovacao E Tecnologias* 11(2) (2021): 897–910.

[10] Oestges, C., and F Quitin, eds., *Inclusive Radio Communications for 5G and Beyond*. London: Academic Press, 2021, pp. 253–293.

[11] Khalifeh, A, K Alakappan, BK Sathish Kumar, JS Prabakaran, and P Nagaradjane. "A Simulation Analysis for LED Spatial Distribution for Indoor Visible Light Communication." *Wireless Personal Communications*, 122(2) (2021): 1867–1890.

[12] Tsao, M, K Yang, S Zoepf, and M Pavone, "Trust but Verify Cryptographic Data Privacy for Mobility Management." *IEEE Transactions on Control of Network Systems* (2021), doi: 10.1109/TCNS.2022.3141027.

[13] Singh, KJ, DS Kapoor, and BS Sohi. "All about human-robot interaction." In *Cognitive Computing for Human-Robot Interaction.* Academic Press, London, 2021, pp. 199–229.

[14] Carou, D. "Aerospace transformation through Industry 4.0 technologies." In *Aerospace and Digitalization*. Springer, Cham, 2021, pp. 17–46.

[15] Bendali-Braham, M, J Weber, G Forestier, L Idoumghar, and PA Muller. "Recent trends in crowd analysis: a review." *Machine Learning with Applications* 4 (2021): 100023.

[16] Kothig, A. *Accessible Integration of Physiological Adaptation in Human-Robot Interaction.* MS thesis. University of Waterloo, 2021.

[17] Zidianakis, E, N Partarakis, S Ntoa, *et al.* "The Invisible Museum: a user-centric platform for creating virtual 3D exhibitions with VR s." *Electronics* 10(3) (2021): 363.

[18] Kondaveeti, H K, N K Kumaravelu, S D Vanambathina, S E Mathe, and S Vappangi. "A systematic literature review on prototyping with Arduino: applications, challenges, advantages, and limitations." *Computer Science Review* 40 (2021): 100364.

[19] Gilmore, S. K. *The Future Railway Control Room: A Practical Framework for Control Room Design Practices*. MS thesis. 2021.

[20] Bhuyan, M H. "A modern review of the spintronic technology: fundamentals, materials, devices, circuits, challenges, and current research trends." *International Journal of Electronics and Communication Engineering* 15(6) (2021): 266–277.

[21] Leon, K S. "Latino criminology: unfucking colonial frameworks in "Latinos and crime" scholarship." *Critical Criminology* 29(1) (2021): 11–35.

[22] Ivashov, S I., L Capineri, T D Bechtel. *et al.* "Design and applications of multi-frequency holographic subsurface radar: review and case histories." *Remote Sensing* 13(17) (2021): 3487.

[23] He, K, Q Cao, G Ren, D Li, and S Zhang. "Map matching for fixed sensor data based on utility theory." *Journal of Advanced Transportation* 2021 (2021). Article ID 5585131, https://doi.org/10.1155/2021/5585131.

[24] Chaudhry, N, and M M Yousaf. "Concurrency control for real-time and mobile transactions: historical view, challenges, and evolution of practices." *Concurrency and Computation: Practice and Experience* (2021): e6549.

[25] Mohammadi, M, K Elfvengren, Q Khadim, and A Mikkola. "The technical-business aspects of two mid-sized manufacturing companies implementing a joint simulation model." In *Real-time Simulation for Sustainable Production: Enhancing User Experience and Creating Business Value.* Routledge, London, 2021, pp. 102–118. Available at https://ebrary.net/164921/business_finance/technical_business_aspects_technical_business_sized_manufacturing_companies_implementing_joint_simul.

[26] Hahn, T, and N Kinski. "Science-industry cooperation: delivering innovations for marine carbon monitoring." *Sea Technology* (2021): 20–23.

[27] Andersen, V J. *Intralogistics System Design with Autonomous Mobile Robots*. MS thesis. NTNU, 2021.

[28] Li, L, W Zhu, and H Hu. "Multivisual animation character 3D model design method based on VR technology." *Complexity* 2021 (2021). Article ID 9988803, https://doi.org/10.1155/2021/9988803.

[29] Dielemans, J. "Ambulance Drones in The Netherlands: A Vision+ Concept Design for 2035." (2021). Available at https://repository.tudelft.nl/islandora/object/uuid%3A2a7e6372-d73c-4d09-bfc1-2009a54e10ec.

[30] Tyagi, AT and S U Aswathy. "Autonomous Intelligent Vehicles (AIV): research statements, open issues, challenges and road for future." *International Journal of Intelligent Networks*, 2 (2021): 83–102, ISSN 2666-6030. https://doi.org/10.1016/j.ijin.2021.07.002.

[31] Mana, M, and A Rachedi. "On the issues of selective jamming in IEEE 802.15. 4-based wireless body area networks." *Peer-to-Peer Networking & Applications* 14(1) (2021).
[32] Rios, J L G, J T Gómez, R K Sharma, F Dressler, and M J F G García. "Wideband OFDM-based Communications in Bus Topology as a Key Enabler for Industry 4.0 Networks." *IEEE Access* 9 (2021): 114167–114178.
[33] Akbari, Y, N Almaadeed, S Al-maadeed, and O Elharrouss. "Applications, databases and open computer vision research from drone videos and images: a survey." *Artificial Intelligence Review* 54(5) (2021): 3887–3938.
[34] Ebrahim, R. *Development of a Cost-Effective Method to Implement Traffic Management Principles in a Small City Environment*. Diss. Stellenbosch: Stellenbosch University, 2021.
[35] Asplund, H, D Astely, P Butovitsch, *et al. 3GPP Physical Layer Solutions for LTE and the Evolution Toward NR*. (2020). Doi: 10.1016/b978-0-12-820046-9.00008-3.
[36] Bellotti, A, J Murphy, L Lin, *et al.* "Paradoxical relationships between active transport and global protein distributions in neurons." *Biophysical Journal* 120(11) (2021): 2085–2101.
[37] Lanagan, S, and K-KR Choo. "On the need for AI to triage encrypted data containers in US law enforcement applications." *Forensic Science International: Digital Investigation* 38 (2021): 301217.
[38] Thiele, LP. "Rise of the Centaurs: the Internet of Things intelligence augmentation." *Towards an International Political Economy of Artificial Intelligence.* Palgrave Macmillan, Cham, 2021, pp. 39–61.
[39] Keramidas, K, F Fosse, A Diaz-Vazquez, *et al.* "Global energy and climate outlook 2020: a new normal beyond Covid-19." Publications Office of the European Union, *Luxembourg*, 2021.
[40] Andrews, RL. "Deep Correlation Learning for Urban Air Quality: Analysis and Prediction in New Zealand." (2021).
[41] Huang, M. *Occupational Polarization Trends in the US Economy and Low-Skilled Worker Career Development*. Diss. Georgetown University, 2021.
[42] Ansari, M, M Ansari, A Asrari, P Fajri, M Rahmani-Andebili and B Ramos. "A collaborative market-based framework to cope with overvoltages caused by massive penetration of rooftop PVs in modern distribution systems." *2020 52nd North American Power Symposium (NAPS)*. IEEE, New York, NY, 2021, 1–5.
[43] Bottarelli, M, P Karadimas, G Epiphaniou, D K B Ismail, and C Maple. "Adaptive and optimum secret key establishment for secure vehicular communications." *IEEE Transactions on Vehicular Technology* 70(3) (2021): 2310–2321.
[44] Abbott, RL. *Aviatar – An Augmented Reality System to Improve Pilot Performance for Unmanned Aerial Systems*. Diss. The University of North Carolina at Charlotte, 2021.
[45] Konstantoudakis, K, D Breitgand, A Doumanoglou. *et al.* "Serverless streaming for emerging media: towards 5G network-driven cost optimization." *Multimedia Tools and Applications* (2021): 1–40. https://doi.org/10.1007/s11042-020-10219-7.
[46] Bai, X, H Jia, and M Xu. "Port congestion and the economics of LPG seaborne transportation." *Maritime Policy & Management* (2021): 1–17.

[47] Verwaaij, D. "TheRoad Ahead: Military Mobility, PESCO, and the Netherlands: A case study on Euro-pean defense Development." (2021).

[48] Sun, Y. *Security and Privacy Solutions for Camera and Camera-Based Authentication*. Diss. The Pennsylvania State University, 2021.

[49] Shklyan, K. "Disillusioned defenders? The integration challenges of American Jewish return migrants in the Israel Defense Forces." *Nations and Nationalism* (2021). Available at: https://ccis.ucsd.edu/_files/journals/nana.12767.pdf.

[50] Torkey, H, G Attiya, and A A Nabi. "An efficient congestion control protocol for wired/wireless networks." *International Journal of Electronics Communication and Computer Engineering* 5(1) (2014): 77.

[51] Hartmann, A. "Greenhouse gas emissions from compacted peat soil." (2021). Available at https://stud.epsilon.slu.se/17294/.

[52] FV Torres, F Josue. *Study on Air Interface Variants and their Harmonization for Beyond 5G Systems*. Diss. Universitat Politècnica de València, 2021.

[53] Yun, H, Y Yu, W Yang, K Lee, and G Kim. "Pano-AVQA: grounded audio-visual question answering on 360deg videos." In *Proceedings of the IEEE/CVF International Conference on Computer Vision*, 2021, 2031–2041.

[54] Priscoli, FD. *Network Control Systems for 5G multi-Connectivity*. Diss. The Sapienza University of Rome.

[55] Chaudhary, V, M Sharma, P Sharma, and D Agarwal. eds. *Deep Learning in Gaming and Animations: Principles and Applications*. CRC Press, London, 2021.

[56] Phung-Duc, T, K Akutsu, K I Kawanishi, O Salameh, and S Wittevrongel. "Queueing models for cognitive wireless networks with sensing time of secondary users." *Annals of Operations Research* 310(2) (2022): 641–660.

[57] Shao, W, A Prabowo, S Zhao, P Koniusz, and F D Salim. "Predicting flight delay with spatio-temporal trajectory convolutional network and airport situational awareness map." *Neurocomputing*, 472 (2021): 280–293.

[58] Pell, S. "Evaluating network technologies for industrial intelligent transportation systems." (2021). Available at https://ir.canterbury.ac.nz/handle/10092/102087.

[59] Alaya-Feki, A B Hadj, A L Cornec, and E Moulines. "Optimization of radio measurements exploitation in wireless mobile networks." *Journal of Communication* 2(7) (2007): 59–67.

[60] Carvalho, J A, A C Mendes, T Brito, and J Lima. "Real cockpit proposal for flight simulation with airbus A32x models: an overview description." In *11th International Conference on Simulation and Modeling Methodologies, Technologies and Applications, Simultech2021*, 2021, 256–263.

[61] Sun, X, S Wandelt, M Husemann, and E Stumpf. "Operational considerations regarding on-demand air mobility: a literature review and research challenges." *Journal of Advanced Transportation* 2021 (2021).

[62] Rios, J L G, J T Gómez, R K Sharma, F Dressler, and M J F G García. "Wideband OFDM-based communications in bus topology as a key enabler for Industry 4.0 networks." *IEEE Access* 9 (2021): 114167–114178.

Chapter 14

Intelligent network access system for environment perception and modeling in Intelligent Transportation Systems

Abstract

The provision of "free tone" or "toll-free" services, for example, is also available on fixed networks through the use of intelligent network (IN) techniques that are peculiar to mobile networks. They will, on the other hand, welcome the ability of an IN architecture (INA) to give tailored services to mobile customers to help them better manage incoming calls. Additionally, the use of IN methods makes it feasible to develop a diverse range of new services. Second-generation cellular systems, like Global System for Mobile Communications (GSM), have architectures that are already capable of supporting IN-type applications. The HLR function, in particular, has a close relationship with the In-Service Control Point (ISP). Mobile customers and service providers will both receive the benefits of technology advancements shortly, which will benefit both parties equally. With the advancement of telecommunication technology and the increasing need for increasingly sophisticated services, attempts to standardize International Intelligent Networks have sprung up to address these issues (IN). Because IN are denigrated by the INs standards, service providers must make their own implementation decisions. These opponents also object to a design that is flexible enough to allow for future expansion to include other IN capability sets (CSs). International IN services provided by standardization organizations such as CCITT/ ITU-T and European Telecommunications Standards Institute (ETSI) enable service providers to implement them INA by providing international IN services. When deploying an intelligent network, the global intelligent network architecture (GINA) should be the starting point, not the endpoint (IN).

Key Words: Intelligent networks (IN), Local number portability (LNP), Internet of Things (IoT), In-service control point (ISP), The public switched telephone network (PSTN)

14.1 Introduction

It is possible to have an intelligent network (IN) that is not reliant on any particular service provider, such as the Internet. As a result, the intelligence of the switch is made available to computer nodes located throughout the network. As a result of this discovery process, network operators will be able to more effectively design and manage their services going forward [1]. Because of the network's flexibility, it is simple and quick to add new features to the system. If and when the Internet of Things (IoT) grows, service providers will encounter several advantages and downsides. With increased network capability to suit customers' continuously changing needs, network intelligence gets more dispersed and intricate as the IN grows in size and complexity. As an example, the networks of traditional operating firms will be interconnected with those of third-party service providers [2]. Local number portability (LNP) causes a plethora of concerns that can only be addressed in an IN environment to comply with government regulations on the subject. To implement standard network topologies such as ITU-T Q.1200, an IN must be implemented. It can be utilized by both stationary and mobile networks. Providing additional value-added services on fixed networks, such as public switched telephone network (PSTN) and integrated services digital network (ISDN) on mobile phones and other portable devices, allows operators to stand out in the marketplace and increase their market share. Nodes on the service layer, rather than solutions based on core switch or equipment intelligence, are responsible for providing the necessary information to users [3]. The switching at this level is completely independent of the core network. The IN nodes are often owned by telecommunications service providers, such as phone companies or mobile phone operators, and are used to transmit data. Because of wireless communications in linked vehicles, a safe and interoperable wireless communications network connecting everything from trucks to trains to traffic lights to cell phones has the potential to revolutionize the way people travel in the United States. New information on how, when, and where people drive their automobiles will be gathered through the use of private signals transmitted between automobiles and infrastructure [4].

In this new data-rich environment, a plethora of new apps will be developed, making roadways safer, less congested, and better for the environment. Consider using apps that will notify you if a car is approaching from behind or if there is ice on the road in front of you while driving. They can also assist you in locating available parking spaces, locating last-minute ride-share partners, and even recommending the best speed to drive at to maximize your vehicle's fuel economy. It is impossible to exaggerate the significance of connected automobile apps and the benefits that they may provide [5]. The PSTN, mobile networks like PSPDN, and ISDN (both narrowband and broadband-based) are all examples of IN applications (B-ISDN). To begin, the adoption of analog telecommunications for the transmission of data marked a watershed moment in human history. This service could not be used in organizations because of the lag it causes when sending and receiving data. A data transfer service was therefore required, one that charged based on the amount of data transferred rather than the amount of time it took to transfer it [6]. It was necessary because it was designed primarily for corporations' use, it was also known as a packet-switched data network. GSM mobile phone technology, which was initially used in 1991, allows for low-speed data transfer. When it comes to running and supplying new services, IN are architectural concepts that have the following characteristics as shown in Figure 14.1.

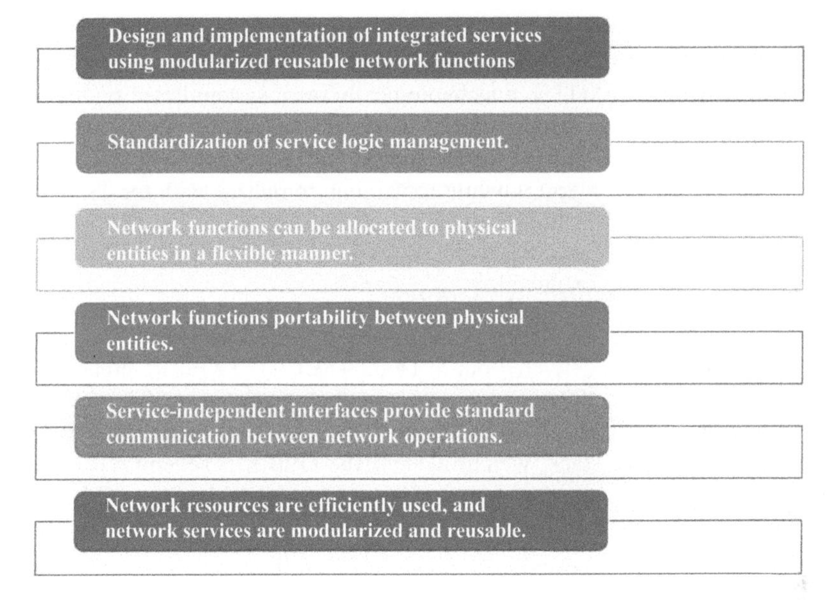

Figure 14.1 IN are architectural concepts that have the following characteristics

Because of the IN's innovative design, it is possible to seamlessly integrate all of these telecommunication services into one system. Physical data transfer in telecommunications and wide-area networks was accomplished through the use of Plesiochronous Digital Hierarchy (PDH) technology. The implementation of the CCITT's Synchronous Digital Hierarchy (SDH) technology resulted in a significant rise in the physical data transfer rates [7]. Asynchronous transfer mode (ATM) is a new technology that was introduced in 1992 that allows for the most efficient use of available bandwidth. Because of the availability of broadband infrastructure, new value-added services, mobile applications, and media material will be made available to the general public. Modem telephony in the United States is being significantly impacted by an increasingly competitive media services industry as well as new technological developments in recent years [8]. The convergence of telecommunications and information technology is driving market developments by expanding customers' access to interactive real-time video and multimedia services, which are becoming more prevalent. This includes digital interactive television, video-on-demand banking, retail, and entertainment services. It also includes electronic publication and electronic publication services. The transmission/access of high-speed, low-cost wide area networks (WBANs), as well as switching/service applications and mobile-enablement support, are all required to make this service function [9].

14.1.1 Universal mobile telecommunications system (UMTS)

International standards for the global telecommunications system, such as the UMTS, are being developed. As part of RACE, the European Union and the ETSI's

SGM5 group have investigated and implemented this technology. In the context of mobile phones, UMTS is a mobile phone system that lets users use their phones at both home and work. UMTS is a technologically open system based on the TMN and IN standards that are used for mobile communications. As a result, the system is capable of converting between ISDN and BISDN via ATM switching and providing broadband mobile access while remaining compliant with the ISDN standard [10].

14.1.2 CCITT signalling system no. 7

Instead of having a distinct data channel for each trunk, this way of transmitting signaling information is delivered out of band. The CCITT OSI Reference Model was used to release the SS7 signaling protocol stack in the year 1980, which was the first time this had happened [11]. In addition to Application Services and User Parts, the SS7 protocol stack, which includes the OSIRM's seven OSIRM layers and is completely digital, also includes the seven OSIRM layers (UP). In the SS7 signaling network structure, the message transfer portion (MTP) and the signal connection control (SCC) portion are components of the network service provider (NSP). Each user part, as well as the intermediate service provider (ISP), is responsible for delivering OSIRM levels 4 through 6. There are numerous moving pieces in the SS7 protocol stack. It is equipped with capabilities to deal with traffic congestion and overcrowding. When a large amount of cargo is transported, there are extra choices for other routes or increased capacity to help alleviate traffic congestion and congestion. As a result of excessive use or a failure, there is a lack of resources available to a network element, resulting in the installed capacity of the network element being reduced.

14.1.3 Network services part

MTP provides message transfer without the need for a network connection, and it supports all three tiers of the SS7 protocol stack. This enables signaling information to be transferred over a network and delivered to the intended receiver without interruption [12]. While systems fail on the network, MTP provides functionalities that allow signaling information to be communicated while the systems fail. During a system or network breakdown, it strives to provide reliable signaling information transfer and delivery over the network, as well as to respond quickly and take the appropriate steps. The signaling data link functionalities of the MTP are made available to the user at the MTP's initial level. A bidirectional signaling data link is a connection that transmits signals in both directions at the same data rate and is capable of transmitting signals in both directions. According to my perception, it adheres as closely as possible to what I consider to be the physical layer of the operating system [13]. MTP Level 2 introduces the signaling connection functionalities, which are described below. The data link layer of the OSI model represents the functionalities of the signaling link. The employment of a signal link in conjunction with a signal data link is a safe way of transporting signals between two signaling locations that are directly connected. At MTP level 3, the user is presented with a visual representation of the signaling network functions. A message can only be delivered to a node using MTP's addressing capabilities, and then the message is dispersed throughout the node using a four-bit service indicator. The use of DPCs and subsystem numbers as an addressing technique increases the capacity of the system in question (SSN). The SSN of a node can be used to determine which SCCP users are present on that node [14].

14.1.4 User part

The user part of the SS7 protocol stack makes use of the SCCP and MTP services provided by the lower layers of the protocol stack. The ISDN-UP, transport control area protocol (TCAP), and OMAP functions are included in the Those of you who work in the networking industry are likely to be familiar with the word TCAP, which stands for TCAP. TCAP makes direct use of the service provided by SCCP. It is for this reason that remote operations protocols and services have been integrated. There is a strong connection between this and the remote operations service element (ROSE) standard [15]. To monitor, coordinate, and regulate all network resources, the SS7 OMAP protocol stack provides the applications protocols and processes required by the protocol stack.

14.1.5 Signalling network structure

When talking about a signaling network, nodes refer to the signaling sites themselves, as well as the signaling lines that connect them [16]. As previously stated, a signaling point (STP) is a level 3 signaling link that transfers signals from one signaling link to another, and it is used to transmit signals between two signaling links (signalling transfer point). Signals that are STPs can also convey level 4 features, such as SCCP and ISDN-UP, as well as other information (signaling points that are STPs). Having "integrated STP functionality" while also having a signaling point that performs both STP and level 4 functions is achievable. In contrast, a stand-alone STP can provide both STP and SCCP capabilities, but not both at the same time. To construct several alternate signaling networks, signaling connections, STPs (both independent and integrated), and signaling points with level 4 protocol functionality can be utilized in conjunction with one another [17]. When using the SS7 Network Services Part protocol, there is no need to be concerned with the underlying signaling network configurations or settings. Any network topology that has unavailability measured in several hours per year or more must have redundancy for the signaling lines to meet the strict availability standards outlined below (e.g., signaling route set unavailability must not exceed 10 min per year). STPs are required to have backups for the majority of the time. The global signaling network will be divided into two functionally distinct tiers: the national tier and the global tier [18]. The outcome is that international and domestic networks can have their unique numbering schemes. Signaling points might be of national or international extent, depending on their location. When a gadget is used for more than one function, it is assigned a different signaling point code for each purpose by a separate signaling network [19].

The rest of the chapter is divided into 10 sections where Section 14.2 describes related work/literature of recent research works by different authors whereas Section 14.3 describes major factors of the connected vehicle system. Section 14.4 describes connected vehicle environment application, Section 14.5 describes connected vehicle safety application, and connected vehicle mobility application is described in Section 14.6. The need for IN described in Section 14.7 and Section 14.8 describes INA. Section 14.9 describes IN standardization and Section 14.10 describe intelligent network functional requirement and at last Section 14.11 describe the summary of the chapter.

14.2 Related/background work

There had previously been proprietary implementations of IN principles, architecture, and protocols by several telecommunications corporations before the ITU-T standardization committee developed them as standards. The IN's original goal was to enhance modern telecommunications networks' basic voice-calling capabilities, particularly the capacity to divert calls when necessary [20]. Providers might build additional services on top of the ones that are already available through a normal phone exchange once this core is in place, thanks to its deployment. ITU-T standards Q.1210 through Q.1219, together known as CS One, published in 1996, went into great detail into the IN. (CS-1). The same set of standards applied to all architectural views, state machines, physical implementation, and protocol implementation. The various telecom providers and operators have all approved of the many variants generated for use around the world. As CS-1 became increasingly popular, so did the creation of CS-2, which had even more improvements [21]. This is in part because the newer ones have more power, but it is also because the previous phone exchanges had problems. CS-2 was far more popular than CS-1, even though both sets of standards were complete at the time. Since a more flexible way to add new complex services to the current network was required, the IN was established. All new functions and/or services must be integrated into the major switch systems before the IN can be built. In recent years, software release cycles have slowed due to the extensive testing needed for network integrity assurance. Due to the IN implementation, services like toll-free numbers and regional number portability were relocated from core switch systems to self-contained nodes. Service providers were able to develop new services and value-added features for their networks without having to contact the core switch maker or wait for a protracted development period because of the open and secure network that was developed in this manner. To give you an idea, the IN technology can instantly convert toll-free numbers to regular PSTN numbers [22]. When compared to its previous function, the IN is now being utilized to build far more complex services like custom local area signaling services (CLASS) and prepaid telephone calls, both of which are under development.

Xu *et al.* (2021): Because of its ability to collect real-time data, the Internet of Vehicles (IoV) is critical in the provision of a wide range of services to consumers.

To facilitate the deployment of a service, acquired data is typically transmitted to a centralized, pricey cloud platform. Real-time services for car users are provided through the deployment of physical resources at roadside units in conjunction with edge computing (EC). Furthermore, although the speed of EC-enabled IoV can be increased through a variety of means, these methods have only a minor impact on the ability to make dynamic judgments in response to real-time queries. Artificial intelligence (AI) can be used to improve the learning capabilities of edge devices, which can then be used to aid in dynamic resource allocation. The use of artificial intelligence to improve EC performance has been extensively researched; nevertheless, explanations that include related concepts or possibilities are difficult to come across. As a result, our investigation into the application of artificial intelligence in IoV edge service improvement has been extensive. Afterward, we will go through what the IoT, electric cars, and artificial intelligence are, as well as how they're connected. To begin, we will examine the IoT edge service frameworks and determine whether artificial intelligence can be used to optimize the deployment of edge servers and offloading services. There are a few concerns that remain with AI-enhanced edge services, which we will examine in more detail at the end of this article [23].

Yingnan Dong *et al.* (2021): While the development of smart cities has improved the aesthetics of our country and drawn the attention of the general public, environmental pollution continues to be a major concern in our country. Now, it is possible to merge environmental awareness with urban information model technology to generate three-dimensional compositions (CIM). A key component of this research is the evaluation of the effectiveness of an intelligent environment system based on CIM. When used in conjunction with CIM technology, technologies like wireless connectivity, micro processing, and other data collecting technologies can be used to create an intelligent environment system. Smart city development, as well as CIM technologies, are employed to achieve this goal. When system non-uniformity and measurement distance inaccuracy are taken into account, it is possible to eliminate the furnace wall temperature perception mistake. When noise and redundant data are removed from the data collected through the use of the Kalman filter technique, the accuracy of the data obtained improves significantly [24].

Mahdi Hashemzadeh *et al.* (2021): It is the new manufacturing plan known as "Industry IIoT," which stands for "Industry 4.0's fourth industrial revolution." As a result of the link between computer-integrated manufacturing (CIM) and AI technologies, a wide range of intelligent industrial services can be developed. The Industrial IoT (IIoT) is predicted to create a large amount of data (IIoT). As a result, new data-driven methodologies are being developed, with intelligence playing a key part in their development. This is where approaches based on machine learning (ML) can be quite useful in this situation. ML, which is a key component of artificial intelligence, can be used to mine massive amounts of data for useful information. Beyond their significance in data mining, knowledge discovery, and the creation of pattern recognition models, they can also serve as a strong computing paradigm for delivering embedded intelligence in industrial applications, among other things. This chapter discusses the benefits and downsides of applying AI, specifically ML, in IIoT systems employing smart industrial IIoT systems, as well as the challenges involved [25].

Chen *et al.* (2021): In mixed-flow traffic networks, it is becoming increasingly difficult to establish the dynamic safety of autonomous driving behavior due to the increase in the number of connected vehicles and the disparity in their intelligence levels. Using a dynamic game model with multi-source information in an intelligent networked environment, this research develops autonomous vehicle dynamics models, forward and backward active safety control, and mixed traffic trajectory optimization planning for self-driving cars. To tackle multiple difficult problems such as perception, judgment, control coordination, and dependability evaluation in an intelligently networked setting, it is necessary to develop new technologies. To ensure that dynamic safety decisions for intelligent networked vehicles are accurate, the V2X actual intelligent transportation system has been integrated with smart laboratory virtual simulation test technologies. This technology makes it possible for autonomous vehicles to make safety decisions under several connected levels of mixed failure to maintain their safety. Continue reading to find out more about this topic. Continue reading to find out more information. In the course of this research, testing costs will be reduced, and game interaction behavior and stress safety response in a road environment will be demonstrated. Additionally, the robustness and application of automatic driving technology will be improved [26].

Mario *et al.* (2021): The purpose of this article is to provide an in-depth look into Construction 4.0 from the aspect of building information modeling (BIM). The origins and applications of Construction 4.0 will be discussed, as well as how its key drivers—integration of BIM with the IoT and big data—will interact in the future (BD). The study's primary goal is to determine the drivers of Construction 4.0 and the amount to which they are triggered by the 4IR, as well as their development and synergy with BIM, as well as the direction in which BIM implementation will take place during the construction phase, among other things. Construction 4.0, as defined by the Fourth Industrial Revolution (4IR), is only possible in certain implementations where BIM, IoT, and BD are the major drivers. The findings of the BIM study reveal indicators of project improvement through the use of IoT and/or BD in conjunction with real-time monitoring, data transfer, and analysis. These integrative approaches to corporate strategy bring all of the aspects mentioned above together in a single package. According to new research, the preconstruction phase of a project is growing and getting more automated as time goes on. These drivers are currently in use in that location [27].

Bruno Guilherme *et al.* (2021): The use of ubiquitous computing systems has increased dramatically in recent years as a result of technological advancements such as mobile computing, more precise sensors, and particular protocols for the IoT. When it comes to research on this topic, context awareness is a prominent way of inquiry. The circumstances of a greenhouse serve as an illustration of the environment in an agricultural context. Recently, several studies have called for the use of sensors to monitor production and/or the use of cameras to capture information on cultivation, to provide farmers with data, reminders, and warnings, among other benefits. This article introduces a computer model for indoor agriculture, named Indoor Plant, which is described in detail. The model examines context histories to provide intelligent generic services like production predictions, challenges in cultivation, and suggestions for greenhouse parameter changes, among other things. It was decided whether or not indoor plant would work based on hydroponic production data from seven-month-long radicchio, lettuce, and arugula cultivations in three different farmer conditions, and the results were promising. To end, findings from the usage of context histories by intelligent service systems are presented in the final section of the essay [28].

Russell *et al.* (2021): Typically, artificial intelligence systems are designed with the notion that they must maximize their behavior to attain a preset and predetermined goal. System designs that follow the conventional method are becoming increasingly popular because they work so well. This document contains an overview of the current state of technology, as well as projections for the next 10 years, and is available for download. General-purpose AI that achieves significant success in the global economy and human roles within it will have a significantly greater impact than is now anticipated. Furthermore, as it becomes increasingly difficult to describe objectives clearly and exhaustively, the standard model will become increasingly unsuitable for practical reasons. This is what I believe will transpire. I believe that a new strategy for building artificial intelligence is required. My suggestion is to have robots that are more durable, controllable, and obedient as they become aware of the true aim. Currently, AI is being developed in several ways [29].

Rahman *et al.* (2021): Through the use of geotagged social media data and interviews with road users, this study evaluates how well the community's shared transportation infrastructure meets them and their needs. It is possible to construct an algorithm that eliminates harmful information regarding how individuals perceive active mobility in the world by combining text categorization with contextual knowledge and contextual knowledge. It is estimated that approximately 75% of the tweets generated by a heuristic-based keyword matching approach are out of context, rendering it ineffective for gathering data from Twitter just through phrase matching. The effectiveness of tweet categorization is evaluated in this study, which makes use of six different text classification algorithms to do so. Following that, content analysis is performed to determine how the keywords are distributed throughout the filtered data that has been generated. We discovered that employing a vectorizer-based logistic regression model with an inverse document frequency was the most successful method for classifying tweets (term frequency). Among the metrics included in this comparison are precision and recall measures, as well as the F1 score (geometric mean of precision and recall) and accuracy metrics. According to the findings of the research, this technique can aid in the collection of more relevant information about walking and bicycling facilities as well as safety concerns. An examination of what people have to say about the biking and pedestrian infrastructure in Washington, DC, reveals what needs to be improved. Through the application of this approach, existing transportation infrastructure may be appraised for quality of service and utilized more efficiently [30].

Tai *et al.* (2021): An innovative XR and deep learning-based IoMT solution for the COVID-19 telemedicine diagnostic are presented in this paper. The solution, which systematically combines VR/AR remote surgical plan/rehearse hardware, customized 5G cloud computing, and deep learning algorithms, provides real-time COVID-19 treatment scheme clues for telemedicine diagnostics. Comparing our new technique to existing perception therapy techniques, we have found that it can greatly increase both performance and overall security. By using 5G transmission, the system was able to acquire 25 clinic data from the 347 positive and 2,270 negative COVID-19 patients in the Red Zone. Afterward, a unique ACGAN-based intelligent prediction algorithm is used to train the new COVID-19 prediction model, which is then used to forecast future events. Furthermore, the Copycat network is used for model stealing and attack for the IoMT to improve the overall security performance of the system. Design of the XR surgical plan and rehearsal framework, which includes all COVID-19 surgical required details that were produced with the guarantee of a real-time reaction, is completed [31].

14.3 Major factors of connected vehicle system

In the automotive industry, a linked car has the capability of connecting wirelessly to other equipment in the surrounding area [32]. The development of the IoT is mainly reliant on the use of linked automobiles. In addition to communicating with other vehicles, mobile devices, and city intersections in both directions, a car that is connected to the internet can communicate with connected entertainment systems that are linked to the driver's cell phone [33]. The IoT, which is used to connect car technologies, has far-reaching implications. The United States Government's Intelligent Transportation Systems strategy incorporates connected automobile technology, and several pilot projects are now underway in various locations across the globe. Communication between automobiles and communication between vehicles and a roadside unit are two of the most important use cases for the IoT automotive in terms of safety (also known as V2X). However, as we will see, IoT-connected automobile technologies have a wide range of other applications in the automotive industry [34]. The United States Department of Transportation (USDOT) evaluated the safety, mobility, and environmental impacts of connected vehicle applications developed using data from four connected vehicle research programs: Vehicle to Infrastructure (V2I) Safety, Dynamic Mobility Applications, Environmental Applications: Real-Time Information Synthesis, and Road-Weather Monitoring. As a result of these initiatives, more than 50 V2I apps have been developed, which are assisting in the reduction of road accidents and deaths, travel time, and pollution. Tests in the field and studies have revealed that these applications have the potential to be widely adopted. According to the findings of several connected vehicle application development programs and studies, V2I apps have the potential to provide significant benefits in terms of safety, mobility, and the environment [35].

How does connected vehicle work?

To trigger essential messages and events, connected vehicles link to a network to enable bidirectional communications between vehicles (cars, trucks, buses, and trains), mobile devices, and infrastructure [36]. Transportation in cities and at intersections is made possible by vehicles equipped with linked vehicle technology, which allows them to communicate continually and receive location information in near real-time. As a result, a pre-programmed response is triggered. The fact page of the federal government provides a succinct and straightforward description of connected automobiles.

The use of in-vehicle or aftermarket gadgets that broadcast critical safety and mobility information will allow cars, trucks, buses, and other vehicles to "speak" to one another. Toll booths, school zones, traffic signals, and other points of interest can all communicate wirelessly with surrounding automobiles if they are equipped with the appropriate technology. Because the information you submit is fully anonymous, no one will be able to monitor your vehicle [37].

Examples of connected vehicles at work in the IoT

Increasing complexity and interconnection of vehicles in the future will allow them to perform a larger range of jobs, which will be made possible by more powerful network infrastructures. For example, the GPS network enables a car to plan a route while considering current traffic conditions to avoid traffic jams, which is an example of a technology that supports connected vehicles and is available now [38]. Lower-latency networks, as well as advancements in connected car technology, will increase the complexity of GPS signals and the usage of these signals by connected cars. Another example is OnStar, which connects a car to a representative of the company so that the driver can seek assistance when necessary. Connected cars increasingly include 4G radios as standard equipment, thereby transforming the vehicle into a mobile hotspot. These are some of the ways that today's connected cars integrate with smart city networks and promote the development of the extremely complex communications required for autonomous driving: Vehicles and infrastructure can now communicate at high speeds, opening up a plethora of new possibilities as shown in Figure 14.2.

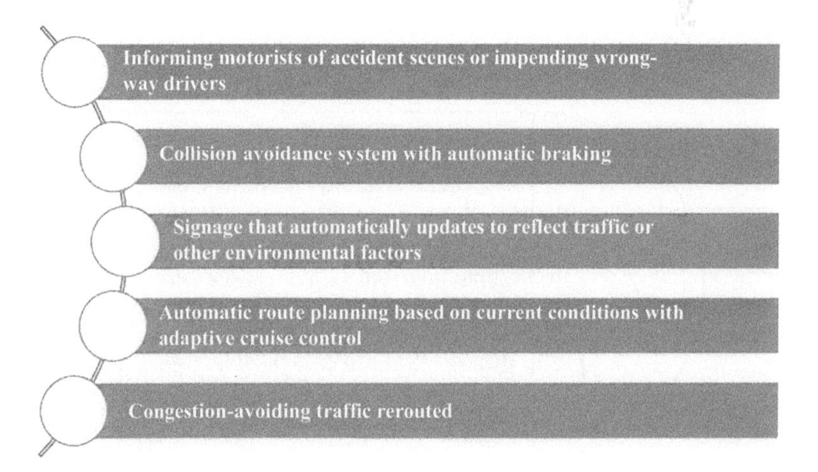

Figure 14.2 Vehicles and infrastructure can now communicate at high speeds

How will 5G networks impact connected vehicle technology?

All future connected cars will be outfitted with 5G receivers and transmitters, which will be standard equipment. Because of the presence of a 5G module, connected autos will be able to communicate in real-time with one another [39]. Two cars traveling in opposing directions may be able to convey information about road conditions based on where they were immediately before communicating. The transmission of driving data by connected vehicles is a fascinating IoT application. Here's an illustration as shown in Figure 14.3

The huge reduction in latency, on the other hand, will allow connected car technology to achieve its primary goal, which is the prevention of accidents and collisions, particularly at crossings in urban environments. As a result, today's autonomous vehicles struggle to maneuver through congested city streets, particularly at intersections and in hazardous environments [40]. It will be possible for the future IoT car to safely navigate through hazardous areas with enough sensors and cameras, as well as the high-speed, low-latency connectivity provided by 5G networks.

Internal on-board use cases for connected vehicle technology

When driving in an internet-connected vehicle, you may do more than just listen to music or make phone calls over the speakers. It is capable of much more when using the smartphone of the driver or passenger [41]. Here's an illustration as shown in Figure 14.4.

The communications backbone for connected vehicle

Intelligent transportation systems (ITS), a smart city and automotive IoT program, will assist linked automobile systems in realizing their full potential. Sensors, cameras, and RFID readers are currently used to monitor intersections and roadways to redirect traffic and compute distances to various locations, among other things [42]. Because of adaptive traffic lighting, emergency services can arrive at accident scenes much faster than in the past. As networks and artificial intelligence capabilities improve, traffic management systems will be able to perform even more in the future. Consider the case where they can track real-time traffic while also looking back in the archives to identify congestion hot spots. If you are looking for better traffic management today, or tomorrow's connected automobile

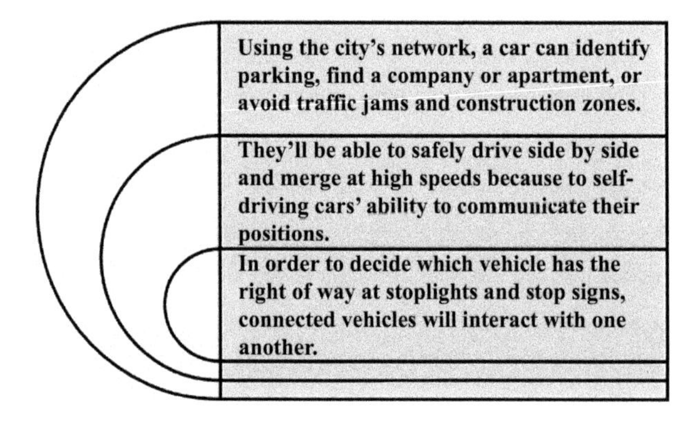

Figure 14.3 IoT application

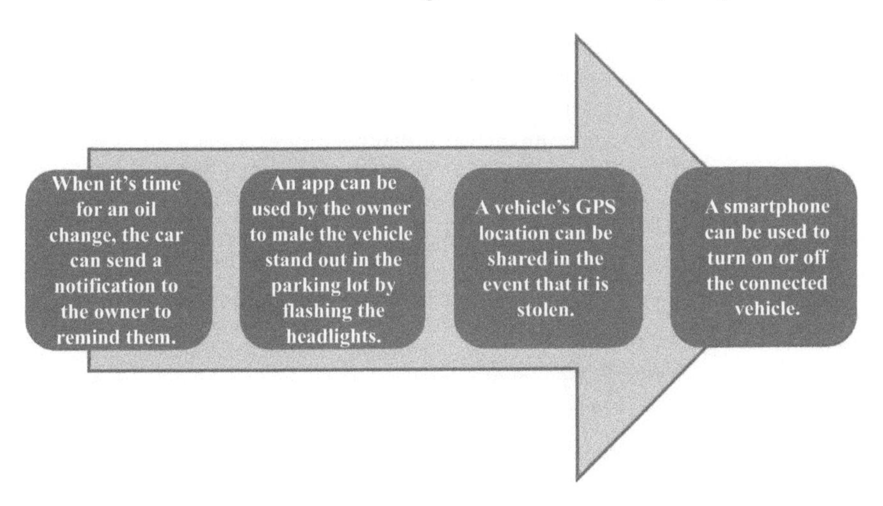

Figure 14.4 Internal on-board use cases for connected vehicle technology

capabilities, smart cities are putting these systems in place right now for both of those goals. Thousands of Digi cellular routers will be put in more than 20,000 junctions by the time this article is printed. When dealing with today's complex traffic management scenarios, gigabit Ethernet speeds, many ports, and computational power is essential, and Digi transportation routers are IoT solutions that can provide these features [43]. These advancements pave the path for future connected automobiles to be equipped with Gigabit Ethernet speeds and many ports, as well as significant processing capacity. It is possible to improve the efficiency of a city's whole traffic management system right away by implementing these solutions now, all while lowering the overall cost and complexity of the city's infrastructure.

14.4 Connected vehicle environment applications

Vehicle connectivity, as a building block for a smart society, is expected to transform the way people get around [44]. To connect vehicles, infrastructure, and passengers, it will be necessary to create wireless communication technologies that are integrated into autos (or in the pockets of the drivers and passengers) as shown in Figure 14.5.

Visual networking index, released by Cisco in March and predicting an average annual growth rate of 37% in what it calls the second-fastest-growing industrial market, says that vehicle connection generates massive amounts of transportation-related data, opening the door to several new applications possibilities (behind linked healthcare). Connected vehicles, which are designed to improve traffic management, public safety, and the environment, will make it feasible to provide new services such as ride-sharing and social networking for automobiles. Increased Internet connection has a variety of benefits, including improved environmental monitoring and infotainment, to name a couple [45]. Because of this, a new architecture and set of standards for developing vehicle application scenarios are being established and standardized.

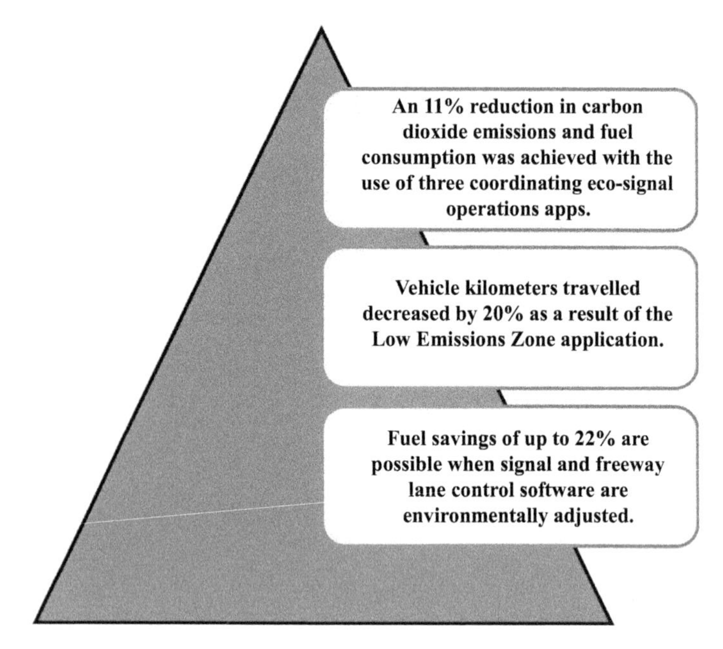

Figure 14.5 Wireless communication technologies

14.5 Connected vehicle safety application

The term "linked auto" refers to vehicles that are connected to their surroundings through the use of apps, services, and other technological innovations. Communication devices (either embedded or portable) are installed into connected vehicles in a variety of configurations. These devices permit communication between vehicles and between vehicles and networks, apps, and services that are located outside the car [46]. These gadgets can either be integrated into a system or carried around with you on your person. A wide range of applications are now accessible, ranging from traffic safety and efficiency to information and entertainment to parking assistance, self-driving automobiles, and GPS (GPS). Cooperative Intelligent Transport Systems (C-ITS) and Advanced Driver Assistance Systems (ADAS), which both provide interactive, real-time support to drivers, are two terms that are frequently used to refer to "connected" automobiles. Using V2V and V2I communications, apps for connected-vehicle safety try to improve situational awareness and prevent traffic accidents by communicating between vehicles and between vehicles and infrastructure (V2I). An ADAS system might include a variety of components such as V2V or V2I systems, vehicle data networks, and sensor technology, to name just a few examples. Adaptive cruise control, automated braking, GPS and traffic warnings, smartphone connectivity, advising the driver of potential risks, and keeping the driver aware of what is in the blind zone are examples of features. It is possible that vehicle-to-vehicle communication technology can help to reduce traffic accidents and congestion by providing critical safety information such as position, speed, and direction between vehicles that are within hearing distance of one another. By implementing this technology, it is possible to improve the effectiveness of front collision warning and blind-spot detection systems, for example. As a result of the shared sensors and awareness data between vehicles, it is envisaged that linked car technology would have a substantial impact on autonomous driving. This will enable cooperative localization and map updates, as well as cooperative maneuvers amongst automated vehicles [47].

Consider the possibility that we live in a world in which automobiles and essential infrastructure are connected and to the rest of us. Because it allows for the establishment of secure, interoperable wireless links between autos, infrastructure, and personal communications devices, connected vehicle technology has the potential to completely transform our transportation system. They are, however, simple precautions, and the driver has complete control at all times when driving. Drivers will be alerted to potential threats via a seat vibration or tone on their visual display. New technology, on the other hand, may be able to alert drivers to potentially dangerous situations. Using limited automated capabilities, such as the junction movement assistance program, drivers can react more swiftly to probable collisions when the car has limited automated capabilities. This program warns vehicles when it is unsafe to enter an intersection without passing through a stop sign [48].

It is difficult to exaggerate the importance of this blind spot warning software, which provides commercial drivers with a virtual perspective of what is going on in their blind area while they are on the road. Drivers will receive an alert if passing a slower-moving vehicle is deemed dangerous using this app. Along with alerting drivers when a train is approaching, connected automobiles can also assist with traffic and safety issues that arise as a result of bad weather. The importance of this is magnified when driving on black ice since even when the weather does not appear to be hazardous, the streets are exceedingly hazardous. Using data from several linked vehicles, it is possible to identify and warn of potential hazards such as slick roads well in advance of the event taking place. Vehicle weather data can be given to traffic management centers (TMCs) so that they can use it to help monitor and regulate the performance of the transportation system, such as by altering traffic signals and speed limitations, for example. It is possible to get up-to-the-minute road weather information by dialing 5-1-1 or listening to highway radio stations. It is also possible to get roadside help and have maintenance cars assigned to the roadside to deliver notifications. When connected vehicles provide drivers with road weather information on their own devices before they leave home, they can help us minimize our carbon footprint and provide environmentally friendly transportation options [49].

When you drive in an eco-lane, you have decided to use a car that is environmentally friendly and to limit your speed to conserve petroleum. Drivers will be informed of how quickly they must travel in these portions using dynamic message signs. Eco lanes are similar to the high-occupancy vehicle lanes that exist today (HOV lanes). Automobiles can communicate with smart traffic signals via vehicle-connected apps to reduce idling and unnecessary stops (VCAs). Adjusting their vehicle speed to pass the next traffic signal on green or slowing down to a standstill can help drivers save fuel, reduce emissions, and lower their costs while also saving money. First responders on the site could be alerted of impending vehicles by using shoulder radios to communicate

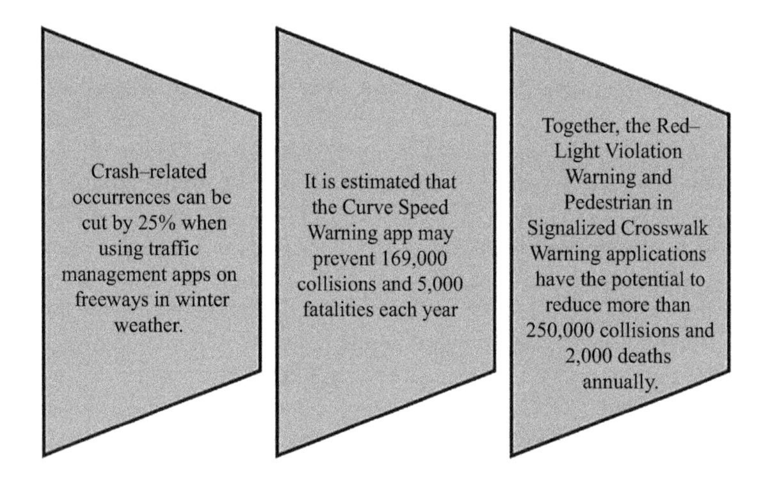

Figure 14.6 Connected vehicle safety application

with one another. In the case of disabled vehicles, car collisions, police activity, and first responders caring for crash victims, the increased knowledge that linked automobiles can supply could be advantageous to all of these groups of individuals. Because of the large number of linked vehicles on the road, it is possible to regulate traffic flow and alleviate congestion at accident sites before things get out of hand. We will be able to optimize traffic flow even more if we can exchange reliable data between ourselves and with smart infrastructure. Increase collaboration amongst public transportation systems and make our neighborhoods and towns even safer and more livable.

In both directions, the application for connection protection allows public transportation operators and customers to speak with one another through a secure connection. Passengers can access real-time transportation information, such as the number of people who will make it to their next destination if a bus is delayed, over the network. To better accommodate clients who, need to catch a train or an airplane, it is likely that bus companies will change their departure times. Transportation-sharing programs that make use of mobile devices and vehicle connectivity improve the logistics of scheduling, reliability, and communication, among other things. This software uses traffic data to locate slow-moving stretches of road and alerts drivers to reduce their speeds to avoid rear-end incidents. It is possible to identify the presence of a person with a disability at a crossing using this app, which alerts drivers to their presence [50]. V2I safety and road weather apps work together to reduce collisions. Figure 14.6 describes connected vehicle safety applications.

14.6 Connected vehicle mobility application

New communications, processing, and storage infrastructure are being developed to meet the varying data volume, veracity, and latency requirements of the connected car ecosystem, which is now in development. For connected vehicles of the future, the detection and avoidance of road hazards will be key safety considerations. For hazard reporting to be actionable, low latency connectivity between car cell towers and data centers is essential for actionability [51]. Autos and fleets of autonomous and remote-operated vehicles will require more and more v2x technology as the technology develops. Vehicles and infrastructure must communicate with one another continuously to maintain track of traffic conditions and critical metadata. Accident investigation, data collecting, and analysis are all part of the job. Traffic congestion, bad weather, and other road hazards necessitate the need for connected vehicles to be responsive and nimble to receive timely notifications. Maintaining a safe environment for vulnerable road users such as pedestrians and bicyclists in highly populated areas is particularly difficult. To facilitate information exchange and system integration, a millisecond response time is essential. This collaborative linked ecosystem takes advantage of surrounding networks as well as cloud-based data processing to improve road safety [52]. To find out more, look into the connected car ecosystem. Intelligent networks can improve current services while also generating new revenue sources in the future. Providers must be able to do the following to achieve these goals as shown in Figure 14.7.

To address the needs of all of their consumers, service providers no longer rely on a small number of vendors. The AIN network and stored program-controlled

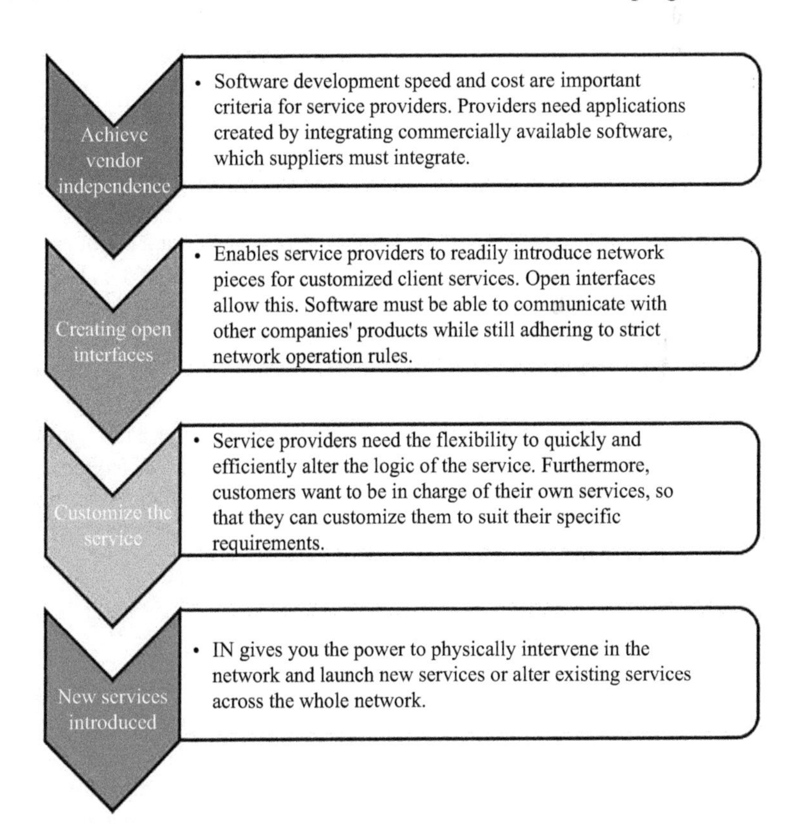

Figure 14.7 Connected vehicle mobility application

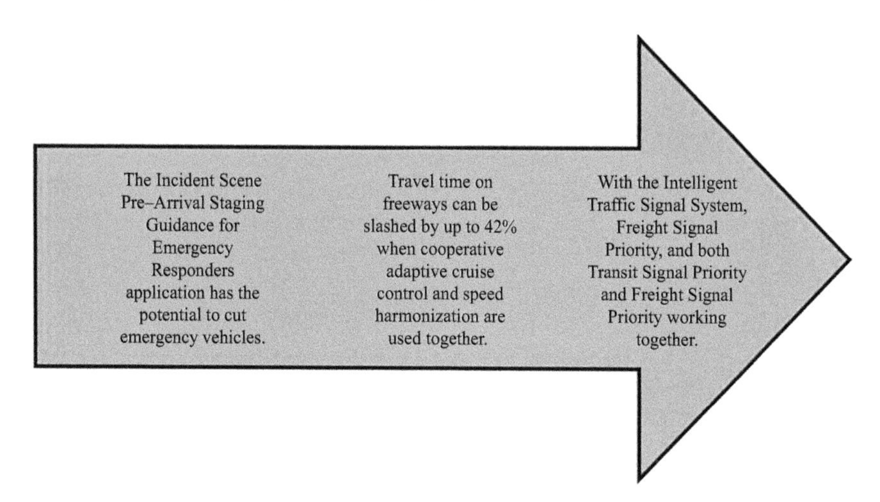

Figure 14.8 V2I apps can be used in combination

switching are developed on top of the SS7 network and stored program-controlled switching. One of the benefits of using AIN technology is that it allows for the separation of service-specific activities and data from other network resources. To avoid reliance on software development and delivery schedules while switching system providers, this capacity should be utilized wherever possible [53]. To better define and adapt their offerings, service providers now have greater flexibility. These capabilities (sometimes referred to as service logic) are programmed and controlled by service providers in the SCP. All information about a client as well as to service is stored in the SCP. The SCP can be accessed by both parties at the same time. The services provided by the IN are not pre-packaged; rather, they are tailored to match the specific needs of each client that engages with them. The process of developing cost-effective services becomes much simpler as the control is transferred to the service provider, which is where it should be. NSPs can test new services before placing them on the market by deploying an SCP that is pre-loaded with service logic and one or more switching systems that have triggering capabilities on the network [54]. To priorities signal timing and reduce travel time and total delay, V2I apps can be used in combination as shown in Figure 14.8.

14.7 The need for IN

In recent years, there has been an explosion of technological innovation, notably in the area of communication networks. Previously, the majority of the telecommunications network was under the control of operators. In recent years, as data sharing and application processing have become more common, the need for standard interfaces between network elements, as well as consumer demand for increasingly sophisticated telecom services have increased dramatically, resulting in a dramatic increase in network element control [55]. Telecommunications networks are now operated, serviced, and in some cases, totally controlled by a single organization. Control and management interfaces with software support are required to combine the control and administration of many services within an operator as well as to be able to provide third-party control and management services. Bellcore began developing the INA that would allow Regional Bell Operating Companies to compete in the newly deregulated telecommunications business as early as almost ten years ago, according to the company. Network operators would be able to introduce, control, and administer new services more efficiently if a central database were to be hosted at a service control point (SCP).

Using IN, service provisioning in a multivendor environment can be made simpler regardless of how the service or network is set up or how many vendors are involved. By the concept of service implementation independence, service providers are free to create their services without having to worry about equipment vendors making service-specific modifications [56]. The ability to assign resources and functionality as they see fit is not bound by the network implementation; instead, equipment manufacturers are free to do so. Until recently, network topologies tended to change in just one direction. Because network operators and service providers must provide customers with services that are deployed independently, IN enables services to be provided without relying on telecommunications networks or equipment vendors, which is advantageous for consumers. Thus, the IN serves as a vehicle for the dissemination and centralization of telecommunications services. This design makes it possible to roll out new customer-centric services in a timely and cost-effective fashion [57].

14.8 INA

IN is a network service control and management architecture for telecommunications networks. In its request for information (RFI) issued in February 1985, the RBOC requested information on the following topics as shown in Figure 14.9.

Because of the lengthy deployment and development process associated with older telecommunications technologies, introducing short-term services was not an option. Because of IN technology, new services may now be launched quickly and seamlessly without interfering with existing ones. An extensive range of industry standards describe the interfaces between different network control points, but IN criticizes every one of these standards. By the interfaces, vendor systems can supply a wide range of diverse products to customers. Afterward, network operators will have the option of adopting any of these products into their existing network infrastructure. Beyond telecommunications providers, IN provides the possibility to integrate new services into the existing network. As a result of IN, new services have been added to the market [58]. The IN reduces the number of SCPs required to choreograph the delivery of exchange services to a bare minimum, reducing costs (SCP). By implementing the CCITI Signalling System No. 7 standard, it is possible to establish connections between network exchanges and service switching points (also known as Service Switching Points). With the implementation of SS7 and the standardization of application interfaces, the market for SCP and SSP

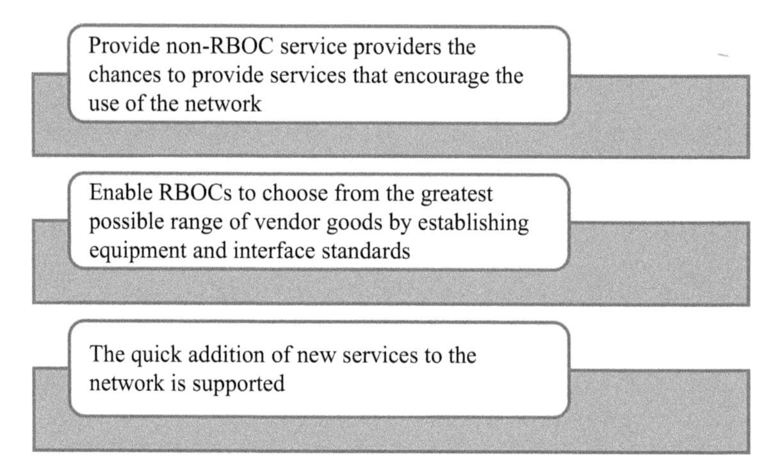

Figure 14.9 The RBOC requested information

software will be accessible to a variety of suppliers (that is, the service control logic and its related data). When new services are introduced or existing services are changed, SSP software cannot be updated to satisfy the short-term requirements of IN (only parameters or trigger updates). A change in parameters or stimuli that has occurred recently will be influenced by the adaptation process. Bellcore initially intended to achieve this goal in two stages, which they designated as phase one and phase two. When the IN/1 definitions were written in 1986, the term "Intelligent Network" was used for the first time. When the IN/2 definitions were written in 1987, it was used for the first time. The definitions for both of these terms were revised in 1989. It was originally planned for publication in 1988, but it was canceled in favor of IN/1+, which was published in 1989. Bellcore discontinued IN/1 + in 1989 due to a lack of multi-vendor participation as well as technical difficulties with the technology. To safeguard the advanced intelligent network (AIN), an MVI was established in 1989 as a backup solution (AIN). In the same period, CCITI and ETSI began working on IN. CCITI. IN/1 marked the introduction of a service-independent architecture. AIN concepts created a service-independent architecture by separating the logic that underpins the switching system from the logic that underpins the service. The CCITI and the ETSI both endorsed these concepts. AlN Release 1 and CCITI CS 1 were both released in 1993, respectively. The specification, development, and control of telecommunication services can be separated from the operation of physical switching networks in an IN.

14.9 IN standardization

14.9.1 IN standards bodies

The CCITT and the ETSI are in charge of defining IN specifications. It is also being worked on by Bellcore, an American non-standards organization that contributes significantly to the American National Standards Institute's Technical Committee [59].

14.9.1.1 European Telecommunications Standards Institute (ETSI)

When the ETSI was founded in 1988, its members included European Telcos (Telecommunications Operating Companies), manufacturers, and end-user organizations. There are two main reasons why the ETSI is present at the event. IN is a member of the second group of countries. ETSI aspires to create European-compatible versions of worldwide standards. Creating European standards in areas where rapid technological improvement is required to keep up with market needs falls under the purview of this organization.

14.9.1.2 CCITT/ITU-T

The Universal Organization for Standardization began working on international IN standards in 1989. Standardization is the responsibility of Study Group XI.4. According to the CCITT or ITU-T (Telecommunication Standardization Sector of the International Telecommunications), the work on the IN standard and its implementation will continue for a short period. The CCITT was the previous name for the Special Interest Group (S/G) of the International Telecommunications Union (ITU). When using a bottom-up strategy, a modest set of criteria is developed first, and these criteria are flexible enough to be modified as the long-term vision becomes more tangible. Collaboration between European Telecommunications Standards Institute (ETSI) and American National Standards Institute (ANSI) is expected to be a critical component in determining how quickly practical IN standardization criteria are developed, to ensure that CCITT recommendations are consistent with their respective programs.

14.9.2 Phased standardization

The CCITT has begun a step-by-step standardized procedure to achieve its objectives, to realize the planned INA. CCITT is also working on developing a long-term CS for each phase of development to gain a better understanding of the IN architecture as a whole (L-TCS). The International Organization for Standardization (ISO) standards are based on CS. In addition to network administration, service processing, and network internet operation, the CSs provide a variety of other capabilities.

14.9.3 Structure of CCIIT IN standards

Each of the other IN standards is constructed on top of the Q.1200 standard, which stands for Q-Series Intelligent Network Recommendations Structure. It is as a result of this that the standards are now sequentially numbering, with one CSx number beginning with 12x and another one following suit with a sequential numbering scheme, such as 12xy, for the CSx recommendation parties, and another one starting with 12x. As a result, recommendation Q.1221 will serve as an introduction to the IN CS2's guiding principles.

14.9.4 Capability Set 1

To counter INA's first move, a coordinated international and European effort was needed. Capability Set 1 (IN Capability Set 1) is a compilation of all of these notions. The CCITT and the ETSI are the two organizations in charge of standardizing CSl technology. The Q.121y series contains a compilation of the CCITT's recommendations. The specifications developed by the CCITT and the ETSI are largely the same. Many major telecom network operators and equipment vendors from around the world are represented in CCITT Study Group XI, working party XU4, which is part of the International Telecommunications Union.

14.10 IN functional requirements

IN functional requirements are required as a result of the demand to provide network capabilities to both clients (service requirements) and network operators to provide network services (network requirements). When a third party makes use of the network's resources, that third party is referred to as a "service user" on the network. Providing a service to its consumers to meet their telecoms requirements is something that an organization does. Other users on the network may be able to deliver or administer services for customers. Third-party service providers include both the companies that supply the services and the services themselves, which are referred to as third-party service providers. Using service criteria will make it easy to determine which services a client is receiving and which are not. These skills have service and network requirements that can be met through the establishment of a network. Figure 14.10 describes IN functional requirements [60].

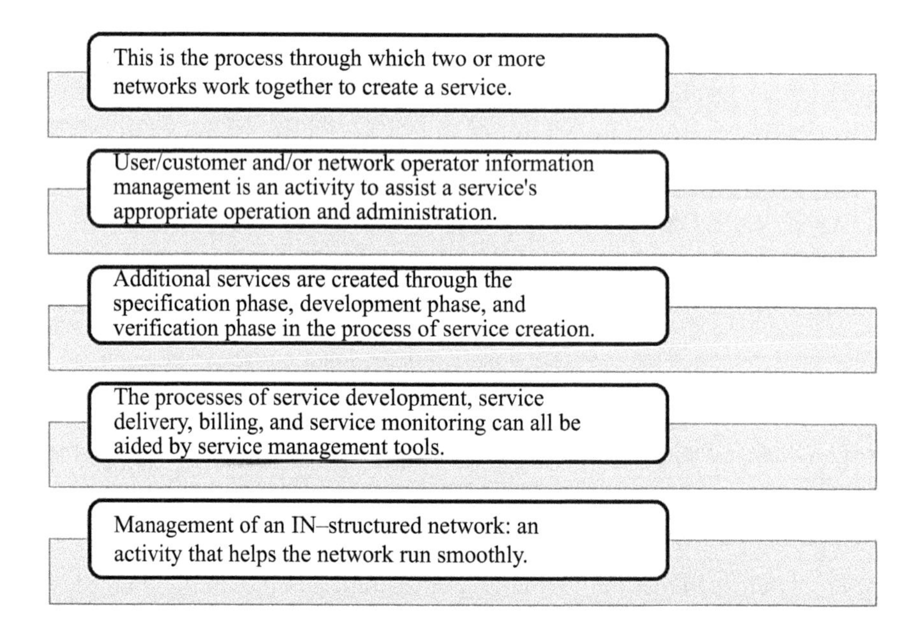

Figure 14.10 IN functional requirements

14.10.1 Service requirements

It is the purpose of IN's effort to debunk a new architectural idea that will allow telecom service providers to meet their customers' present and future service needs rapidly, economically, and without relying on a single vendor to achieve this. This includes enhancing the quality of network service operations and management while simultaneously minimizing the costs associated with these operations and management. When it comes to establishing the IN architecture, [Q1201] meets the following general service requirements as shown in Figure 14.11.

To provide consumers with service creation services, service requirements for service creation must be met by network capabilities that network operators can supply to them. Networks must be equipped with the appropriate capabilities to provide consumers with service management services. An IN-structured network's service requirements for service processing refer to the network capabilities required to provide basic and additional services from the perspective of a consumer. In partnership with the IN, we hope to provide new services efficiently while simultaneously upgrading existing services. Because the provision of services is apparent from that perspective, the consumer has no way of knowing if the service is supplied IN or OUT of the country. Identifying service and access processing requirements are doable and feasible. You can deliver the following core services, as well as additional services, by utilizing IN service capabilities [Ql20l] as shown in Figure 14.12.

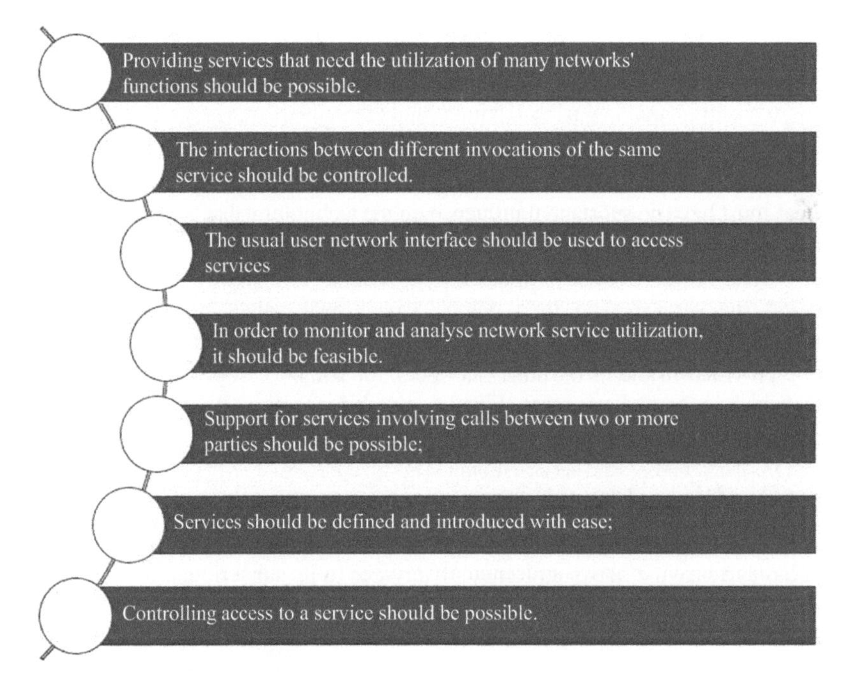

Figure 14.11 Service requirements

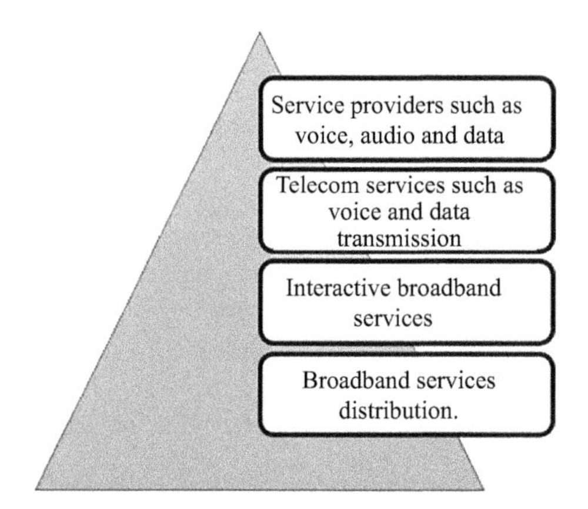

Figure 14.12 IN service capabilities

It is anticipated that the IN-access capabilities will be usable by all telecommunications networks, including Public Switched Telephone Network (PSTN), Integrated Services Digital Network (ISDN), narrowband and broadband, packet-switched public data networks (PDNs), and mobile networks. To name a few examples, there are PSTNs, ISDNs, and mobile phone networks. The fact that only PSTN, Public Land Mobile Network (PLMN), and ISDN are permitted in IN CS1 suggests that IN should allow service providers to define their services without consideration to equipment provider-specific improvements in the PSTN, PLMN, and ISDN technologies. The development of CS 1 for high-value commercial service providers and enterprises has a priority on the flexibility of routing, pricing, and user engagement, among other things. We will go into greater detail regarding the benchmark services and features in a subsequent section. The CCTIT, on the other hand, is not responsible for standardizing these products or services. In addition to being technically feasible and intelligible, the services must have no substantial influence on the technology that is now in use. Type A services are the responsibility of CS 1. Within the framework of CS 1 standards, the type A category possessed several distinct advantages. They start with a broad choice of tried-and-true services and then deliver exceptional value. Because of the well-understood connection control linkages between network components, these services can be provided in the time frame necessary for the IN CSl-product development in 1993, which is around one year. Finally, the shift to more rapid service delivery is advantageous to both the service provider and the equipment manufacturer alike.

14.10.2 Network requirements

Service creation network needs are the network capabilities that a network operator must have to provide new supplemental services to its subscribers. It is necessary to define, build, and test services as part of the service creation process. The network capabilities required by a network operator to successfully manage services are referred to as network requirements for service management. Overall network requirements of IN are stated in [Q1201] as shown in Figure 14.13.

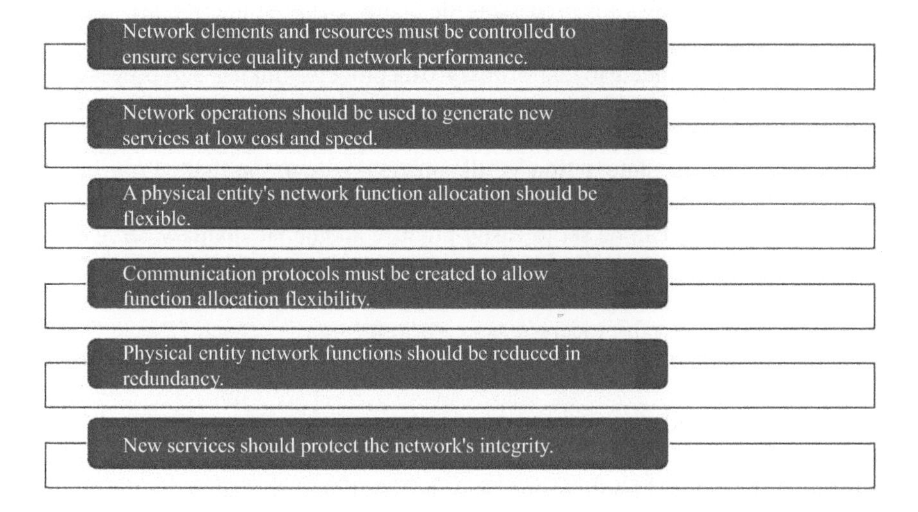

Figure 14.13 Network requirements

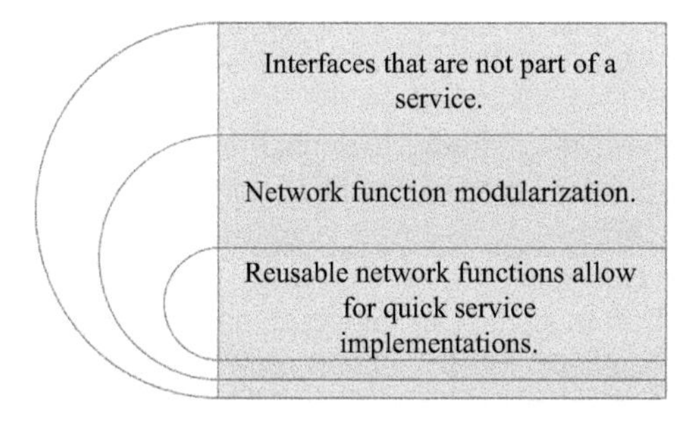

Figure 14.14 IN intends to overcome this impediment

The network capabilities required by an IN-structured network to provide basic and supplemental services are referred to as network requirements for service processing (or simply network requirements). For traditional "non-IN" network operators to be able to swiftly design and deploy new supplementary services, service processing requires the usage of network resources. The IN intends to overcome this impediment in the following ways as shown in Figure 14.14.

So, the in-service processing model was born. This technique's three basic components are call processes, "hooks" to connect with IN service logic, and "coded" services. Here are the important components:

- Services that do not require extra capabilities should be able to use the basic call method. Decomposing the core call mechanism into service-independent sub-processes allows for processing flexibility (without interference from the outside during execution).

- Hooks will be added to the primary call process, linking sub-processes to service logic. In-service logic can interact with "hooks." The fundamental call method must be constantly monitored for circumstances that cause an IN-service logic interaction. Call processing can be halted during an interaction.
- An environment can be set to easily produce additional auxiliary services. IN logic "programmed" can introduce new services. It uses "hooks" to connect in-service logic to the basic call procedure. The IN-service logic regulates the fundamental call sub-processes.

14.11 Conclusion

If the development and implementation of an intelligent grid are to be successful, it will be necessary to consider social and cultural issues. Future intelligent energy systems will be difficult to implement due to the increasing number of interdependent individuals that will act as decision-makers and stakeholders in the system. Smart grid systems have been thoroughly investigated in terms of their objectives and capabilities, but in reality, they are only a small element of the larger "smart system" of the IoT, which includes other smart grid systems. Various current studies are focused on the technical functionality of intelligent networks; nevertheless, they should examine "a system inside a system," with many self-propelled components responding to a wide range of economic and environmental concerns beyond basic operational considerations. For infrastructure systems to continue to fulfill evolving user needs while also serving new public goals, it will be important to maintain, expand, and innovate them in the future.

Investments in production capacity with a controlled degree of flexibility, for example, could give the flexibility required to respond to the volatility and unpredictability associated with renewable energy sources (especially gas-fired units). Because of the widespread storage capacity of their batteries, electric vehicles (EVs) boost the adaptability of the system, even more, when they are introduced. Furthermore, commercial organizations are becoming key actors in other value chain sectors, such as the creation of smart grids, and they must justify their investments by offering strong economic justifications for their decisions. It appears as a result that investment in intelligent networks under the market paradigm differs in several respects from the traditional public monopoly approach. When it comes to tomorrow's energy markets, smart grids will have a significant impact on both our professional and personal life. We may require some time to become acclimated to their presence and operation, as well as their interference with our human lives, which are frequently guided by emotions and vices such as lethargy before we can deal with them effectively. Computers can process large amounts of data accurately, and algorithms are impervious to human emotion, yet understanding the cold logic of computers can be difficult at first.

References

[1] Lv, Z, L Qiao, and S Verma. "AI-enabled IoT-edge data analytics for connected living." *ACM Transactions on Internet Technology* 21(4) (2021): 1–20.

[2] Hu, HQ, J Zhang, and CH Jin. "Research into the design and evaluation of information technology in a learning environment." In *Computational Social Science.* CRC Press, London, 2021, pp. 590–595.

[3] Fitzek, FHP, S C Li, S Speidel, and T Strufe. "Tactile Internet with human-in-the-loop: new frontiers of transdisciplinary research." In *Tactile Internet.* Academic Press, London, 2021, pp. 1–19.

[4] Chen, Q, Y Xie, S Guo, J Bai, and Q Shu. "Sensing system of environmental perception technologies for driverless vehicle: a review of the state of the art and challenges." *Sensors and Actuators A: Physical* 319(2021): 112566.

[5] Zhang, X, H Duan, and S Qu. "Design of "Internet^{+}"-based intelligent greenhouse control system." *2021 IEEE Asia-Pacific Conference on Image Processing, Electronics and Computers (IPEC).* IEEE, New York, NY, 2021, 187–190. Doi: 10.1109/IPEC51340.2021.9421134.

[6] Rehman, A, T Sadad, T Saba, A Hussain, and U Tariq. "Real-time diagnosis system of COVID-19 using X-ray images and deep learning." *It Professional* 23(4) (2021): 57–62.

[7] Mahendran, JK, D T Barry, A K Nivedha, and S M Bhandarkar. "Computer vision-based assistance system for the visually impaired using mobile edge artificial intelligence." In *Proceedings of the IEEE/CVF Conference on Computer Vision and Pattern Recognition*, 2021, 2418–2427.

[8] Tan, L, *et al.* "Toward real-time and efficient cardiovascular monitoring for COVID-19 patients by 5G-enabled wearable medical devices: a deep learning approach." *Neural Computing and Applications* (2021): 1–14.

[9] Browne, M. "Artificial intelligence data-driven Internet of Things systems, real-time process monitoring, and sustainable industrial value creation in smart networked factories." *Journal of Self-Governance and Management Economics* 9(2) (2021): 21–31.

[10] Chen, J. "Visual design of landscape architecture based on high-density three-dimensional Internet of Things." *Complexity* Volume 2021 (2021). Article ID 5534338, https://doi.org/10.1155/2021/5534338.

[11] Suler, P, L Palmer, and S Bilan. "Internet of Things sensing networks, digitized mass production, and sustainable organizational performance in cyber-physical system-based smart factories." *Journal of Self-Governance and Management Economics* 9(2) (2021): 42–51.

[12] Zgank, A. "IoT-based bee swarm activity acoustic classification using deep neural networks." *Sensors* 21(3) (2021): 676.

[13] Mo, Y, S Ma, H Gong, Z Chen, J Zhang, and D Tao. "Terra: a smart and sensible digital twin framework for robust robot deployment in challenging environments." *IEEE Internet of Things Journal* 8(18) (2021): 14039–14050.

[14] Saniuk, S, A Saniuk, and D Cagáňová. "Cyber Industry Networks as an environment of the industry 4.0 implementation." *Wireless Networks* 27(3) (2021): 1649–1655.

[15] Ma, J, H Xie, K Song, and H Liu. "A Bayesian driver agent model for autonomous vehicles system based on knowledge-aware and real-time data." *Sensors* 21(2) (2021): 331.

[16] Hussain, A, T Ali, F Althobiani, *et al.* "Security framework for IoT based real-time health applications." *Electronics* 10(6) (2021): 719.

[17] Zhao, X, S Yongchareon, and N-W Cho. "Enabling situational awareness of business processes." *Business Process Management Journal* 27(3) (2021): 779–795.

[18] Wang, Z, W Feng, J Ye, J Yang, and C Liu. "A study on intelligent manufacturing industrial internet for injection molding industry based on digital twin." *Complexity* 2021 (2021).

[19] Liu, J, Y Duan, Y Wu, R Chen, L Chen, and G Chen. "Information flow perception modeling and optimization of Internet of Things for cloud services." *Future Generation Computer Systems* 115 (2021): 671–679.

[20] Alsufyani, A, Y Alotaibi, A O Almagrabi, S A Alghamdi, and N Alsufyani. "Optimized intelligent data management framework for a cyber-physical system for computational applications." *Complex & Intelligent Systems* (2021): 1–13.

[21] Ali, ES, M Bakri Hassan, and RA Saeed. "Machine learning technologies on Internet of Vehicles." In *Intelligent Technologies for Internet of Vehicles.* Springer, Cham, 2021, pp. 225–252.

[22] Wisetsri, W, K Mangalasserri, L P L Cavaliere, "The impact of marketing practices on NGO performance: the pestel model effect." *Age* 53 (2021): 62.

[23] Xu, X, H Li, W Xu, Z Liu, L Yao, and F Dai. "Artificial intelligence for edge service optimization in internet of vehicles: a survey." *Tsinghua Science and Technology* 27(2) (2021): 270–287.

[24] Wang, J, X Sheng, and Y Dong. "Intelligent environment system based on CIM." *Journal of Physics: Conference Series*. 1881. 4. IOP Publishing, 2021.

[25] Banaie, F, and M Hashemzadeh. "Complementing IIoT services through AI: feasibility and suitability." *AI-Enabled Threat Detection and Security Analysis for Industrial IoT.* Springer, Cham, 2021, pp. 7–19.

[26] Chen, X, X Min, N Li, "Dynamic safety measurement-control technology for intelligent connected vehicles based on digital twin system." *Vibroengineering Procedia* 37 (2021): 78–85.

[27] Begić, H, and M Galić. "A systematic review of construction 4.0 in the context of the BIM 4.0 Premise." *Buildings* 11(8) (2021): 337.

[28] Martini, BG, G A Helfer, J L V Barbosa, *et al.* "IndoorPlant: a model for intelligent services in indoor agriculture based on context histories." *Sensors* 21(5) (2021): 1631.

[29] Russell, S. "The history and future of AI." *Oxford Review of Economic Policy* 37(3) (2021): 509–520.

[30] Rahman, R, K Redwan Shabab, K Chandra Roy, M H Zaki, and S Hasan. "Real-Time Twitter data mining approach to infer user perception toward active mobility." *Transportation Research Record* 2675(9) (2021): 947–960.

[31] Tai, Y, B Gao, Q Li, Z Yu, C Zhu, and V Chang. "Trustworthy and intelligent covid-19 diagnostic iomt through xr and deep learning-based clinic data access." *IEEE Internet of Things Journal* (2021).

[32] Wu, C, F Ge, G Shang, *et al.* "Research on visual perception of intelligent robots based on ADMS." *Journal of Physics: Conference Series*. 1748(2) IOP Publishing, 2021.

[33] Kovacova, M, and E Lewis. "Smart factory performance, cognitive automation, and industrial big data analytics in sustainable manufacturing Internet of Things." *Journal of Self-Governance and Management Economics* 9(3) (2021): 9–21.

[34] Varlamov, O. "Brains" for Robots: application of the Mivar expert systems for implementation of autonomous intelligent robots." *Big Data Research* (2021): 100241.

[35] Huang, X, F Wild, and D Whitelock. "Design dimensions for holographic intelligent agents: a comparative analysis." (2021). Available at http://ceur-ws.org/Vol-2902/paper2.pdf.

[36] Durana, P, N Perkins, and K Valaskova. "Artificial intelligence data-driven internet of things systems, real-time advanced analytics, and cyber-physical production networks in sustainable smart manufacturing." *Economics, Management, and Financial Markets* 16 (2021): 20–30.

[37] Lv, Z, L Qiao, A Kumar Singh, and Q Wang. "AI-empowered IoT security for smart cities." *ACM Transactions on Internet Technology* 21(4) (2021): 1–21.

[38] Karunachandra, RTHSK, and HMKKMB Herath. "Binocular Vision-Based Intelligent 3-D Perception for Robotics Application." (2021).

[39] Mishra, S. and A K Tyagi. "The role of machine learning techniques in Internet of Things-based cloud applications." In Pal, S, De, D, Buyya, R. (eds) *Artificial Intelligence-based Internet of Things Systems. Internet of Things (Technology, Communications and Computing)*. 2022, Springer, Cham. https://doi.org/10.1007/978-3-030-87059-1_4.

[40] Gan, B, C Zhang, Y Chen, and Y C Chen. "Research on role modeling and behavior control of virtual reality animation interactive system in the Internet of Things." *Journal of Real-Time Image Processing* 18(4) (2021): 1069–1083.

[41] Zhou, S, S T Ng, Y Yang, and J F Xu. "Integrating computer vision and traffic modeling for near-real-time signal timing optimization of multiple intersections." *Sustainable Cities and Society* 68 (2021): 102775.

[42] Sun, X, and K Shu. "Application research of perception data fusion system of agricultural product supply chain based on Internet of things." *EURASIP Journal on Wireless Communications and Networking* 2021(1) (2021): 1–18.

[43] Lei, G, R Yao, Y Zhao, and Y Zheng. "Detection and modeling of unstructured roads in forest areas based on visual-2D Lidar data fusion." *Forests* 12(7) (2021): 820.

[44] Xie, Y, K Lian, Q Liu, C Zhang, and H Liu. "Digital twin for cutting tool: modeling, application and service strategy." *Journal of Manufacturing Systems* 58 (2021): 305–312.

[45] Caiado, RGG, L F Scavarda, L O Gavião, P Ivson, D L de Mattos Nascimento, and J A Garza-Reyes. "A fuzzy rule-based industry 4.0 maturity model for operations and supply chain management." *International Journal of Production Economics* 231 (2021): 107883.

[46] Wang, J. "Development trend of intelligent transportation in the era of Big Data." *Journal of Physics: Conference Series*. 1881. 4. IOP Publishing, 2021.

[47] Mendiboure, L, MA Chalouf, and F Krief. "Toward new intelligent architectures for the Internet of Vehicles." *Intelligent Network Management and Control: Intelligent Security, Multi-criteria Optimization, Cloud Computing, Internet of Vehicles, Intelligent Radio* (2021): 193–215.

[48] Javaid, M, A Haleem, R P Singh, and R Suman. "Significance of Quality 4.0 towards comprehensive enhancement in the manufacturing sector." *Sensors International* 2 (2021): 100109.
[49] Jhong, S-Y, Y Y Chen, C H Hsia, S C Lin, K H Hsu, and C F Lai. "Nighttime object detection system with a lightweight deep network for the internet of vehicles." *Journal of Real-Time Image Processing* 18(4) (2021): 1141–1155.
[50] Xu, C, and G Zhu. "Intelligent manufacturing lie group machine learning: real-time and efficient inspection system based on fog computing." *Journal of Intelligent Manufacturing* 32(1) (2021): 237–249.
[51] Haidine, A, F Z Salmam, A Aqqal, and A Dahbi. "Artificial intelligence and machine learning in 5G and beyond: a survey and perspectives." *Moving Broadband Mobile Communications Forward: Intelligent Technologies for 5G and Beyond* (2021): 47.
[52] Monks, I, R A Stewart, O Sahin, and R J Keller. "Taxonomy and model for valuing the contribution of digital water meters to sustainability objectives." *Journal of Environmental Management* 293 (2021): 112846.
[53] Janeera, DA, S S R Gnanamalar, K C Ramya, and A G A Kumar. "Internet of Things and artificial intelligence-enabled secure autonomous vehicles for smart cities." In M Kathiresh and R Neelaveni. (eds) *Automotive Embedded Systems.* EAI/Springer Innovations in Communication and Computing. Springer, Cham, 2021, https://doi.org/10.1007/978-3-030-59897-6_11.
[54] Rehman, A, K Haseeb, T Saba, and H Kolivand. "M-SMDM: a model of security measures using Green Internet of Things with cloud integrated data management for smart cities." *Environmental Technology & Innovation* 24 (2021): 101802.
[55] Zhang, X, and X Ming. "A comprehensive industrial practice for Industrial Internet Platform (IIP): general model, reference architecture, and industrial verification." *Computers & Industrial Engineering* 158 (2021): 107426.
[56] Watkins, D. "Real-time big data analytics, smart industrial value creation, and robotic wireless sensor networks on Internet of Things-based decision support systems." *Economics, Management, and Financial Markets* 16(1) (2021): 31–41.
[57] Madhav, A V S, and A K Tyagi. "The world with future technologies (post-COVID-19): open issues, challenges, and the road ahead." In Tyagi, A K, Abraham, A, and Kaklauskas, A. (eds) *Intelligent Interactive Multimedia Systems for e-Healthcare Applications*. Springer, Singapore. https://doi.org/10.1007/978-981-16-6542-4_22.
[58] Madduru, P. "Artificial intelligence as a service in distributed multi-access edge computing on 5g extracting data using IOT and including AR/VR for real-time reporting." *Information Technology In Industry* 9(1) (2021): 912–931.
[59] Anjum, M, A Voinov, F Taghikhah, and S F Pileggi. "Discussoo: towards an intelligent tool for multi-scale participatory modeling." *Environmental Modelling & Software* 140 (2021): 105044.
[60] Khan, MA. "Intelligent environment enabling autonomous driving." *IEEE Access* 9 (2021): 32997–33017.

Chapter 15

Multi-channel operations, coexistence, and spectrum sharing for vehicular communications

Abstract

With recent advancements in wireless communications, new computer networking research topics have sprung up to provide data network access to regions where wired solutions are impractical or impossible to implement. Vehicle traffic is receiving more academic and industry attention as a result of the sheer variety and significance of related applications, which include everything from road safety to traffic control to mobile entertainment. Autonomous Vehicular Ad Hoc Networks (VANETs) are self-organizing networks made up of autos that are on the go (mobile ad hoc networks). Specialized networking solutions must be designed and tested using computer simulations to determine whether or not they are feasible for the creation of VANETs. It is extremely difficult to create realistic non-uniform vehicle and velocity distributions and unique connection dynamics in VANET simulations, which pose substantial obstacles. However, although the efficient use of other service channels has gotten less attention, resource allocations for safety-related applications on the common control channel have attracted a great deal of attention. Platooning of trucks, autonomous driving, and other intelligent transportation system (ITS) safety applications are all expected to exist side by side with non-ITS mobility technologies in the future, necessitating the usage of various communication channels and sharing of the ITS spectrum. Through the redirection of traffic to alternate paths, multi-channel operations seek to reduce the communication stress placed on individual channels and networks. It is discussed here how ITS legislation and procedures in the United States and Europe deal with coexistence and multi-channel communication. ITS frequency allocations and access restrictions are discussed first, followed by explanations of the protocols that are currently available in standards and research and development.

Keywords: Keywords Roadside unit (RSU), Vehicular communication (VC), Basic safety messages (BSMs), Onboard unit (OBU), Time-division multiple access (TDMA)

15.1 Introduction

A vital contributor to industrial advancement and a novel driving experience, dedicated short-range communication (DSRC) road safety technologies were developed by the cooperative intelligent transportation systems (C-ITS), which was founded by the Department of Defense [1]. Risks on the road can be detected and anticipated using only one well-known safety channel at a time, such as cooperative awareness (CAM) or decentralized environmental notification (DENM), which are used in the European Union (EU), and basic safety messages (BSMs), which are used in the United States, and other well-known safety channels that are not used in the EU. Because this channel has a limited capacity, the vast bulk of scientific, standardized, and industrial research and development has been committed to the development of smart cooperative communication and network solutions that alleviate traffic congestion. Cooperative Intelligent Transport Systems (C-ITS) applications, on the other hand, can make use of a variety of channels [2]. C-ITS apps have been underutilized although they were made accessible early and have the potential to develop from C-ITS applications. There were only one of these channels selected as the common control channel (CCH) in the EU and only one of these channels selected as the control channel in the United States when it came to Day One deployments of C-ITS applications like road hazard alert and intersection collision alert [3]. Multi-channel solutions that make use of direct, smart, and dynamic information are expected to become increasingly important shortly for the equitable and efficient utilization of the full 5.9 GHz C-ITS spectrum. Multi-channel devices are used for three different purposes. They make better use of all available channels' resources by redirecting certain traffic to nearby channels [4]. The provision of services can be done in real time across a variety of channels, and clients can choose to use the service that has been given to them on whichever channel they want. By dynamically switching channels when detrimental interference is detected, they enable diverse technologies to coexist in the same frequency ranges while maintaining high performance. This is an option that they make available. According to the findings of the study, the standards used to characterize these systems have changed at different rates in the United States and the EU [5]. This chapter was written while the European Telecommunications Standards Institute (ETSI) standards were still in the process of being developed. There are no restrictions placed on the use of C-ITS channels; however, the radiation and application access restrictions imposed on each channel make this possible. A channel can be anything from a command-and-control channel to a channel that is available to all C-ITS services. We propose to explore the various channel allocation mechanisms utilized in the United States and the EU for C-ITS applications. During this session, we will cover the distinctions and similarities between the two and how they will be put to use in various forms of media [6]. Figure 15.1 shows multi-channel operations, coexistence, and spectrum sharing for vehicular communication (VC).

The world of DSRC technology is a competitive one, and not all DSRC gadgets are created equal. There are no restrictions on how a system can be configured:

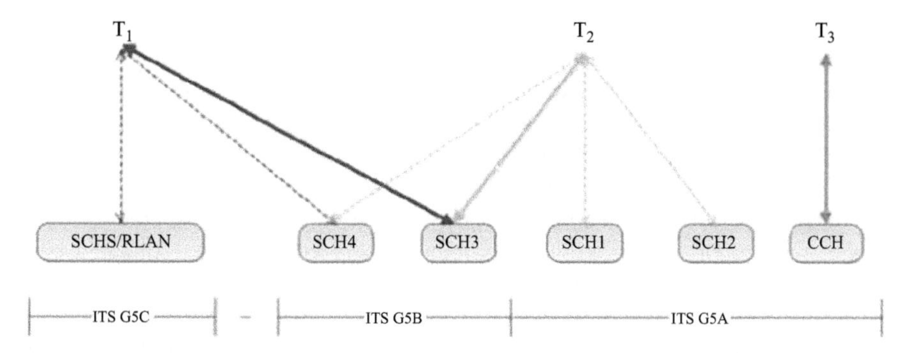

Figure 15.1 Multi-channel operations, coexistence, and spectrum sharing for VC [7]

it can be static or mobile, switchable or non-switching, with a high number of transceivers or none at all, or any combination of these characteristics [8]. We will be concentrating on the various ITS stations in the United States and the EU. When it comes to accessing multiple channels, the onboard unit (OBU) and the roadside unit (RSU) have distinctly different capabilities from one another. Achieving efficient use of all available channels necessitates the development of mechanisms that allow users to switch between channels in real time or slow motion, monitor channel load, offload traffic to other channels, and establish mechanisms that allow C-ITS service providers (SPs) and service consumers (SCs) to meet on a common channel for the consumption of services on DSRC devices that support one or more transceivers [9], with a focus on multi-channel switching methods and their applications. Finally, spectral resources are in short supply in the telecommunications industry. The use of a frequency band such as 70 MHz, which has been set aside but is not fully utilized, attracts the attention of other technologies [10].

The rest of the chapter is framed as follows: Section 15.2 presents the literature survey, i.e., related work done recently by a different author, Section 15.3 describes vehicular mobility, while Section 15.4 the integrated simulation of mobility and communications. Section 15.5 describes frequency allocation. ITS station types and restrictions are described in Section 15.6. Section 15.7 explains the multi-channel operation. Section 15.8 describes the spectrum sharing for VC, whereas Section 15.9 describes the coexistence issues and future challenges. Section 15.10 provides the summary of the complete chapter.

15.2 Related/background work

The use of cognitive radio (CR) technology may have an impact on the way spectrum access networks are designed and operated. Spectrum management technology allows for the modification and control of the spectrum access network to meet the needs of users in CR networks [11]. An optimization problem is a problem that researchers create to find the best possible solution that increases the usefulness of the solution to the intended audience. When it comes to CR networks, game theory is a beneficial spectrum management method, particularly for rate and power control. The use of a game theory-based spectrum allocation technique for CR networks has been proposed. It is important to highlight that these studies did not take into consideration the pricing challenges that CR network players face when selling their spectrum [12].

Irfan *et al.* (2021): Ad hoc wireless networks are becoming more prevalent due to the proliferation of new wireless communication protocols and technologies, as well as increased application and user needs. The most extensively utilized medium access technique for deterministic ad hoc networks is the time-division multiple access (TDMA) protocol. However, like traffic and the number of wireless devices increase, protocol scaling issues arise. These protocols cannot make full use of the spatial spectrum because disclosing nodes is still a problem. The deficit will be alleviated by moving away from fixed allocations and toward free-form spectrum sharing. To improve spectral efficiency and allow the coexistence of numerous networks that are physically adjacent to one another, more distributed scheduling approaches for dynamic spectrum sharing are required. By using control messages sent between nearby nodes in one-hop networks, this research develops a dynamic distributed multi-channel TDMA (DDMC-TDMA) slot management protocol for allocating, removing, and allocating new slots between nodes. Because it is used for slot management in a large-scale, high-density ad hoc network, the topology of the network is unimportant in this scenario. On the ns-3 simulator, many topologies and situations were tested using DDMC-TDMA. According to the simulation results, DDMC-TDMA may achieve near-optimal spectrum usage by addressing both the hidden node and the exposed node issues. There was no loss in performance during testing, even in large-scale, high-density networks. As a result, it permits the use of untrusted wireless networks while yet operating within the same domain [13].

Keyvan *et al.* (2021): Vehicle communications networks are critical for the security and convenience of future C-ITS. It was the IEEE 802.11p DSRC protocol stack that was used for vehicle-to-everything connection in the United States and Japan. Other countries, such as Australia, have put the protocol stack through its paces. Cellular vehicle-to-everything side connection, a novel device-to-device cellular technology that connects automobiles to everything on the network, has emerged with the growth of cellular technologies. If these two technologies cannot coexist, there will be a lot of debate over which one will dominate the C-ITS landscape. Both technologies are constrained in terms of performance by different factors. DSRC and cellular vehicle-to-everything technologies are compared to see how much spectrum they can share in a shared environment. Researchers found that the ITS band can simultaneously send DSRC and CVTE signals for DSRC, proving the study's hypothesis. A descriptive function model is used to assess transmission and reception methods on hybrid vehicle-to-everything platforms [14].

Albinsaid *et al.* (2021): To make better use of radio spectrum, one alternative is to use dynamic spectrum access (DSA). To provide the best possible service to licensee network users, a DSA controller must avoid interfering with the incumbent network's communications. It is difficult for incumbents and licensees to divide the radio spectrum in a distributed DSA network because there is no centralized controller to do so. Power distribution techniques cannot be developed using optimization-driven methodologies in this type of network. In this study, we provide a distributed DSA-based communication framework for multi-agent reinforcement learning (RL), in which the average signal-to-noise ratio value serves as an incentive to use the DSA-based framework for communication. As agents, we take advantage of multiple-cell multi-user multiple-input multiple-output licensee networks. These RL algorithms use DSA's physical layer parameters to learn how to obtain the optimal power allocation policies for accessing the spectrum in a distributed manner without relying on a central DSA controller for interference management with the incumbent. These algorithms include Q-learning, deep deterministic policy gradient, and twin delayed deep deterministic (TD3). These trade-offs are recognized and analyzed in terms of algorithmic performance, temporal complexity, and the potential to scale up DSA networks [15].

Mihai Damian *et al.* (2021): Wireless technology has huge promise for increasing transportation and communication network safety and efficiency. End-users and the respective government agencies and organizations have a strong desire for realistic solutions. In a Vehicular Ad Hoc Network (VANET), using a uniform protocol for vehicle-to-vehicle and vehicle-to-road communications would be beneficial to avoid collisions and accidents. Warning messages can be sent from vehicle-to-vehicle or from vehicle-to-infrastructure using alternate multi-hop routing strategies. To improve communication dependability, this system uses two non-overlapping channels for vehicle-to-vehicle communications. It also aims to achieve low latency communication to improve communication dependability. We will be able to employ common communication protocols and coding methodology if we incorporate the contributions from this project. These protocols and coding methodologies will be customized to the circumstances of the situation. In this book, a communication protocol is employed to emphasize and imitate the exchanges that take place between road infrastructure and its users. The findings of this study demonstrate that the communication protocols vehicle-to-roadside and vehicle-to-vehicle (roadside-to-vehicle) are mutually beneficial and that the shared node significantly enhances the transmission process in low-latency situations, making it perfect for the development of traffic safety systems in general [16].

Merkebu *et al.* (2021): It has become increasingly common to employ machine learning (ML) techniques to solve difficult wireless network issues in recent years. This peaceful coexistence has been made possible by the use of a convolutional neural network (CNN) and Q-learning-based ML techniques. There are currently two ways to determine when a co-located technology transmits: sniffing Wi-Fi packets or utilizing a central coordinator that can communicate with the co-located networks to discuss their status and requirements via a collaboration protocol that is currently under development. It is necessary to follow the steps listed below. In coexistence schemes, the employment of such technologies for traffic status sensing saves costs and complexity, but it also increases traffic overhead and reaction time. Both are unwanted consequences. An over-the-air collection of medium occupation statistics about Wi-Fi frames without decoding is used to analyze the saturation state of a network without decoding to address this problem. Wi-Fi network saturation levels are calculated using these variables and a CNN model based on artificial neural networks. Using ML to identify if a Wi-Fi network is saturated is an option, according to the findings of this research [17].

Ali *et al.* (2021): To manage traffic efficiently and reduce the frequency of traffic accidents, ITS relies on vehicle communications. Mobile communications have grown in popularity over time. Each vehicle must communicate periodic awareness messages throughout the surrounding region to support the use of safety applications, as required by the ETSI ITS-G5 and IEEE 1609 standard families for safety. Increased vehicle density necessitates the exchange of more messages, which could lead to radio channel congestion and degradation of safety-critical services. There is a decentralized congestion control (DCC) mechanism in place at the ETSI to reduce channel congestion using transmission characteristics (such as transmission rate, transmit power, and data rate). The DCC mechanism's performance varies based on the method employed to manage congestion. Researchers studied DCC standardization processes, as well as current techniques and algorithms for congestion reduction, as part of this project [18].

Dong *et al.* (2021): Co-channel interference, which happens when many devices use the same radio channel at the same time, has the potential to significantly degrade network performance. Although coexistence issues have been investigated in several studies, no comprehensive review has been published that summarizes and compares current research results and challenges in IEEE 802.15.4 networks, Bluetooth, and wireless local area networks (WLANs). These types of networks are examined in depth in this study, and we provide a detailed assessment of each. Many essential parameters, such as the packet reception ratio and latency, as well as scalability and energy efficiency, are discussed in terms of current techniques, analyses, and simulation models, among other considerations. Even though both Bluetooth and IEEE 802.15.4 are Wi-Fi-enabled personal area networks (PANs), there are some variations between them (wireless PANs). WLAN interference, on the other hand, does not affect WLAN. Both the IEEE 802.15.4 and Bluetooth networks may coexist peacefully side by side, although they are very similar. The paper concludes with a discussion of research objectives and challenges for the future, including deep learning and reinforcement learning-based algorithms for detecting, minimizing, and reducing wireless PAN and WLAN cochannel interference signals and signals [19].

Aygün *et al.* (2021): Using electricity as a power source, engineers were able to design and build new types of aerial vehicles and transportation systems. Because they can be used for so many different things, DROIDS are becoming more popular among businesses. Soon, you will see more of them in the sky. New ideas in the business suggest that a flying taxi may carry passengers for hundreds of kilometers in the air. UTM systems are already being created to deal with future airspace complexity that will necessitate previously unimagined communication requirements. Voice-oriented communications will become increasingly difficult to meet as commercial flight numbers rise, and future alternatives such as single-pilot operations would require even stronger connectivity to keep up with the rising demand. In this study, we examine the viability of allowing aerial vehicle networking applications using current and future communication technologies. Classification and analysis of connectivity requirements for each aerial vehicle use case is the initial step, which is completed in stages. Using the data from this study, we hope to provide an overview of recent findings in the literature on wireless communication standard implementation's benefits and drawbacks. We plan to cover every aspect of wireless communication technology using more than 500 pieces of relevant research. A review of numerous network configurations led us to compile a list of open-source testbed platforms for upcoming scientists. Despite extensive research into cellular technologies to enable connectivity for aerial platforms, this study showed that only a single wireless technology can match the stringent connectivity requirements of aerial use cases, particularly piloting-related use cases. With the emergence of new investigations into multi-technology heterogeneous network architectures, it became apparent that further work was needed to enable reliable and real-time communication in the sky. Future research should focus on aerial networks that can accommodate a variety of quality-of-service demands for aerial applications [20].

Chen *et al.* (2021) Wireless networks that can be configured to meet the needs of a constantly changing environment are more cost-effective and stable in the long run. This study presents a strategy for DSA that is both efficient and effective, based on CR and primary-prioritized recurrent deep reinforcement learning (RDRL). Because of the Markov state of the spectrum, priority queueing is given to primary and secondary users. Secondary users' spectrum access options provide us the best of both worlds in terms of fair spectrum access and throughput under a wide range of ideal criteria. Secondary users can alter their parameters to select the most efficient access method, thanks to our learning-based DSA technology. Our learning-based DSA technology makes this possible. Prioritized experience replay combined with a recurrent neural network improves the dueling deep Q-convergence network's (DQN) performance, as proven in this research (dueling DQN). RDRL, as proposed, beats the present dueling DQN and DQN schemes in terms of convergence speed and channel throughput, according to the findings of comprehensive testing [21].

Dong-hyu Kim *et al.* (2021): As a result of market-based standardization, there have previously been disagreements about standards both inside and between industries. When it comes to market supremacy, it is important to have standards that are competitive with one another. Contests between standards developed by different committees or standard-setting organizations have received some attention, but competitions between standards established by separate committees or standard-setting organizations have received less attention. Through the use of the standardization mode and the heterogeneity of the participants, this study can identify a variety of different sorts of standards conflicts. Fighting over committee standards is a new sort of conflict that has emerged as a result of system convergence: battles over the definition of what constitutes a standard. The dynamics of these disagreements are illustrated by two examples. Research is being done on the various communication protocols for vehicle-to-everything (V2X) and V2X-to-everything, which are used in the charging of electric vehicles and the installation of smart meters. Communication interference may arise if committee standards for the same frequency bands are not satisfied, which will have a negative influence on the performance of smart systems. A wide range of social incompatibilities exists between a large number of players and coalitions from varied businesses, all of whom have an interest in sustaining a specific norm, thus complicating the already difficult situation. Following the publication of the findings, there are repercussions for heterogeneity, players, and government involvement in standardization conflicts [22].

15.3 Vehicular mobility

The random waypoint model was the first ad hoc network model to be created and was introduced in 1995. At each convincing point, referred to as a waypoint, each random walk (RW) node selects an arbitrary destination and a random velocity from a pool of possibilities [23]. With each iteration of the simulation, the value of the simulation variable decreases steadily until it achieves a steady-state value. The "VELOCITY DECAY" parameter, which was a significant concern in this particular model, was corrected. Navidi and Camp discovered that the distribution is self-reliant in terms of position and speed when operating in the static regime. The application of the Palm strategy assisted in the resolution of a disagreement [24]. Using this random direction model, you may execute fast, staggered motions while optionally waiting between each component of the walk that is performed. Throughout the movement of the knot, the speed and direction remain unchanged. A simulation was performed to estimate the likelihood that the availability of a link would be influenced by traffic. No matter what model you use, you must deal with the issue of "PROBABLY PROVIDE LINK AVAILABILITY."

Further, Sedar and his colleagues proposed a model based on proximity nodes and created speed constraints for nodes that were within a specific distance of one another [25]. Because they are more inclined to travel, the nodes that were constructed near the point of attraction were given a higher probability of being visited. Because of this, it is feasible that realistic driving scenarios will be created in the future. When examining automobile mobility in a specific metro region, the random walk model can be used to uncover social factors that influence driving behavior [26]. According to this mobility paradigm, cars have the option of either driving to neighboring locations or remaining in their current location for an extended period. One of its weaknesses is its inability to change direction fast or recall prior moves. A consequence of this is that neither of these mobility models can come close to precisely simulating the motion of a real-world vehicle. It is possible to construct a grid-like layout for the city of Manhattan utilizing the Manhattan mobility model, which is characterized by chaotic traffic [27]. Figure 15.2 shows different types of VC.

During intersections, the traffic light model algorithm is used to manage the movement of mobile nodes. The establishment of distinct nodes is restricted as a

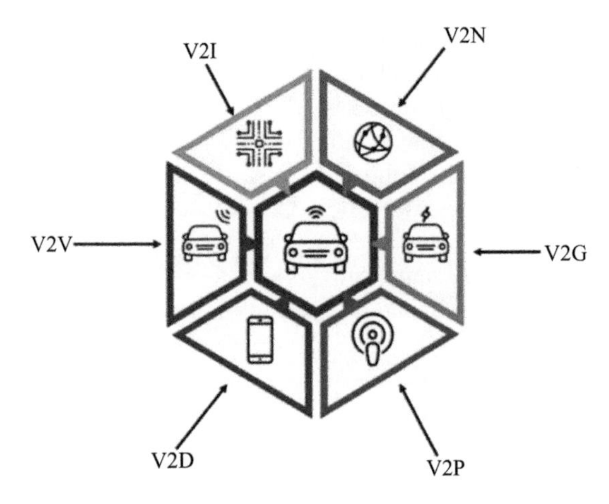

Figure 15.2 Vehicular communication

result of the road's directional characteristics [28]. To accomplish this, it contains both vertical (north and south) and horizontal (east and west) elements. In particular, it is vital to note that Manhattan's mobile node line-based mobility model is still a static model that cannot change the simulation pace of the user. For a realistic model to be accurate, it must account for variations in speed caused by environmental conditions. To mimic a PCS, the Gauss–Markov mobility model was developed (Personal Communication System). It is the stochastic or Gauss–Markov process, which governs the variation in node speed across time, that determines how the network performs [29]. With the random waypoint mechanism, it is possible to generate nodes with varying speeds and directions at random. If the interval is left at its default value of zero, stopping and turning will be a difficulty for the driver. With this model, it is possible to determine how quickly and in which direction a node is migrating. It follows as a result that nodes move in a specific direction and at a specific rate that has been determined throughout time. Expect to see more of the same when that length of time has passed [30].

15.3.1 Generating mobility model for VANETs

As shown in Figure 15.3, the classification of VANETs and mobility models, also known as Fiore's classification, can be divided into four categories.

15.3.1.1 Synthetic model

To achieve a genuine physical effect, mathematical models must be used in conjunction with synthetic mobility models [31]. Figure 15.4 shows definitions of the five models that fall under the synthetic model's jurisdiction.

15.3.1.2 Survey-based model

For another way of saying it, a survey model selected which variables would be included in the final results of the survey. Surveys are used to design mobility

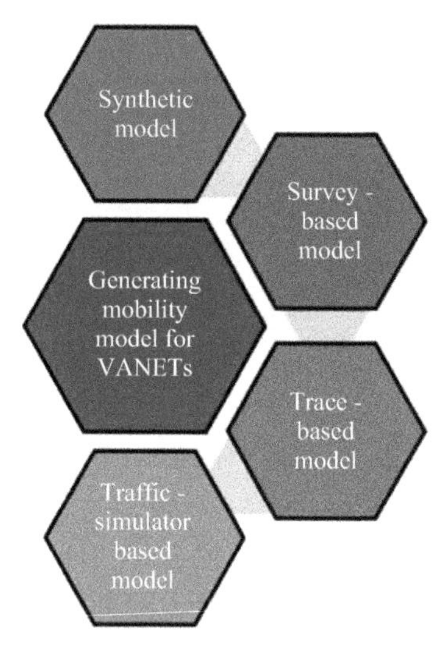

Figure 15.3 Classification of generating mobility model for VANETs

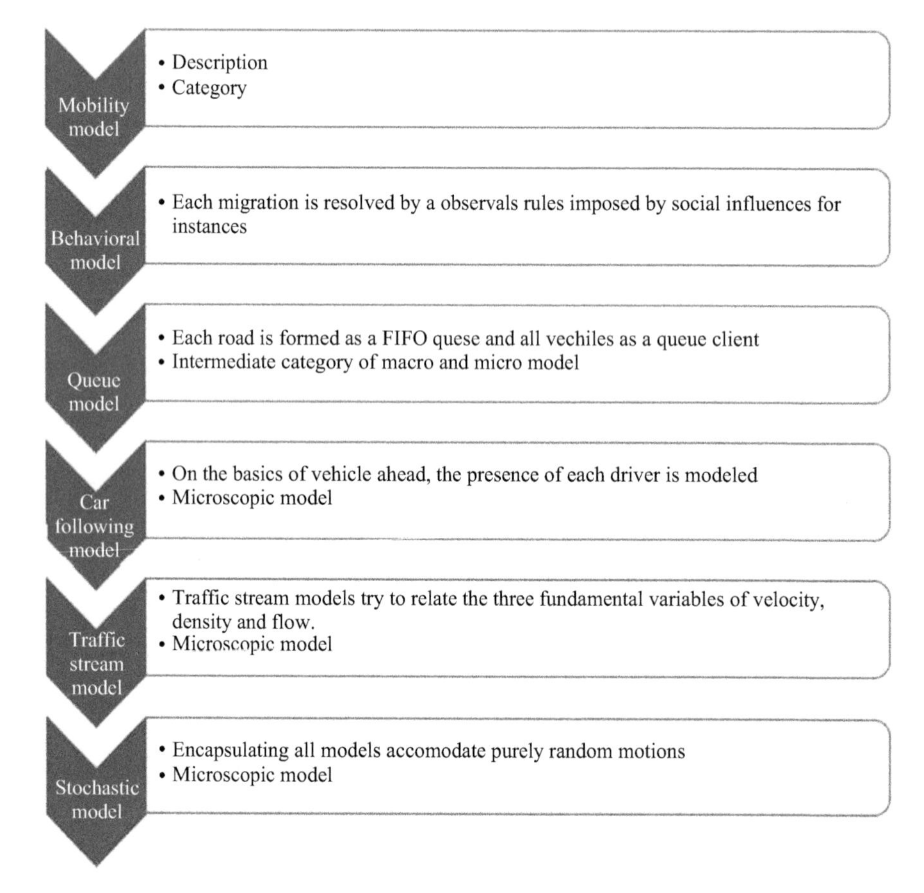

Figure 15.4 Classification of synthetic model [32]

models, which are then implemented. Polls and other forms of study are the most reliable sources of knowledge about macroscopic movement, according to the National Science Foundation [33]. By analyzing statistical data, one can create a mobility model that closely reflects real-world traffic patterns, which can be either pseudo-random or deterministic.

15.3.1.3 Trace-based model

Because the trace-based model is generated from actual mobility traces, the model's mobility pattern is accurate. Traces can be used to determine where a user is, or where a user track is located, on a computer network [34]. When it comes to movement patterns, the most accurate data is collected by employing a trace-based technique. Traces are utilized in the simulation to represent the movement of the vehicle. Traces in a network reveal where nodes have gone and when they have been there. Because of the network's decentralized design, it is difficult to collect real-time traces from all nodes at the same time. The most difficult element of the process is extrapolating patterns from traces that are not visible in the traces themselves. It is feasible to generate confident assumptions about movement patterns that do not present in the traces by employing complex mathematical models. The constraint is typically associated with the type of campaign that is being evaluated [35]. In the case of bus networks, extrapolation models cannot be used to estimate personal automobile traffic because movement traces have already been collected.

15.3.1.4 Traffic-simulator-based model

Micro-traffic, energy consumption, pollution, and noise levels can all be assessed in a city using a traffic-simulator-based model [36]. Traffic engineering simulators like CORSIM, PARAMICS, and VISSIM are now being developed in cities around the world. Incompatibility of the traces and the lack of a finalized interface means these simulators cannot be used for network simulators right away. Additionally, because some traffic simulators are commercial goods, a license from the maker may be required to use them. The end-user may have access to validated traffic patterns by attaching a traffic analyzer to network simulator input files, but this puts them in charge of obtaining the level of detail that no mobility model can achieve on its own [37]. For this reason, it is extremely difficult to calibrate traffic simulators without also making numerous other changes, which is a significant disadvantage.

15.4 Integrated simulation of mobility and communications

Simple car-following models mixed with traffic management create congestion in the system. To account for the ratio of vehicles equipped with radios that are actively talking with one another, the model uses a ratio called enabled vehicular penetration [38]. Vehicle traffic mobility model integration for testing new applications is now feasible, and data transmitted can dynamically alter the routes taken by participating nodes during a run (such as traffic information). STRAW incorporates street patterns into the design to generate a road map for a specific geographic area. STRAW is required to collect data on various road segments, such as the type of road (e.g., residential road or split highway), the start and finish places, the name of the street, and so forth [39]. It must also construct a "point list," which is a list of spots along the segment that are not in a straight line (e.g., detour). In addition, STRAW guarantees that vehicles have at least one clear lane to move in either direction when traveling. Figure 15.5 shows modeling and simulation of vehicular networks communications.

To establish where the automobiles will start their journey on the field, we use a random street layout technique. Until the traffic light changes, the simulation drives a car in a random street lane. The simulation continues indefinitely until the light changes [41]. If there is another vehicle in the same lane as the new car, it will almost definitely be parked behind it. When this happens, all of the nearby vehicles are immediately put on a halt (i.e., assigned a speed of 0). The TIGER data files of the United States Census Bureau are the foundation of our street layouts for the time being. Some files in TIGER data packets organized by state county contain information about numerous geographic elements such as the location and size of schools, parks, highways, and other landmarks. By using these data, we can learn about various aspects of roads and street names and "classes," for example, as well as geographic locations, forms, and capacities [42].

15.4.1 Intra-segment mobility

To keep up with the current driver's pace, as soon as the simulation starts, nodes will try to accelerate continuously. A "car-following model" describes this type of

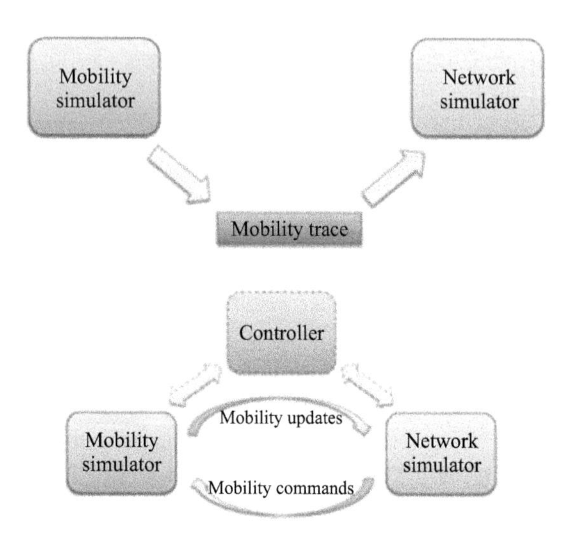

Figure 15.5 Modeling and simulation of vehicular networks communications [40]

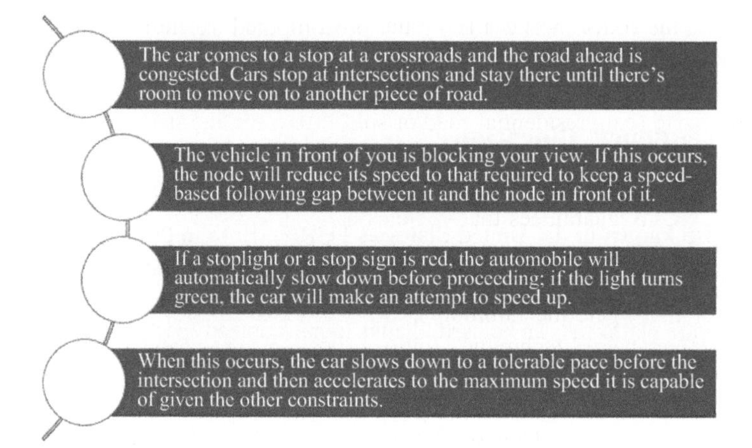

Figure 15.6 Inter-segment mobility

model. Nodes' maximum speed is set to the present road's stated limit plus the mean of a Gaussian distribution of values in this scenario [43]. The following regulations will alter the car's speed as shown in Figure 15.6.

15.4.2 Inter-segment mobility

It is modeled after traditional traffic-management technologies in the first version of the admission control system [44]. When using our traffic simulator, it is a breeze to use stop signs and timed traffic signals at the appropriate times. Eventually, the model will have triggered lighting and guarded turns, among other features. Because lane changing is not yet supported, attempts to make a turn are not taken into consideration. To ensure that a car has enough space before it enters a crossroad, an additional type of admission control can be implemented. Because the simulator does not have access to real-world traffic control data, the type of traffic control assigned to each intersection is determined by the type of road segments existing at that intersection [45]. There is a stop sign that restricts access when two neighborhood roads come together, as well as a timed stoplight that restricts access when a minor road and a state highway come together.

15.4.3 Route management and execution

It is determined via a component known as route management and execution what the overall simulation path for each vehicle will be. Mobility is incorporated into our strategy through the use of origin–destination (OD) pairings and simple segment mobility (simple STRAW) (STRAW OD). Each intersection in the preceding model, which was used to calculate the vehicle's next leg of the route, was determined by chance [46]. Rather than taking into account the likelihood that a car will make a right turn at each given intersection, this model only considers the probability that it will do so to estimate the next segment it will take. As a result of these calculations, the decision is made by the precomputed shortest path between the starting point and the destination.

15.5 Frequency allocation

Vehicle-to-infrastructure communication (V2V) is a key pillar of the C-ITS system concept (also known in the United States as DSRC). Both in the United States and Europe, a licensed spectrum has been set aside just for this. Both in the United States and Europe, these specifics are the same or almost equivalent. In both zones, the same hardware platforms can be used, thanks to harmonized spectrum allotment [47].

15.5.1 US allocation

Spectrum policy in the United States is under the authority of the Federal Communications Commission (FCC). To make room for ICT, the FCC set aside 75 MHz of the 5.850–5.925 GHz frequency range (also known as the 5.9 GHz channel) in 1999 due to the DSRC service (IT). Following discussions with the ITS community and the U.S. Department of Transportation, the FCC issued spectrum licenses and service guidelines in 2003. A total of seven non-overlapping 10 MHz channels is required, as is an unoccupied 5 MHz band at the low end of the spectrum [48]. These standards are strict. There are restrictions to public (government-managed) devices that are not present in private (privately owned) devices, and there are differences between the two types of devices in terms of these constraints. The majority of important applications will use transmit power in the 10–20 dB range instead of that high level, even though power limits normally allow for an EIPR of 33 dBm or more. Reciprocal interfering must be prevented by decreasing transmit power by the transmission range [49]. There are seven 10 MHz bands in all that make up the CCH system, and channel 178 is one of them. After Wireless Access in Vehicular Environments (WAVE) was introduced in 2010, the new IEEE 802.11p-2010 standard was developed to replace it. As a result, the FCC's rules have remained largely intact for years. For example, IEEE 802.11p enables IEEE 802.11 compliance beyond the basic service set, which is an important breakthrough in terms of new functionality [50].

15.5.2 Allocation in the EU

To ensure that all EU member states have equal access to the spectral areas designated to them, the European Commission is in charge of enforcing European spectrum regulations [51]. A portion of this frequency, known as ITS-G5A, was designated by the European Commission (ECC) for use in ITS safety communications in 2008, along with an additional 20 MHz in the 5.905–5.925 GHz spectrum for possible future usage (often known as ITS-G5D). ITS non-safety communications, including data transmissions, are allowed to use 20 MHz of the 5.855–5.875 GHz spectrum, according to an ECC Recommendation issued in 2008. Six of the ECC spectrum distribution's service channels (SCHs) and one CCH are under the jurisdiction of the Federal Communications Commission (FCC). When it comes to availability and utilization, EU allotments are slightly different from those of the FCCs. Only communications related to IT road safety will be allowed on the ITS-G5A frequency band's CCH, SCH1, and SCH2 frequencies [52]. Data from the ITS non-safety system is sent over the ITS-G5B frequency band by SCH3 and SCH4. Other future ITS enhancements will be made possible by the inclusion of newer, lower-frequency bands, which will be used in conjunction with the G5D ITS frequency bands. EC's decision means that these frequencies will not be made available to the general public shortly. Certain bands may not work properly if you live in a specific part of Europe. A special U.S. channel approved by the FCC carries BSMs instead of CCH-encoded CAMs and DENMs, which are sent over both channels in Europe. In this instance, an important distinction has been made. The ECC's ITS-G5A spectrum is not linked to a specific technology, which is another difference between the approaches of the two regulating agencies [53]. A 10 MHz orthogonal frequency division multi-access safety channel is required for the transmission of traffic for other technologies (such as third-generation partnership project long-term evolution (3GPP LTE) and Wi-Fi-Giga) to work inside the ITS-G5A range. Due to the greater spectrum power limits on SCH2, any inter-vehicle communication (IVC) is restricted to short-range transmissions because of interference from neighboring channels (exactly 23 dBm EIRP on the 10 MHz channel). The ITS-G5C band's spectral power limits apply to this band as well as the radio LAN (RLAN) bands (e.g., Wi-Fi-5).

15.6 ITS station types and restrictions

The purpose of a station is determined by whether it is mobile or fixed. An ITS station's ability to access different types of channels, how it executes multi-channel operations, and how rigorous the station may be about sending are all determined by the number of transceivers that the station has at its disposal [54–56]. DSRC devices are subject to access limitations in the United States, and the European Union's ITS station is discussed in this section (ITS-S).

License-required DSRC devices can only use the 5.9-GHz portion of the band. Each type of equipment must have a particular type of licensing according to the FCC's guidelines. OBUs can also function in regions where cars and pedestrians are allowed to travel together. By Part 95 of the FCC's rules, OBUs must obtain an explicit license. RSUs, on the other hand, must be stationary at all times and can only operate on a single site or within a specified geographic zone. RSU operations are governed by FCC guidelines under Part 90. The 5.9 GHz frequency can be used by OBUs to communicate with one another and with RSUs. RSUs connect with OBUs exclusively using the 5.9 GHz frequency. Signal preemption on channel 184 comes in handy for emergency vehicles. To avoid interfering with other traffic, an emergency vehicle might ask a signal controller to adjust the signal condition. This preemption request can travel for many kilometers when powered up to 40 dBm, allowing the signal controller plenty of time to safely clear the intersection before the next vehicle comes through. Although they have additional identifiers, channels 172 and 184 are still regarded as SCHs. In the European Union ITS-S, the ETSI oversees both stationary RSUs and mobile OBUs. This section discusses the ITS-G5 band's ITS-S architecture. Every one of the ITS-G5 transceivers assigned to one or more ITS channels is found in an ITS-S station. In the ITS-S classification, transportation information systems can be divided into three subcategories: safety, traffic efficiency, and commercial applications. The ETSI specification includes information on the supported ITS transceivers, as well as the number and kind of ITS-G5 transceivers that can be used. The ETSI ITS Day 2 application window has been reopened as of the date of this writing. Rather than focusing on a specific approach, we offer an approach that may be applied broadly. To meet ITS application requirements, single-channel IVC communication on the ITS-G5 may or may not be required. Because these devices lack a scan mode, a base channel must be provided for every ITS-G5 transceiver arrangement. Unwanted communications on this base channel may be expected by ITS-G5 transceivers.

15.7 Multi-channel operations

Multi-channel transceivers are wireless transceivers that can operate on many wireless channels at the same time [57–59]. Channel swiping or channel swiping and swiping are other names for this technique. There are numerous DSRC (US) and C-ITS (EU) communication channels available; therefore, rather than assigning a device statically to each channel or avoiding the usage of specific channels entirely, it is advised that multiple channels be used. As described in this section, multi-channel operation in the 5.9 GHz range can be made much more user-friendly.

15.7.1 US regulations

Because there are so many possibilities on the market, individual apps and services may be linked to different channels. Certain channel assignments may be required at specific times and locations based on features such as road geometry, traffic flow, RSUs, and other sources of interference in the vicinity; however, when these variables change, a different set of channel assignments may be required. There will be times when using all of those channels will be necessary for a vehicle or infrastructure equipment, but it is doubtful that a device will be required to use more than two or three of them simultaneously for extended durations in the future (hundreds of milliseconds). A device having nine radios, one for each accessible channel, is theoretically feasible. The effort and money spent were completely in vain. On the other hand, the IEEE 1609 standard series has you covered.

To communicate with one another, all DSRC devices need to learn how to use Wi-Fi scanning to find other DSRC devices. When it comes to BSM, the channel is wide open (channel 172). There is no reason to believe that the other channels cannot be used for other purposes except emergency communication. High-level announcements are used to broadcast service information via IEEE 1609 channel switching when networks are switched. Many distinct interval types exist, including the service channel interval. There is a phrase for this period called the "CCH interval". It is better when devices can find one another on their own rather than having to be planned ahead of time, which is called "reconfiguration." By looking at the Web Services Agreement (WAS), the SP can determine which SCHs are made available to the public. As soon as the user wants to make use of an advertised service, the user must switch from the CCH-tuned radio to one that has been tuned to the specified SCH interval. As soon as the following CCH interval arrives, the gadget turns the radio back on and resumes listening to promoted services over the CCH. See how the single radio paradigm lets you access any or all available channels, one channel at a time, but still just needing to carry one radio? You can turn on and off individual radios on a device without affecting the others. One radio will be permanently tuned to channel 172 for emergency communications and the other will use a switching paradigm to access other services in most operational base units (OBUs). This layout is likely to be favored by a large number of people.

According to IEEE 1609 specifications, a user device has the option to control how much it participates in services by implementing an internal mechanism. As services are requested by the device's upper layers, this management function keeps track of those requests. If a received WSA contains a line item from the service request list, management functions begin channel switching actions to address the issue. Management can use a priority setting in a request list to priorities which services are supplied in cases when several services are offered at the same time. The standards allow the use of primitives to update this service request list. The allocation of PSID values to application regions is documented by the IEEE 1609.12 identifier allocations standard. This registration role, for example, could be taken up by the IEEE Registration Authority in the future. A PSID is assigned by the IEEE 1609 Working Group in the same numbering space as an ITS application identifier (ITS-AID) maintained by ISO and ETSI. The PSID's size might vary from one to four bytes depending on the application. This data is formatted according to IEEE 1609.3. Protocols like IEEE 1609.3s wave short message protocol and WAVE security services make use of the PSID value (IEEE 1609.2). As a result of its adaptability, it can be used in numerous situations. A provider service context (PSC) field may be included in the WSA for each advertised PSID to provide more information about the service supplied. In the PSC field, 1–31 bytes can be used as a range. It is up to the entity in charge of preserving the PSID data to decide which PSC field format should be utilized for each PSID value. This high-level switching mechanism, in which a device tunes to the CCH at every CCH interval and the SCH at every SCH interval, is known as alternate channel access according to IEEE 1609. Using a single-channel access key allows a device to use continuous channel access to stay permanently tuned to a single channel. Immediate and prolonged channel access options are also provided in the specifications. An advertised service is immediately switched on by the user device upon demand, therefore there is no need to wait for a new service cycle period to begin (SCH). To keep the user device tuned to a specific SCH for a longer period, you can extend the duration of channel access. When a user's smartphone is out of reach for a lengthy period, CCH receives a notification. As long as you use these additional techniques, you will be on your way to seeking aid sooner. When both CCH and SCH are operational, an SP can continue to provide the service as long as it can do so during the SCH interval. Based on this data, the user device can make decisions, such as whether to start quick access or whether to allow delayed access to begin. Continuous, alternating, instantaneous, and prolonged channel access are some of the options available to a device. At different times of the day, different devices may operate in different modes.

Using DSRC services has never been easier, thanks to IEEE 1609 channel switching technology. Each sync interval lasts 100 milliseconds and is measured in milliseconds because it is synchronized with the GPS second boundary (msec). The CCH period and the SCH interval are the two periods that make up a sync interval. Before the SCH interval, there is a period known as the CCH period. Because they are mentioned in the standard, these default durations have been employed in the vast majority of tests thus far. The standard allows any other length value as long as the sync interval is less than one second. To allow everything to be reset, there is a guard interval after each CCH and SCH interval has finished and before the next one starts again. Relocating radio resources will be used to achieve this goal. It is possible that if you do not pay close attention to the beginning of a scheduled SCH (or CCH) interval, you will meet an unexpectedly high volume of traffic awaiting transfer. Channel access techniques utilized by the IEEE 802.11 medium access control protocol, when traffic is not divided evenly throughout the interval as it would be under normal circumstances, can cause an excessive amount of frame collisions. A busy channel will be reported by En queue packet interval if CCH or SCH is not begun, which can assist alleviate the situation somewhat.

15.7.2 EU regulations: ETSI ITS

The ETSI's proposal for a multi-channel operation has a striking resemblance to the IEEE 1609.4 and IEEE 1609.3 standards in terms of structure and content. Because this T1 transceiver does not allow multiple channels, it is not possible to operate in a multi-channel configuration. Among the many benefits of this approach is that it enhances the CCH's capacity to speak on safety-related matters. While the ITS-S is being converted to an SCH, the capacity of the CCH channel is severely limited by the synchronous channel switching (SCH) described. Given that, probably, the CCH will not be able to support safety-related applications to their full potential, ITS-S must be able to receive information from the CCH at all times. Two transceivers are necessary for an ITS-S to be able to provide ITS services as well as other services. In comparison to having an ITS-G5A transceiver that functions on the other two SCHs, having an ITS-G5 transceiver that runs on one of the SCHs permanently tuned to it is a greater challenge. Because of this, ETSI was uninterested in information technology service management. To alleviate channel congestion caused by ITS Day One applications on CCH, ETSI concentrated on measures to reduce channel congestion.

15.7.2.1 Multi-channel congestion control

Even while we try our best to reflect the most recent changes, we cannot guarantee that specifics would not be changed shortly. Anyone interested in learning more about this problem should reference the relevant standard, as recommended. This chapter was the first time the ETSI channel switching standard had received any attention from scholars outside of Europe. Depending on the channel settings, services can be provided on up to four separate SCHs. Service management and IEEE 1609 have a lot in common when it comes to safety. SAM is used whenever an ITS SP publishes an announcement about a new service so that it can determine which SCHs are accessible through the service. They seem genuinely interested. First, clients and providers must get together to advertise the service before they can get together to receive it. As long as services and customers are on board with using SAM rendezvous as a channel for ITS communication, the service can be delivered. The service announcement channel (SAC) is the channel through which an ITS transceiver transmits service announcement messages (SAMs) (SACH). To facilitate multi-channel operations, ITS-Ss using T2 transceivers, as shown in Figure 15.4, must monitor ITS-G5A frequency channels. Because of this, even whether safety-related messages are relayed or offloaded, all T2 ITS-S on ITS-G5A must be using the same secondary SCH to ensure that all safety-related messages are conveyed. Only SCH is allowed, just like in IEEE 1609.4.

15.7.2.2 Channel switching principles

Even while we try our best to reflect the most recent changes, we cannot guarantee that specifics would not be changed shortly. Anyone interested in learning more about this problem should reference the relevant standard, as recommended. This chapter was the first time the ETSI channel switching standard had received any attention from scholars outside of Europe. Depending on the channel settings, services can be provided on up to four separate SCHs. Service management and IEEE 1609 have a lot in common when it comes to safety. SAM is used whenever an ITS SP publishes an announcement about a new service so that it can determine which SCHs are accessible through the service. They seem genuinely interested. First, clients and providers must get together to advertise the service before they can get together to receive it. As long as services and customers are on board with using SAM rendezvous as a channel for ITS communication, the service can be delivered. The SAC is the channel through which an ITS transceiver transmits SAMs (SACH). To facilitate multi-channel operations, ITS-Ss using T2 transceivers must monitor ITS-G5A frequency channels. Because of this, even whether safety-related messages are relayed or offloaded, all T2 ITS-S on ITS-G5A must be using the same secondary SCH to ensure that all safety-related messages are conveyed. Only synchronous channel switching (SCH) is allowed, just like in IEEE 1609.4.

15.8 Spectrum sharing for VCs

To enhance traffic safety and efficiency, vehicle networks, which allow vehicles to communicate with one another and with roadside infrastructure, should be created.

As a result, it will have a major influence on the future development of the smart city [60]. DSRC and LTE vehicles are two prominent concepts for automotive communication technologies and standards (LTE-V). With this system, the FCC-allocated frequency for ITS is used to minimize interference from the previous system (ITS). Vehicle data is sent from base stations to individual vehicles using the LTE permitted spectrum. DSRC and LTE-V networks would be in direct competition for spectrum if they were to use the ITS band, which would increase the number of resources available for vehicular network services. In a multi-operator scenario, where different operators have implemented different technological standards, network providers must compete for vehicle users. It is critical to figure out how to compete in the ITS band on the DSRC and LTE-V networks of different operators while maintaining QoS and coexisting in the ITS band on an equal footing. For the time being, the research is concentrating on vehicular user access in a multi-vehicle network, in particular mechanism for a lower handoff network that the driver can select to save time while switching between it and the other networks. Vehicle entrants will have a higher success rate if they can employ a more efficient entry method. To provide suitable network access to vehicle users, an optimal access strategy must be developed. It is becoming a big topic in networking circles to talk about traffic unloading in a diverse network environment [61]. To reduce the burden on mobile networks, it uses game theory-based traffic offloading, which selects Wi-Fi transfers from vehicles with the greatest deals. Cellular networks can be made more efficient by leveraging vehicle-to-vehicle communication to help with transmission between cars and the base station. Network coexistence difficulties can be addressed and network congestion reduced through the usage of unlicensed LTE and Wi-Fi spectrum sharing. A hybrid approach will be used to transport cellular data traffic over unlicensed bands, including techniques such as traffic unloading and resource sharing. To optimize U-LTE network protocol parameters at the medium access control layer and the physical layer and to maximize throughput, we need a cross-layer proportional fairness framework. Researchers are attempting to improve the listen-before-talk mechanism and design a network parameter modification technique to boost the capacity of U-LTE and Wi-Fi networks. They do not account for clients who travel by automobile, even though all of these studies assume that a single network operator operates both the U-LTE and Wi-Fi networks [62].

15.9 Coexistence issues and future challenges

Unlicensed National Information Infrastructure (U-NII) devices should not operate in the 5 GHz band, according to a 2013 FCC Notice of Proposed Rulemaking (NPRM). IEEE 802.11 protocols, more commonly known as "Wi-Fi," are used by many U-NII devices [63]. A lack of readily available spectrum has long been cited by Wi-Fi devices as a major cause for the FCC's NPRM. By using this technique, the 5 GHz spectrum is divided into numerous smaller bands, each of which has a unique frequency. Because of this, the FCC has previously approved the use of UNI I II A B C D and UNI I III for

U-NII operation. In the NPRM, it was recommended that U-frequency NII's operation be altered. In the 5 GHz band, the term "spectrum sharing" is frequently used since U-NII devices use a spectrum that is generally reserved for licensed devices operating in these subbands [64]. Part 15 of the FCC's regulations prohibit U-NII operations from interfering with legally authorized transmissions in a harmful way. According to IEEE 802.11ac2013, wireless networks can use channel bandwidths of 80 MHz or 160 MHz. The U-NII-4 frequency, discovered by chance, piqued the interest of Wi-Fi users worldwide. They complained about having to share the spectrum in the DSRC frequency band. When U-NII devices began transmitting at this frequency, it alarmed DSRC stakeholders who were worried about the agency's ability to carry out its mission of saving human lives [65]. According to a similar proposal, Wi-Fi RLANs should be permitted to operate in the European ITSG5 spectrum. To begin an inquiry, the European Commission contacted the European Conference of Postal and Telecommunications Administrations (CEPT) as soon as it became aware of the matter in September of the same year. Assigned to Engineering Group 24 of the CEPT, they took on the challenge (SE-24: Short Range Devices). Because the CEN DSRC cannot coexist with other technologies that use its particular frequency, it is an important flaw in the European approach that must be fixed as soon as feasible. The DSRC contacted the Wi-Fi community at the beginning of 2013 to start a discussion about whether or not it is viable to regulate Wi-Fi sharing. In August of that year, the IEEE 802.11 WG organized a Tiger Team to study the coexistence of DSRC and Wi-Fi. There were two different ways that Wi-Fi members of the Tiger Team came up with to share information. The DSRC protocol may identify the DSRC device since it uses the IEEE 802.11 CSMA/CA MAC protocol, which listens before it communicates. The CCA proposal is known as the Clear Channel Assessment (CCA). IEEE 802.11ac devices want to use two 20 MHz channels, which is what the DSRC group is proposing. This is a requirement of the rechannelization strategy. Rather than using 5,925 GHz as the boundary, U-upper NII-4's border needs to be drawn at 5,895 GHz instead. The DSRC community has rejected the idea of rechannelization for various factors. According to information shared by Tiger Team with the FCC and other government officials, the CCA concept could help enable sharing without getting in the way. Wi-Fi developers have been urged by members of the DSRC community to keep working on the CCA proposal so that it can be tested on DSR devices in the future. All stakeholders have agreed that before a sharing solution can be allowed for distribution, it must be thoroughly examined. If U-NII devices can share the DSRC band with other devices, the FCC is likely to approve them. The Tiger Team is taking these suggestions under consideration after receiving input from the DSRC community.

15.10 Conclusion

In this chapter, we looked at multi-channel VC systems from the perspectives of North American and European researchers at their most fundamental level. Multi-channel switching techniques are required because the United States has seven DSRC channels, and that operating them all cheaply with a variety of DSRC transceivers is problematic. Although they serve similar objectives and have numerous parallels, their differences can be traced back to the different spectrum allocations in the United States and the European Union. Because just three ITS-G5 channels have been assigned in Europe since 2008, multi-channel switching is no longer a concern. As a result, the ETSI has concentrated on rerouting traffic to less congested routes. We began by discussing channel allocations and various ITS stations before moving on to the topic of channel switching (e.g., RSUs and OBUs). In our final session, we discussed the ideas of channel switching defined by IEEE 1609.4 and IEEE 1609.3, as well as the ETSI and International Telecommunications Union recommendations for switching between different types of channels. The unloading or relaying of some of the ETSI CCH safety-critical traffic to a secondary SCH can assist in controlling the load on the primary CCH, as demonstrated by an example. The asynchronous multi-channel service management system that met current industry standards and was flexible enough to satisfy the needs of both SPs and customers was exhibited and tested. It was a complete and utter success for us. The approaching requirement for DSRC coexistence and spectrum sharing with other technologies on ITS Day Two will make the development of viable multi-channel solutions even more crucial (i.e., Wi-Fi-Giga or LTE-A). It is quite unlikely that a single form of transportation will make use of all of the ITS bands. Cognitive ideas were employed in the early stages of the process to dynamically move traffic between channels based on the presence or absence of other types of traffic. Because of the ongoing improvement and expansion of the processes outlined in this chapter, a greater range of IT services, ranging from safety-related ones such as traffic management, will be supported by them. Congestion control and multi-channel dynamic spectrum sharing were among the first technology goals set by the DSRC/ITS-G5 group (MCDSS). In the future, it is envisaged that these aims will take precedence above all others.

References

[1] Szott, S, K Kosek-Szott, P Gawłowicz, *et al.* "WiFi Meets ML: A Survey on Improving IEEE 802.11 Performance with Machine Learning." *arXiv preprint arXiv*:2109.04786 (2021).

[2] Touati, H, A Chriki, H Snoussi, and F Kamoun. "Cognitive radio and dynamic TDMA for efficient UAVs swarm communications." *Computer Networks* 196 (2021): 108264.

[3] Mamadou, A M, and G Chalhoub. "Enhancing the CSMA/CA of IEEE 802.15. 4 for better coexistence with IEEE 802.11." *Wireless Networks* 27(6) (2021): 3903–3914.

[4] Bagchi, S, and J Y Siddiqui. "Spectrum sensing for cognitive radio using a filter bank approach." *Intelligent Multi-modal Data Processing* (2021): 205–230.

[5] Zhou, T, D Qin, X Nie, X Li, L Zhang, and C Li. "Coalitional game-based user association integrated with open loop power control for green communications in uplink HCNs." *Wireless Personal Communications* 120(4) (2021): 3117–3133.

[6] López-Raventós, Á, and B Bellalta. "IEEE 802.11 be Multi-Link Operation: When the Best Could Be to Use Only a Single Interface." *arXiv preprint arXiv*:2105.10199 (2021).

[7] Härri, J, and J Kenney. "Multi-channel operations, coexistence and spectrum sharing for vehicular communications." *Vehicular ad hoc Networks*. Springer, Cham, 2015. 193–218.

[8] Mukherjee, P, and T De. "Content independent location-based clustering for 5G device to device communications." *Wireless Personal Communications* 118(4) (2021): 2573–2599.

[9] Amjad, M S. *Towards low latency and bandwidth-efficient communication in wireless systems*. Diss. Technische Universitaet Berlin, Germany, 2021.

[10] Kang, B, J Yang, J Paek, and S Bahk. "Atomic: adaptive transmission power and message interval control for C-V2X Mode 4." *IEEE Access* 9 (2021): 12309–12321.

[11] Le, K N. "Systems of wireless relays under imperfect-and-outdated-CSI severity." *Digital Signal Processing* 117 (2021): 103154.

[12] Pei, E, L Zhou, B Deng, X Lu, Y Li, and Z Zhang. "A Q-learning based energy threshold optimization algorithm in LAA networks." *IEEE Transactions on Vehicular Technology* 70(7) (2021): 7037–7049.

[13] Jabandžić, I, S Giannoulis, R Mennes, F A De Figueiredo, M Claeys, and I Moerman. "A dynamic distributed multi-channel TDMA slot management protocol for ad hoc networks." *IEEE Access* 9 (2021): 61864–61886.

[14] Ansari, K. "Joint use of DSRC and C-V2X for V2X communications in the 5.9 GHz ITS band." *IET Intelligent Transport Systems* (2021).

[15] Albinsaid, H, K Singh, S Biswas and C-P. Li. "Multi-agent reinforcement learning based distributed dynamic spectrum access." *IEEE Transactions on Cognitive Communications and Networking* (2021). Doi: 10.1109/TCCN.2021.3120996.

[16] Zadobrischi, E, and M Damian. "Vehicular communications utility in road safety applications: a step toward self-aware intelligent traffic systems." *Symmetry* 13(3) (2021): 438.

[17] Girmay, M, A Shahid, V Maglogiannis, D Naudts, and I Moerman. "Machine learning-enabled Wi-Fi saturation sensing for fair coexistence in unlicensed spectrum." *IEEE Access* 9 (2021): 42959–42974.

[18] Balador, A, E Cinque, M Pratesi, F Valentini, C Bai, A A Gómez, and M Mohammadi. "Survey on decentralized congestion control methods for vehicular communication." *Vehicular Communications* 33 (2022): 100394.

[19] Chen, D, Y Zhuang, J Huai, *et al.* "Coexistence and interference mitigation for WPANs and WLANs from traditional approaches to deep learning: a review." *IEEE Sensors Journal* 21(22) (2021): 25561–25589.

[20] Baltaci, A, E Dinc, M Ozger, A Alabbasi, C Cavdar and D Schupke. "A survey of wireless networks for future aerial communications (FACOM)." *IEEE Communications Surveys & Tutorials* 23(4) (2021): 2833–2884.

[21] Chen, M, A Liu, W Liu, K Ota, M Dong and N N Xiong. "RDRL: a recurrent deep reinforcement learning scheme for dynamic spectrum access in

reconfigurable wireless networks." *IEEE Transactions on Network Science and Engineering* 9(2) (2022): 364–376.

[22] Eom, D H Lee, and D-hyu Kim. "Committee standards battles in the era of convergence: implications for smart systems." *International Journal of Information Management* 60 (2021): 102380.

[23] Kosek-Szott, K, A L Valvo, S Szott, P Gallo, and I Tinnirello. "Downlink channel access performance of NR-U: impact of numerology and mini-slots on coexistence with Wi-Fi in the 5 GHz band." *Computer Networks* 195 (2021): 108188.

[24] Nair, M M, A K Tyagi, and N Sreenath, "The future with Industry 4.0 at the Core of Society 5.0: open issues, future opportunities and challenges," 2021 International Conference on Computer Communication and Informatics (ICCCI), 2021, pp. 1–7, doi: 10.1109/ICCCI50826.2021.9402498.

[25] Sedar, R, C Kalalas, F Vázquez-Gallego, and J Alonso-Zarate. "Intelligent transport system as an example of a wireless IoT system." *Wireless Networks and Industrial IoT.* Springer, Cham, 2021. 243–262.

[26] Zaidawi, D J, and S B Sadkhan. "Blind spectrum sensing algorithms in CRNs: a brief overview." *2021 7th International Engineering Conference "Research & Innovation amid Global Pandemic"(IEC)*. IEEE, 2021, 78–83, doi: 10.1109/IEC52205.2021.9476142.

[27] Lee, K-E, J G Park, and S-J Yoo. "Intelligent cognitive radio ad-hoc network: planning, learning, and dynamic configuration." *Electronics* 10(3) (2021): 254.

[28] Nurcahyani, I, and J W Lee. "Role of machine learning in resource allocation strategy over vehicular networks: a survey." *Sensors* 21(19) (2021): 6542.

[29] Jiang, W, and W Yu. "Multi-Agent Reinforcement Learning based Joint Cooperative Spectrum Sensing and Channel Access for Cognitive UAV Networks." *arXiv preprint arXiv*:2103.08181 (2021).

[30] Abbasi, M, A Shahraki, M J Piran, and A Taherkordi. "Deep reinforcement learning for QoS provisioning at the MAC layer: a survey." *Engineering Applications of Artificial Intelligence* 102 (2021): 104234.

[31] Liu, B, W Han, E Wang, *et al.* "An efficient message dissemination scheme for cooperative drivings via multi-agent hierarchical attention reinforcement learning." *2021 IEEE 41st International Conference on Distributed Computing Systems (ICDCS)*. IEEE, 2021, 326–336.

[32] Verma, S, R Paulus, and A Agrawal. "A survey on vehicular mobility models and mobility management." *Journal of Emerging Technologies and Innovative Research* 5(12) (2018): 196–204.

[33] Isnawati, A F, and M A Afandi. "Game theoretical power control in heterogeneous network."*2021 9th International Conference on Information and Communication Technology (ICoICT)*. IEEE, 2021, 149–154, doi: 10.1109/ICoICT52021.2021.9527439.

[34] Alwarafy, A, M Abdallah, B S Ciftler, A Al-Fuqaha, and M Hamdi. "Deep Reinforcement Learning for Radio Resource Allocation and Management in Next Generation Heterogeneous Wireless Networks: A Survey." *arXiv preprint arXiv*:2106.00574 (2021).

[35] Singhal, C, and V Patil. "HCR-WSN: hybrid MIMO cognitive radio system for wireless sensor network." *Computer Communications* 169 (2021): 11–25.
[36] Cody, T, and P A Beling. "Heterogeneous transfer in deep learning for spectrogram classification in cognitive communications." *2021 IEEE Cognitive Communications for Aerospace Applications Workshop (CCAAW)*. IEEE, 2021.
[37] Hamza, B J, and T H Yousif. "Multiple transceivers based WiMAX mesh network to optimize routing algorithm." *Journal of Physics: Conference Series*. IOP Publishing,2021. 1
[38] Malathy, S, P Jayarajan, M H D Hindia, *et al.* "Routing constraints in the device-to-device communication for beyond IoT 5G networks: a review." *Wireless Networks* 27(5) (2021): 3207–3231.
[39] Obite, F, Aliyu D U, and E Okafor. "An overview of deep reinforcement learning for spectrum sensing in cognitive radio networks." *Digital Signal Processing* (2021): 103014.
[40] Ros, F J, Juan A M, and Pedro M R. "A survey on modeling and simulation of vehicular networks: communications, mobility, and tools." *Computer Communications* 43 (2014): 1–15.
[41] Maldonado, R, C Rosa, and K I Pedersen. "Multi-link techniques for new radio-unlicensed URLLC in hostile environments." *2021 IEEE 93rd Vehicular Technology Conference (VTC2021-Spring)*. IEEE, 2021, 1–6, doi: 10.1109/VTC2021-Spring51267.2021.9449023.
[42] Glossner, J, S Murphy, and D Iancu. "An Overview of the Drone Open-Source Ecosystem." *arXiv preprint arXiv*:2110.02260 (2021).
[43] Gao, D, S Wang, Y Liu, W Jiang, Z Li, and T He. "Spoofing-jamming attack based on cross-technology communication for wireless networks." *Computer Communications* 177 (2021): 86–95.
[44] Loginov, V, E Khorov, A Lyakhov and I Akyildiz. "CR-LBT: listen-before-talk with collision resolution for 5G NR-U networks." *IEEE Transactions on Mobile Computing* (2021), doi: 10.1109/TMC.2021.3055028.
[45] Hasan, M M, M A Rahman, A Sedigh, *et al.* "Search and rescue operation in flooded areas: a survey on emerging sensor networking-enabled IoT-oriented technologies and applications." *Cognitive Systems Research* 67 (2021): 104–123.
[46] Panahi, F H, Farzad H P, and T Ohtsuki. "Spectrum-Aware Energy Efficiency Analysis in K-tier 5G HetNets." *Electronics* 10(7) (2021): 839.
[47] Srivastava, A, and G Kaur. "Resource management for traffic imbalance problem in green cognitive radio networks." *Physical Communication* 48 (2021): 101437.
[48] Dash, B Kumar, and J Peng. "Performance study of zigbee networks in an apartment-based indoor environment." *Proceedings of Sixth International Congress on Information and Communication Technology*. Springer, Singapore, 2022.
[49] Alonso, R M, D Plets, M Deruyck, L Martens, G G Nieto, and W Joseph. "Multi-objective optimization of cognitive radio networks." *Computer Networks* 184 (2021): 107651.
[50] Amulya, S. "Survey on improving QoS of cognitive sensor networks using spectrum availability based routing techniques." *Turkish Journal of Computer and Mathematics Education (TURCOMAT)* 12(11) (2021): 2206–2225.

[51] Girmay, M, V Maglogiannis, D Naudts, A Shahid, and I Moerman. "Coexistence scheme for uncoordinated LTE and WiFi networks using experience replay based Q-learning." *Sensors* 21(21) (2021): 6977.

[52] Zhdanovskiy, V D, V A Loginov, and A I Lyakhov. "A study on the impact of out-of-band emissions on performance of 5G new radio-unlicensed (NR-U) networks." *Journal of Communications Technology and Electronics* 66(6) (2021): 784–795.

[53] Leena, K. "Efficient channel allocation for cognitive radio Internet of Things." *Turkish Journal of Computer and Mathematics Education (TURCOMAT)* 12(10) (2021): 3476–3482.

[54] Shih, P-kan. Bayesian Nonparametric Reinforcement Learning in LTE and Wi-Fi Coexistence. Diss. Arizona State University, 2021.

[55] Huang, Y, Y Shen, J Wang, and X Zhang. "A platoon-centric multi-channel access scheme for hybrid traffic." *IEEE Transactions on Vehicular Technology* 70(6) (2021): 5404–5418.

[56] Gallego-Madrid, J, R Sanchez-Iborra, P M Ruiz, and A F Skarmeta. "Machine learning-based zero-touch network and service management: a survey." *Digital Communications and Networks* (2021), https://doi.org/10.1016/j.dcan.2021.09.001.

[57] Zhen, X, H Shan, G Yu, Y Cheng, and L X Cai. "Joint resource allocation over licensed and unlicensed spectrum in U-LTE networks." *Wireless Networks* 27(2) (2021): 1089–1102.

[58] Salameh, H Bany, S Shtyyat, and Y Jararweh. "Adaptive variable-size virtual clustering for control channel assignment in dynamic access networks: design and simulations." *Simulation Modelling Practice and Theory* 106 (2021): 102197.

[59] Obadi, A B, P J Soh, O Aldayel, M H Al-Doori, M Mercuri, and D Schreurs. "A survey on vital signs detection using radar techniques and processing with FPGA implementation." *IEEE Circuits and Systems Magazine* 21(1) (2021): 41–74.

[60] Baswade, A M, M Reddy, B R Tamma, and V Sathya. "Performance analysis of spatially distributed LTE-U/NR-U and Wi-Fi networks: an analytical model for coexistence study." *Journal of Network and Computer Applications* 191 (2021): 103157.

[61] Rahul, A R, S R Sabuj, M Akbar, H S Jo, and M A Hossain. "An optimization-based approach to enhance the throughput and energy efficiency for cognitive unmanned aerial vehicle networks." *Wireless Networks* 27(1) (2021): 475–493.

[62] Casetti, C Ettore, D M Malinverno, and D F Raviglione. "Implementation of standard-compliant ETSI ITS-G5 Networking and Transport Layers on ns-3." (2021).

[63] Abbas, T, M Boban, J L Calvo, Y Zang, and M Nilsson. "Spectrum and channel modeling." *Cellular V2X for connected automated driving* (2021): 91–136.

[64] Jeong, S, Y Baek, and S H. Son. "Distributed urban platooning towards high flexibility, adaptability, and stability." *Sensors* 21(8) (2021): 2684.

[65] Lee, K. E., J. G. Park, and S. J. Yoo. "Intelligent cognitive radio ad-hoc network: planning, learning, and dynamic configuration." *Electronics* 10 (2021): 254.

Chapter 16

Simulation tools and techniques for vehicular communications and its applications

Abstract

A wide range of technologies have been adopted to improve transportation by making it more convenient and efficient for passengers. However, recent academic–industry collaborations demonstrate that the paradigm for intelligent transportation systems is on the verge of shifting. Vehicles will be outfitted with computing, communication, and sensor systems, among other things. It is because of them that new and more diversified transportation systems that increase safety and efficiency while also providing entertainment will be made feasible in the future. An examination of a variety of concepts, methods, and technology connected to vehicle communication systems is the focus of this article. The use of simulation trials to evaluate system performance continues to be the most common method of evaluating inter-vehicle communication (IVC) system performance, even though field operational testing has already commenced. The strategy, tactics, and models that they use will be distinct from those used by other professions if they are to be effective. This chapter addresses how to prioritize scalability and applicability when selecting models, as well as the interactions that occur between the models themselves. When it comes to IVC simulations, the trade-off between scalability and applicability is extensively addressed in the models as the granularity of the simulation rises. Now we will speak about some of the most widely used simulation frameworks in IVC, and we will examine how they compare to the technique we'll be utilizing in the future. Vehicles in network simulation (Veins), iTETRIS, and VSimRTI are three well-known free IVC simulation systems, and they are all available online. Specifically, we address the appropriate selection of models and the granularity of their granularity about IVC application requirements to provide recommendations for effective and scalable simulations of IVC applications and to provide an overview of their relevant support in each toolkit.

Key Words: Keywords Simulation tools, Vehicular communication, Intelligent automation, Driverless, Efficiency

16.1 Introduction

Whether you are a vehicle driver, a bus passenger, a train passenger, or a bike, it is impossible to fathom a future without transportation options [1]. These modifications are being implemented right now to keep up with the rising demand for products and services in the marketplace. When it comes to transportation, it includes everything necessary to get people where they want to go for the least amount of money feasible. Direct expenses like gasoline and maintenance, as well as indirect costs such as discomfort, pollution, and accidents, are covered by the insurance. Because of physical, economic, and environmental restrictions, extending the current road infrastructure to meet the increased traffic levels would be prohibitively expensive [2]. In other words, the current urban traffic management challenge is more about taking an environmentally friendly approach to managing the system's available network and constituent entities, while also improving quality by providing timely information, tailored travel-related services, and high-quality guidance to the entities that use these roads [3]. This educational center will act as a catalyst for the development of a diverse variety of innovative transportation-related goods and services. Sussman compares transportation to a common local information overlay system (CLIOS). It is challenging because there are numerous interconnected subsystems, feedback loops connect them, and they all function on various time scales with a certain amount of uncertainty, among other factors [4]. Because the implications of this phenomenon are large, long-lasting, and geographically broad, it is impossible to exaggerate its significance. In addition to the environmental system, this system is linked to several other systems as well [5].

Given the complexity of technology and the significant social ramifications of its use, openness means that it interacts with other essential societal systems, such as sociopolitical-economic-environmental systems. It becomes important to use more comprehensive quantitative models, technologically advanced processes, and qualitative analytical frameworks as institutions grow more intricate and sensitive in real-time and across all geographic scales. Because a model must incorporate all of the components of reality, it is critical to have a simulation platform that allows users to abstract and replicate the components of reality [6]. Because tests on the real urban traffic system are not practicable, models (such as analytical, simulation, and prototype designs) must be utilized to better understand the behavior of the system (the actual process of building the model provides awareness of the real system). Models like this allow us to quantify variables such as productivity, analyze options, pick the most optimal choice, and assess the impact of various policies on a given population [7].

With the increase in mobility of people and goods, there is an increase in traffic congestion, deaths, and injuries. More than a dozen attempts to alleviate these difficulties have been undertaken in the previous ten years, and a few of these efforts have resulted in solutions that we now employ. Example applications include transmitting emergency information on the FM radio band, interrupting user-tuned reception for a brief period, and displaying warnings about changing circumstances on variable message signs along motorways that are spaced several hundred meters apart or strategically placed along the highway [8]. Meanwhile, vehicles are becoming increasingly technologically advanced, with a growing number of driver aid and safety technologies being included. Anti-lock braking systems, navigation systems, compasses, parking radars, and cameras are some of the most common autonomous sensor technologies found in vehicles. Passive safety measures protect passengers and the vehicle by preventing them from being exposed to potentially dangerous driving conditions [9]. The driver may personalize her driving experience and remain up to date on the state of the vehicle by utilizing a variety of onboard controls and information sources. They can alert the driver of approaching risks and help to reduce the severity of those threats by delivering appropriate information based on real-time views of the road, topography, and vehicle location. Fleet management and traffic data collecting are utilizing several various technologies (such as roadside cameras) and more complicated systems to go beyond the limits of today's basic technology. Intelligent transportation systems (ITS) are in a strong position to make substantial progress as a result of recent technological advancements in mobile computing, wireless communication, and remote sensing [10].

As a result of a large number of processors and sensors that have been mounted aboard, automobiles are already very complex computing systems. Wireless connection, processing, and sensing are just a few of the new features that have been introduced. In real-time, autonomous systems gather and share data with other (theoretically neighboring) systems, allowing them to improve their performance [11]. Cars that are part of a network are included in this category. While radar and vision alone cannot provide a complete picture of a situation, they may be used in conjunction with other technologies to provide a more complete picture outside of the line of sight. Detecting potentially dangerous conditions over a large region and in a short period necessitate the use of infrastructure and vehicles working together. When developing dependable and efficient driving assistance systems, it is critical to employ proper virtual reality (VR) designs, which are critical for both road safety and transit efficiency. The information on this page has been updated over the past few years to offer an overview of venture capital (VC) systems [12].

There are state-of-the-art and technical data from a variety of R&D efforts included in this collection. Instead of concentrating just on the architecture, it would be preferable to collect current commercial and academic knowledge in a clear and quantifiable manner. Beginning with a brief introduction to VR technology, their significance in smart transportation systems, and some of the most notable projects that have been performed thus far [13]. Applications with virtualization capabilities, onboard equipment, wireless data transfer technologies, and virtualization networking protocols will all be discussed. Finally, we will talk about what the future holds for virtualization systems in general.

In automobiles, new processing, networking, sensors, and user interfaces will be implemented. Drivers and passengers will be able to take advantage of new and integrated services, as well as applications aimed at improving transportation safety and efficiency [14]. Wired Internet connections and specialized roadside infrastructure units, as well as existing or future wireless infrastructure (such as cellular), are all expected to play a key part in the future of transportation (roadside units (RSUs)). It should also be feasible to connect wirelessly to user-portable devices using onboard equipment. In conjunction with VC systems, the introduction of new information sources and the improvement in the quality of current information that is important to driving safety and transportation efficiency are both achieved over time [15]. Some systems have the capability of sending out warnings when specific circumstances are met, which is an essential safety feature. RSUs, with the assistance of other cars, broadcast the same warnings about unsafe or congested traffic conditions, whether they are disseminated locally or nationally. Many efforts, including standardization working groups and industry consortia, have been launched across the world to further the development of such VC systems and related technologies [16]. From all of the coordinated actions and initiatives, it is apparent that collaboration is crucial for success. TCITS of the European Telecommunications Standards Institute (ETSI) and the International Organization for Standardization (ISO) TC204 WG16 have collaborated on cooperative system projects in Europe, especially the Come Safety initiative (ITS Communications). The result is wide acceptance across Europe, as well as in a few countries outside of Europe, for the reasons stated above [17]. To put it another way, VC-specific wireless transmission and medium access technologies are the most important enablers of the VC revolution. Furthermore, networking technologies allow for the conceptual sharing of data among devices that are dispersed around the world (vehicles, RSUs, and other servers). They assist with the previously stated applications as needed by mediating data extraction from network activities (mostly location- and timestamped data) and creating up sessions between two VC system components [18].

The remainder of this chapter is organized as follows. Section 16.2 describes the related work/background of simulation tools and techniques for vehicular and its application/ ITS/AIV and then in Section 16.3, there is a brief study of typical inter-vehicle communication (IVC) applications. In Section 16.4, we discuss connected vehicle environment application whereas in Section 16.5, we discuss wireless communication and networking. In Section 16.6, we further move on to a study of mobility modeling for ITS/AIV and in Section 16.7, we discuss network and mobility simulation tools, Section 16.8 is a brief study of integrated IVC simulated toolkits. Section 16.9 is the summary of the chapter and it also describes the future outlook for researchers.

16.2 Related/background work

In the field of ITS, a collection of innovative technologies is available for building a safe network in which motorized cars, bicycles, and pedestrians may coexist in peace [19]. To minimize traffic congestion, a variety of approaches are available, including safety (which reduces the number of collisions and traffic disputes, as well as traffic law breaches), mobility (which reduces the amount of time spent traveling, as well as travel budgets), and congestion (to increase the LoS of the road facility). Transportation systems that are both environmentally friendly and sustainable are essential in the twenty-first century. Automobiles and infrastructure will become more "intelligent" and efficient in the future, thanks to the integration of existing computer, communications, surveillance, and traffic management technology [20]. A reduction in emissions, as well as increased efficiency in road operations, would help the environment. The successful monitoring and control of the city's transportation infrastructure in real-time are dependent on the deployment of cutting-edge technologies. The sensors and control technologies that characterize an ITS have been covering highways, cars, and passengers for a few years now, and this is expected to continue [21]. In terms of boosting network speed while still keeping an eye on congestion, safety, and user and environmental comfort, this represents a significant step forward. Traffic flow management and control operations, as well as the provision of traffic information to drivers and other users, are the two major categories in which current ITS advancements are being made (with traffic information services). While on the road, the next generation of technological advancements will place a strong focus on technology that assists drivers in making decisions and operating their vehicles safely [22].

Yichuang Sun *et al.* (2021): The usage of passive UHF radio frequency identification (RFID) for AV and ETC systems may become more common in the future (RFID). Biometric identification is highly successful in many situations when both an RF radiation zone and an anti-collision procedure are used in conjunction with one another, as is the case in many circumstances. When applied to a typical AVI/ETC application scenario, the impacts of scene shape and vehicle speed on identification rate are investigated and simulated in detail. The identification zone is initially computed using a ray-tracing approach that is based on the theory of reflection. Finally, the communication process is split into three pieces, with the probability distributions for each phase being estimated separately from the other two sections. Optimization of tag speed and antenna inclination angles, according to numerical models, has a significant influence on the speed with which tags may be detected [23].

Dowd *et al.* (2021): As a result of the introduction of unmanned aerial vehicles (UAVs) and many studies demonstrating that cameras can track and monitor traffic beginning in 2000, cameras have become common sensors on UAVs. A similar pattern of development is occurring in the fields of dedicated short-range communication (DSRC) and vehicle-to-everything cell phone (C-V2X) communication technologies. Our ultimate objective is to develop a single piece of software that can simulate both ground and air traffic flows simultaneously. This environment and PTV Vissim are being developed in collaboration to assist transportation professionals in the usage of PTV Vissim while modeling air and ground traffic. The simulation system is built on custom Python packages named PyPTV and Vissim event-based scripts, which are written in Python. As a starting point, we will talk about software design and methodology, as well as techniques for speeding up simulations. As the data demonstrates, a simulation environment may be used for a broad variety of tasks and projects. Videos demonstrate how simulations are put into action [24].

Farhat Tasnim *et al.* (2021): Because of the increase in the number of intelligent cars on the road, enabling secure communication between them via a variety of networks has become a significant problem. Through the analysis of real-time data, vehicular communication attempts to increase communication security while at the same time lowering traffic congestion costs. To keep up with the growth in-vehicle systems, there are various ways to interact, including sending and receiving messages as well as transmitting data, which raises concerns about security and privacy. Healthcare, transportation management, academics, and genetic engineering are just a few of the fields in which artificial intelligence (AI) has made significant strides in recent years. Because of this, blockchain has demonstrated potential in industries such as banking and government, where security is paramount. Researchers have employed AI and blockchain technology, both of which were recently created, both individually and in combination, to address security problems associated with automotive networks. Malicious attacks such as Sybil, DoS, and MITM result in data manipulation, data outflow, message delay, and traffic congestion, among other things (man in the middle). Our team looked through the most recent research to find out what's new in the world of automobile ad hoc networks and vehicle social networks. While emphasizing AI and blockchain-based solutions, we have also identified new research topics that should be investigated further [25].

Mohamed Ben *et al.* (2021): The cloud computing idea was designed to allow people to share services when they are on the road. When cloud computing and Ad Hoc Vehicular Networks (VANET) come together, a new type of cloud computing known as the vehicular cloud (VC) is developed, and it is referred to as the VC. Consumer vehicles get services from the VC through provider cars, which are owned by the VC and operated by its employees. These services have a huge influence on everything from Internet access to data storage to data applications and everything in between. As a result of the high-speed mobility of automobiles, people must seek out services in their immediate vicinity. Customers may find it difficult to choose the most appropriate selection from among the many accessible possibilities. Using the RSU directory and the cluster head (CH) directory, this study provides a novel technique that allows the VC to be formed and the visibility of supplier vehicle resources in the VC to be increased at the same time. In a mobile cloud, the automobiles that make up the cloud are all traveling along a single road at the same time. The mist generated by the other cars, on the other hand, is rather apparent. Our technique, as proven by the simulation results, reduces wait times while improving the proportion of hits received by the system [26].

Sahaya Beni *et al.* (2021): Autonomous cars in vehicular critical energy infrastructures are now possible even without the intervention of a person, thanks to the development of fifth-generation networks (5G) targeted at vehicle-to-everything (V2X) communication and the proliferation of 5G networks (CEI). Despite technical improvements in autonomous cars, safety-critical messages (SCMs) continue to play a critical role in minimizing collisions, preventing injuries, and saving lives by providing drivers with critical information (AVs). In this research, we present the Migrating Consignment Region (MiCR) based on SDN clustering, which is based on the Federated K-means technique for spreading SCMs to AVs via 5G-V2X communication and is based on SDN clustering (SDN). In contrast to alternative methods, which create clusters for each instance, MiCR maintains moving clusters for spreading SCMs to AVS with exceptionally high reliability and low latency while maintaining moving clusters. The suggested MiCR technique has been evaluated in comparison to other alternatives by simulating real-time highway route maps. The result is a decrease in network overload and SCM delivery ratio over prior approaches, as well as a decrease in latency and disseminating efficiency [27].

Martin *et al.* (2021): The use of autonomous guided vehicles has recently been proven to be extremely effective in the completion of material handling duties (automated guided vehicles (AGVs)). While this is true, the cost of installing AGVs is expensive because the workplace layout must be altered to provide room for mobility areas. Recently, industrial systems have evolved to make use of mobile robots that are more adaptive and cooperative, referred to as autonomous intelligent robots, which are more adaptable and cooperative (AIVs). Because they are more intelligent, next-generation mobile robots do not require limited zones and are capable of interacting with other mobility obstacles on the job site, such as human operators. While ensuring that the AIV transportation responsibilities are completed on time, the goal of this study is to determine the appropriate size of the AIV fleet in a dynamic and unpredictable environment that includes human beings. To simulate the complexity of the Internet of Things, Big Data, and sensors as real-time data sources, an AIV trip time prediction simulation has been developed [28].

Fragapane *et al.* (2021): Automobile mobile robots are already being used in manufacturing and warehousing operations, as well as cross-docks and terminals (AMRs). Today's technology and control software enable them to operate autonomously, even while operating in highly dynamic environments. AMR systems, as opposed to AGV systems, decentralize the decision-making process, whereas AGV systems, which are administered by a central unit for all AGVs, decentralize the decision-making process when utilizing an AMR system instead. Because of its decentralized decision-making structure, the system can respond quickly and flexibly to changes in its internal state and environment. As a result of these transitions, many old systems and decision-making processes have been changed. These works on AMRs in intralogistics planning and control have been identified and categorized in this research. We perform a detailed literature analysis to illustrate how technological advances in AMR impact planning and control decision-making processes. This contributes to the corpus of knowledge and assists managers in realizing their full potential. The last section includes research recommendations for further investigation [29].

Cao *et al.* (2021): In the vast majority of existing map-matching algorithms, GPS data is utilized to create the processes that are used to match maps. Because of the limited sample size and the high incidence of position errors in GPS data, it is inappropriate for use in ITSs. However, the advantages of AVI data exceed the problems of GPS data by a wide margin. Because of the sparsity issue, existing map matching algorithms are unable to be used with AVI data. As a result, in this work, a novel map matching approach for sparse AVI data is created, which is designated as AVI-MM. The connected observation pairs are used as input for the first phase, and the AVI trajectory is broken down into its parts for the second step. In the following step, a list of candidate sub-paths is created for each set of sensor pairs, and a matching probability is assigned to each of these candidate sub-paths. In the last stage, sub-paths with a high likelihood of matching are linked together. As a result of the development of a candidate set generation algorithm, it is possible to generate a large number of feasible and appealing candidates from sparse AVI observations. This is because different sub-paths are capable of connecting them all. By combining a spatial–temporal analysis with an examination of drivers' route-choosing behavior, it is possible to define the matching probability and construct the right path consistently and accurately. To assess and evaluate the effectiveness of the proposed approach, field tests are conducted. If accuracy is important, the suggested AVI-MM methodology outperforms three existing benchmark approaches while consuming a fraction of the computational resources [30].

Leszczyna *et al.* (2021): To ensure that critical cyber assets are adequately secured from attacks, it is necessary to do frequent cybersecurity assessments. Over the years, the cybersecurity industry has developed a variety of various techniques for analyzing risks. One such methodology is the threat model. Following the completion of the literature search described in this study, it was revealed that there are no reviews of their goods available. As a result, the primary goal of this research was to close this knowledge gap by identifying and thoroughly analyzing scientific literature-based cybersecurity evaluation methodologies. It was established that an organized inquiry approach, as well as a set of assessment criteria, would be used in the investigation. As a consequence of this investigation, a total of 32 possibilities are provided. It is one of the major issues to determine whether or not the techniques can be applied effectively under practical circumstances. These concerns have been identified, as well as potential solutions that may be developed to solve them. While reading the article, you may come across some research that may be utilized in combination with the content throughout the assessment process. Finally, a section outlines areas that still require improvement and makes recommendations for new research and development initiatives in the future. The goal is to aid academics and practitioners in their assessments by offering recommendations on which technique to use and by highlighting potential research areas in the process [31].

Serhan Tanyel *et al.* (2021): Because space is scarce in city centers to develop new infrastructure, the utilization of parking garage capacity is a key source of worry. Using IPGSs, which limit parking capacity, to cope with shortages like these is a viable option. When it comes to the occupancy conditions of three surface parking lots, it is recommended that an IPGS model be used to compare the actual occupancy conditions of three surface parking lots with those of a conventional parking utility system. Comparisons are made using a multi-agent simulation program to determine occupancy ratios, wasted Value of Time (VoT), and hazardous gas emissions [32].

16.3 IVC applications for ITS/ AIV

Vehicle communication (VC) systems will aid in the improvement of transportation safety, efficiency, and the overall quality of the user services provided by automobiles. When it comes to developing new systems, the first and second categories are the most significant sources of inspiration to draw from. VC and IT settings are frequently compatible with it, and it may also function as an engine for the third-category market in many cases. VC systems have been the subject of several initiatives, standardization organizations, and consortia across the world, all of which have created and built applications for them [33]. A large list of prospective applications was developed in the beginning, looking to the future while also considering current transportation requirements and capabilities, and then investigating how venture capitalists may take on and expand support for each of these applications. In Figure 16.1, it can be seen that the vast majority of applications may be categorized into one of the three groups that were previously discussed [34].

Although the names are not identical across projects (e.g., C2C-CC and VSC-A), they are comparable to those of other standardization efforts (e.g., ETSI and IEEE) rather than being unique to each project (e.g., SAFESPOT, CVIS, COM2REACT, and SEVECOM). For the sake of ensuring that each application has all it needs, we have separated the material into the following four categories as shown in Figure 16.2.

As a result, we want to make certain that we have the most up-to-date understanding possible of test locations and testbeds for certain application scenarios before moving further. These criteria can be adjusted even more when individual systems are tested in greater depth over time. They prioritize transportation safety over other considerations such as cost or performance since the first eight applications are time-sensitive and based on a whim. Pre-crash sensing needs the greatest amount of delay out of the whole group [35]. Note that these latency numbers are not connected to any specific reliability requirements at this moment, which is a side benefit. It is necessary to specify end-to-end delays from start to

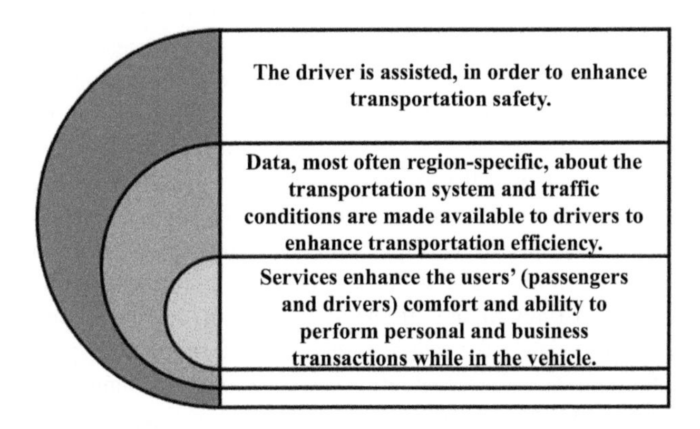

Figure 16.1 IVC applications fall into three categories

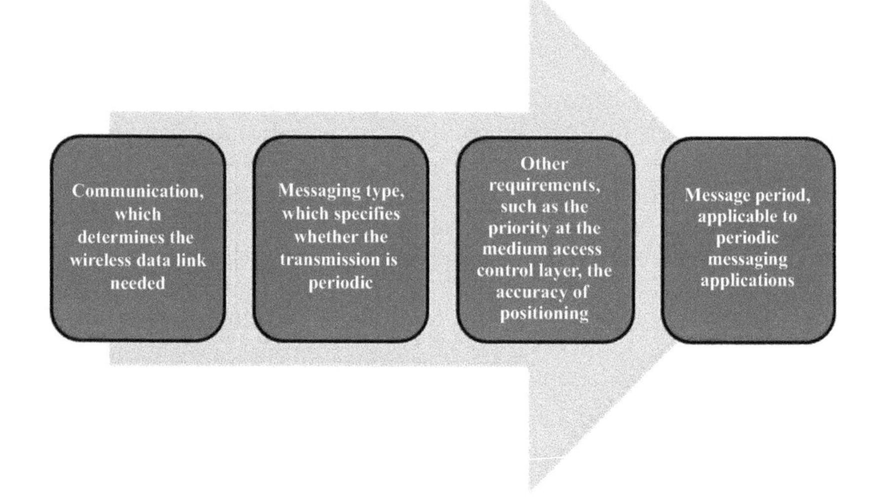

Figure 16.2 The material into the following four categories

finish (i.e., for V2V-enabled safety applications, maximum V2V application layer delays). After an event has happened, the data would reveal how long it takes for the receiving vehicle to recognize the event (such as a vehicle detection or a collision danger) because all processing at every layer would be included in the data collection. It is required to define multihop communication definitions. It is also feasible to make adjustments to these standards as the situation necessitates. As part of pre-crash sensing, a very short communication range is observed via the lens of a particular perspective (which corresponds to small distances between vehicles). The usage of recommended ranges for certain applications can be beneficial when dealing with incoming messages such as beacons, and protocol designers can profit from this if they utilize them appropriately [36].

Even though all vehicles and RSUs require position awareness, the precision offered by GNSS-based or other localization systems is less precise than the accuracy mentioned above. Consequently, the device's computing power is put to the test. Ad hoc communication with infrastructure is used in the following four applications that are all related to improving transportation efficiency (RSUs, generic or specialized such as the toll collection units). It has a lower priority, lower beaconing rates, and latencies that can be up to five times greater than those of safety-related applications. In addition, protocols such as V2V can be utilized to interact with one another. Unlike broadcast safety applications, unicast V2V communication is distinct in that it may incorporate cellular connectivity. This makes it an excellent choice for public safety applications. Finally, numbers 13–16 provide services to users rather than allowing V2V data exchange (passengers and drivers). They will also require access to the Internet at some point. As a result of their connection to the internet through ad hoc V2I (when the car is within range of an access point), cellular, or other networks, IP addressable hosts and servers are required for automobiles (such as Wi-Fi).

When latency needs are low, cooperative awareness takes a backseat to a more traditional approach [37]. Other concerns, such as digital rights management (DRM), exist that are not particular to venture capital but affect everyone (i.e., mechanisms and policies that specify and enforce access control to the obtained content). The explanation above demonstrates a connection between communication and networking protocols as well as their associated applications. The fact that we are working together does not exclude us from having some technical issues to sort out before we can go forward. Future initiatives, such as those detailed in the preceding sections, are expected to continue to use V2V and V2I communication techniques. Open VC design allows for data exchange; however, sophisticated data interchange is not required at this time. VC architecture allows for data sharing because of this agnostic method, you may create an app without or with very little venture capital funding at this moment, which is an exciting prospect. The highly automated vehicles for intelligent transportation (HAVE-IT) project, for example, anticipates that VC will be commercially available. So that the driver and the driving system may share work equally effectively, we aim to build a co-pilot with an electric design that is both redundant and fail-safe. Have-IT is laying the groundwork for future cooperative data fusion while maintaining a local perspective of the world [38].

Because the application plays such a significant part in the design of IVC systems, there is a strong bi-polarity to the communication design when developing IVC systems. IVC applications define the requirements, while access technologies and the shared mobile wireless medium provide the necessary capabilities to meet those requirements. It is impossible to develop a communication solution that satisfies all of the requirements of an IVC application due to an engineering misunderstanding. This has an influence on the protocol design at many levels, as well as the simulation toolkits that are used to evaluate the protocols in question. IVC applications have been divided into three groups based on their size, infrastructure support, and communication sensitivity to maintain some sense of order in an otherwise chaotic environment [39]. Figure 16.3 shows IVC applications have been classified into three classes.

Applications in this category tend to be broad in scope, with a primary goal of increasing throughput as their primary goal. The classification has an impact on the models and toolkits that are required to analyze the performance of IVC applications (or conformance). Every time a decision is made on an IVC model, the decision is always a trade-off between performance and accuracy. It is necessary to mimic just those behaviors that have an impact on the IVC application because of the restricted simulation capability.

16.4 Connected vehicle environmental application

Because of their capacity to be both autonomous and linked, cars in a connected vehicle ecosystem may continue to function independently of one another and from

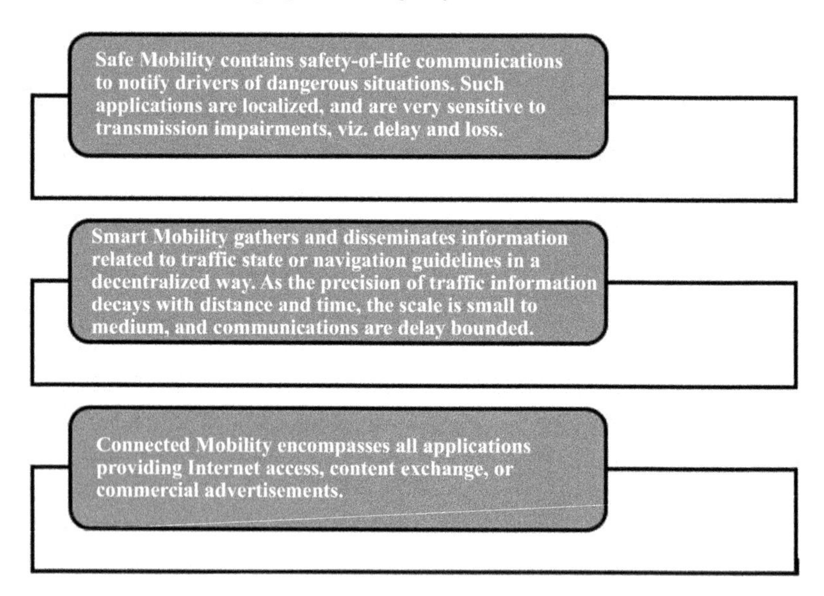

Figure 16.3 IVC applications have been classified into three classes

the infrastructure in the same way they did previously [40]. They are generally equivalent when it comes to typical ITS components, such as rights-of-way along the road and the employment of its own devices to regulate traffic flow, although they have major infrastructural variations. In a connected vehicle environment, on the other hand, agencies now have access to network data generated by in-car devices and transferred over communications channels that they do not have control over. Moreover, they can disseminate information to all travel agencies through the use of technologies and networks that they do not control or own. Many organizations and travelers may be unfamiliar with this new hybrid environment, which is understandable. Even while it may take some time to get used to, the end effect should be better data and faster decisions in the long run. These notions, although they are novel, are built on a strong foundation of previous knowledge. To communicate effectively in the connected car ecosystem, it is necessary to develop a common language and set of standards. It has been a long time since the rules governing how drivers should be notified were updated and clarified. Octagonal stop signs were needed by the transportation sector in the 1930s for the sake of safety in the workplace. Because of the shape's ability to broadcast crucial safety information in real-time, drivers all around the world are now aware of what to do in an emergency [41]. New concepts are being developed in the connected automobile environment that will be known to travel companies and travelers in the future, according to the researchers. Curve speed warning is a simple example of a distributed or independent application that you may learn from. In the connected vehicle industry, there is common awareness that not all cars are created equal and that situations and conditions might vary over time. Oncoming cars are now warned to slow down when a yellow warning sign is placed in their line of sight, allowing them to safely negotiate the curve. As an example of an application concept, a central system might be used to assign a location to all of the region's dangerous curves, which would be a central system. Other means of transmitting data to vehicles will be covered in greater depth later in this chapter, under the heading Communications. Once the data has been transferred to mobile devices, the mobile app may be able to detect whether or not the car in which it is running is approaching one of those corners and whether or not a warning is necessary. For large vehicles, such as school buses, a loud warning signal is necessary. Although it is possible that a warning is justified for a small vehicle traveling below the acceptable speed limit because of changes in the weather and the surface becoming more slippery, it is unlikely. Often, while thinking about connected vehicles, it is helpful to think about how the underlying data may be separated from how it is shown to the user [42].

16.5 Wireless communications and networking

To assess the IVC protocol and application performance, the tools that are utilized must be capable of supporting a wide range of simulation models. Signal propagation models must consider CSMA behavior, congestion control, and even multi-channel operations while developing their models of transmission [43]. Channel access models must take the CSMA into mind when designing channel access models. In addition to facilities-layer protocols such as BSM or CAM, IVC-specific dissemination techniques such as beaconing or geocasting may be necessary. In the following sections, we will talk about both domains and show you some of the most prevalent models used in this industry today. Wireless communication technologies are expected to be utilized extensively in future ITSs, according to industry analysts and researchers. For both short- and long-range applications, radio modem communication on UHF and VHF frequencies is widely utilized in the ITS industry. IEEE 802.11 technologies, including WAVE and the DSRC standard proposed by the Intelligent Transportation Society of America and the United States Mobile ad hoc and mesh networks, have the potential to enhance the capabilities of these technologies. Communications across long distances rely on high-speed infrastructure networks, such as 5G, to function properly [44]. Long-distance communication systems need a more substantial and expensive infrastructure than shorter-distance protocols, as compared to shorter-distance protocols.

16.5.1 Computational technologies

Electronics for automobiles have evolved in recent years, resulting in the employment of fewer but more powerful computer processors in their construction and operation. As real-time operating systems and hardware memory management become more widely used, smaller and more costly CPU modules are becoming more popular among computer users [45]. All of these technologies, including models-based process control, AI, and ubiquitous computing, are now possible on

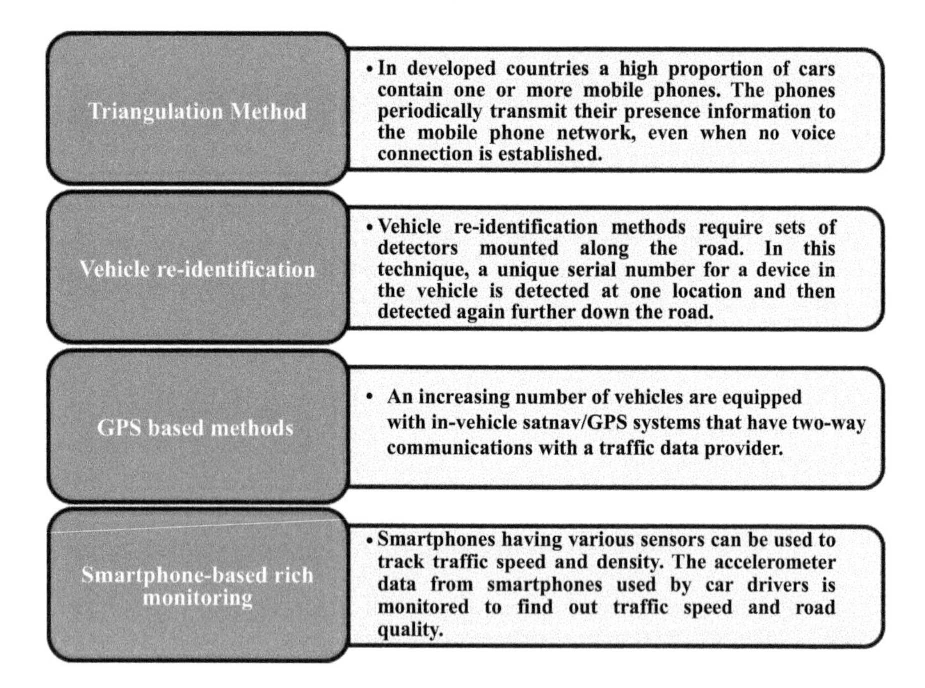

Figure 16.4 The four primary methods of gathering raw data

the new embedded system platforms. AI will be required in large quantities for futuristic transportation systems to be successful.

16.5.2 Floating car data/floating cellular data

The data from other means of transportation was obtained using a floating vehicle or a probe, as described above. Figure 16.4 illustrates the four primary methods of gathering raw data that were employed.

As seen in Figure 16.5, there are several advantages to adopting floating car data technology over standard traffic measurement methods.

16.5.3 Sensing

Progress in telecommunications and information technology, such as the development of ultramodern/state of the art chip designs, RFID, and low-cost intelligent beacon sensing technologies, have increased the technical capabilities that will allow ITSs to improve motorist safety around the world in the coming years [46]. ITS sensing systems take advantage of networked systems based on vehicles and infrastructure, such as intelligent vehicle technology, to improve their performance. When it comes to installing indestructible infrastructure sensors fast, there are two options: manual distribution during preventative road construction maintenance, and sensor injection gear (such as in-road reflectors). A variety of vehicle detection systems are available to keep track of vehicles operating in high-risk areas around

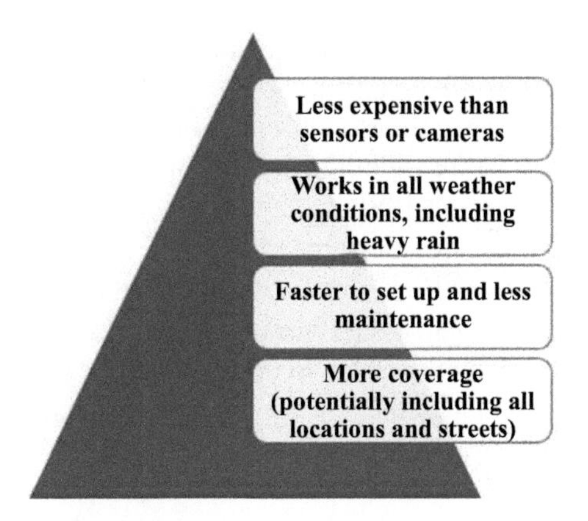

Figure 16.5 Floating vehicle data technology outperforms conventional traffic monitoring techniques

the world. These systems include the deployment of electronic beacons on infrastructure and vehicles for identification communications, as well as the use of video automatic number plate recognition and vehicle magnetic signature detection technology.

16.5.4 Inductive loop detection

A car can be recognized by installing inductive loops in the roadbed, which identify the vehicle by detecting the magnetic field produced by the loop. Detectors are available in a range of shapes and sizes. Many different approaches are available for calculating how many vehicles have traveled through the loop in a given length of time [47]. A more advanced sensor is used to assess the parameters of the vehicle, such as speed and distance. The usage of loops is suitable for both slow- and fast-moving vehicles, and they may be implemented in a single lane or over several lanes.

16.5.5 Video vehicle detection

Video cameras may also be used to identify vehicles since they can automatically monitor traffic flow and record any incidents that occur on the road [48]. Non-intrusive technology, such as automated number plate recognition (ANPR), refers to technology that does not need the placement of components directly into the road surface or roadbed. Images captured by cameras are sent to processing units, which monitor how the video picture changes when cars pass in front of it. The cameras are typically installed on or near utility poles or other high-profile structures that run parallel to or above the road and are visible from a distance. A typical video detection system requires an initial configuration to teach the processor the default backdrop picture, which is often a still image. For example, normal practice calls for measuring the distance between lane markers or the height of the camera above the road surface. Video detection processors are capable of identifying up to eight cameras at the same time, depending on the brand and kind of processor. Data from traffic camera systems is frequently used to calculate traffic flow rates, vehicle counts, and lane occupancy, for example. Various other outputs, including gap, headway, stopped-vehicle recognition, and wrong-way vehicle alerts, are also available from certain systems [49].

16.5.6 Bluetooth detection

It is a solid and inexpensive alternative to broadcast a vehicle's location while it is in motion, and Bluetooth may be used to do so. Roadside sensors detect Bluetooth devices in passing automobiles that are connected to the internet [50]. When these sensors are linked together, it is possible to calculate the travel duration as well as the data for the origin and destination matrices. In Figure 16.6, which illustrates the differences between different traffic measurement technologies, the comparison of different traffic measurement technology is shown.

In conjunction with the rising prevalence of Bluetooth devices in vehicles and the availability of additional information, the quantity of data collected over time becomes increasingly precise and significant for reasons such as predicting trip time and distance, among others [51]. It is also possible to estimate traffic density using audio cues such as tyre and engine noise, honks, and air turbulence noise when they are combined. To analyze traffic conditions, a roadside microphone collects a wide spectrum of vehicle sounds. It will then be possible to employ techniques for audio signal processing. When compared to the other approaches we have examined, the accuracy of this system is on par with them.

16.5.7 Radar detection

Situated on the side of the road, side-of-the-road radars monitor both moving and stopped or idling vehicles. The installation of radar systems, like video systems, requires the radar to learn about its environment to discriminate between automobiles and other objects [52]. It can also be utilized in low-light environments. To count cars and estimate traffic density, traffic flow radar scans all lanes in a restricted zone using a "side-fire" approach, which counts vehicles and estimates traffic density. 360° radar sensors scan all lanes of a road over an extended length of time to identify stopped vehicles (SVD) and to automatically detect incidents when they occur. When utilized at a distance of more than 10 miles, radar is believed to be more effective

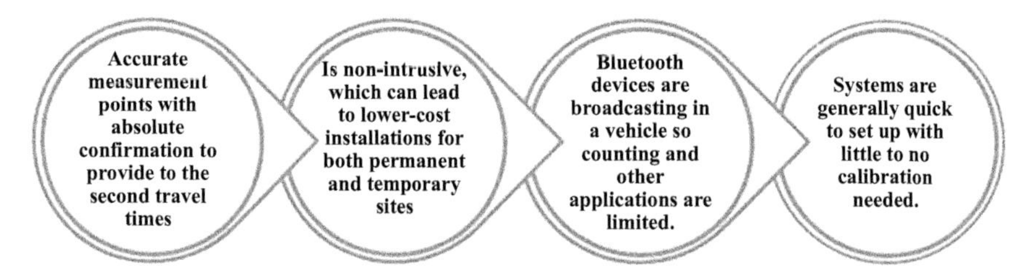

Figure 16.6 Differences between different traffic measurement technologies

than other technologies in terms of efficacy. SVD radars are anticipated to be deployed on the Smart highways in the United Kingdom.

16.5.8 Information fusion from multiple traffic sensing modalities

The intelligent integration of data from different sensing technologies may be used to provide a more exact assessment of the traffic situation. When combined with other technologies, the use of auditory, visual, and sensor data acquired by roadside sensors can provide a significant safety improvement.

16.5.9 Signal propagation and fading

The physical layer influences on the application layer such as throughput, latency, and bit error rate are typically taken into consideration in simulations concentrating on the application layer. Because the first two may be easily simulated in any discrete event simulation, we will limit our discussion to bit error rate (BER) calculations in the next sections [53]. Following the fine-graininess of the simulated applications, as seen in Figure 16.7, there are three alternative fading models to consider and study further.

16.5.10 Channel access

While a large number of emitters must share the same radio channel resources, protocols must find out how to do so in an efficient manner. The need that channels access to be strictly dispersed, in contrast to personal communications, where a centralized method is sometimes employed, makes IVC one of the most distinctive features of the technology (WLAN, 4G). To be effective, distribution channel

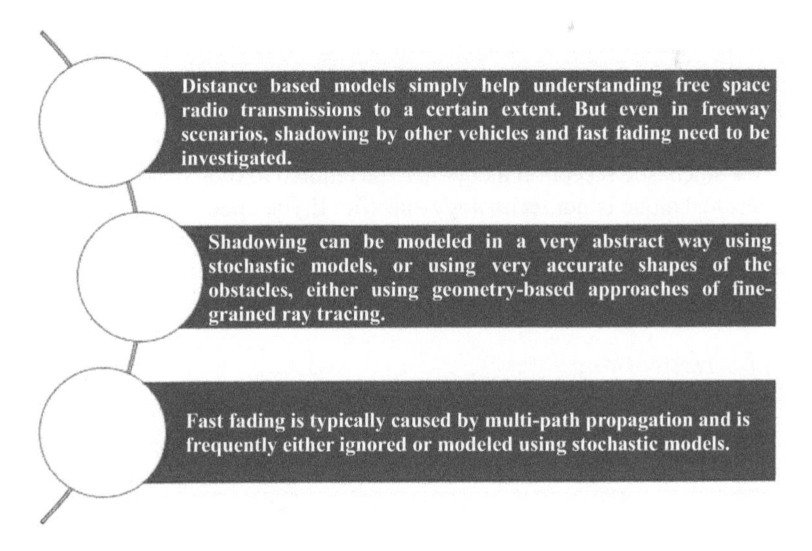

Figure 16.7 Three fading models

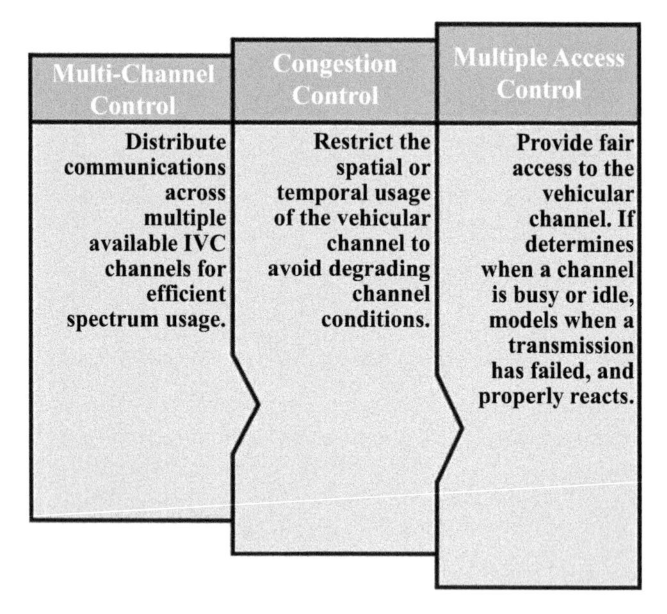

Figure 16.8 IVC primarily fulfill three characteristics

access techniques for IVC must primarily fulfill three characteristics, which are illustrated in Figure 16.8.

When it comes to access control, one of the most straightforward solutions is a packet reception probability model that is independent of channel circumstances (idle/busy). This probability model can be represented by a deterministic function (e.g., the unit disc model) or a stochastic process [54]. Example: A declining reception probability may be simulated by taking into consideration the broadcast range, the number of neighbors in range, and other factors such as the transmission rate or packet size, in addition to the transmit rate. To accommodate the communication characteristics of LTE or an ITS G5 technology, depending on the application, the stochastic reception model may be tailored to those characteristics. As a result, this technique is not technology-specific. If you choose an idealized model, the results of your simulations may be extremely enlightening while still being finished in a relatively short amount of time. The most significant long-term benefits of this technology will accrue to connected and autonomous cars.

16.5.11 Networking

IVC information is propagated using network protocols, which may be carried out through vehicle networks, the back-end, or Internet-based services. Aside from that, the networking layer aspires to offer users a comprehensive picture of their IVC configuration. Rather than providing a comprehensive discussion of vehicular networking protocols, the objective of this section is to illustrate the connection that may be established between the IVC application class and the IVC networking

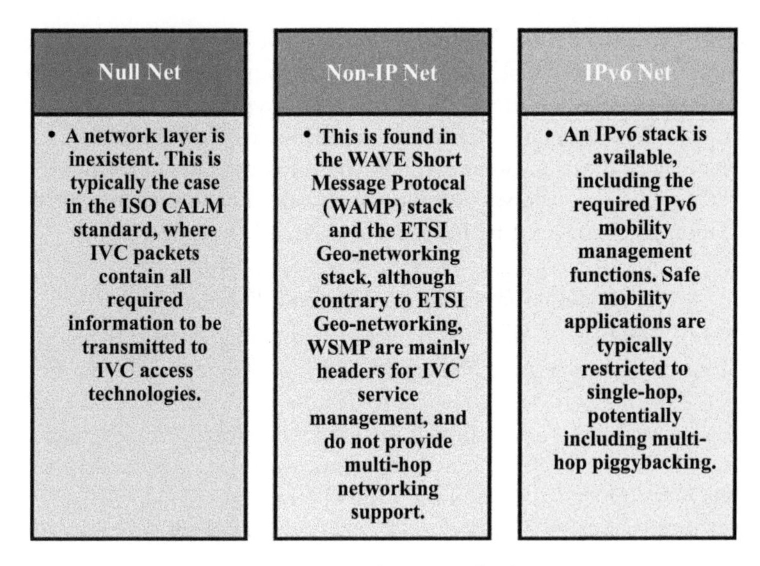

Figure 16.9 Network protocols three types

stack [55]. Figure 16.9 depicts three different types of network protocols that can be utilized independently of whatever IVC standard is being implemented.

As a result, it is doubtful that a networking stack will be necessary. To facilitate safety-related IVC, the application/facilities layer messages in CAM and BSM, for example, contain the required addressing information [56]. The decentralized emergency notification message (DENM) network is under the control of the European Telecommunications Standards Institute (DENM). Even if a networking layer is not required for the vast majority of safe mobility apps, a Facilities layer stack is essential to ensure that the applications conform to ETSI standards. To build intelligent mobility applications, data must be transferred both horizontally (between individual cars) and vertically (between several vehicles) (to back-end or Internet services). If this is the case, the IVC protocols specify the type of networking stack to be utilized in the environment. Because the Internet Engineering Task Force (IETF) does not support IPv6 multi-hop for ITS applications over a vehicular network, a non-IP stack is necessary to transfer smart mobility data between cars [57]. If smart mobility traffic is mostly vertical, cars must be connected to the Internet to function. Having an IPv6 network is important for assessing IVC applications. This is especially true for applications that require cars to be connected to their backend or the Internet. Because of the high volume of vehicle traffic on the IPv6 backbone, IPv6 mobility and traffic control may be necessary.

16.6 Mobility modeling for ITS/AIV

As previously said, models help us to get a better grasp of reality by abstracting it away from its physical manifestation [58]. The development of a basic platform that includes two essential tools, namely a GIST capable of dealing with space and a traffic microsimulation tool, is required for a comprehensive approach to the integration and administration of ITS systems. ITS systems are becoming increasingly complex. To help municipalities in the deployment of comprehensive, integrated, modular, and action-oriented ITSs, the intelligent urban traffic operations system (IUTOS), was developed. For the framework to be successful, the modeling and simulation platform must be in place, as seen in Figure 16.10.

16.6.1 Geographical information system for transportation

Due to a large number of static and dynamic spatial characteristics present in the transportation data, it is a great choice for GIS modeling applications. Geospatial information systems (GISs) provide considerable analytical and revisualization capabilities by enabling precise digital mapping and storing of geographical data with standard Database Management Systems (DBMS). Although Geographic Information Systems in Transportation (GIS T) has been around since 1960, it has only lately seen major advancements and the publication of a dedicated bibliography in the academic area of GIS T [59]. There are several challenges presented by the current ITS environment, including network representation, clear communication, the integration of new technologies, interoperability, and analytical and dynamic modeling, to name a few. The network data model that underpins GIST, the fundamental foundation for most transportation analysis projects, is well-suited

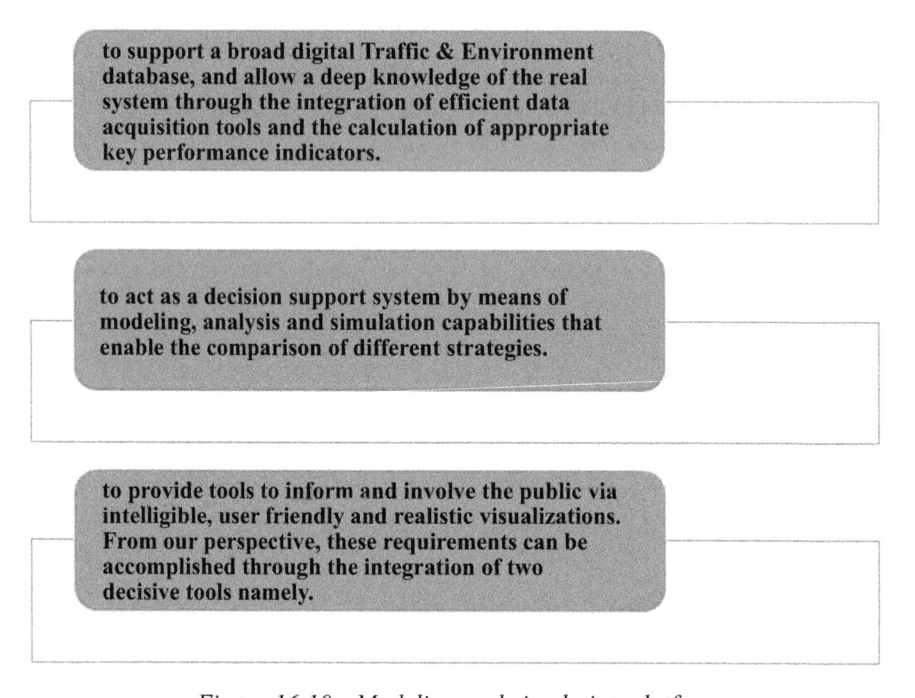

Figure 16.10 Modeling and simulation platform

to several ITS applications, such as vehicle location and routing, collision warning and guidance, and advanced trip planning, among others. Topological representation, in combination with dynamic segmentation/linear reference techniques, and matrix operations, is a powerful tool for ITS modeling and simulation. The integrative abilities of a GIS technician are essential for an ITS holistic approach. The fact that it incorporates a variety of systems (such as land use and transportation systems) and different themes from the same subject (such as road base networks, inventories, traffic operations, and traffic impacts), and that it incorporates data from a variety of sources and styles (such as databases, census files, picture files, and GPS data points) at a variety of resolutions (such as intersections, segments, and traffic analysis zones), makes spatial modeling a complicated endeavor.

16.6.2 Traffic microsimulation

DTA techniques have been created in response to contemporary ITS settings, which necessitate the assessment of short-term operational management schemes, as well as the management and control of online traffic flows and flows of people. As a result, dynamic traffic assignment (DTA) methods have advanced in recent years. When designing the DTA, factors such as mode and route selection, traveler behavior, network performance as a result of accidents and poor weather, lane closures, road construction, and other special events are all taken into consideration [60]. Microsimulation models are currently included in the most popular dynamic traffic models, and they are used to simulate driver behavior relating to car following, gap acceptance, and lane choice limitations. Animated microsimulation traffic simulations created by the AIMSUN software (source: www.aimsun.com). Individual vehicle movements can be represented second-by-second or sub-second-by-second, and how vehicles move and interact can be specified, allowing for the evaluation of urban street traffic performance to be conducted (traffic, transit, pedestrians, etc.). By utilizing microsimulation technologies that give realistic graphical dynamic displays, it is feasible to gain the support of many stakeholders involved in ITS initiatives. They have proved to be effective in traffic studies since they are capable of reproducing individual vehicle movements over a wide range of transportation networks. As a result, they have made considerable strides in recent years, and their application in transportation operations has increased as a result. They are particularly useful for analyzing scenarios and giving short-term forecasts, as well as for supporting real-time decisions.

16.6.3 Overview and constraints

To construct a mobility model, three primary building elements are required: motion restrictions, mobility needs, and traffic demands [61]. A full mobility model is made up of these three components working in conjunction with one another. The planned mobility is explained by the mobility and traffic needs, but the intended mobility is restricted by motion restrictions. In this case, the granularity of these two blocks is what determines how accurate a model is. When developing an application, it is necessary to take into account three sorts of motion limitations. Figure 16.11

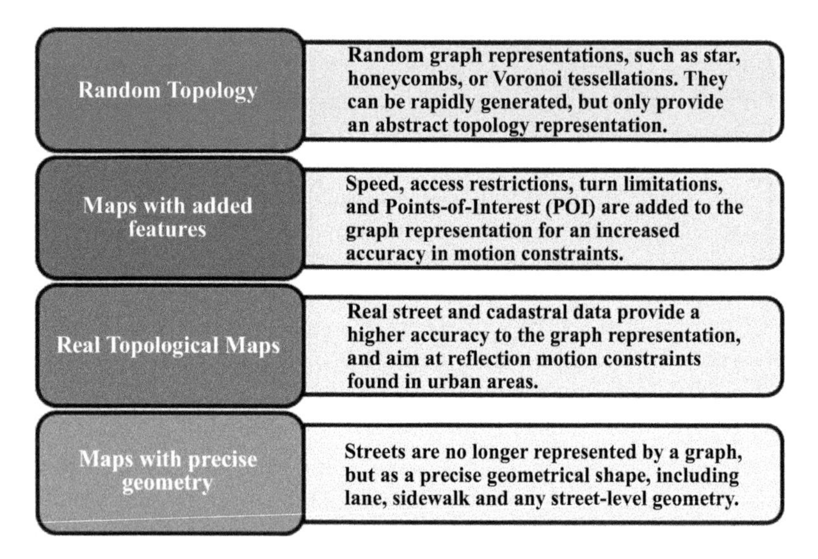

Figure 16.11 Substantial influence on mobility patterns

demonstrates, on the other hand, that increasing motion restriction accuracy has a substantial influence on mobility patterns. Remember to keep this in mind while designing the motion constraint block.

This makes it possible to more accurately identify cars and people inside a network, as well as to keep track of their interactions with one another. These and other considerations have resulted in the classification of mobility needs into a variety of categories [62]. Historically, random models have been the most common and straightforward method of representing the mobility of communication networks. Because of their simplicity and stochastic qualities when used to represent vehicle movement, they make up for their lack of precision with their stochastic properties. Microscopically detailed models of traffic flow are necessary for bigger evaluations because they correctly depict how a vehicle interacts with other vehicles, traffic regulations, and street topologies, among other things. Whenever mesoscopic models are used for large-scale assessment, they provide a great balance between accuracy and simulation performance. Behavioral models, rather than flow or traffic equations, can be used to demonstrate patterns in vehicle movements in situations when strictly physical principles fail to represent the movement of the vehicle in question. When we talk about traffic demands, wc are talking about the broad patterns of car movement. Each junction is approached using the Random Turns method, which results in a random turn being made at each intersection. Using origin-destination (O-D) matrices, it is feasible to depict the beginning and destination locations, as well as the optimal paths between them, in greater detail than previously conceivable. It is possible to get population models for a whole city or a wide region, which may be used to fine-tune O-D Matrices and preferences in routes and mobility options, among other things. When it comes to vehicle mobility models, three fundamental building blocks are enhanced by the inclusion of new features and enhancements at the expense of increased cost and complexity. A lot of factors impact vehicle mobility as shown in the figure, including its capacity to move about and the quantity of traffic on the road [63].

16.6.4 Towards more realistic mobility modeling

One of the difficulties is developing a traffic demand model that properly reflects vehicle mobility at both the macroscopical and microscopic levels of analysis. As an example, demonstrate how just a flow model is required for testing traffic safety applications to mimic realistic microscopic vehicle-to-vehicle (V2V) interactions (e.g., lane departure warning systems). Traffic models are necessary for large-scale mobility patterns such as traffic efficiency analysis and traffic flow forecasting. Models for traffic flow and distribution are built using physical equations to ensure the safety and efficiency of the traffic flow and distribution system. Cars tend to behave in the same manner under any set of conditions. The reality is that drivers are not robots; their responses to numerous stimuli vary depending on the circumstances they find themselves in [64]. This is demonstrated by the patterns that appear, which are characterized by reactions that do not perfectly follow laws but are instead tailored to the conditions in which they occur. This type of model is commonly used in accident simulation, for example, where it is frequently used for simulation. Researchers are paying growing attention to large-scale calibrated urban mobility models based on trace/survey models, which are becoming increasingly popular. Because of their complexity, as well as the geographical and temporal constraints of the calibration data, these models' accuracy is restricted, although they are quite close to the original. Make sure you pick the right mobility model for your application and protocol needs since these define which mobility model should be utilized and how accurate it should be. Although node mobility is currently supported by all major network simulators, the sophistication of the mobility models used differs significantly across them. A significant influence on the outcomes of mobile ad hoc network simulations is exerted by the quality of the mobile models included in the simulation. A detailed study has demonstrated the influence of mobility models on IVC simulation results as well as the limitations of simple random mobility models, demonstrating the relevance of these models.

16.7 Network and mobility simulation tools—analysis and research views

In the following sections, we will present and briefly discuss some of the most often used simulators for IVC application and protocol performance evaluation. For a comprehensive examination of the vast majority of the tools and techniques, now accessible [65,66].

16.7.1 The ns-3 network simulator

ns-3 is a discrete-event simulation environment that was created to replace the widely used ns-2 network simulator, which was developed in the 1990s. A key difference between ns-2 and ns-3 is the structure and modularity, which is intended to make the system more scalable and adaptable. For simulation purposes, ns-3 makes use of optional Python scripts rather than the C++ and OTcl scripts that were previously utilized in ns-2. The basic architecture of C++, on the other hand, remains object-oriented. There have been few instances of ns-2 tons-3 models being converted to C++. Although it is exclusively available for commercial use, the underlying architecture was designed for network virtualization and real-time testbed integration. The most significant aspect of ns-3 is the object aggregation model that it employs. Connections between nodes, applications, and protocol stacks can be established at any point in time. The "weak base class" problem in C++ causes every time the base class is reused in a new configuration to need its modification. As a potential solution, the following method has been suggested.

Through the management of access between many aggregated items, the aggregation paradigm significantly simplifies the automated memory control method. It is smaller in contrast to ns-2, but it can handle more models, such as IEEE 802.11, 3GPP LTE, IPv6, and various MANET routing protocols, in addition to IPv6, than ns-2. Of particular note are the OFDM symbol processing, space-time channel modeling, and extraordinarily precise reception model that are all included in the physical layer implementation of ns-3, which allows it to more accurately simulate wireless transmissions at the signal level. The fact that so many processing resources are required means that this can only be used in limited circumstances. ns-3 has become even more helpful as a result of the development of ETSI ITS compliant IVC protocol stacks such as iTETRIS and COLOMBO1. CAM, DENM, and service announcement message (SAM) are among the communication-related facilities that have been included in the system, as have geo-routing protocols (geounicast, geobroadcasting, and geo-anycast) and delay tolerant networks (DTN) capabilities. It is feasible to gain access to heterogeneous access technologies, such as ETSI IT-G5, for example. With the iTETRIS management software, you may perform multi-channel operations and choose from a variety of technical alternatives. ITS G5 channels can only be unloaded by one of these devices; the other offloads traffic to a variety of access technologies and the network stack as required by IVC applications. Also provided are DCC channel load monitors at the management layer and data plane, both of which are ETSI-compliant, as well as DCC flow control capabilities in the data plane.

16.7.2 The OMNeT++ simulation environment

Everyone can now use OMNeT++ version 4 for free, as long as you do not intend to profit from it in any way. OMNEST is a commercially licensed version of the same simulation environment that is sold by Simulcraft, Inc. under the OMNEST brand name, which is licensed from the business. To utilize OMNeT++, you will need an integrated development environment (IDE), a simulation kernel, and an execution environment. When assembling and customizing simulations graphically, it takes advantage of Eclipse's simulation capabilities rather than manipulating plain text files directly. Execution environments are available in two varieties. Batch executions will take place on dedicated computers in a command-line-based environment, without the need for intervention. When utilizing a graphical user interface (GUI), it is easier to interact with the various components of the simulator. OMNeT++ distinguishes between code that behaves in a unique way and code that behaves in a descriptively descriptive way. A connection is established between the OMNeT++ kernel and all of the behavioral code, all of which is written in C++ (i.e., code detailing how basic modules handle and send messages, as well as how channels handle messages). Each msg and ndf file contain a message definition (i.e., code specifying the module/channel and message structure) as well as a network description, which is both kept in the same directory. The initialization file is responsible for handling all module run-time settings.

A C++ program contains all of the behavioral code for OMNeT++ components, which makes it simple to combine them with other libraries and debug them using commercial tools such as Visual Studio. Consequently, it is ideal for both quick prototyping and the development of high-quality software applications for a variety of applications. In this kernel, several distinct module libraries are utilized to simulate various sorts of protocol stacks from different vendors. Channel modeling and signal processing that is based on realistic data are major elements of the MiXiM module library, for example. Signal power levels change over time and frequency when seen as three-dimensional entities at a given moment in time and space. When it comes to signal simulation, MiXiM is responsible for modeling the propagation and interference of a wide range of signals. There is nothing further that the model builder has to do for the model to work properly. It is not necessary to have accurate models of common upper-layer Internet protocols for IVC modeling; as a result, MiXiM is well-suited for IVC modeling. Overlay networks and peer-to-peer networking are the primary focus of the Oversim module library extension, whereas the INET Framework is concerned with correct IPv4 and IPv6 representations, as well as Internet transport layer protocols and applications at the application layer.

Simu LTE is a module library that simulates cellular networks that use LTE and LTE advanced technology. The IEEE 802.11p DSRC and IEEE 1609.4 WAVE path loss models are among the IVC-specific route loss models featured in OMNET++, as is the IEEE 802.11p DSRC. auxiliary models such as driver behavior and emission calculation are also incorporated in the simulation. ITS G5 protocol stack and the DCC protocol stack, both developed by the International Telecommunications Standards Institute, have been deployed. The OMNeT++ framework is capable of running simulations that have been developed with these module libraries.

16.7.3 JiST/SWANS

It is built on the JiST platform and is known as SWANS (scalable wireless network simulator). JiST design, for example, allows users to execute typical Java network applications over simulated networks while conserving memory. This is one of the many advantages of JiST architecture. JiST/SWANS source code is now available to anybody who wants to use it thanks to the Creative Commons license provided by the Cornell Research Foundation. Ulm University was the site of the development of several JiST/SWANS modifications and upgrades, notably for mobile ad hoc networks. Configuration files, for example, are used by the DUCKS software to create a fully functional simulation research environment. Aside from that, the DUCKS program provides a flexible means of archiving study findings that may be customized. A configuration file is used to indicate how results should be saved and where they should be saved.

16.7.4 The SUMO simulation environment

Modern simulation systems employ traffic microsimulation models, which are found in a broad variety of applications. When adopting free and open-source software modeling frameworks, researchers may more readily compare their findings, which is beneficial for both the researchers and the public. The Simulation of Urban MObility (SUMO) small road traffic simulation environment, which is the most often used simulation environment, will be briefly discussed in this part. Because of the widespread use of simulations in the scientific community, it is simple to compare different simulations. Thanks to the fact that SUMO is Free and Open-Source Software licensed under the GNU General Public License, you may simulate multi-modal traffic on city-scale networks with this software (version 2 or later). SUMO simulations, whether or not they are accompanied by an OpenGL-based GUI, let people interact directly with the simulation while it is being run. As part of its development for realistic simulation of numerous different vehicles, the SUMO model was expanded to incorporate both an adaptation of the Kraußand tiny vehicle mobility model and more complex mobility models such as the Intelligent Driver Model (IDM) model. Some vehicles have a static route allocated to them, while others have a dynamic route that has been specifically developed for them to go on. Others just drive according to a predetermined timetable. In SUMO, every road can have an infinite number of lanes, regardless of its length. Individual lanes can take on any shape or size because there are no design restrictions.

16.7.5 PTV-VISSIM- An Advanced Traffic Simulation Software

Pedestrians and automobiles are based on psycho-physical drivers, whereas cyclists and pedestrians are based on social forces, according to the concept. Traffic simulations occur in a range of forms and sizes, depending on the need. They will probably include scenes involving automobiles, trucks, buses, and two-wheelers (such as bicycles and motorcycles) (bus, tram, underground). VISSIM includes a pedestrian model as well as an automobile model. When it comes to private traffic, OD matrices are used to design automobile journeys, whereas schedules are used when it comes to public transit. Transport of data throughout the network is accomplished through the use of preset routes. The APIs provided by VISSIM are useful when it comes to introducing new features or improving the functioning of current ones. The creation of new driver-behavior models is made possible through the use of a DLL interface.

While the simulation is underway, a COM interface may be used to allow read and write access to the simulation data. In this way, the simulation can interact with other tools while it is running, as long as they are linked to it.

16.8 Integrated Inter-Vehicle Communication (IVC) simulation toolkits

While mobility is simply regarded as an annoyance when it comes to wireless communication apps, it is the application in IVC applications, which means that it must be dynamically updated to function properly. Another issue is that a single simulator cannot handle all of the different IVC application models that are required due to the vast range of specialized knowledge that is required, especially for standard compliance, making it impossible to use a single simulator for everything. Federating the necessary simulators in an IVC simulation framework is a successful technique since each community (mobility, application, communication, and so on) often relies on its unique simulators, which are federated in an IVC simulation framework [67–69]. Traditionally, the synthetic traces from traffic simulators have been retrieved as mobility files and then incorporated into the network simulator using a particular parser designed for that purpose. Particularly relevant is the interplay between communication and mobility in the context of the Internet of Things. Although vehicle communication or networking is not required to achieve mobility, this technique is nevertheless popular. Apps developed by IVC will alter mobility to prevent accidents or minimize traffic congestion. Two (or more) simulators are connected in a two-way fashion to exchange mobility data or to affect the control of one another. This technique is presently the most widely used approach for evaluating IVC software assessments. An enormous amount of effort has been devoted to the development of integrated IVC simulation frameworks as a result. Efforts to bring these programmers together have culminated in the last several years as a result of consolidation efforts.

16.8.1 Veins

As an extension to OMNeT++, Veins is an open-source simulation framework that was developed on top of it. In the discrete event situation, on the other hand, the OMNeT++ simulation kernel is in charge of all simulation control and data collecting activities. Veins provide a modular framework that makes use of SUMO to mimic vehicle movement in real-time to model unique applications for a variety of industries. To keep the MiXiM model suite's discrete event simulation of wireless channels distinct from the Internet protocol simulation and cellular network communication simulation, separate model libraries for Internet protocol simulation and cellular network communication simulation have been developed. In conjunction with the IVC model suite of the OMNET++, it is possible to build a bespoke data generation and dissemination protocol for traffic safety, efficiency, and infotainment. Because all of Veins' modules are built and linked together into a single executable, it is feasible to execute a GUI application or batch simulation from the command line. A simulation in this simulator is a virtual implementation of an application with automated vehicles/ of vehicles with intelligent nature. The SUMO Road traffic simulator is used to replicate the movement of vehicles on the road. The simulation now in use starts a second instance, which is managed by the first instance of the simulation. Veins can make use of SUMO-integrated object subscriptions to request push alerts and updates from the simulation when vehicles are being constructed or their circumstances change. Combining realistic channel and access models with behavior and mobility input enables the gathering of a wide variety of characteristics, such as the assessment of junction collision avoidance solutions, that would otherwise be impossible to collect. On the Veins website, you may find further information, the entire source code, a beginner's guide, and associated publications.

16.8.2 iTETRIS

At the heart of the iTETRIS system, the control system (iCS) is the interface that communicates with the rest of the system. Because there are no restrictions on the number of simulator instances that may be utilized, it can be used with a broad range of simulators. ns-3 Making Use of Cutting-Edge Technology in the Transportation and Networking System Management of the Upper-Layer Interface the SUMO Pollutant Model Lighting Control is used to simulate a road network. It is possible to configure, synchronize, and control the iCS application layer proxy interface for the ETSI ITS G5 facility layer lower layers via a human–machine interface (HMI). With the integration of ITS Facilities IVC application features into the iCS, such as station management and local dynamic maps, its function becomes even more critical. When using the IVC application module, the result container may be used to store and transport generic data between several instances of the module. An open API set, controlled by the iCS subscription module, regulates data flow across the ns-3, SUMO, and IVC applications. With this one-of-a-kind capability, IVC programmers may simulate dynamic interactions between mobility and networking realistically. It is iTETRIS IVC application simulator that stands out in this regard. By extracting the logic from the primary simulators and transferring it to a secondary simulator, the assessment of ITS applications is made easier to do. The architecture of the IVC simulator is divided into two parts: the top and the bottom. The application logic of the IVC and the ICS are intricately linked in a variety of ways. Everything about relationships is taken care of at the first layer. By utilizing a block named Payload Storage to store IVC data on the application simulator, this data may be saved locally on the application simulator. Because unlike other methods of increasing system scalability, this approach does not require real data to be included in the payloads of the simulated packets, which allows it to be used to enhance system scalability. The logic for the IVC application is located within the second layer. It is at this layer where IVC application logics are implemented using an extendable higher-layer node architecture similar to that of ns-3, with the ITS controller serving as an interface and coordinator. The application simulator and other iTETRIS components, such as the IVC application logics and the IVC application simulator, make model integration and interaction during simulation execution simpler. The ability to have these two applications running at the same time is essential in situations where, for example, one IVC software is in charge of traffic monitoring while the other is in charge of delivering personalized advice services. The iTETRIS community website3, which also provides information on the iTETRIS platform, has links to iCS and IVC application simulators.

16.8.3 VSimRTI

VSimRTI goes above and beyond the call of duty when it comes to segmenting the various components. M&S High-Level Architecture (IEEE, Standard) is an all-encompassing architecture that links multiple simulators, each of which focuses on a distinct area (HLA). There are no worries when it comes to management when you use VSimRTI because it takes care of everything. Optimistic synchronization, for example, is concerned with the performance of high-performance simulators. Because common VSimRTI interfaces are available, it is feasible to integrate and interchange simulators with relative ease. With the introduction of simulators, it is now feasible to simulate vehicle traffic, emissions, wireless communication (cellular and ad hoc), user behavior, and mobility application modeling more realistically than was previously possible. The variety of simulators that may be created using VSimRTI can be used almost immediately. In addition to traffic simulation, communication simulation is provided by the programmers ns-3, OMNeT++, JiST/SWANS, and a mobile phone communication simulator (see below). Electric vehicle simulations are also being included in VSimRTI as part of an ongoing update. This V2X application simulator may be used to simulate V2X apps. Applications can request sensor data or communicate with communication modules by interacting with a simulated vehicle-like environment. To produce a format that can be utilized by real-world automobile components, data from traffic, communication, and other related simulations are employed. Java logic is used to control the operation of the program. If you want to fine-tune the behavior of an application in the VSimRTI application simulator, you may change the CAM transmission rate or the circumstances under which a DENM is broadcast. Additional information may be found on the VSimRTI website.

16.9 Conclusion and future work

This chapter describes the simulation modeling requirements for three types of IVC applications: safe, smart, and linked. There are various granularity levels available and categorized according to their relevance to different IVC application types. They were considered. Each of these aspects must be balanced. To achieve this objective, we had to disclose which models were essential for IVC simulations. In addition, finding simulators with all the required models and granularity for IVC application development is challenging. The most recent simulation models and methods were also evaluated and presented. Simulators that include all essential models at different granularity levels have enhanced the quality and comparability of ITS simulation studies. As a result, many modern studies of simulation performance use similar methods and tightly integrated simulation components. These methods are common. Veins, iTETRIS, and VSimRTI are three separate yet complementary toolkits. In this case, it is tough to choose between the three options. Much will rely on the protocols or apps used to find the best option. Using only one of these toolkits may significantly enhance the research's credibility and repeatability. One factor is the availability of well-established IVC models. Examples include realistic mobility models that may be dynamically linked to IVC applications, as well as propagation and channel access models. He discusses ITES and urban transportation networks, focusing on the Portuguese experience. So that these intelligent solutions may be properly explored, municipalities must be led by an integrated framework. An urban transportation modeling and simulation platform that can handle the system's spatial and dynamic characteristics must be incorporated. The criteria need transportation GIS and a traffic simulation program. Recent concerted initiatives have seen notable achievements and forward momentum. On the other hand, a broad overview of the present state of the art is provided. Despite this, implementing VC procedures will be challenging. Large-scale field experiments will be needed to fully confirm the system's dependability. Aside from data lines and networking in general, additional applications, particularly those with strict requirements, are involved. Even if tough circumstances never arise, ensuring efficient and effective operation is essential (e.g., as the size of VC networks scales up). When developing architecture and standards for secure VC and privacy-enhancing technologies, security should not be ignored. Secure VC systems may be as effective as non-secure ones. Trustworthy VC systems might be implemented if individuals realized how vital security is. The project's financial, legal, and organizational elements should also be considered. Commercialization requires product innovation. Without widespread deployment (in cars and/or infrastructure), drivers will see few advantages. It's a no-brainer to invest in anything long term. Understanding how venture capital works are critical. Clearly define who is in charge of what and how actions are done. Standardization may be achieved through initiatives, industrial consortia, standardization committees, and working groups. Future solutions to the issue may be revolutionary. Work on autonomous vehicles, new sensor technologies, and driver assistance systems may be very helpful.

References

[1] Liu, B, Han, C, Liu, X, and Li, W. "Vehicle artificial intelligence system based on intelligent image analysis and 5G network." *International Journal of Wireless Information Networks* (2021): 1–17.

[2] Martínez-Díaz, M, and F Soriguera. "Short-term prediction of freeway travel times by fusing input-output vehicle counts and GPS tracking data." *Transportation Letters* 13(3) (2021): 193–200.

[3] Böhm, F, Dietz, M, Preindl, T, Pernul, G. "Augmented reality and the digital twin: state-of-the-art and perspectives for cybersecurity." *Journal of Cybersecurity and Privacy* 1(3) (2021): 519–538.

[4] Zhou, C, Xiong, R, Zeng, H, *et al.* "Aerial locating method design for civil aviation RFI: UAV monitoring platform and ground terminal system." *Journal of Intelligent & Robotic Systems* 103(2) (2021): 1–13.

[5] Oza, P, Abhyankar, N, Seth, A, and Sharma, G. "Road Incidents and Closures Classification in a Connected Vehicular Environment". 10.13140/RG.2.2.31982.74561.

[6] Mohammadi, M, Elfvengren, K, Khadim, Q, and Mikkola, A. "The technical-business aspects of two mid-sized manufacturing companies implementing a joint simulation model." *Real-time Simulation for Sustainable Production*. Routledge, London, 2021, pp. 102–118.

[7] Caicedo, P and C Roviro. *Model-based real-time monitoring of large-scale urban traffic networks for decision making*. Diss. Universidad Nacional de Colombia.

[8] Kim, H Kyoung, Y Chung, and M Kim. "Effect of enhanced ADAS camera capability on traffic state estimation." *Sensors* 21(6) (2021): 1996.

[9] Li, L, W Zhu, and H Hu. "Multivisual animation character 3D model design method based on VR technology." *Complexity* 2021 Volume 2021, Article ID 9988803, https://doi.org/10.1155/2021/9988803.

[10] Moreira, L Henrique, and CG Ralha. "Evaluation of decision-making strategies for robots in intralogistics problems using multi-agent planning." In 2021 IEEE Congress on Evolutionary Computation (CEC). IEEE, New York, NY, 2021.

[11] Nishibori, S, T Murase, and Y Tadokoro. "Periodic networked imaging with nanoscale sensor nodes via two-layered time-division access." *IEEE Internet of Things Journal* (2021).

[12] Cheng, H, Q Hu, X Zhang, Z Yu, Y Yang and N Xiong. "Trusted resource allocation based on smart contracts for blockchain-enabled Internet of Things." *IEEE Internet of Things Journal* (2021).

[13] Gilmore, SK. The Future Railway Control Room: A Practical Framework for Control Room Design Practices. *MS thesis*. 2021.

[14] Rahman, H, Md, and M Abdel-Aty. "Application of connected and automated vehicles in a large-scale network by considering vehicle-to-vehicle and vehicle-to-infrastructure technology." *Transportation Research Record* 2675(1)(2021): 93–113.

[15] Bai, J, W Yu, Z Xiao, *et al.* "Two-stream spatial-temporal graph convolutional networks for driver drowsiness detection." *IEEE Transactions on Cybernetics* (2021).

[16] Jesus, GT, and MFC Júnior. "The roles of modeling and simulation in assessing spacecraft Integration Readiness Levels." 2021 IEEE International Systems Conference (SysCon). IEEE, New York, NY, 2021.

[17] Deb, PK, A Mukherjee, and S Misra. "XiA: send-it-anyway Q-routing for 6G-enabled UAV-LEO communications." *IEEE Transactions on Network Science and Engineering* 8(4) (2021): 2722–2731. Doi: 10.1109/TNSE.2021.3086484.

[18] Khan, S Mahmud, and AD Patire. "Third-party data fusion to estimate freeway performance measures." *Transportation Research Record* (2021): 03611981211024240.

[19] Ierardi, C. *Distributed Estimation Techniques Forcyber-Physical Systems*. Diss. 2021.

[20] Gatien, S, A Khan, and J Gales. "Technology and human factor considerations in adapting airport landside facilities and operations to autonomous and connected vehicles." *International Conference on Applied Human Factors and Ergonomics*. Springer, Cham, 2021.

[21] Poltronieri, F, M Tortonesi, A Morelli, C Stefanelli, and N Suri. "A value-of-information-based management framework for fog services." *International Journal of Network Management* (2021): e2156.

[22] Rahman, MH, Mohamed A-A, and Y Wu. "A multi-vehicle communication system to assess the safety and mobility of connected and automated vehicles." *Transportation Research Part C: Emerging Technologies* 124 (2021): 102887.

[23] She, K, and Y Sun. "Impacts of scene geometry and vehicle speed on the performance of RFID based AVI/ETC system." *2021 International Wireless Communications and Mobile Computing (IWCMC)*. IEEE, New York, NY, 2021.

[24] Dowd, G, O Kavas-Torris, L Guvenc, and B Aksun-Guvenc. "Simulation environment for visualizing connected ground and air traffic." *Transportation Research Record* 2675(6) (2021): 15–22.

[25] Progga, FT, H Shahriar, C Zhang, and M Valero. "Securing vehicular network using AI and blockchain-based approaches." *Artificial Intelligence and Blockchain for Future Cybersecurity Applications*. Springer, Cham, 2021, pp. 31–44.

[26] Bezziane, MB, A Korichi, M Fekair, and A Nadjet. "FR-VC: a novel approach to finding resources in the vehicular cloud." *Innovative and Intelligent Technology-Based Services for Smart Environments – Smart Sensing and Artificial Intelligence*. CRC Press, London, 2021, pp. 149–155.

[27] Prathiba, SB, G Raja, A K Bashir, A A AlZubi and B Gupta. "SDN-assisted safety message dissemination framework for vehicular critical energy infrastructure." *IEEE Transactions on Industrial Informatics* 18(5) (2022): 3510–3518.

[28] Martin, L, M González-Romo, M Sahnoun, B Bettayeb, N He and J Gao. "Effect of human-robot interaction on the fleet size of AIV transporters in FMS." In 2021 1st International Conference on Cyber Management and Engineering (CyMaEn). IEEE, New York, NY, 2021.

[29] Fragapane, G, R De Koster, F Sgarbossa, J O Strandhagen. "Planning and control of autonomous mobile robots for intralogistics: literature review and research agenda." *European Journal of Operational Research* (2021).

[30] Cao, Q, G Ren, D Li, H Li and J Ma. "Map matching for sparse automatic vehicle identification data." *IEEE Transactions on Intelligent Transportation Systems* (2021). Doi: 10.1109/TITS.2021.3058123.

[31] Leszczyna, R. "Review of cybersecurity assessment methods: applicability perspective." *Computers & Security* (2021): 102376.

[32] Dogaroglu, B, SP Caliskanelli, and S Tanyel. "Comparison of intelligent parking guidance system and conventional system with regard to capacity utilisation." *Sustainable Cities and Society* 74 (2021): 103152.

[33] Rios, JLG, J T Gómez, R K Sharma, F Dressler and M J F-G García. "Wideband OFDM-based communications in bus topology as a key enabler for Industry 4.0 networks." *IEEE Access* 9 (2021): 114167–114178.

[34] Basso, F, A Cifuentes, R Pezoa, and M Varas. "A vehicle-by-vehicle approach to assess the impact of variable message signs on driving behavior." *Transportation Research Part C: Emerging Technologies* 125 (2021): 103015.

[35] He, Z, J Wang, and C Song. "A review of mobile robot motion planning methods: from classical motion planning workflows to reinforcement learning-based architectures." arXiv preprint arXiv:2108.13619 (2021).

[36] Abdeldayem, O M., A Dabbish, M M Habashy, *et al.* "Viral outbreaks detection and surveillance using wastewater-based epidemiology, viral air sampling, and machine learning techniques: A comprehensive review and outlook." *Science of The Total Environment* (2021): 149834.

[37] Ray, AM, A Sarkar, A Obaid, and S Pandiaraj. "IoT security using steganography." In *Multidisciplinary Approach to Modern Digital Steganography*. IGI Global, Hershey, PA, 2021, pp. 191–210.

[38] Witrisal, K, A-H Carles, G Stefan, *et al.* "Localization and tracking." In *Inclusive Radio Communications for 5G and Beyond*. Academic Press, London, 2021, pp. 253–293.

[39] Lalla, E, M Konstantinidis, C Czakler. *et al.* "Remote science activities during the AMADEE-18 Mars analog mission: preparation and execution during a simulated planetary surface mission." *Journal of Space Safety Engineering* 8(1) (2021): 75–85.

[40] Alharbi, A, M Yamin, and G Halikias. "Smart technologies for comprehensive traffic control and management." In *2021 8th International Conference on Computing for Sustainable Global Development (INDIACom)*. IEEE, New York, NY, 2021.

[41] Cui, J, W Macke, H Yedidsion, A Goyal, D Urielli and P Stone. "Scalable Multiagent Driving Policies For Reducing Traffic Congestion." arXiv preprint arXiv:2103.00058 (2021).

[42] Asha, AZ, C Smith, G Freeman, S Crump, S Somanath, L Oehlberg, and E Sharlin. "Developing an improved ACT-R model for pilot situation awareness measurement." *IEEE Access* 9 (2021): 122113–122124.

[43] Chen, H, S Liu, L Pang, X Wanyan and Y Fang. "Developing an improved ACT-R model for pilot situation awareness measurement." *IEEE Access* 9 (2021): 122113–122124.

[44] Brandín, R, and S Abrishami. "Information traceability platforms for asset data lifecycle: blockchain-based technologies." *Smart and Sustainable Built Environment* (2021).

[45] Hildmann, H, K Eledlebi, F Saffre, and A Isakovic. "The swarm is more than the sum of its drones." *Development and Future of Internet of Drones (IoD): Insights, Trends and Road Ahead* 332 (2021): 1.

[46] Chen, Y-T, E Sun, M-F Chang, and Y-B Lin. "Pragmatic real-time logistics management with traffic IoT infrastructure: big data predictive analytics of

freight travel time for Logistics 4.0." *International Journal of Production Economics* 238 (2021): 108157.

[47] Yang, L, D Sun, and H Ruan. "Research on acquisition and tracking algorithm of global satellite positioning receiver based on UWB." *Wireless Communications and Mobile Computing* 2021 (2021).

[48] Fernando R, V Marco, A M-M Jose, *et al.* "Spectroscopic Study of Terrestrial Analogues to Support Rover Missions to Mars – A Raman-Centred Review." (2021). https://doi.org/10.1016/j.aca.2021.339003.

[49] Aldinucci, M, A Giovanni, A Antonio, *et al.* "The Italian research on HPC key technologies across EuroHPC." In *Proceedings of the 18th ACM International Conference on Computing Frontiers*, Association for Computing Machinery, New York, NY, USA, 178–184. Doi: https://doi.org/10.1145/3457388.3458508.

[50] Bertsekas, D. "Lessons from AlphaZero for Optimal, Model Predictive, and Adaptive Control." arXiv preprint arXiv:2108.10315 (2021).

[51] Kondaveeti, HK, and S E Mathe. "A systematic literature review on prototyping with Arduino: applications, challenges, advantages, and limitations." *Computer Science Review* 40 (2021): 100364.

[52] Chebi, H, and A Benaissa. "Novel approach by fuzzy logic to deal with dynamic analysis of shadow elimination and occlusion detection in video sequences of high-density scenes." *IETE Journal of Research* (2021): 1–12.

[53] Carou, D. "Aerospace transformation through industry 4.0 technologies." In *Aerospace and Digitalization*. Springer, Cham, 2021, pp. 17–46.

[54] Bouzekri, E, C Martinie, P Palanque, K Atwood, and C Gris. "Should I add recommendations to my warning system? The RCRAFT framework can answer this and other questions about supporting the assessment of automation designs." In IFIP Conference on Human-Computer Interaction. Springer, Cham, 2021.

[55] Mufti, N, and SAA Shah. "Automatic number plate recognition: a detailed survey of relevant algorithms." *Sensors* 21(9) (2021): 3028.

[56] Bogdoll, D, S Orf, L Tottel, JM Zoller. "Taxonomy and Survey on Remote Human Input Systems for Driving Automation Systems." arXiv preprint arXiv:2109.08599 (2021).

[57] Gholami, A, W Daobin, R Seyed, Z T Davoodi. "An adaptive neural fuzzy inference system model for freeway travel time estimation based on existing detector facilities." *Case Studies on Transport Policy* (2021).

[58] Vitali, G, M Francia, M Golfarelli, and M Canavari. "Crop management with the IoT: an interdisciplinary survey." *Agronomy* 11(1) (2021): 181.

[59] Goswami, SS, and DK Behera. "Solving material handling equipment selection problems in an industry with the help of entropy integrated COPRAS and ARAS MCDM techniques." *Process Integration and Optimization for Sustainability* (2021): 1–27.

[60] Huang, A, L Yang, C Tianjian. "StarFL: hybrid federated learning architecture for smart urban computing." *ACM Transactions on Intelligent Systems and Technology (TIST)* 12(4) (2021): 1–23.

[61] Singhal, A, R Roy, D Mittal, and P Dahiya. “Optimal tuning using global neighborhood algorithm for cruise control system.” In 2021 7th International Conference on Advanced Computing and Communication Systems (ICACCS). Vol. 1. IEEE, New Yor, NY, 2021.
[62] Dömeke, A. *A Media Caching Approach Utilizing Social Groups Information in 5G Edge Networks*. Diss. Bilkent University, 2021.
[63] Li, J, Y Ma, P Li, and A Butz. “A journey through nature: exploring virtual restorative environments as a means to relax in confined spaces.” *Creativity and Cognition*. 2021.
[64] Peng, Y, WA. Scales, and D Lin. “GNSS-based hardware-in-the-loop simulations of spacecraft formation flying with the global ionospheric model TIEGCM.” *GPS Solutions* 25(2) (2021): 1–14.
[65] Han, H, Z Jun, X Zehui, *et al.* “Reconfigurable intelligent surface aided power control for physical-layer broadcasting.” *IEEE Transactions on Communications* (2021).
[66] Kulikov, A, A Loskutov, A Kurkin, *et al.* “Development and operation modes of hydrogen fuel cell generation system for remote consumers’ power supply.” *Sustainability* 13(16) (2021): 9355.
[67] Majid, I, R Sabatini, K A Kramer, *et al.* “Restructuring avionics engineering curricula to meet contemporary requirements and future challenges.” *IEEE Aerospace and Electronic Systems Magazine* 36(4) (2021): 46–58.
[68] Politi, E, I Panagiotopoulos, I Varlamis, and G Dimitrakopoulos. “A survey of UAS technologies to enable beyond visual line of sight (BVLOS) operations.” *VEHITS*. 2021.
[69] Islas-Cota, E, J. Octavio Gutierrez-Garcia, O A Christian, R Luis-Felipe. “A systematic review of intelligent assistants.” *Future Generation Computer Systems* 128 (2022): 45–62.
[70] Abbas, Q, and A Alsheddy. “Driver fatigue detection systems using multi-sensors, smartphone, and cloud-based computing platforms: a comparative analysis.” *Sensors* 21(1) (2021): 56.

Chapter 17

Security, privacy, and trust for Autonomous Intelligent Vehicles/Intelligent Transportation Systems

Abstract

The world has been evolving at lightning speed with new technologies being invented every other day and the transportation sector is surely one of the fields taking the fast track towards innovation, advanced technology and inventions. Intelligent transport system (ITS) is one of the enhanced applications of the transportation sector which provides intelligent and smart network/framework services by holistically addressing the various aspects of the sector like safety, traffic control, interconnected networks, etc. Autonomous intelligent vehicles (AIV) are majorly considered as the crucial components of any ITS framework. This chapter addresses one of the most sought-after challenges of the field which is linked to the security, privacy, and trust/reliability aspects of an intelligent and smart framework. The fact that such systems extract large amounts of data from the users and nodes within the system is one of the major highlights which may lead to various consequences and external attacks if due importance is not given. Furthermore, exposing the data acquired to other organizations or data breaches which leads to confidential data of the users reaching malicious users puts the privacy and safety of the users at stake. This chapter will elucidate and elaborate on the various safety and security aspects of ITS/AIV and will also give insights on the perspective or trust and reliability of the users being important factors that drive the successful implementation of the system.

Key Words: Keywords Security, Privacy, Trust, Autonomous intelligent vehicles, Intelligent transportation system, automation

17.1 Introduction

Advancing technologies and evolving tech strategies have surely been a boost for a plethora of fields and domains in the society. However, when it comes to the transportation sector which involves manual driving, the statistics of accidents and mishaps in this field are extremely high; the main cause rooting down to the human errors involved during the decisions made while on road [1]. The World Health Organization (WHO) states that around 1.3 million people are affected by such mishaps each year and are victims of death for the same reason. Another striking fact put up by the WHO team is that the crashes and accidents caused on road contributes to around 3% of the nations' gross domestic product (GDP) and exposes children and youngsters to a highly vulnerable chance of losing their lives [2]. This is clear evidence of the fact that it is high time to engage factors and parameters that would increase the accuracy of the decisions made for vehicles while on road. What better to fit in to this model when the automation and intelligent devices have been throbbing extensively in the various sectors [3].

Concepts like smart cities and smart societies have been emerging in various parts of the globe and an integral portion of all of these is the evolution in the transportation sector through technological and scientific aspects. Furthermore, in a generation which completely relies on automation and technology for a living, intelligent transportation system with autonomous intelligent vehicles (AIVs) is the boomer. In any developing city, mobilization with ease and comfort is of top priority and hence we see an increasing incline towards intelligent and smart transport frameworks. In fact, the uniqueness of ITS lies in its varied application and distinct features expanding over and beyond the usual traffic control and road management. It is a field which has turned out to be an interdisciplinary aspect spanning over ideations including travel information system, public transport system, vehicle control systems, rural transportation systems, and so on. To top it off, ITS being adopted in various applications have turned out to be more data centric and involves large scale data consumption and transmission for efficiently carrying out tasks and analyses. Evolving ITS applications include computational hardware, tracking systems, sensors and communication technologies, data processing, digitalization, etc. When there is so much data handling at stake, the security, privacy, and reliability of the operations carried out with ITS are thoroughly questioned before it is widely accepted in the community [4]. Despite the rapidly evolving tech strategies and portals, there are quite a few security issues and concerns to be addressed with regards to the safety observed while streaming the data, encryption standards used for preserving informational privacy, storage policies, and much more.

Issues like confidentiality, integrity, authenticity and identification, behavioral privacy, fault tolerance, and availability are few of the many parameters which are extensively used for identifying and analyzing the security and privacy aspect of smart and intelligent vehicles and transportation systems. Figure 17.1 illustrates the work flow of the basic ITS framework and engages how the architectural functioning of the system usually works. However, in order to ensure a reliable system, further components and modules of safety, security, and privacy aspects need to be incorporated. The main motive of this work is to showcase the various aspects and perspectives of the security and privacy sector of AIVs and ITS [5]. Section 15.2 of this chapter elucidates some of the existing works and researches in this field while Section 15.3 focuses on the safety approaches and parameters for such automated vehicles and its systems. Section 15.4 walks the readers through the various architectures and frameworks which are possible and Section 15.5 gives insights on the essence of security and privacy in these fields. Following this Section 15.6 talks about the possible risks and security values of Vehicular Ad Hoc Networks (VANETs) and ITS, while

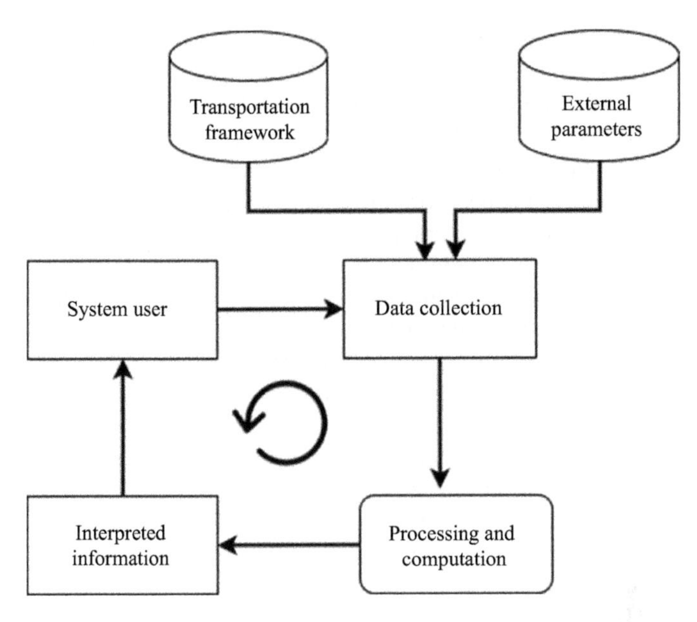

Figure 17.1 Basic workflow of an ITS framework

Section 15.7 throws light on the security and privacy challenges and Section 15.8 explains the effect of trust and reliability in the same field. Section 15.9 talks about potential threats if intelligent networks followed by the conclusion in Section 15.10.

17.2 Related/background work

ITS and AIV being a field which has immense amount of transformation and aspects, there are numerous techniques which help in preserving the privacy and security of the developed frameworks. The authors of [6] have researched extensively and have put forth a paper which portrays the different categories and classes of privacy issues in ITS and have also discussed about possible integrated networks like Controller Area Network, Flex-Ray, etc. for developing resilience to possible attacks and data breaches. In the work put forth by [7], the researchers have very rightly cited a scheme for preserving security and privacy of the data and users utilized in the ITS frameworks. It makes use of a unique technology which helps in preserving the data during information exchange and transmission and on absence of such a scheme, it has been proven that there are high chances of numerous anomalies and mishaps. The strategy used here is the emergent intelligence (EI) for gathering, dispatching, and interpreting the information to enhance the robustness, flexibility, autonomy, etc. which is put into action by utilizing the Crypto++ package. One of the other major security concerns is the ability to track and trace a person if the data reaches the hands of a malicious user.

The research carried out by the authors of [8] have showcased that this can be avoided by making use of pseudonym pools which helps vehicles to switch easily between various identities. As per the research, this scheme is further extended by making use of time slotted pseudonym pool which has a static size which in turn offers benefits of reducing the storage required and encapsulates the users' privacy and security in a better way. In [9], the authors have put forward a collection of strategies to combat the issues of privacy and security which include masking the collected data and information while transmission, using reliable and trustworthy third-party applications wherever required and by adhering to the rules and norms put forth by the government with regards to security and privacy. One of the other interesting perspectives toward this sensitive topic is to curate cooperative intelligent transport systems (C-ITS) which would not only enhance conventional smart transportation techniques but would also address the privacy and security concerns of for the developed architecture. The framework proposed by them integrates blockchain and deep learning techniques for achieving the same. Blockchain will be used to handle the safe and secure data transmission between various modules and functional groups within the system and elsewhere. Further, using enhanced Proof of Work strategy, the integrity and core of the data is strengthened so as to combat any poisoning attacks. With regard to the deep learning module, long-short term memory-auto encoder (LSTM-AE) is used for converting and marshaling the data from C-ITS into a varied format for preventing any possible attacks [10].

The authors of [11] have opened up possible streams to be considered when addressing the security and privacy aspects of AIV and ITS. Parameters like secrecy, authentication, access control, policy enforcement, trust management, and secure middleware are the major points explained and described in the paper. The above-mentioned techniques and strategies are some of the current frameworks which have been proposed with regards to the security and privacy of data and information in ITS/AIV. The next section will talk about the various safety approaches and methods in the field of ITS/AIV.

17.3 AV/ITS/AIV safety approaches

Safety on road and safety of transportation systems are extremely essential in a world where communities and societies are evolving and are heavily dependent on vehicles for facilitating their mobilization needs and requirements. However, just like how very coin has two sides, the transportation sector has also faced a significant increase in the number of accidents, casualties, and dangers caused by illicit driving practices. To combat this and to increase the efficiency and automation and to bring in an intelligent quotient to the transport sector, researchers and scientists have been integrating various technological aspects to uplift this sector in making it smart and automated. Unfortunately, these techniques and strategies too possess various bottlenecks, the main area of concern lies in the field of security, privacy and trust. Since technological integration and automated frameworks in this sector call for extensive and massive amount of data and information pertaining to its users, it is extremely essential to ensure safe, secure, and reliable data transmission. Apart from security of data, there are numerous other aspects and perspectives to be considered for a holistic safe and secure framework. One possible aspect of safety and security in this sector deals with that of the traffic related issues and concerns. Since AIV/ITS make use of networks and interconnected nodes and objects for extracting relevant information, one of the best ways to securely lock this layer is to have a robust and strongly connected network along with efficient sensors so as to achieve high level of accuracy and precision with respect to the traffic related data being extracted. Another possible aspect which can be incorporated to further improvise the safety environment is to develop a system in such a way that it has a track of all criminal records in and around the positioned place of travel. This would be a great technique for authorities to get frequent updates and notifications on the whereabouts of a loos criminal and can be used for tracking them down [11]. Another aspect for measuring the safety and reliability of an ITS/AIV is with regards to the generic road safety. There are a number of components and factors which can affect the reliability and safety of smart and automated vehicles on road. Collision detection and warning are a highly accountable parameter in this class, especially in conditions where drivers indulge in rash driving. In certain countries across the globe, reckless drivers hardly follow the rules and regulation and the traffic control system which not only exposes them to a high-risk factor but also exposes the fellow safe drivers to serious risk.

Considering the safety aspect as a modular system, the developers and researchers of automated intelligent vehicles and intelligent transportation systems will mainly consider the following aspects of safety standards during the curation phase [12,13]:

- The developers of such an extensively autonomous system should define a particular threshold for evaluating and comparing the safety performance of the systems and frameworks. If the safety creates an environment which has a much higher performance statement, then further automations for driverless systems can be developed in the near future.
- Proper testing and operational checks are to be conducted before implementing and executing the system or framework in real time as it would give a clear idea of the possible mishaps and technical glitch.
- Another approach to tackle any possible bottlenecks caused by the safety standards of the system is to have trial runs with a few public users to ensure that the valuable inputs provided by them are also included for further enhancement. Furthermore, considering the user safety is an important aspect when developing such extensive systems.

The above-mentioned points and statements indicate the major safety and security approaches for a vehicle or network curated in the field of ITS. The next section elaborates on the various architectures of AI and ITS.

17.4 Differences over architecture of AV/ITS/AIV and raised issues

Identifying the right architecture for the development of any technological system is highly important to ensure that the various components are able to cohesively work and function together. There are various types of architectures available for an ITS system in general. One of the architectures is the artificial societies, computational experiments, and parallel execution (ACP) which mainly consists of three steps—portrayal through AI, interpretations, and analysis by computational experiments, and finally, managing and authorizing the work through parallel execution of tasks. One of the major reasons for the usage of this particular ACP architecture is the two striking features it offers—inseparability and unpredictability. The fact that systems which adopt this architecture would be inseparable indicates how intrinsically the different functional modules of the system would be tied together irrespective of the complexity and globality of resources involved [14]. The architecture being unpredictable indicates that the globalized outlook of the individual components cannot be analyzed and can only be examined on a larger scale or level. Figure 17.2 shown gives a brief idea of how the ACP approach has evolved rooting down to the natural sciences, social sciences, and artificial sciences leading to the generation of different societal worlds/communities. Having said this, the main motive of using ACP for developing automated systems and frameworks in the transportation sector is because of the flexibility, timeliness and effectiveness it offers. Figure 17.3

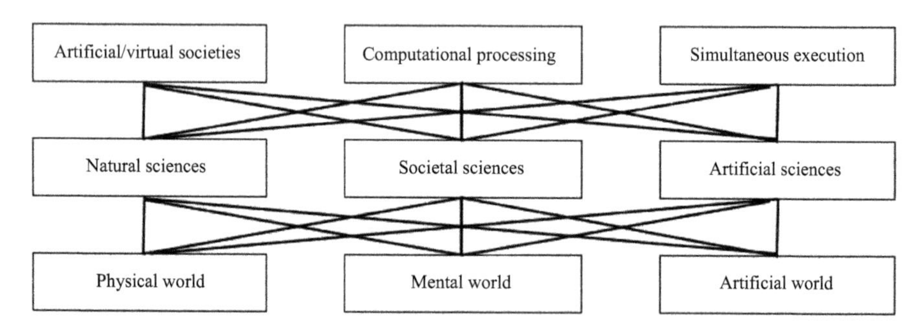

Figure 17.2 Philosophical and scientific linkage of ACP approach

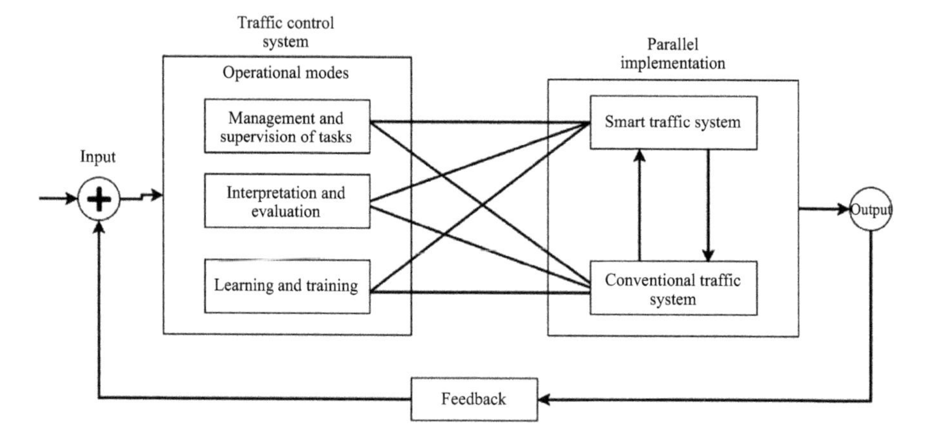

Figure 17.3 Parallel ACP approach framework

demonstrates the working of the ACP model showing the step wise development of the input provided from the nodes in the transport system which passes through the control center and is then processed parallelly to generate the required output.

Another possible architecture for ITS/AIV is the artificial transportation system (ATS) architecture which was introduced way before the ACP architecture. One of the main highlights of the ATS architecture is to avail and utilize the large-scale data available with ease and efficiency. In other words, ATS is the foundational model for the ACP architecture wherein, ATS extends its features of artificial societies and computerized traffic simulation methods. Figure 17.4 gives an insight into the major components involved in an ATS modeled transportation system [15].

One of the other types of architecture for such automated systems in the transportation sector is the ontology-based system architecture which consists of three major layers including those of ontology layer, reasoning layer and query layer. The ontology layer is the foundational layer which functions on the OWL-RDF language in order to capture the basic essence of the road traffic and other information pertaining to the vehicles on the road. The upper most layer i.e., the query layer, contains the various agents which have direct access to the information obtained. Figure 17.5 gives the workflow of the system architecture of the ontology-based system [16]. On considering the above-mentioned architectures for the development of ITS/AIV/AV, it can be closely observed that each model has its unique features and characteristics that make it utilizable in its own way.

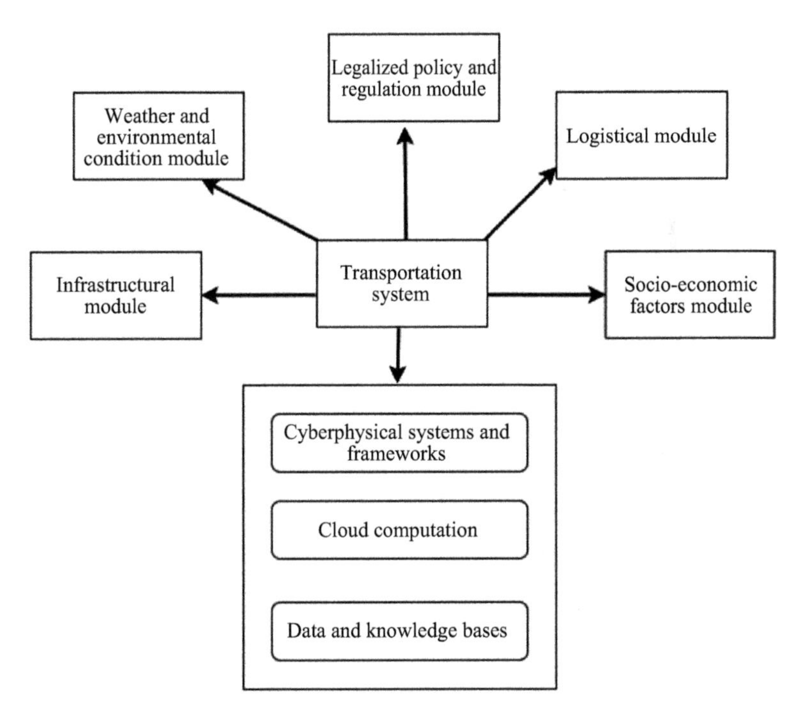

Figure 17.4 Scope, architectural framework and services of ATS

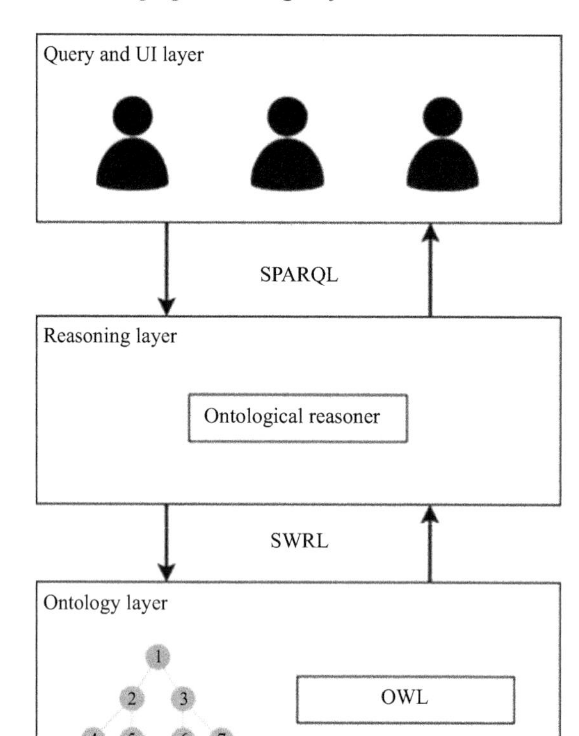

Figure 17.5 Workflow of OWL-based architecture

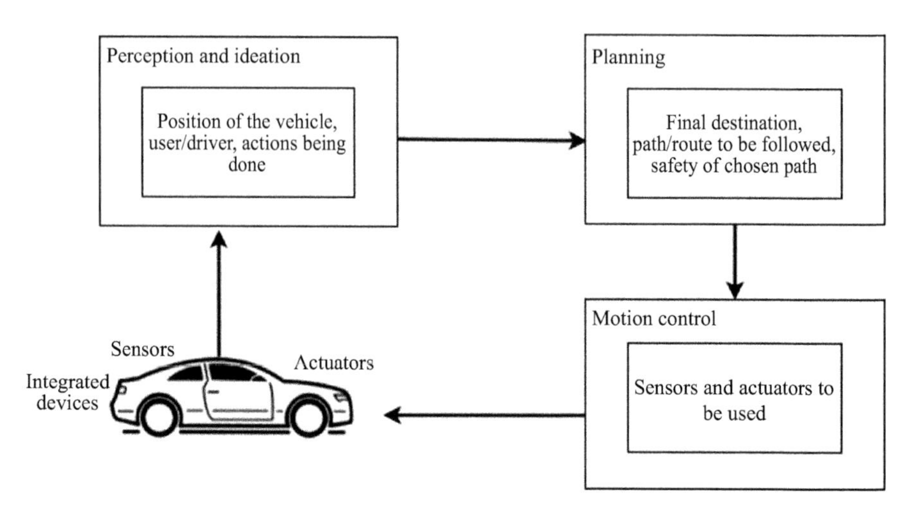

Figure 17.6 Layers and functionalities in the AIV/AV model

However, each of them has its own drawbacks and bottle necks making it vulnerable from various aspects. One of the major issues raised with regards to all of these architectures is the issue of security and trust. With such robust architectures, reliability and privacy preservation of the large-scale data being collected is often put t stake exposing the system to a plethora of attacks and consequential acts. From similar aspects, if we concentrate solely on the architecture of AIVs or automated vehicles, most of them are likely to possess a robotic architecture which consists of three major components: perception, planning, and motion control. These components are used to analyze and maneuver the vehicle forward while also helping the vehicles make decisions and necessary conclusions as and when required. Figure 17.6 represents the various systemic components needed for an AIV/AV and elaborates on how each component complements each other [17].

17.5 What security, privacy and trust mean in ITS/AIV?

In general, the term security refers to the state of being free and out of any possible dangerous situations or threats. Privacy is usually defined as a particular state in which one is not being judged/observed by the others and is left at their own will. Trust often resembles reliability or the ability of someone to completely rely on someone else or an object with regards to a particular aspect. These terms are often used interchangeably but as defined above have significant different characteristics. Security, privacy, and trust in the field of ITS/AIV relate to the same definition given above with regards to the users and the system being proposed. In ITS, safety mainly refers to a situation wherein the vehicle must be kept in such a way that it aims towards the protection of human life/users at all times or in other words, the system/vehicle must ensure the prevention of mishaps which may harm its users or bring harm to others within the transport network [15]. Apart from regular vehicle accessories essential for the sustenance of a vehicle or intelligent transportation system, components which can handle critical analyses and which can be relied upon for fault tolerance are highly critical.

In such intelligent automated systems, it is essential to maintain a safety integrity level (SIL) which would provide a threshold value for determining how safe the system or framework is from aspects like those of development, programming, operations, ability to handle glitches, etc. When it comes to security in ITS/AIV, it mainly focuses on preventing any sort of external attacks or breaches on the existing system/network. In other words, the developers should maintain a consistent track on creating a cyber-safe framework in the intelligent transportation field. Not only should the system be secure from thefts but should also be resilient against any sort of malfunctioning, communication gaps, and software bugs. Furthermore, information security also falls under the category of security of AIV/ITS wherein the main focus throws light on the confidentiality and integrity of the information and data that is being handled by the system [17]. Similarly, privacy is one of the most important and critical elements in today's world especially in generations where the entire community is highly cautious about the information and other details being shared and intruded by others. One of the commonly used techniques for privacy is that of the policy-based privacy preservation technique for efficient and optimized results. Another way to combat the concerns and issues raised due to privacy would be to develop a policy which would assure the users and others relying on the system that the data or the information collected would not be exposed to other organizations or parties, would not be broadcasted or shared which would guarantee the users that the intelligent and smart transportation system is safeguarding the privacy.

17.6 Security risk analysis of a trust model for VANETs/ ITS

VANET is a type of mobile ad hoc networks with a strong feature of being trained with already defined routes and maps consisting of vehicular nodes and a framework along with numerous road side units (RSUs) and on-board units (OBUs). When it comes to fields like VANETs and ITS, most of the existing security measures do not provide stable and enhanced controls for counteracting the attacks within the model. This calls for efficient security mechanisms in VANETs and ITS to control, configure and gather the safety and security services. A possible way to overcome the pressing concern of security is to use a trust-based model for the same which ideally works around the concept of establishing a trustworthy relation between the nodes and vehicles in the network. The model basically calculates the trust metrics and quantity of every vehicular node individually against malicious and malignant coefficients of the nodes. It is usually carried out by direct and indirect computations in a centralized or decentralized manner.

The node which possesses the highest trust metric is the one which takes up the role of the leader. This model when implemented in VANETs/ITS helps in boosting the security quotient and efficiency to the next level and also helps in classifying vehicles/nodes. In order to assess the model developed against the security parameter, two commonly used mechanisms are those of SecRAM and ETSI TVRA. The SecRAM assessments covers various aspects like developing the contextual symbol and identifying relative objectives, finding out threats and their probability, analyzing, and interpreting the possible consequential losses, examining the risk factors, etc. On the other hand, the ETSI TVRA method is used in computing the risk factor of every possible threat which attacks the framework and recognizes the risks by excluding and cornering the vulnerabilities and other salient exposable features. With the help of both of these models, the main motive is to uplift the systemic framework by picking out the security aspects, revealing the plausible threats opposing the system, and mitigating plans of actions for fault tolerances and rollbacks [18,19]. One of the other trust-based models for analyzing the security risks is an entity-centric trust model which is weighted in nature. In this model, the trust value being computed is completely dependent on the essence of direct trust and recommendation basis. The model being weighted is one of the best ways to develop and scrutinize the trust elements from a holistic aspect. Hence, it can be concluded that using the models mentioned above, the application paradigms used in developing VANETS/AIV/ITS can ensure that the applications work effectively with ease and flexibility using dynamic factors, weighted values, direct trust, and other recommended parameters. Moreover, studies state that the above-mentioned models are simple yet complex enough to carry out the operations in the smart and intelligent transport environments smoothly [20].

17.7 ITS security and privacy challenges

As seen in the previous sections of this paper, security and privacy are two of the most potent features to be saliently developed while building any framework that is intelligent and smart in the transportation sector. There are a few major aspects of the system where the security and privacy pose a threat/challenge to the system. The first aspect is that of the communication model which involves the transmission of data from one vehicular node to another, data extracted from sensors, data being transmitted to the ground station, etc. When sending over messages to the various nodes, RSUs, OBUs, etc., there are high chances of attacks from within/outside the system by malicious hackers and crackers or any malicious software [19]. Further, if any important or crucial messages and details are transmitted without strong encryption techniques and strategies, it can cause data leakages and data breaches holding the users details and private information at stake.

When it comes to AIV/AV, one of the important aspects of security is to ensure that the smart vehicles are able to sense and detect the surrounding environment accurately and precisely in applications which involve automated driving and maneuvering. In such cases, it is highly important to ensure that at most care is given to secure the lives of the users availing the services. Similarly, the smart vehicles must be able to capture the right amount of speed with which the vehicle should move to ensure the right speed management and cooperative navigation techniques. When it comes to the possible attacks on the ITS network and nodes, there are various types and forms of attacks which in turn affects the privacy of the data and information as well. The different possible attacks on an ITS framework are as given below [21]:

- Man in the middle attack: It is one of the basic types of attacks in which the individual/party attacking the system has the power to intercept and breach the data and information being transmitted and shared between any two nodes. For example, the attacker can intercept and modify the positional data of a particular vehicle being transmitted from one vehicular node to that of the other.
- Routing attack: In this attack, the routing protocols implemented in the ITS framework is maliciously infused and this prohibits the messages and data packets from reaching the desired destination during transit.
- Timing attack: As the name suggests, this attack aims toward delaying the delivery of packages so as to dwindle the operational execution of various functional modules in the system leading to collision and other unnecessary consequences.

- Spoofing: The attackers broadcast and spam every node with bogus and unwanted data to result in extensive reaction from the system.
- Denial of service attack: It is one of the other classic cyber-attacks wherein the attacker hampers the availability of the resources in the system framework and this can be highly dangerous as it can often affect the critical features of the framework.

The above-mentioned points and types of attacks sum up the major security and privacy challenges and concerns faced in ITS.

17.8 Understanding trust in automation and AVs/ AIVs

This is one of those topics which completely correlates on how factors and influences which are way beyond the natural human–system interaction can encourage the users to build out a relation of trust and reliability with the intelligent and smart vehicles. It is one of the essential aspects in the intelligent transportation sector which is being heavily researched upon so as to curate a holistic interpersonal trust component. As mentioned earlier, trust is what defines and terms the feeling of an attitude which promotes complete dependability and reliability either on a system or another individual/party [22]. Over the years, there has been a mixed collective response to the acceptance of AIV/AV where a few studies found to have received positive responses while the other few gathered a slightly negative response. In some cases, the users had mixed opinions too where they were able to strongly point out that these systems have potential benefits and a few risky elements too. Another key analyzation and interpretation are that people usually tend to prefer lower levels of AIV/AV's when compared to three-tiered ones mainly because of the significant number of crashes likely to be observed as the tiers increase. This drives us to the motive of taking up the initiative to educate the audience and users of the system with regards to the happenings and workings of the system to be able to identify and correlate to the essence of trusting/relying on it.

17.9 Potential threats of intelligent network-vehicle systems

Intelligent network vehicle is a new type of evolved smart vehicle system which is integrated with necessary sensors, tech gadgets, controllers, etc. along with efficient decisive and control skills. The network framework which is developed is majorly dependent on vehicles, roads, and related components using a plethora of communication techniques and strategies. In fact, with rapid technological advancement over the years, ITS and intelligent network vehicles seem to amalgamate and work closely with each other at the core [23]. Just like any other interconnected network, this system of network also has possibilities of threats and system crashes. If an attacker wishes to breach the data or information being transmitted from one node to the other in the network system, they would have to acquire a good gain over the communication buses and the connected devices. Once the attacker acquires control and intercepts the communication line or gains control over any of the network nodes, the attacker can easily carry out his mission of attacking the system. However, this can be overcome by incorporating special service centers (SCs) to each of the individually connected nodes in the network. Another possible threat in the network vehicle system is the transmission of bogus data across various nodes which would in turn lead to mayhem as the nodes keep working extensively with every data packet being received.

A few of the other possible threats the systemic network is exposed to include loopholes in gathering and collecting data, malfunctioning the hardware and other connected devices, DoS attacks, masquerade attacks, etc. [24]. Another class of attack is the black-hole attack wherein the attacker would intercept and remove packets to be transmitted from one node to another which ultimately creates a hole in which no more movement of packets are observed. Similarly, replay attack is another typical attack which is considered to be a variation of man in the middle attack. In this, the transmission of data packets is delayed causing a complete turnover in the execution of tasks and functions by the nodes. Impersonation attack is one of the other types of attacks where the target node of the attacker will take up the role of an RSU so as to trick the users and extract authentication details for accessing confidential data for misuse. On similar grounds, threats like malware, falsified-information attacks, timing-attacks, etc., are some of the other major security concerns in the intelligent network vehicle systems [25]. Further, popular issues toward ITS/ AIV and recommended solutions can be found in the [26–31].

17.10 Conclusion and future scope

In this chapter, we have thrown light on the essence of trust, security and privacy for an intelligent transportation system and AIV. The various threats and risks exposed by the system have been elucidated. This chapter also gives various aspects of safety approaches and security strategies to safeguard the systems and the network from potential threats and issues. ITS and AIV/AV being a field which is highly dwelled upon, the topic considered for research in this paper can be utilized for further studies in preserving the security and sustaining privacy of the information and data extracted from the interconnected network of systems. Furthermore, this paper has also described the possibilities of a trust model for analyzing the above-mentioned reliability parameters for an AV/AIV. As a future scope, mentioned issues like scalability, security, privacy, trust, etc., can be considered for further research work for researchers/ scientist around the globe.

References

[1] Pal R, Ghosh A, Kumar R., et al.. Public health crisis of road traffic accidents in India: Risk factor assessment and recommendations on prevention on the behalf of the Academy of Family Physicians of India. *Journal of Family Medicine and Primary Care* 2019;8(3):775.

[2] Website Link: https://www.who.int/news-room/fact-sheets/detail/road-traffic-injuries

[3] Tokody D, Mezei IJ, and Schuster, G. An overview of autonomous intelligent vehicle systems. In *Vehicle and Automotive Engineering*, Springer, New York, NY, 2017, pp. 287–307.

[4] Jaiswal AK, Krishna S, Agarwal A, Ghosh A, and Pal R. Alcohol and road safety: Investigation and legal aspects. *Journal of Medical Science* 2018;11: 154–60.

[5] Yu K, Lin L, Alazab M, Tan L, and Gu B. Deep learning-based traffic safety solution for a mixture of autonomous and manual vehicles in a 5G-enabled intelligent transportation system. in *IEEE Transactions on Intelligent Transportation Systems* 2021;22(7):4337–4347, doi: 10.1109/TITS.2020.3042504.

[6] Hahn DA, Munir A, and Behzadan V. Security and privacy issues in intelligent transportation systems: classification and challenges. *IEEE Intell. Transp. Syst. Mag.* 2021;13(1):181–196.

[7] Chavhan S, Gupta D, Garg S, Khanna A, Choi BJ, and Hossain MS. Privacy and security management in intelligent transportation system. *IEEE Access* 2020;8:148677–148688.

[8] Eckhoff D, German R, Sommer C, Dressler F, and Gansen T. Slotswap: strong and affordable location privacy in intelligent transportation systems. *IEEE Communications Magazine* 2011;49(11):126–133.

[9] Fries RN, Gahrooei MR, Chowdhury M, and Conway AJ. Meeting privacy challenges while advancing intelligent transportation systems. *Transportation Research Part C: Emerging Technologies* 2012;25:34–45.

[10] Kumar R, Kumar P, Tripathi R, Gupta GP, Kumar N, and Hassan MM. A privacy-preserving-based secure framework using blockchain-enabled deep-learning in cooperative intelligent transport system. *IEEE Transactions on Intelligent Transportation Systems*, doi: 10.1109/TITS.2021.3098636.

[11] de Souza AM, Pedrosa LLC, Botega LC, and Villas L. Its safe: an intelligent transportation system for improving safety and traffic efficiency. In 2018 IEEE 87th Vehicular Technology Conference (VTC Spring), 2018, pp. 1–7, doi: 10.1109/VTCSpring.2018.8417760.

[12] Botega L, Oliveira ACM, Perira VA, Saran JF, Villas LA, and Araujo RB. Quality-aware human-driven information fusion model. In 20th International Conference on Information Fusion, 2017.

[13] de Souza AM, Yokoyama RS, Maia G, Loureiro A, and Villas L., "Real-time path planning to prevent traffic jam through an intelligent transportation system." In IEEE Symposium on Computers and Communication (ISCC '16), 2016, pp. 726–731.

[14] Sanchez-Medina JJ, Galn-Moreno MJ, and Rubio-Royo E. Traffic signal optimization in 'La Almozara' district in Saragossa under congestion conditions using genetic algorithms traffic microsimulation and cluster computing. *IEEE Transactions on Intelligent Transportation Systems* 2010; 11(1):132–141.

[15] Wang F. Parallel control and management for intelligent transportation systems: concepts, architectures, and applications. *IEEE Transactions on Intelligent Transportation Systems* 2010;11(3):630–638, doi: 10.1109/TITS.2010.2060218.
[16] Fernandes LC, Souza JR, Pessin G, *et al.* CaRINA intelligent robotic car: architectural design and applications. *Journal of Systems Architecture* 2014;60:372–392.
[17] Garcia O, Vitor GB, Ferreira JV, *et al.* The VILMA intelligent vehicle: an architectural design for cooperative control between driver and automated system. *Journal of Modern Transportation 2018*;26:220–229, https://doi.org/10.1007/s40534-018-0160-3
[18] Tokody D, Albini A, Ady L, Rajnai Z, and Pongrácz F. Safety and security through the design of autonomous intelligent vehicle systems and intelligent infrastructure in the smart city. *Interdisciplinary Description of Complex Systems: INDECS* 2018;16(3-A):384–396.
[19] Hasrouny H, Bassil C, Samhat AE, and Laouiti A. Security risk analysis of a trust model for secure group leader-based communication in VANET. In *Vehicular Ad-Hoc Networks for Smart Cities*. Springer, Singapore, 2017, pp. 71–83.
[20] Yao X, Zhang X, Ning H, and Li P. Using trust model to ensure reliable data acquisition in VANETs. *Ad Hoc Networks* 2017;55:107–118.
[21] Mecheva T. and Kakanakov N. Cybersecurity in intelligent transportation systems. *Computers* 2020;9(4):83.
[22] Raats K, Fors V, and Pink S. Understanding Trust in Automated Vehicles. In Proceedings of the 31st Australian Conference on Human-Computer-Interaction (OZCHI'19). Association for Computing Machinery, New York, NY, USA, 2019, pp. 352–358.
[23] Zou B, Gao M, and Cui X. Research on information security framework of intelligent connected vehicle. In Proceedings of the 2017 International Conference on Cryptography, Security and Privacy. 2017, pp. 91–95.
[24] Ben Othmane L, Al-Fuqaha A, ben Hamida E, and van den Brand M.. Towards extended safety in connected vehicles. In 16th international IEEE conference on intelligent transportation systems (ITSC 2013). IEEE, New York, NY, 2013, pp. 652–657.
[25] Dibaei M, Zheng X, Jiang K, et al. Attacks and defences on intelligent connected vehicles: a survey. *Digital Communications and Networks* 2020;6(4):399–421, ISSN 2352-8648.
[26] Tyagi AM and Aswathy SU. Autonomous intelligent vehicles (AIV): research statements, open issues, challenges and road for future. *International Journal of Intelligent Networks* 2021;2:83–102, ISSN 2666-6030. https://doi.org/10.1016/j.ijin.2021.07.002.
[27] Tyagi AK and Sreenath N. Location privacy preserving techniques for location based services over road networks. In Proceeding of IEEE/ International Conference on Communication and Signal Processing (ICCSP), Tamilnadu, India, 2015, pp. 1319–1326, ISBN: 978-1-4799-8080-2.

[28] Krishna AM and Tyagi AK. Intrusion detection in intelligent transportation system and its applications using blockchain technology. In 2020 International Conference on Emerging Trends in Information Technology and Engineering (ic-ETITE), 2020, pp. 1–8, doi: 10.1109/ic-ETITE47903.2020.332.

[29] Varsha R, *et al.* Deep learning based blockchain solution for preserving privacy in future vehicles. *International Journal of Hybrid Intelligent System* 2020;16(4):223–236.

[30] Sravanthi K, Vijay B, and Amit AK. Preserving privacy techniques for autonomous vehicles. *International Journal of Emerging Trends in Engineering Research* 2020;8:5180–5190. 10.30534/ijeter/2020/48892020.

[31] Mohan Krishna A, Tyagi AK, and Prasad SVAV. Preserving privacy in future vehicles of tomorrow. *JCR* 2020;7(19):6675–6684. doi: 10.31838/jcr.07.19.768.

Chapter 18

Future research trends, applications, and standards in emerging vehicle technology

Abstract

The transportation sector has been undergoing revolutionary changes over the last few years with so many technically advanced integrations including those of smart vehicles, automation, connected vehicle networks, and so on. This Chapter gives insights and analyses on the evolution of the transportation industry with respect to the incorporation of various tech formulas. Such evolving concepts prove to be a relief to the numerous curbs and challenges that are encountered like traffic- related problems and safety and reliability concerns in connected vehicular networks. Off late, one of the highly improvising domains in the transportation domain is the Internet of Vehicles (IoV) which deals with interconnection and communication of vehicular nodes in the framework and Vehicular Ad Hoc Networks (VANETs). In order to ensure that such frameworks can work robustly, a variety ofous routing protocols which are involved in data transmission exist. Another important factor to be assessed when dwelling into such fields is to monitor the performance metrics rigorously and constantly analyze the improvisational components for further growth and improvement. In order to curb traffic management- related issues, real-time information extraction systems are often implemented so as to extract wanted details and values from the vehicles of the connected framework to ensure that the control station can manage the traffic accordingly. Further, with the evolution of wireless network connections, technologies like 5G on integrating with IoV can break barriers and enhance the efficiency to a great extendsextent. This chapter provides a holistic perspective of emerging vehicle technologies along with their trends, applications, and standards.

Key Words: Keywords Future research trends, Vehicular applications, Standards for driverless vehicles, Future smart era for vehicles

18.1 Introduction

The vehicle industry has been one of the highly impacted sectors with advancing technologies over the last few decades. Be it with respect to automation, mobilization, interconnectivity, and what not. This transformation is solely because of the radical transformation and change in the choices and options made by society and the community with regards to the automotive sector. With advancing science and technology, there has been a sheer increase in the demand and call for the adoption of smarter, intelligent, and safer options of transport to boost the exchange of goods and offerings across cities and towns [1]. The main aim of technologies and strategies in this particular sector is to enhance and improvise the productivity features of the automobiles and provide a smooth transitioning experience during the different stages and phases of its development – innovation and research, design, and implementation. The key point of observation here is the fact that in order to bring in such grave innovations and optimistic changes, a number of parameters and components are required for their key contributions to elevate the execution of the ideas and deep-rooted strategies. If we consider the example of navigation, it has a plethora of variations – real time, dynamic, previously trained and stored, etc. One of the major components required for the implementation of these navigational strategies is information and data. In fact, the developments in the technological aspect of the vehicle sector have made to being one of the top ten automation industry trends and technologies that haves been evolving at a rapid rate. Statistical surveys and researches have stated that one of the major contributions to the Indian GDP is the transportation sector, and with its consistent evolution and transformation, it has been gaining a lot of attention and focus from a global outlook as well. Most importantly, strict safety norms and emission control is followed at par in order to ensure a secure and safe environment to the greatest extentd [2].

Starting from the invention of a wheel centuries ago, all the way to transitioning towards thea driverless car, advancements in vehicular technologies are fast approaching. Numerous companies and corporations of the transportation sector are focused on introducing vehicles and automobiles with their own tail of innovation, automation, and unique style. However, with constant improvement in the mechanization aspect, vehicles are going to be far more than just automated. Companies and researchers across the world have been working on devising techniques and strategies which would have many more vehicular capabilities apart from just being able to take their hands away from the steering wheel. They have pushed themselves towards designing vehicles which that could share characteristics to that of vehicle-to-vehicle technology for exploiting benefits which would consist of interconnecting vehicles for further interactions [3].

As observed in Figure 18.1, it was in the 19th nineteenth century that the very first car was invented and sparked up the introduction to the automobile industry followed by the introduction of cars which that were produced using an assembly line. Then came the vehicles which were accompanied by electric ignition starters, cigarette lighters, built- in radio features, coil spring suspension, car keys, hydraulic power steering, air conditioning, modernized seat belts and electric windows, connected cars and global positioning systems (GPS) navigations, hybrid vehicles, rear- view cameras and sensors, automated parking, autopilot vehicles, and even self-driving cars.

The structure of the work is as follows. After the introduction in Section 18.1, Section 18.2 deals with related and background work of the existing systems while

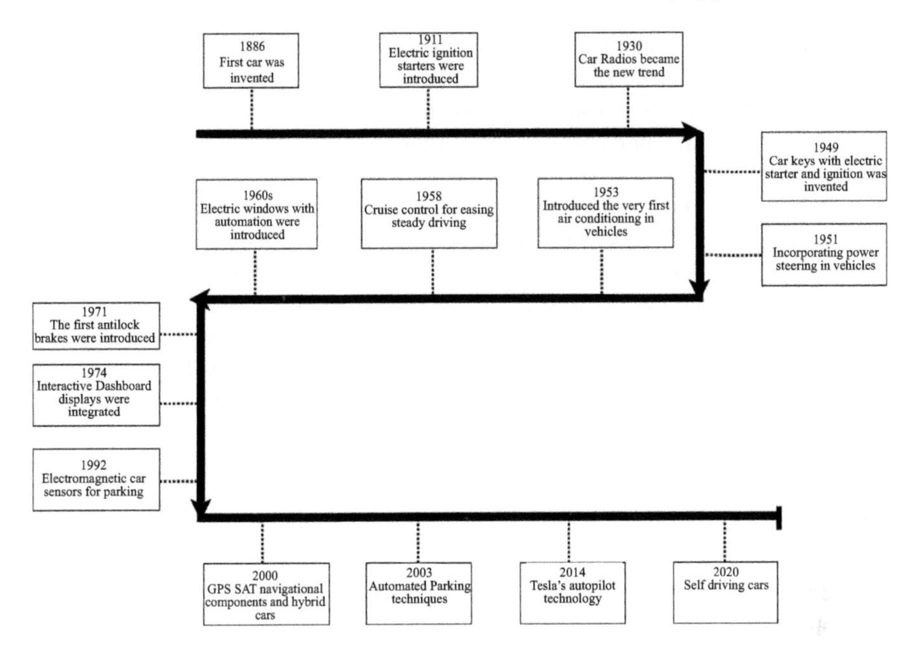

Figure 18.1 Timeline for the evolution of vehicular technologies over the years

Section 18.3 deals with Vehicular Ad Hoc Network (VANET)- based applications and the requirements for running networks smoothly. Section 18.4 deals with the measurement of performance of connected vehicle and applications and Section 18.5 portrays the content related to real- time traffic information systems. Following this, Section 18.6 highlights the environmental impacts of automation of vehicles and vehicular networks, and. Section 18.7 deals with the development of vehicle networking communication standards and Section 18.8 throws light on 5G and the Iinternet of Vvehicles (IoV). This chapter ends with a conclusion in Section 18.9 and also cites the references used for the research in this work.

18.2 Related/ background work

With many researchers, engineers, and scientists taking interest in the evolving and transforming technologies for revolutionizing the vehicle sector, there are plenty of researches and surveys being carried out from all across the globe. The authors of [4] have proposed the usage of electric vehicles for logistics and mobility and how would it impact the environment. The paper portrays the various challenges which are to be overcome when using an electric vehicle, and the fact that there i's a growing demand for these in the public space, further adds to the skill [5]. One major challenge which they had to work on was with regards to the charging stations, provided they are able to chare the vehicles without too much of a hassle and within a lesser amount of time. In order to combat this issue, the authors have devised a swap model which ensures distributed network optimization for better efficiency. Researchers of [6] have put forth some of the possible emerging technologies in the latest of years. The main aim of their work is to showcase their solutions and propositions on how to make vehicles smarter and intelligent by curbing the challenges faced in the fields of safety, fuel consumption, communication strategies, and multimedia transformation. The striking technological advancements for enhancing vehicular developments which they have proposed include providing on-the-fly mobile internet services which could interconnect vehicles, utilizing sensors and control systems for auto navigational possibilities, and so on. The paper suggests that smart vehicles are more of an extension to the intelligent interconnected vehicles which encapsulate challenges from various aspects in order to curb any mishap that may arise [6].

One of the other prominent works in the field of emerging technology belongs to the authors of [7] with the key focus of their paper being self-driving automated vehicles. The paper also focuses on the surveys conducted by them catering to the needs of older people on adapting to these new technologies when trying to exploit the facilities provided [8]. Further, the authors have put forth the idea of maintaining a built-in health monitoring system which that could provide a substantially strong basis in identifying the medical conditions and vital parameters of the drivers and the passengers of the vehicle [9]. One of the other common enhancing ventures in the field of vehicle technology is the integration of Internet of Things (IoT) technology, which has turned out to be one of the highly utilized and researched strategies off late. And this popularity of IoT is solely accredited to the heavy users of the internet which has taken a leap towards advancement such that it can interconnect numerous nodes, devices, and gadgets together. The authors of [10] have majorly focused on the development of mobile ad hoc networks (MANETs) and Vehicle Ad Hoc Networks (VANETs) and have walked through the various algorithms and processes involved in adopting the same [11]. These network systems would surely be one of the best ways to indulge in communication and interaction between various traffic-controlled intersections and junctions and would be a great step forward in solving accident-related issues and problems [12].

The research put forth in [13] showcases a variety of possible upcoming technologies in the transportation sector including those of energy- storing panels, airbag lenticulation, dashboards integrated with augmented and virtual reality, and cars which can interact and transmit essential details with each other on roads. Of these, the one idea which is likely to get a maximum reach in our society belongs to that of energy- storing body panels attached in vehicles wherein the body panels can store energy during the drives and gets recharged with the help of normal batteries at a much faster rate. Similarly, the next tech research done by them with regards to the integration of airbags that can stop vehicles is the usage of airbags which would be activated and functionalized whenever the sensors predict a possible damage to the car [14]. One of the other proposed systems in the field of smart vehicles is the usage of Wi-Fi in VANETs so as to interconnect the various vehicles for the exchange of information regarding the surrounding status of the road and traffic [15]. The intercommunication between vehicles is carried out through V2V (vVehicle- to-v Vehicle (V2V) networks and are is highly used in cooperative applications and services as stated by authors in [16]. The above-mentioned systems and techniques are some of the existing systems which have been implemented or which have been proposed by researchers in this particular field.

18.3 VANET-based applications and requirements

VANET which stands for a Vehicular Ad Hoc Network is a form of wireless network that focuses on the task of multi-hop and is completely dependent on topology and the mobility and scalability of the nodes being affected. In the VANETs, the nodes involved are the vehicles that would be involved in motion across roads. The most important point to be noted is that this type of Vehicular Ad Hoc Network makes use of two types of communication strategies to maintain an efficient and consistent interaction – vehicle-to-vehicle and vehicle-to-infrastructure (V2V and V2I) [17]. VANET is often classified under the category of a MANET, wherein the mobile nodes are replaced with vehicles. V2V communication strategy makes use of direct communication and interaction between vehicles without any added interferences. However, in V2I, there is a wireless communication and transfer of data amidst the vehicles and the interconnected infrastructure.

V2I is a form of dual communication strategy which that supports and complements the safety and security of data, scalability, and mobility. These communication strategies are highly essential in the upcoming era of emerging vehicular technologies because they a're best suited for enhancing applications that call for automation. Along with the latest developments and establishments in the field of VANETs, there has been a significant number of emerging applications revolving around this specialized case of MANETs [18]. One of the major requirements in VANET is the need and necessity for an efficient routing protocol. Though it may not seem practically viable to instantiate protocols which that will be suitable for all types of VANET applications, various trials have been initiated to develop protocols which that could be suited for specific applications and domains with ease and flexibility. For most of the applications which that involve safety- related depiction, the broadcasting-forwarding mechanism of packets using interpolated CCA (chose cipher-text attack) is implemented [19]. Another possible routing approach is to exploit the swarming protocol which is based on the impact of content delivery in VANETs. With regards to applications which that focus on the premium comfortable side, SODAD (segment-oriented data abstraction and dissemination) can be utilized which is highly useful in creating scalable information centers which are not centralized. MOPR (Movement- prediction oriented- based prediction routing) is one of the other possible routing mechanisms which is efficient enough to predict the location of vehicles in the near future and can estimate if a certain route is likely to be damaged/tweaked or not [20].

Some of the major featured highlights and requirements of VANETs are as follows:

- Dynamic topological framework – In order to maintain the network communication and integration of nodes between the vehicles moving with high speed, such dynamic topologies are a must [21].

- Prediction and mobility of modelling – Prediction of the accurate position of the vehicles can be quite challenging and to overcome this, trained models are used which lay its their foundation on the previously processed roadmaps [21].
- Environment for interaction – After developing prediction models, the next step is to strengthen the communication and interaction process between the nodes in the network. Necessary steps are to be taken in order to ensure efficient data transmission during the mobilization of nodes [22].
- Delays and constraints – One of the most important points is the timely delivery of messages without any data loss or cross- talks [22].
- Communicating with onboard sensors – Sensors form an extremely integral part of the VANET building process and they will be able to extract and read data. Further, they are also responsible for transmitting these details to the central board for further analysis, and hence sensors are often utilized for the formation of interconnections and routing protocols [22].

When it comes to routing protocols, there are a variety ofvarious options for the same. They can be broadly classified into two major categories: topology- based and geographic- based routing norms. The former classification of routing protocols used the data and information extracted from the links existing inside the VANET in order to transfer the data packets to/from the desired locations. Topology- based routing protocols have further divisions as well. The first type is the proactive or table-driven protocol which are is instantiated from the shortest path algorithms and these protocols are quite organized as they store data in structured tabular format. This allows for reduced latencies and delays and eliminates the necessity for route discovery or overhead costs. Fisheye state routing is a commonly used type of proactive protocol which reduces the bandwidth being consumed to a drastic extent. The next type of topological routing is reactive or on-demand routing protocols [23]. As the name suggests, these protocols initiate the discovery path of routes only when a particular vehicle/node wishes to interact with the other and is extremely essential in reducing bogus and unnecessary traffic [24].

The comparison shown in Table 18.1 elucidates the difference between topology-based and geography-based routing protocols with respect to the essential

Table 18.1 Comparing the two prominent types of routing protocols in VANETs

Protocols	Topology-based	Geography-based
Forwarding concept	Wireless technique with multiple hop forwarding	Wireless technique with multiple hop forwarding
Recovery strategy	Multiple hop forwarding concept	Flooding technique
Requirement of virtual map	No	No
Requirement of virtual infrastructure	No	No
Possibility of traffic flow	Yes	Yes
Application domain	Urban areas and cities	Highways and fast roads

parameters of the type of strategy followed for the protocol, traffic- related queries, domain of application, and so on. One of the popular routing protocols under this category is the ad hoc on-demand distance vector (AODV) protocol which indulges in creating a routing path when a node is in need of the same. Further, the dynamic source routing (DSR) protocol is a commonly used on- demand protocol which that works in two phases – the route identification phase and the route maintenance phase. Temporally ordered routing (TOR) protocol is another deviation of the on-demand category which makes use of acyclic graphs for broadly dispatching the packets and the data across the framework. Moving on to the geographic routing protocols, they follow a mechanism which that ensures that every node in the network is aware of the positioning of the subsequent neighboring nodes across the network [25]. This strategy is a great way for enhancing the scalability and mobility of the proposed framework. A delay-t Tolerant network (DTN) is a type of geographic routing protocol which makes use of the carry and forward technique for transmitting data which is conducted by making use of the parameters present with respect to the neighboring nodes [26]. One of the other major types of geographical protocol is the Beacon protocol which focuses on transmitting short messages at frequent intervals and periods of time. Following this, the next type is the overlay protocol where every node in the framework is connected through virtual/logical interconnections which are founded on the existing network. There are many more such routing protocols including those of vehicle- assisted data delivery (VADD), geographical opportunistic routing (GeOpps), greedy perimeter stateless routing (GPSR), etc. all of which are variants of the geographical routing protocol [27]. Figures 18.2 and 18.3 display the classification and distribution of different protocols based on the categories it falls under. The diagram is highly useful in analyzing the various types of routing protocols that exist within the unicast mode of

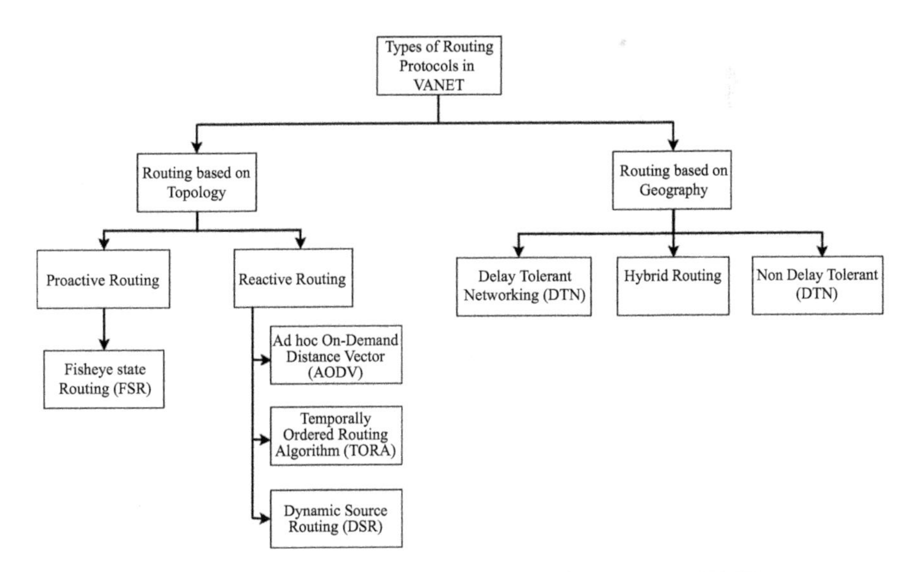

Figure 18.2 Routing protocols classification in VANETs

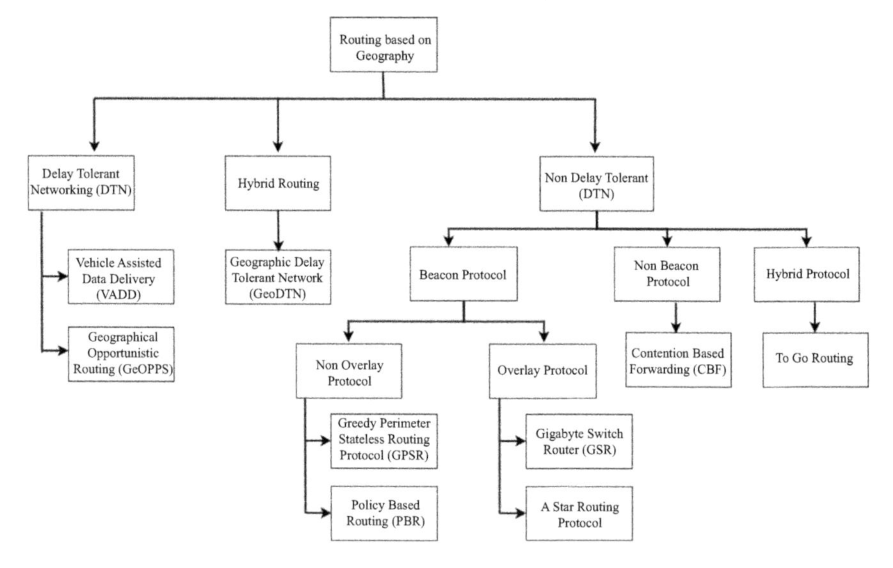

Figure 18.3 Expanded version of geographical routing protocol

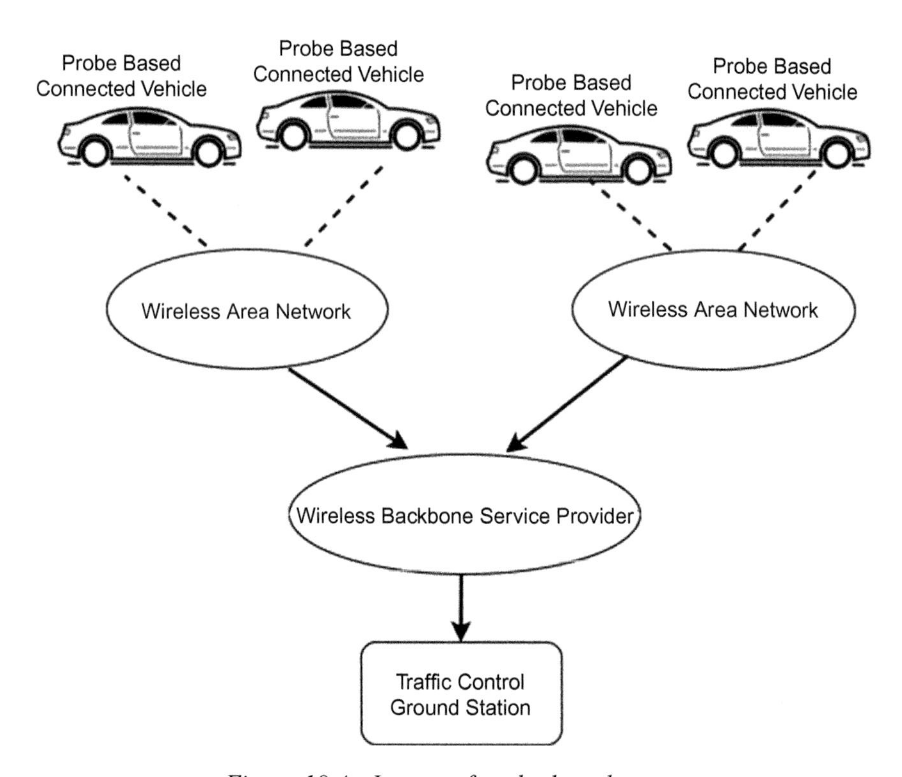

Figure 18.4 Layout of probe-based system

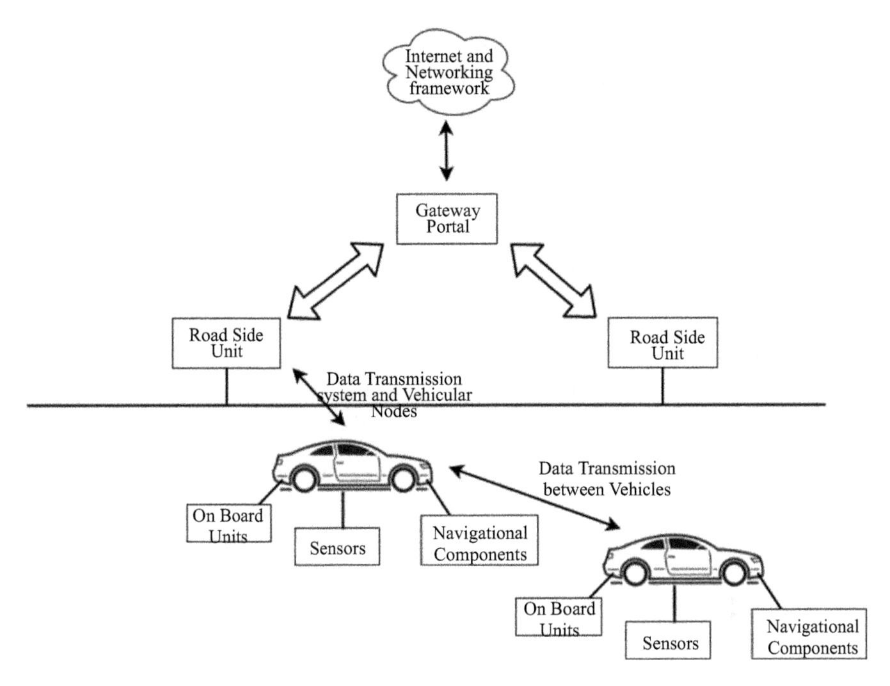

Figure 18.5 Communication architecture for interconnected vehicles

routing and data transmission which is a highly required component when it comes to VANETs.

18.4 Measurement of the performance of connected vehicle and applications

When dealing with automated and smart vehicular applications, it is extremely essential to measure the various parameters that would affect the performance and efficiency of the system. Such applications in the transportation sector are commonly termed as CAV (connected and/or automated vehicles (CAV) [27]. To ensure the smooth functional execution of such dynamic and fast- evolving concepts, it is necessary to curate a performance- related taxonomy of its own to monitor the efficiency of the connected vehicles and corresponding applications. The right blend of sensors, routing, and data transmission techniques will enhance the mobility and, safety and reduces the environmental hazards. Therefore, these three components are surely required for monitoring the overall performance and efficiency of the system [28].

- Safety: This is one of the major factors which requires significant contribution during the research and development stage so as to ensure CAVs develop with safe and secure motives and objectives. Providing proper notifications and warnings in case of any warning triggers, incorporating distancing mechanisms for vehicles, etc. are a few of the possible techniques which come under the safety parameter of performance evaluation [29].

- Mobility: One of the other crucial components which that affects the performance of CAVs is the extent of mobilization that i's possible as it is dependent on the operational efficiency of the mode of transportation. For example, coordinating traffic signals automatically and platooning [30].
- Environmental impacts: This is one of those parameters that is are integral when it comes towards developing applications that restrict and constrain the amount of pollution it causes to the environment. On considering the environmental contributions of an application or connected/automated vehicle CAV that i's being developed, it surely does boost the economic value of the system and also implies that we are aiming for a sustainable and green life ahead [31].

The above-mentioned metrics are the major measures which are used for computing the performance and impact of the connected and automated vehicleCAV applications to analyze the benefits and bottlenecks of the curated system. The reason as to why these parameters are important is because they provide a holistic aggregate of various aspects involved in development.

18.5 Real-time traffic information estimation

The real- time traffic information system is an integral and important component in any of the intelligent transportation systems (ITS). ITS is one of the highly evolving and revolutionary concepts in the vehicular domains with the main aim of stepping up the game in the transportation sector by introducing a plethora of innovative and advanced strategies to improvise traffic- related issues and to integrate intelligent and smart techniques in vehicles [32]. There are numerous techniques which are used to extract the information pertaining to traffic and related context. Of these, the most prominent and important one is the probe vehicular technique. In order to collect traffic- related data real time, this technique is not domain- specific and can be stretched out to numerous applications including those of accident detection, management of traffic autonomously, retrieving information for mechanizing the flexibility of mobilizing traffic, and for supervising the road conditions [33]. The pProbe technique itself has various types, one of which is the cellular- based probe system which is wireless in nature. This is one of the most promising strategies when it comes to extracting data from the real- time feed and is usually implemented in urban areas and city lines. This technique is in fact existent in countries like the USA due to the reliability and efficiency of data transfer with respect to traffic- related information.

This concept is implemented by utilizing vehicles which that have built- in global positioning systems (GPS) by incorporating additional components like a global system for mobile communication (GSM) and general packet radio service (GPRS) [34]. These components are highly useful in getting details of the location, position, and speed of the interconnected vehicles on the road. Using such valuable data, the control systems can easily get an accurate and precise view of the traffic situation on various roads and paths. One of the other commonly used techniques is the VANET- based probe systems which makes use of the ALOHA mechanism. The protocol used in this technique is used for transferring traffic managements reports and statements to the ground control station [35]. However, the major difference between the VANET- based and cellular- based probe techniques is the use of a wired network framework for the former and wireless network framework for the latter. The Figures 18.4 and 18.5 shown below explains the connection and interpolation of the data being transmitted from different nodes wirelessly using the probe techniques mentioned above in order to facilitate the extraction of data. The data received from the vehicular nodes are accessed through wide- area networks (WANs) which are then transmitted to the traffic management system through essential routers and wireless network service providers [36].

18.6 Environmental impact

One of the highly unlikeable yet unavoidable side effects of advancing technologies in the vehicular sector is the heavy impact of vehicles on the environment which may lead to increasing pollution rates which in turn harms the surroundings. The extent of advancement and enhancement is to such an extent that there are existing systems and researches going on with regards to driverless vehicles, automated paring systems, communication and transmission of data between vehicles and what not but amidst these, the entire globe is striving towards addressing and fighting against the environmental changes and effects caused due to the same. With increasing facilities and services in vehicles, people would indulge in more comfortable traveling and would further expect more demanding features for the purpose of entertainment and recreational activities in vehicles, especially when considering driverless cars and similar technological incorporations. This would in turn lead to more amount of travel time which will ultimately increase the environmental adversities further. Another possible impact is the chance of mere accidents in case of any technical glitch in the traffic management system or in case of any latencies involved in transferring crucial data between vehicles and to the control station. Further, such mishaps would also occur in cases where the implemented algorithms or mechanisms are faulty or provide less precise/accurate results. However, on looking at the positive aspects, we surely do have quite a few beneficial advantages which would impact the environment positively. With efficient VANETs and IoV in place along with properly trained and automated traffic management systems, the usual traffic and congestion can be effectively mitigated and controlled which would result in saving a lot of time as well as energy. Similarly, the concept of platooning can be observed in connected automated vehicles which implies that the interconnected vehicles of a particular framework would be moving with respect to each other in a uniform manner. This ultimately leads to increased roadway bandwidths, lower traffic jams, and so on. Further, techniques like eco-friendly vehicular systems can be introduced for efficient utilization of fuel and energy consumption and optimal acceleration mannerisms. Hence, it can be concluded that emerging vehicular technologies is very similar to two sides of the same coin as it has positive impacts as well as adversities on the environment.

18.7 The development of vehicle networking communication standards

Vehicular networking is of extreme importance when it comes to emerging and advancing technologies in the future application of the transportation domain. In fact, the efficient and effective networking techniques have an array of applications in a number of domains including those of safety and security on roads, traffic control and management, information retrieval, etc. Vehicular networking technologies and communication standards help in allowing the implementation of emerging vehicular technologies with a widely acceptable standard and proper framework. In simpler words, vehicular network communications and standards for the foundation for developing an effective and impactful ITS system or VANETs because they require stable, multifaceted, and versatile strategies for transmitting data in a safe and secure way. Using the inter-vehicular communication techniques, the vehicular nodes of the framework will be able to interact with each other and also respond to the controlling base station, thus promoting sound communication methodologies and a well- structureds networking framework in the transportation domain [37]. When it comes to applications like VANETs and ITS which are some of the emerging vehicular concepts, these usually possess on board units (OBUs) integrated with the vehicular nodes for incorporation of various interaction-based algorithms and mechanisms. The data which is extracted from these sensors and gadgets are then transmitted to the corresponding road side units (RSUs) and can even be forwarded to neighboring vehicles based on the application purpose. For such extensive network communications and robust features, highly advanced wireless networking capabilities are a must so that scalability and mobilization grow hand- in- hand with tech advancements. In such a situation, the main parameter to be developed to full potential is the synchronization factor and related services.

Figure 18.6 visualizes the various communication layers and overview of how exactly a vehicular communication scenario would look like if implemented in real time between RSUs, vehicles, base stations, and mobile/remote control centers. Another essential turning point in the vehicular domain has been the standardization so as to acquire proper transport management and adapt to diverse and varying standards across the world in the vehicular industry. There are numerous such standards that exist across the world which allows for the harmonic and peaceful compatibility of such networks through the normalized standards. Few of the well-known vehicular communication standards and involved organizations include Cooperative ITS (C-ITS), European Telecommunications Standards Institute (ETSI), European Committee for Standardization (CEN), International Telecommunication Union (ITU), Society of Automotive Engineers (SAE), and so on [38].

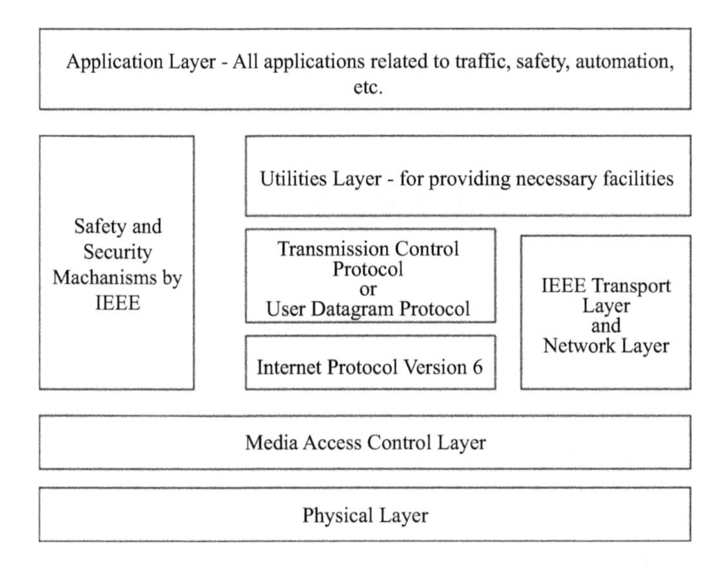

Figure 18.6 Communication lLayers followed for protocol standardization during communication

18.8 5G and Internet of Vehicles

With the rapid advancement in technology, the world has currently reached the very fifth generation of wireless network 5G, surpassing the preceding 1G, 2G, 3G, and 4G technologies. 5G is surely a revolutionary concept which that mainly aims towards connecting literally everything across the globe including gadgets, machines, devices, and people. The main advantage of this revolutionary concept is the extremely high data speed of gigabytes per seconds, very less latency- related issues, high reliability, extensive capacity for the uplifting of massive number of connected devices and nodes, and an extremely user-friendly experience [39]. 5G is no't essentially a technology developed by a single individual or researcher; it is more often compared to a framework in the mobile ecosystem wherein many companies and researchers from across the world joined hands to bring 5G to life. This networking technique has extremely wide and large bandwidths with massive spectrums spectra enabling it to deliver with speed and efficiency as well as broaden its service offerings across various domains [38]. One of the other mind- gobbling innovations and ongoing researches is in the field of Internet of Vehicles (IoV), which is often known to be the driving factor for the common VANETs in the world [40]. IoV has two main components under its wings – networking of vehicles in the associated framework and intelligent and smart vehicles. IoV is strongly integrated with deep learning, cognitive sciences, artificial intelligence, etc. in order to combat the hurdles and challenges that hinder the possible growth and expansion in the particular field [41–43].

However, the magical blend of IoV and 5G is a true blessing because of the strong benefits and upliftment obtained by combining the advantages of these two

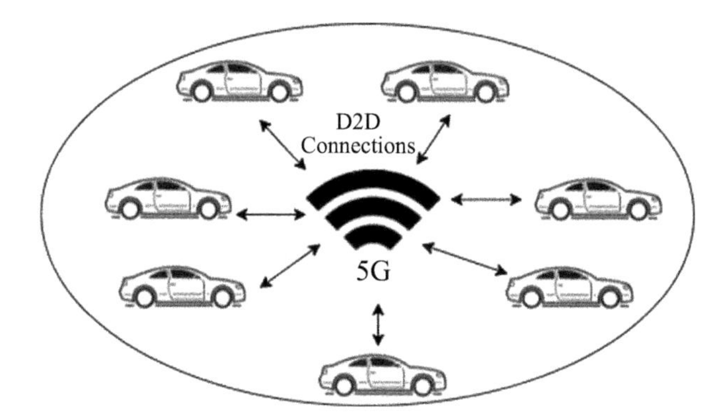

Figure 18.7 Communication architecture for D2D network using 5G in the Internet of Vehicles

emerging tech concepts. 5G and IoV proves to be highly useful in facilitating network frameworks for communication and interaction, signal processing schemes, cloud computation, virtualization of network frameworks, etc. The addition of 5G technology to IoV has turned out to be a boosting asset for ITS with regards to curating communication performances, extracting data, reliable data transfers, processing, computing, and what not. In other words, the combination of both IoV and 5G complements each other and supports the vehicular infrastructure from a holistic aspect. The network architecture used for 5G while integrating with IoV is a sub category of beam division multiple access (BDMA) which makes use of orthogonal beams, dividing the beam of rays from the antenna into varying locations of mobile station access points. One of the highly pressing needs for 5G technology and IoV is the existence of smart terminals which are often exposed to exponentially growing data traffic. The communication technique used in this approach is the device- to- device (D2D) strategy as shown in Figure 18.7 below. IoV uses a similar architecture for enhanced communication with the help of D2D and this, in turn, gives rise to increased spectral performance, user-friendly experience, and expanding communication applications. Therefore, it can be observed and reiterated that the combination of 5G and IoV in the emerging vehicular technology domain is a great way to ensure secure, trust worthy, and enhanced transportation advancements.

18.9 Conclusion

As per the analysis and research done on future research trends, applications, and standards on emerging vehicle technology, this chapter elucidates various parameters, components, and factors that contribute significantly to the evolution of the transportation sector. VANETs, IoV, and ITS are all some of the significant innovations in this particular domain and continues to be researched upon for improvisations in the future. When it comes to setting up network frameworks that need to be robust, efficient, and highly dynamic, a number of different aspects need to be considered while developing them, including those of routing protocols, nodes and devices, and standardization. Further, with so many sideline technologies advancing at a rapid rate, integration of tech concepts like 5G, deep learning, artificial intelligence, etc. along with such frameworks are a sheer boost and takes the implementation to the next level. On the whole, there are a plethora of comprehensive surveys carried out with regards to the data- centric and interconnected vehicular frameworks in the research domain. Apart from the above-mentioned technologies, concepts like IoT, blockchain, big data, etc. are also paving their way into gaining spotlights in integration with proposed vehicular frameworks. This research topic has an expanded future scope for research and implementation which are bounded by a few constraints and endless opportunities. The chapter provides great deeper insights and meaningful data in the field of emerging vehicular technology by providing a blend of various advancing tech strategies with vehicles, the impact of them on the environment along with illustrations, comparisons and diagrams for better understanding, and analysis of complementing data.

References

[1] Bajpai, J.N., 2016. Emerging vehicle technologies & the search for urban mobility solutions. *Urban, Planning and Transport Research*, 4(1), 83–100.

[2] Rajeev Singh, Website Link: https://assets.kpmg/content/dam/kpmg/pdf/2015/08/ACMA.pdf

[3] Toral, R., 2018. Evolving autonomous vehicle technology and the erosion of privacy. *University of Miami Business Law ReviewU. Miami Bus. L. Rev.*, 27, 153.

[4] Juan, A.A., Mendez, C.A., Faulin, J., De Armas, J., and Grasman, S.E., 2016. Electric vehicles in logistics and transportation: A a survey on emerging environmental, strategic, and operational challenges. *Energies*, 9(2), 86.

[5] Yang, J. and Sun, H., 2015. Battery swap station location-routing problem with capacitated electric vehicles, *Computers & Operations Research Comput. Oper. Res.* 2015, 55, 217–232.

[6] Shin, J., Bhat, C.R., You, D., Garikapati, V.M., and Pendyala, R.M., 2015. Consumer preferences and willingness to pay for advanced vehicle technology options and fuel types. *Transportation Research Part C: Emerging Technologies*, 60, 511–524.

[7] Srinivasan, R., Sharmili, A., Saravanan, S. and Jayaprakash, D., 2016, December. Smart vehicles with everything. In *2016 2nd International Conference on Contemporary Computing and Informatics (IC3I)* 400–403., IEEE, pp. 400–403.

[8] Johnson, E. N. and Pritchett, A. R. (1995). Experimental, Study of Vertical Flight Path Mode Awareness, ICAT, Report No. *ASL*-95-3.

[9] Lindenberger, U. and Baltes, P. B. (1994). Sensory functioning and intelligence in old age: A a strong connection. *Psychology and Aging* 9, 3, 339−355.

[10] Keertikumar, M., Shubham, M., and Banakar, R.M., 2015, October. Evolution of IoT in smart vehicles: An an overview. In *2015 International Conference on Green Computing and Internet of Things (ICGCIoT).*, 804–809. IEEE, pp. 804–809.
[11] Stankovic, 2014. "Research directions for the Internet of Things." IEEE Internet of Things Journal., Vol. 1, No. (1), Feb. 2014, pp. 3–9.
[12] Prabhakar, R. and Kamal Kant, A., "Comparative study of VANET and MANET routing protocols." Proceedings of the International Conference on Advance Computing and Communication Technologies (ACCT), 2011, pp. 1–7.
[13] Madhav, A.V.S., and Tyagi, A.K. 2022. The world with future technologies (post-COVID-19): open issues, challenges, and the road ahead. In: Tyagi, A. K., Abraham, A., Kaklauskas, A. (eds) *Intelligent Interactive Multimedia Systems for e-Healthcare Applications*. Springer, Singapore. https://doi.org/10.1007/978-981-16-6542-4_22.
[14] Liu, J. and Yin, Y., 2019. An integrated method for sustainable energy storing node optimization selection in China. *Energy Conversion and Management*, 199, 112049.
[15] Al-Hader, M., Rodzi, A., Sharif, A.R., and Ahmad, N., 2009. "Smart city components architecture." Proceedings of the International Conference on Modeling and Simulation Technology, Sep. 2009, pp. 93–97.
[16] Devare, A., Hande, A., Jha, A., Sanap, S., and Gawade, S., 2016. A survey on Internet of things for smart vehicles. *International Journal of Innovative Research in Science, Engineering and Technology*, 5(2), 1212–1217.
[17] Hakim, B., Abderrezak, R., 2015. "Modeling tools to evaluate the performance of wireless multi-hop networks." In Editor(s): Mohammad S. Obaidat, Petros Nicopolitidis, Faouzi Zarai, (eds.), Modeling and Simulation of Computer Networks and Systems, Morgan Kaufmann, Burlington, MA, 2015, Pages pp. 653–682 (Chapter 23), ISBN 9780128008874.
[18] Aijaz, A., Bochow, B., Dötzer, F., Festag, A., Gerlach, M., Kroh, R. and Leinmüller, T., 2005. "Attacks on inter vehicle communication systems – an analysis." The Network on Wheels Project, Tech. Rep.
[19] T. Taleb, T., M. Ochi, A. Jamalipour, N. Kato, and Y. Nemoto, "An efficient vehicle-heading based routing protocol for VANET networks," . in Proceedings of the. IEEE WCNC, Las Vegas, NV, Apr. 2006, pp. 2199–2204.
[20] Taleb, T., Sakhaee, E., Jamalipour, A., Hashimoto, K., Kato, N. and Nemoto, Y., 2007. A stable routing protocol to support ITS services in VANET networks. *IEEE Transactions on Vehicular technology*, 56(6), 3337–3347.
[21] Lee, K.C., Lee, U. and Gerla, M., 2009. "Survey of routing protocols in vehicular ad hoc networks." in Advances in Vehicular Ad-Hoc Networks: Developments and Challenges, IGI Global, London, Oct, . 2009.
[22] Kumar, V., Mishra, S. and Chand, N., 2013. Applications of VANETs: present & future. *Communications and Network*, 5(01), 12.
[23] Paul, B., Ibrahim, M., Bikas, M. and Naser, A., 2012. Vanet routing protocols: pros and cons. arXiv preprint arXiv:1204.1201.

[24] Cheng, P.-C., Weng, J.-T., Tung, L.-C., Lee, K. C., Gerla M., and Härri J., (2008). "GeoDTN+NAV: a hybrid geographic and DTN routing with navigation assistance in urban vehicular networks,". Proceedings of the 1st International Symposium on Vehicular Computing Systems (ISVCS'08), Dublin, Irland, July.

[25] Schnaufer, S. and Effelsberg, W., 2008. "Position-based unicast routing for city scenarios." 2008 International Symposium on a World of Wireless, Mobile and Multimedia Networks, 2008. WoWMoM 2008, 23–26 June 2008, pp. 1–8.

[26] Lee, K.C., Lee, U., and Gerla, M., 2009. "TO-GO:TOpology-assist geo-opportunistic routing in urban vehicular grids." Sixth International Conference on Wireless On-Demand Network Systems and Services, 2009. WONS 2009, 2–4 Feb. 2009. Sixth International Conference on, vol., no., pp. 11–18, 2–4 Feb. 2009.

[27] D Tian, G Wu, K Boriboonsomsin, and Barth, Matthew. 2018. Barth, Performance measurement evaluation framework and co-benefit/: tradeoff analysis for connected and automated vehicles (CAV) Applications: a survey, . IEEE Intelligent Transportation Systems Magazine (IF3.419), 2018-01-01, DOI:10.1109/mits.2018.2842020.

[28] K. Li, T. Chen, Y. Luo, and J. Wang, 2012. "Intelligent environment-friendly vehicles: concept and case studies." *IEEE Transactions on Intelligent Transportation SystemsIEEE Trans. Intell. Transp. Syst.*, vol. 13, no. 1, pp. 318–328, Mar. 2012.

[29] M. Zargayouna, F. Balbo, and K. Ndiaye, 2016. "Generic model for resource allocation in transportation application to urban parking management," . Transportation Research Part C: Emerging TechnologiesTransp. Res. C, Emerg. Technol., vol. 71, pp. 538–554, Oct. 2016.

[30] Amit Kumar, T. and, S U Aswathy, S.U., 2021. Autonomous intelligent vehicles (AIV): Research research statements, open issues, challenges and road for future, . International Journal of Intelligent Networks, Volume 2, 2021, Pages 83–102, ISSN 2666-6030. https://doi.org/10.1016/j.ijin.2021.07.002.

[31] V. Milanés, J. Godoy, J. Villagrá, and J. Pérez, 2011. "Automated on-ramp merging system for congested traffic situations," . IEEE Transactions on Intelligent Transportation SystemsIEEE Trans. Intell. Transp. Syst., 12, 2. 500–508, June 2011.

[32] Schäfer, R.P., Thiessenhusen, K.U., Brockfeld, E., and Wagner, P., 2002. A traffic information system by means of real-time floating-car data. ITS World Congress 11. Available at https://elib.dlr.de/6499/1/chicago_final.pdf.

[33] J. Zhu, and S. Roy, 2003. "MAC for dedicated short-range communications in intelligent transport system," . IEEE Communications Magazine, 41, 60–67, Dec. 2003.

[34] Akyildiz, I. F., X. Wang, X., and W. Wang, W., 2005. "Wireless mesh networks: a survey," . *Computer Networks Journal (Elsevier)*, 47(4), 445–4872005.

[35] W.H. Lee, S.S. Tseng, and W.Y. Shieh, 2010. 2010. Collaborative real-time traffic information generation and sharing framework for the intelligent transportation system. *Information Sciences*, 180(1), 2010, 62–70.

[36] Standard Specification for Telecommunications and Information Exchange Between Roadside and Vehicle Systems – 5 GHz Band Dedicated Short Range Communications (DSRC) Medium Access Control (MAC) and Physical Layer (PHY) Specifications, ASTM E2213-03, Sep. 2003.

[37] Anand Paul, A., Naveen Chilamkurti, N., Alfred Daniel, A., Seungmin and Rho, C., 2017. Chapter 1 - Introduction: intelligent vehicular communications, . Editor(s):In Anand Paul, Naveen Chilamkurti, Alfred Daniel, Seungmin Rho (eds.), Intelligent Vehicular Networks and Communications, Elsevier, New York, NY2017, Pages pp. 1–20 (Chapter 1), ISBN 9780128092668, https://doi.org/10.1016/B978-0-12-809266-8.00001-6.

[38] Paul, A., Chilamkurti, N., Daniel, A., and Rho, C.Anand Paul, Naveen Chilamkurti, Alfred Daniel, Seungmin Rho, 2017. Chapter 4 - Evaluation of vehicular network models, . In Editor(s): Anand Paul, Naveen Chilamkurti, Alfred Daniel, Seungmin Rho (eds.), Intelligent Vehicular Networks and Communications, Elsevier, New York, NY, 2017, Pages pp. 77–112 (Chapter 4), ISBN 9780128092668, https://doi.org/10.1016/B978-0-12-809266-8.00004-1.

[39] M. Wellens, B. Westphal, and P. Mahonen, 2007. "Performance evaluation of IEEE 802.11-based WLANs in vehicular scenarios," . Iin: Vehicular IEEE 65th Technology Conference, 2007. VTC2007-Spring. *IEEE 65th*, 2007, pp. 1167–1171.

[40] Y. Yuan, R. Lei, L. Xue, and Z. Xingshe, 2013. "Delay analysis and study of IEEE 802.11p based DSRC safety communication in a highway environment." Iin *INFOCOM, 2013 Proceedings IEEE*, 2013, pp. 1591–1599.

[41] Z. Qingwen, Z. Yanmin, C. Chao, Z. Hongzi, and L. Bo, 2013. "When 3G mMeets VANET: 3G-assisted data delivery in VANETs," . *Sensors Journal, IEEE*, 48, *10*, (13) 3575–3584.

[42] A. K. V, A. K. Tyagi and S. P. Kumar, "Blockchain technology for securing Internet of Vehicle: issues and challenges." *2022 International Conference on Computer Communication and Informatics (ICCCI)*, 2022, pp. 1–6.

[43] R. Varsha, M. N. Meghna, M. N. Siddharth, and A. Tyagi, 2020. "Deep learning based blockchain solution for preserving privacy in future vehicles." *International Journal of Hybrid Intelligent System*, 16, 4, 223–236.

Chapter 19
Conclusion

This book has been written to learn more about Vehicular Ad Hoc Networks (VANETs) [1] and their related emerging technologies for road vehicle automation. Toward this, we have discussed many chapters in this book. The summary of each chapter can be explained as follows:

Chapter 1 describes how VANET requires a distinct set of routing protocols than a mobile ad hoc network (MANET) to give unfettered access to nodes. A vehicle communication network connects cars and allows information to be delivered and received to improve traffic flow, road safety, and overall network efficiency. It refers to the capability of a vehicle node connecting to another communication node via wireless communication in the radio spectrum. While VANET stands for inter-vehicular communication (IVC), it focuses on spontaneous networking rather than using infrastructures like highway side units and cellular networks, as in 2015. The VANET's route protocols are detailed and grouped into five categories: topology-based routing, position-based routing, cluster-based rutting, geo-positioning routing, and radio transmission routing. In other cases, serial connections without crossing other pieces are the best option. Consider a vehicle that breaks down and warns approaching motorists. To avoid a collision, nearby motorists should approach the notice as quickly as feasible.

Over the last decade, academics and the industry have worked to improve wireless communication in cars. The use of network coding methods in VANETs may be contributing to this trend. Second, the SERV class contains supplemental services such as quality of service (QoS), security, and location. To ensure VANET security, the system must meet specific criteria. Threats or attacks in VANETs may not match these criteria. The urge for privacy is addressed differently in each country. Several countries require drivers to be identifiable by their license plates to avoid crime. Others may impose a mandatory privacy policy on their systems. Also, the need for anonymity is a significant issue in public acceptance of VANET implementation. Today's more capable onboard apps may keep a wide range of personal data and traces that can be exploited to reveal people's habits, movements, and whereabouts. The above dangers must be addressed before using VANET communication architecture. VANETs are a vital component of an efficient, safe, educational, and entertaining transportation system. Nowadays, people spend a lot of time in cars or other vehicles.

The roadside unit (RSU) can act as an access point or router for storing and distributing data [2]. It can also keep and distribute data. Vehicles must upload and download data from RSUs. Later it will demonstrate advanced VANET-based traffic management and monitoring applications, platooning, junctions, vulnerable road users (VRUs), and other related capabilities. Due to the current scenario, various future VANET-based applications have been considered. These apps focus on essential aspects of transportation, including safety, efficiency, long-term sustainability, and user comfort. Several applications are built on the Internet of Things (IoT), smart cities, and smart grids [3]. This chapter discusses the future transport architecture for VANET applications and offers an overview of the present transport infrastructure. If developed appropriately, future VANET apps may suggest a fully automated and cooperative transportation system.

Next, Chapter 2 discusses the medium access control (MAC) layer of VANETs/ hybrid MAC protocols in VANETs. Also, it discusses background details required to understand the internal structure of VANET (including layers). Understanding a vehicle structure is crucial, but it is not mandatory to understand a driver or passenger. In this chapter, we explain several protocols which have been used in VANETs to make an efficient and secure communication. Also, standardization of layers including protocols has been explained from a user's perspective, which will help VANETs network to make qualitative. At last, several possibilities toward the development of new protocols for VANETs have been included in this chapter.

Further, Chapter 3 describes about IoT devices and its used in ad hoc network or VANET that provides driving safety and comfort to the drivers/ passengers as an Intelligent Transportation system (ITS). VANET vehicles have several purposes like safety, navigation, traffic statistics, etc. They also act as relay nodes for packets sent to other vehicles. It might also be a database. Several data transport methods rely on VANET receivers. A reactive receiver system called VIRTUS was developed to enable video streaming over VANET. VIRTUS provides high-speed data communication between automobiles without roadside infrastructure. A routing protocol specifies the interaction of two communication components. It includes establishing a route, communicating information, and maintaining or recovering a course. Wireless communication's primary goal is to reduce communication time while utilizing minimal network resources. The optimum route is determined by the delay and connectivity of each road section. Based on weight data, traffic segments may be chosen one by one for optimal routing. The data packet will then be delivered along the route using a reliable greedy approach. After a virtual sensor network (VSN) is implemented, the VSN nodes occasionally send ads to their neighbors, identifying their identity and hops from the sink. Initially, all nodes are placed apart from the sink node. The matching sink node is added to the routing tables and publicity nodes. Urban VANET (UVANET) provides many routing techniques to transfer data efficiently. Because topology varies often, UVANET routing systems are challenging to implement. UVANET employs position-based routing protocols that simply offer data about the destination and its surroundings if no road maintenance is required.

VANET gives information on the channel state in various situations, which can help with routing decisions. Each strategy is chosen for a particular set of circumstances. Receiver-based VANET data transmission methods VIRTUS is a reactive receiver-based system for video streaming over VANET. VIRTUS offers high-speed car-to-car connectivity. The receiving nodes interact based on your present and future positions. Intelligent transportation system (ITS) can use VANET technology, which is exciting. Many academics have sought to make VANET routing protocols more dependable, scalable, and efficient. Because VANET has specific qualities like restricted mobility, high knot speed, and dynamic topology, implementing a routing protocol to assure specific QoS is one of the primary problems for automotive networks. Because the VANET network is active, it is difficult to disrupt the route between source and target nodes. Connecting Applications (based on IoT devices) are unreliable, so vehicles over road are also unreliable. In the automobile, ITS addresses traffic and road safety issues. Determining the closest forwarding node to the destination reduces expectations and improves routing efficiency.

However, transmission timeouts increase the possibility of several routes. Excessive transmission may also increase end-to-end delay. Thus, present and future simulations will evaluate the proposed system's performance in various circumstances. The protocol design and limiting concerns were thoroughly explored. Existing protocols and new protocols were compared, opening up new avenues for VANET research. Future VANET technologies will enable the Internet of Vehicles (IoV) and vehicle cloud (VC) architecture. Furthermore, the growth of long-term evolution (LTE) device-to-device technologies will open up additional study options in the future.

Next, Chapter 4 discusses VANET forwarding and remote networking. Recent advances in in-vehicle connection and dedicated short-range communications (DSRC) technologies have sparked interest in VANETs. Vehicle-to-vehicle (V2V) and vehicle-to-infrastructure (V2I) are two ways to interact in automobiles having onboard units (OBUs). Emergency vehicle warnings, curve speed warnings, pedestrian crossing warnings, and more. This technology is vital to ITS since it makes driving more fun while simultaneously boosting traffic safety. Because VANETs employ open wireless media, they are vulnerable to a high number of self-driving automobiles.

Many attacks, such as monitoring the attacker, stealing or altering information, or sending messages fraudulently, compromise vehicle privacy [4–8,16]. The wrong note on the road might trigger a severe vehicle accident. The traffic control center misread traffic flow information, so attackers may pose as ambulances and approach traffic signals for assistance. Channel eavesdroppers and relay cars can obtain the message and vehicle privacy data. No stone is left unturned in VANET security. To make the network safe, rogue nodes must be identified. VANETs employ broadcasting or multi-hop to communicate with other vehicles. Users can save their lives by using data from other cars. Nodes in dynamic vehicular networks that deceive surrounding cars about traffic congestion or accidents may generate a traffic standstill. VANET is no longer a distant dream, thanks to considerable investments from areas including government, automotive, and navigation safety. VANET applications, sites, and prospects quickly expand, spanning many service kinds, demands, and aims. Developing wireless networks, ensuring message delivery, and detecting events are just a few new and exciting available research subjects. Communication in vehicles is increasingly being researched and contested. Ultimately, these potential technologies would be integrated into global vehicular communication programs and standards.

VANETs, enabled by driverless automobiles and V2V and V2I connections, can produce a more efficient and secure transportation system [9,10,16]. The European Union (EU) will shortly develop an application (android and web) for driving assistance apps. This chapter discusses VANETs as a possible future transportation solution. Next, sophisticated VANET-based traffic management and control software for platooning, junctions, and VRUs are introduced. The present situation is evaluating VANET-based apps. Safety, efficiency, sustainability, and comfort are all factors considered while creating transportation apps. Smart cities, smart grids, and the IoT are emerging technologies. This chapter achieves a twin objective by providing an overview of present transportation systems and future VANET-based applications. 4G is a crucial objective to enhance mobile data transmission speeds while retaining high data transfer rates. Since 2010, 4G networks worldwide have used the LTE standard. 4G will speed up data transfer, increase mobility, and allow seamless network switching. To put it another way, 4G looks to be lacking in every regard. Using a cellular network allows you to use existing infrastructure ™ participation of base stations in mobile environments delays data transfers. DSRC has eliminated the need for a communication base station.

Next, Chapter 5 presents some efforts on "UAV Relay in VANETs Against Smart Jamming with Reinforcement Learning/ Deep learning" which describes that VANETs provide V2V and V2I communications, enabling autonomous driving and increasing onboard device utilization while improving transmission safety. The VANET is vulnerable owing to OBU mobility and a vast dynamic network with RSUs. Automatic automobiles have grown in popularity in recent decades, particularly in Europe. The internet links all cars and allows data to be delivered and received. A hacker trying to steal data from a car-to-car link may cause an accident. Most machine learning algorithms work effectively with datasets of a few hundred attributes or columns. The sheer number of characteristics in an unstructured dataset like one created from an image makes this method impractical. Reinforcement learning is how deep learning algorithms learn more about a picture as it moves through the layers of a neural network. First, a strategy exists for transporting data from one location to another. Once installed, the protocol broadcasts data to all automobiles utilizing it to share the same information.

The G-hop and GOEA algorithms may be used in real-world systems using mobile edge computing and software-defined radios. The G-hop broadcast protocol uses mobility groups to guide autos to the relay, thus its name. Among the benefits include faster data transmission, better coverage, and lower energy use. Despite improvements in orthogonal frequency-division multiple access, LTE networks are still vulnerable to active radio node assaults. An attacker initially sends a signal to confuse or distract the target (a jamming node) from gathering personal information. Denial-of-service assaults on communication networks are sometimes referred to as Layer-1 attacks. It has been a long time since wireless communication networks were jammed. Because short-time Fourier transform (STFT) employs the same rectangular window for all sites, the time-frequency resolution will be similar. Wavelet may accomplish multi-resolution on multiple signal locations since they have different window widths than STFTs.

The adaptive key is essential because sophisticated jammers modify the signal's properties. The Ant system states that this strategy will fail if jamming is discovered before the agents have generated a tabu list. In addition to being incorrect in assessing network characteristics, jamming detection methods are prone to faults. While jamming is easier to implement, detecting and combating it is more challenging. When planning your trip, consider that the deep reinforcement learning (DRL) framework necessitates training, so the DRL paradigm's applicability in real-world systems has been questioned. The stochastic model simplifies the natural design while ignoring subtle tendencies. The DRL framework's training and performance evaluation demand more efficient simulation data generation. The needed mathematical complexity and synchronization are pretty high. Intrusion detection can help prevent mishaps and keep the public safe. The intrusion detection system successfully identifies assaults by analyzing and categorizing VANET signals. Previous attempts to identify intrusions on the VANET using artificial intelligence and machine learning were unsuccessful.

Next, Chapter 6 discusses vehicular cloud computing [13,14] and its related terms in it. Cloud-based vehicles are the vehicles that are being used nowadays by the automobile industry. A brief history of its existence and scope have been explained in this chapter. Today, cloud computing is used almost in all applications to provide efficient access to storage and reliable delivery of data anywhere, anywhere. Cloud computing is of three types (in general): public, private, and hybrid. Such characteristics and applications toward cloud computing (especially on vehicles) have been explained in this chapter. Also, the Internet protocol addressing like IPv4, IP6 importance in the previous decade and limitations in the current era also is explained (including benefits and demerits). How IPv6 is good for Industry 4.0 (as a necessity)/cloud-based vehicles and Society 5.0 is discussed in this chapter (including opportunities and challenges with IPv6).

Further, Chapter 7 discusses advanced vehicle motion control and LTE/5G/6G for vehicular communications. This chapter describes many of these advances are driven by commercial pressures. They may have substantial implications on how we live in the future. The fifth and sixth generation (5G and 6G) broadband cellular networks are expected to transform existing society and technology profoundly. While the market is enthused about 5G, it may also research and influence customer behavior. Until the 6G standard is established, it is only helpful as a conceptual guide. Some applications, such as multiple view holograms, may need data speeds that 5G cannot provide. Deploying 6G networks may be required in the future. Most vehicle communication systems are meant to make driving safer while lowering the financial damage caused by accidents.

Contrary to popular belief, the terms VANET and IVC are equivalent when referring to vehicle-to-vehicle communications. There are several wireless protocols for radio communication between cars, vehicles and infrastructure, and infrastructure and infrastructure. The primary goal of these communication standards is to increase road safety, traffic efficiency, and driver and passenger comfort. Cars are the third most popular option after homes and jobs. The term "information and entertainment" (infotainment) embraces both traditional and emergent Internet uses (e.g., content download, media streaming, VoIP, web browsing, social networking, blog uploading, gaming, cloud access). As connected cars grow, 5G will need new technologies like spectrum sensing and direct device identification node coordination. 4G LTE currently allows direct device detection and communication without infrastructure. 5G would require broader coverage areas to enable new proximity-based services and data access.

6G's timing and phase synchronization requirements are expected to be greater than 5G's. Wireless systems enhance communication speeds by increasing spectrum efficiency and transmission capacity. These tactics are familiar to wireless users. Many technologies can help improve the efficiency of the B5G and 6G spectrum. They are looking at employing this GHz/THz frequency range to extend their accessible spectrum. 5G's performance will determine the future of connected automobiles. Over-the-air updates are crucial for keeping vehicle data and systems current, but they require a fast connection between the car and a nearby base station. 5G is still a work in progress, but providing cars with extra flash storage allows data to be routed to a central point for processing. If numerous base stations transmit at the same time, more coverage is necessary. 5G will enable future automotive functionalities that are not now conceivable. While the technology boosts production, it also improves safety. It will be thrilling and stimulating for drivers to stay connected in so many places when 5G is accessible in automobiles. New, high-impact 5G use cases will speed up its adoption. Reduced user input latency may lead to new user behaviors and new ways to influence them. Better self-monitoring abilities may help us better leverage current user behavior and social impact components.

Further, Chapter 8 discusses the importance of computer vision in autonomous intelligent vehicles (AIV) [9–12,15]. Note that autonomous vehicles (AVs) are a subset of AIV. Computer vision is a superset of artificial intelligence that contains the concept of machine learning, deep learning, etc. These learning techniques when enabled or used in vehicles make vehicles intelligent and on the other side using the IoT, vehicles become automated. Hence, learning techniques for AIV like artificial intelligence-based AIV, machine learning in AIV, long short-term memory/deep learning in AIV, and robot-based AIV have been explained in this chapter. Note that AIV, AVs, connected vehicles, hybrid vehicles are different types of vehicles. Using learning techniques and smart devices in vehicles make vehicles smarter and driving convenient and hassle-free for the driver. In this chapter, future possibilities for computer vision toward AIV (including research and challenges). Driverless cars/vehicles are the necessity of the future with a smaller number of road accidents, less emission of CO_2/greenhouses gases.

Then, Chapter 9 discusses biometric and blockchain-based vehicular technology in detail. It starts with the background work required to explain security and its necessity to vehicles. Then the scope of biometric technology is explained in detail and shows that today biometric concept is the most secure concept because each individual contains a unique thump impression (in this world) which is difficult to compromise. But if it is comprised by any third party then a new concept of blockchain technology is used combinedly to enhance the security of vehicular systems. Here, blockchain technology is a distributed ledger technology that is used to provide trust and secure concepts over the communication networks (via string information using a hash key in blocks, where each block is interlinked to each other). Hence, this chapter discusses how biometric and blockchain technology can provide the highest level of security to systems. Further, a few important issues like scalability, trust issues, privacy and security, etc. have been explained in this chapter in biometric – blockchain-enabled vehicles.

Next, Chapter 10 discusses IoT-based vehicular technology (IoV) and its scope in the near future. IoVs are vehicles that use smart devices to make themselves intelligent and smart (automated)/interactive with the end-users. These types of vehicles make user's/driver's/passengers' travel journey comfortable and hassle-free. Few issues like road accidents, jamming problem in next upcoming routes, etc., have been notified by smart devices to these vehicles. With this, the user can take appropriate step/action to change his travel journey or route to reach his/her destination. In this chapter, the working of IoVs, their architectures, benefits toward smart generation/future generation have been explained in detail. Also, how the IoVs are better and more efficient than ITS and autonomous intelligent vehicles are discussed. Further, several open issues like security, privacy, trust, and open challenges toward integration of the IoT with other vehicles have been discussed. In the last, a few applications of IoV and IoT-based pure electric vehicles are explained in this chapter.

Further, Chapter 11 describes the introduction to an ITS that minimizes traffic concerns to improve traffic efficiency. It enhances traffic information, local convenience, real-time operational data, seating access, and passenger safety and comfort. ITS is widely used in various nations. It also informs passengers about the bus's position and the density of people. The system is so well designed that passengers and even drivers are unaware of the delay. ITS, or intelligent transportation system, occasionally uses communications and IT to alleviate traffic congestion. ITS is a massive change in various elements of transportation. ITS is an integration of various IoTs and sensor technologies which work together to deliver efficient, reliable services to the end users at surface. ITS is a complex technology system meant to ensure efficiency and safety. ITS is used globally to improve traffic flow and minimize travel time. In rising nations, rural-to-urban migration has been unequal. Many emerging countries have urbanized without much motorization or growth. But even though few people can afford cars, congestion in multi-modal transportation networks is developing rapidly. They pollute the air, endanger public safety, and exacerbate social inequality. Due to its importance in the globe, the transportation network is constantly evolving.

An ITS is a set of transport infrastructure and communication technologies designed to reduce traffic congestion and environmental impact. It helps drivers make educated choices. As a result, traffic signal and traffic management systems use more IT than operational technology. Technology and available options for ITS systems are numerous. Recent security threats include ransomware. This dangerous application encrypts computer data and demands payment for decryption and data recovery. If the attacking nodes are unavailable but the attacking party is situated in both, it must attenuate the signal or change the position data. As an IoT subtype, the ITS may be constructed utilizing similar methodologies and concepts. It can be used in ITS. With ITS's complexity comes a need for proactive protection. There is a strategic application of ITS cyber security in numerous fields. Many articles describe the future look of ITS; however, few fundamental discoveries can be applied by ITS. Blockchain, which keeps track of prior transactions, can reduce well-planned opportunistic behavior. Reliable news creates an economic climate of low transaction costs, crucial to economic efficiency and advancement.

Smart contracts (based on blockchain concept) and IT can help increase confidence and eliminate uncertainty in behavioral transactions. Distributed and decentralized accounting systems reduce middle-man costs. Governments and audit agencies' access to blockchain company accounts boosts corporate profit and fiscal management while decreasing tax evasion. To create an ITS that meets the demands of many users, all of these features must be integrated. Intelligent ITS development should include priorities safety, accuracy, and cost-effectiveness. ITS is a significant problem for the future of rising countries like India. People and goods must be transported conveniently to both urban and interurban locations in this Internet era. Safer, more competitive, and more cohesive intelligent mobility centers on ITS.

Then, Chapter 12 discusses AV and its localization, navigation/tracking/traces in this current smart era. The difference between autonomous and automated vehicles has been explained in this chapter. Further importance of autonomous vehicle technology scope is discussed over other types of existed vehicles like a connected vehicle, intelligent vehicles, etc. Then few promises and perils of AV technology (including effects of AV technology on safety and crashes) have been explained. Further few other effects of AV technologies on mobility for those unable to drive are explained in detail. Note that an autonomous vehicle/general vehicle either runs on petrol or CNG or LPG, etc. But few automobile companies are giving an option to the user to use electric vehicles, but such vehicles are very costly and out of reach for the general public. So, the energy and emission implications of AVs have been explained in this chapter. Then developing costs and several disadvantages for implementing AIV in today's scenario/era have been explained. Further, a few other safety applications for VANETs/ITS are explained in this chapter. In the last, other applications like traffic condition sensing application for AIV, and opportunities and challenges for the future towards AVs have been included in this work in detail.

Next, Chapter 13 presents a few views on realistic vehicular mobility/channel models for vehicular communications for AIV or combining these traits results in difficult-to-model propagation situations. Due to the dynamic environment, low antenna heights, and significant vehicle mobility, signal statistics rise on both local and large scale very rapidly in an urban setting. The propagation characteristics show that the signal interacts with the built-up environment traveling from the transmitter to the receiver. Rays bounce off buildings and other barriers in cities, increasing the total number of rays reaching the receiver. Before implementing vehicular protocols and applications, a proper channel model must be designed. Consequently, we recommend a channel model depending on the protocol/application being tested, the available geographic data, and the simulation time limits.

Infotainment includes ITS applications like map consulting and Internet access. As vehicle sensors proliferate, the ITS application category "sensors on wheels" has grown. The data acquired by car sensors may be aggregated and sent to traffic control centers to help regulate traffic flow. ITS data transfer requires the establishment of MANETs. The premise of broad sense stationarity is no longer valid due to fast environmental change. Correlations between path parameters are prevalent, contrary to the popular idea of uncorrelated scatterers. The correct simulation of the signal propagation environment is critical to a successful VANET protocol and application evaluation. To accurately show VANET channels, static and moving objects (such as buildings and plants) must be included (other vehicles on the road). Sorting objects is accessible; explaining the ecosystem they create is another store. Car communication occurs in three environments: highways, suburbs, and rural areas. Aside from traffic speed and density, roadside items such as trees and bushes can significantly affect signal transmission. A model's capacity to realistically portray many circumstances is a significant reason for its inclusion in this category. The models in this section have been tested on real-world data and may be downloaded.

The models are based on their propagation mechanism, implementation method, and channel characteristics. There is a need for models that can be employed in large-scale simulations of the vehicular network due to the limits of current models. Testing protocol and other application assessment alternatives are used in vehicles as communication channels. Raytracing is commonly employed to generate the appearance of depth. Geometry-based stochastic models (GBSMs) include geometry-based dynamical models. A GBSM model's parameter values can be predicted, whereas the discrete element model's parameter values may be characterized statistically. Unlike geometrical stochastic models, non-geometrical stochastic models focus on high-level behavioral aspects. They are frequently referred to as tapped-delay lines since their original representation is based on echoes. Wireless local area networks are the most frequent type. Wireless connection simulation models for unicast transport protocols over multiple wireless channels are discussed in this chapter.

Further, Chapter 14 describes the intelligent network access system for ITS, i.e., for environment perception, modeling and growth of IoTs in vehicular networking. Also, several advantages and disadvantages are also included in this chapter with respect to intelligent network access system. As intelligent networks (IN) develop in size and complexity to meet changing client requirements, network intelligence becomes more distributed and sophisticated. Users need information from service layer nodes, not from a core switch or equipment intelligence. This degree of switching is independent of the leading network. Telecom businesses, such as phone companies or mobile phone operators, own the IN nodes and utilize them to send data.

A secure and interoperable wireless communications network integrating trucks, trains, traffic lights, and cell phones might transform American travel. Private signals sent between cars and infrastructure will collect new data on how, when, and where individuals travel. Consumers' access to interactive real-time video and multimedia services is increasing as telecommunications and information technology merge. Video-on-demand banking, retail, and entertainment services are examples of this. It also encompasses electronic publications and services IN moving services like toll-free and regional number portability from core switch systems to standalone nodes. The open and secure network established this way allowed service providers to develop new services and value-added features for their networks without contacting the core switch maker or waiting for a long development time. This technology can instantaneously transform toll-free numbers into ordinary PSTN numbers. The growth of the IoT is dependent on connected cars. An automobile connected to the internet may interact with other vehicles, mobile devices, and city junctions in both directions.

The IoT is used to link automotive technology. The US government's ITS goal includes linked car technology, and multiple pilot programmers are presently ongoing globally. Cars, trucks, buses, and other vehicles will be able to "talk" with each other via in-vehicle or aftermarket devices that broadcast essential safety and mobility information. Those outfitted with the right equipment may wirelessly connect with nearby cars, including toll booths. Vehicle connection is intended to revolutionize transportation in an intelligent society. To link vehicles, infrastructure, and passengers, automakers will need to develop wireless communication technology (or in the pockets of the drivers and passengers). ADAS systems may incorporate V2V or V2I systems, vehicle data networks, and sensor technologies, to mention a few. The driver is advised of possible threats and is kept aware of what is in the blind zone by systems such as adaptive cruise control and automatic braking. Communication between cars may decrease traffic accidents and congestion by giving crucial safety information such as location, speed, and direction between vehicles within hearing distance. In addition to fundamental operational difficulties, current research should explore "a system inside a system," with multiple self-propelled components responding to a wide variety of economic and environmental concerns. Maintaining, expanding, and innovating infrastructure systems will meet changing user requirements and new public aims.

Then, Chapter 15 describes multi-channel operations, coexistence, and spectrum sharing for vehicular communications. Risks on the road can be recognized and predicted using just one well-known safety channel at a time, such as cooperative awareness messages or decentralized environmental notification messages in the EU, and basic safety messages (BSMs) in the United States. Most scientific, standardized, and industry research and development have been devoted to brilliant cooperative communication and network solutions that reduce traffic congestion. Conversely, C-ITS applications can utilize several channels.

Despite their early availability, C-ITS apps have been underused. In the EU and the United States, only one of these channels was chosen as the control channel (CCH) for Day One deployments of C-ITS applications, including road hazard alert and intersection collision alert. This chapter was prepared before the ETSI standards were finalized. The usage of C-ITS channels is not restricted, but the radiation and application access limits set on each track allow it. A channel might be a command-and-control channel or one shared by all C-ITS services. We propose C-ITS channel allocation processes to investigate the United States and EU. This chapter discusses the differences and similarities during this lesson and employs them in various mediums. Spectrum management technology allows users in cognitive radio networks to modify and regulate the spectrum access network.

An optimization issue is a problem created by researchers to discover the best possible answer for the target audience. Game theory can help manage spectrum in cognitive radio networks, especially for rate and power control. The random waypoint model was introduced in 1995 as the first ad hoc network model. Each RW node chooses an arbitrary destination and a random velocity from a pool of options at each waypoint. The simulation variable's value drops slowly with each simulation iteration until it reaches a steady state. The random waypoint technique may produce random nodes with varying speeds and directions. Stopping and turning will be difficult if the interval is left at zero. This model can predict a node's migration speed and direction. As a result, nodes move in a particular direction and at a specific rate. The data supplied can dynamically vary the paths taken by participating nodes throughout a run (such as traffic information).

These characteristics are essentially the same in both the United States and Europe. Because of the shared spectrum, the same hardware platforms may be utilized in both zones. The United States has seven DSRC channels, and running them all inexpensively with various DSRC transceivers requires multi-channel switching. Their disparities stem from the different spectrum allocations in the United States and the EU. Since 2008, only three ITS-G5 channels have been assigned throughout Europe, eliminating multi-channel switching.

Next, Chapter 16 describes simulation tools and techniques for vehicular communications and their applications/ITS/AIV are now being adopted to meet increased market demand for goods and services. Getting people where they want to go for the least amount of money possible is included. The insurance covers direct and indirect expenditures like gasoline, maintenance, pollution, and accidents. Extending the present road system would be extremely expensive due to physical, economic, and environmental constraints. The contemporary urban traffic management problem is providing timely information, targeted travel-related services, and high-quality assistance to the organizations that utilize these roadways. This educational center will catalyze the development of new transportation products and services. It is difficult because there are many interrelated subsystems, feedback loops, and they all operate on different time scales with some uncertainty.

There are several ways to alleviate traffic congestion, including safety (which minimizes crashes, traffic conflicts, and law violations), mobility (which reduces travel time and costs), and congestion (to increase the line-of-sight of the road facility). Sustainable transportation systems are vital in the twenty-first century. Technology integration will make future cars and infrastructure more "intelligent" and efficient. Reduced emissions and improved road operations would assist the environment. Modern technologies are required to effectively monitor and regulate the city's transportation infrastructure in real time. It is projected that the sensors and control systems that make up an ITS will continue to encompass roadways, automobiles, and passengers. This is a big step forward in network speed while keeping an eye on congestion, safety, and user and environmental comfort. It is typically compatible with VC and IT setups, and it may often be used as a third-category market engine. Several efforts, standardization groups, and consortia have established and constructed applications for VC systems globally. Initially, a long list of potential applications was compiled, looking to the future while considering existing transportation needs and capabilities and how venture capitalists may assist these applications. There are three categories of IVC applications: safe, intelligent, and connected.

Various granularity levels are available, grouped by IVC application type. They were weighed. These facets must be balanced. To reach this goal, we have to reveal the models required for IVC simulations. It is also challenging to locate simulators with all the essential models and granularity for IVC app development. The latest simulation models and approaches were also reviewed. Simulators with all key models at various granularities have improved ITS simulation study quality and comparability. As a result, many recent simulation studies employ comparable approaches and simulation components. This is standard. iTETRIS, Veins, and VSimRTI are three complementary toolkits. Choosing between the three possibilities is difficult here. There are several protocols or programmers to discover the most satisfactory solution. A single toolbox can considerably improve study credibility and reproducibility—the existence of well-established IVC models. Realistic mobility models, propagation, and channel access models are some examples.

Next, Chapter 17 discusses several issues like security, privacy, and trust raised in AIV, AV, and ITS. First, we discuss some related/background work and then further few safety approaches have been explained toward AV/ITS/AIV from the end-user perspectives. Safety approaches are necessary to provide comfort and reliability to its consumers. In this chapter, differences over the architecture of AV/ITS/AIV are explained (including raised issues in respective vehicles). Further, several security risk analyses for a trust model (for VANETs/ITS) have been explained in detail. If a risk is overcome from a model, then it became more comfortable or safe to use. Further, a few ITS security and privacy challenges are discussed in detail. Also, several potential threats are discussed mitigated in IN-based vehicle systems/IoTs-based cloud vehicles/AIVs. Note that securing is an AIV/ITS/AV/IN (intelligent vehicle) is always a primary concern and proving better security builds more trust among users and improves the user's better experience/satisfaction toward automobiles industries.

In the last, Chapter 18 discusses future research trends, applications, and standards in emerging vehicle technology in detail. This chapter explains some background for future vehicles (using 5G, called IoV) or vehicles of tomorrow [1,9–12], and then explains more applications to grow and the requirements for future vehicles, and also few possibilities like measuring the performance of connected vehicles (with respect to their applications), real-time traffic information estimation, and their impact on environmental/our earth in near future. We need to work on the development of vehicle networking communication standards to improve vehicle communication with other applications. Note that in the future driverless car, electric vehicles, and vehicles with 5G and IoV using 6G will be a possibility to make people's life convenient and easier to live (by having the intelligence to take effective add quick decisions in emergencies). We hope that future vehicles of tomorrow will help us to live longer and stronger via reduced emission of CO2 gases or other gases. In the last, we invite all researchers and scientific communities to come forward and contribute in any way to protect our earth against pollution/for a sustainable earth.

Disclaimer. Papers on Vehicular Ad Hoc Network/ ITS provided in this conclusion chapter are only given as examples for future reference (for readers/ researchers). To leave any citation or link is not intentional.

References

[1] A K Tyagi, S Kumari, T F Fernandez, and C Aravindan. P3 block: privacy preserved, trusted smart parking allotment for future vehicles of tomorrow. In: O Gervasi, *et al.* (eds) *Computational Science and Its Applications – ICCSA 2020. Lecture Notes in Computer Science*, Springer, Cham, vol. 12254, 2020. https://doi.org/10.1007/978-3-030-58817-5_56

[2] A K Tyagi and S Niladhuri. ISPAS: an intelligent, smart parking allotment system for travelling vehicles in urban areas. *International Journal of Security and Its Applications* 11(12) (2017): 45–66, ISSN: 1738-9976 IJSIA, SERSC Australia.

[3] M M Nair and A K Tyagi. Privacy: history, statistics, policy, laws, preservation and threat analysis. *Journal of Information Assurance & Security* 16(1) (2021): 24–34.

[4] K Sravanthi, V K Burugari, and A Tyagi. Preserving privacy techniques for autonomous vehicles. *International Journal of Emerging Trends in Engineering Research (IJETER)* 8 (2020): 5180–5190, doi: 10.30534/ijeter/2020/48892020

[5] A M Krishna, A K Tyagi, and S V A V Prasad. Preserving privacy in future vehicles of tomorrow. *JCR* 7(19) (2020): 6675–6684, doi: 10.31838/jcr.07.19.768

[6] A K Tyagi and N Sreenath. Preserving location privacy in location based services against sybil attacks. *International Journal of Security and Its Applications* 9(12) (2015) 189–210, ISSN: 1738-9976 (Print), ISSN: 2207-9629 (Online).

[7] A K Tyagi and N Sreenath. A comparative study on privacy preserving techniques for location based services. *British Journal of Mathematics and Computer Science* 10(4) (2015) 1–25, ISSN: 2231-0851.

[8] A K Tyagi and N Sreenath. Location privacy preserving techniques for location based services over road networks. In: *Proceeding of IEEE/ International*

Conference on Communication and Signal Processing (ICCSP), Tamilnadu, India, 2015, pp. 1319–1326, ISBN: 978-1-4799-8080-2.

[9] M Krishna and A K Tyagi. Intrusion detection in intelligent transportation system and its applications using blockchain technology. *2020 International Conference on Emerging Trends in Information Technology and Engineering (ic-ETITE)*, 2020, pp. 1–8, doi: 10.1109/ic-ETITE47903.2020.332.

[10] A K Tyagi and N Sreenath. Vehicular ad hoc networks: new challenges in carpooling and parking services. In: *Proceeding of International Conference on Computational Intelligence and Communication (CIC), International Journal of Computer Science and Information Security (IJCSIS), Pondicherry, India*, 14 (2016). pp. 13–24.

[11] R Varsha *et al.* 'Deep learning based blockchain solution for preserving privacy in future vehicles'. *International Journal of Hybrid Intelligent System*, 16(4) (2020) 223–236.

[12] A K Tyagi and S U Aswathy. Autonomous intelligent vehicles (AIV): research statements, open issues, challenges and road for future. *International Journal of Intelligent Networks*, 2 (2021) 83-102, ISSN 2666-6030. https://doi.org/10.1016/j.ijin.2021.07.002.

[13] A K Tyagi and N Sreenath. Providing trust enabled services in vehicular cloud computing. In *Proceedings of the International Conference on Informatics and Analytics (ICIA-16)*. Association for Computing Machinery, New York, NY, USA, 3 (2016), pp. 1–10. Doi: https://doi.org/10.1145/2980258.2980263.

[14] A K Tyagi and N. Sreenath. Providing trust enabled services in vehicular cloud computing, *2016 International Conference on Research Advances in Integrated Navigation Systems (RAINS)*, 2016, pp. 1–7, doi: 10.1109/RAINS.2016.7764391.

[15] A K Tyagi, D Agarwal and N Sreenath. SecVT: securing the vehicles of tomorrow using blockchain technology, *2022 International Conference on Computer Communication and Informatics (ICCCI)*, 2022, pp. 1–6, doi: 10.1109/ICCCI54379.2022.9740965.

[16] A K V, A K Tyagi and S P Kumar. Blockchain technology for securing Internet of Vehicle: issues and challenges, *2022 International Conference on Computer Communication and Informatics (ICCCI)*, 2022, pp. 1–6, doi: 10.1109/ICCCI54379.2022.9740856.

Index